MATHEMATICAL INTEREST THEORY

MATHEMATICAL INTEREST THEORY

James W. Daniel

Leslie Jane Federer Vaaler

Department of Mathematics
The University of Texas at Austin

PEARSON
Prentice
Hall

Upper Saddle River, New Jersey 07458

Library of Congress Cataloging-in-Publication Data

Daniel, James W.
   Mathematical interest theory / James W Daniel, Leslie Jane Federer Vaaler.
      p. cm.
   Includes index.
   ISBN 0-13-147285-2
      1. Interest rates—Mathematical models.   2. Interest rate futures—Mathematical models.
   3. Risk management—Mathematical models.   I. Vaaler, Leslie Jane Federer.   II. Title.

   HB539.D33 2007
   332.801'513—dc22

                                                                              2006042586

Editorial Director: *Marcia J. Horton*
Senior Editor: *Holly Stark*
Production Editor: *Raegan Keida*
Assistant Managing Editor: *Bayani Mendoza de Leon*
Senior Managing Editor: *Linda Mihatov Behrens*
Executive Managing Editor: *Kathleen Schiaparelli*
Manufacturing Buyer: *Maura Zaldivar*
Marketing Manager: *Tim Galligan*
Marketing Assistant: *Jennifer de Leeuwerk*
Editorial Assistant: *Jennifer Urban*
Art Director: *Jayne Conte*
Director of Creative Services: *Paul Belfanti*
Cover Designer: *Bruce Kenselaar*
Manager, Cover Visual Research & Permissions: *Karen Sanatar*
Cover Image: *Getty Images, Inc.*

© 2007 Pearson Education, Inc.
Pearson Prentice Hall
Pearson Education, Inc.
Upper Saddle River, New Jersey 07458

Pearson Prentice Hall™ is a trademark of Pearson Education, Inc.

Printed in the United States of America
10 9 8 7 6 5 4 3 2

ISBN 0-13-147285-2

Pearson Education LTD., *London*
Pearson Education Australia PTY, Limited, *Sydney*
Pearson Education Singapore, Pte. Ltd
Pearson Education North Asia Ltd, *Hong Kong*
Pearson Education Canada, Ltd., *Toronto*
Pearson Educacion de Mexico, S.A. de C.V.
Pearson Education - Japan, *Tokyo*
Pearson Education Malaysia, Pte. Ltd

To my wife, Ann Daniel; my teacher, Paul Mielke;
and my actuarial guide, Gene Wisdom.

– James W. Daniel

Investors receive interest.
To Abigail Louise Vaaler and Douglas Quinn Vaaler,
the best interest ever.
To Jeffrey David Vaaler, my co-investor.

– Leslie Jane Federer Vaaler

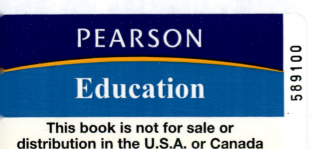

Contents

APPENDICES

Preface

To students

We hope you enjoy this book and find it useful. Those who understand interest theory can be informed borrowers, making intelligent choices about mortgages and other loans, and they can also be wise investors. We anticipate that the majority of our readers will be interested in exploring actuarial careers. However, among our enthusiastic students, we have counted many who just felt that this was material they would like to understand; they found it interesting and applicable. Interest rates affect us all! The authors have frequently been told that the course for which the book was designed was a student's "favorite" in his or her college career — and this was not just by students in the University of Texas' actuarial program! With this in mind, we have tried to write a book that will be appealing to any student who wishes to become familiar with how investments grow over time, and who appreciates this being carried out in a mathematically precise manner.

If you are embarking upon a career as an actuary, a mastery of interest theory will be very important for you. This is reflected in the fact that the Society of Actuaries (SOA) and the Casualty Actuarial Society (CAS) each require their Associates to pass a jointly administered exam including interest theory. Our text is designed to meet the needs of individuals hoping to master this material. Should you work in other financial areas, we believe that an exposure to the rigorous introduction presented in this text will be advantageous.

We were trained as mathematicians, specializing in numerical analysis and algebraic number theory, respectively. We learned interest theory by teaching the subject. Not being insiders in the world of finance, we were like you, the readers for whom this book has been written. Sure, we have a great deal more mathematical sophistication and experience than we would expect of our student readers but, like you, we found ourselves searching for the larger context in which to place the material. It is one thing to be adept at handling equations, another to understand why it is useful to make these calculations. We hope that our struggles to see the big picture, and the questions posed to us by our students over the twenty terms that we have taught interest theory, have helped us to write a book in which each concept is well motivated and described in a manner that leaves the reader feeling satisfied.

While we have strived to write a student-friendly book, helpful to those trying to learn interest theory on their own, as well as to students in organized classes, we want to warn you that this is not easy material. In fact, the level of difficulty may be deceptive. The majority of the problems describe a financial arrangement in words — producing the dreaded "word problem." To solve the problems, you must extract all the information contained in each sentence. Sometimes this is accomplished by writing equations, but at other times it would be best to use a diagram or list. It is at this stage that many students of interest theory have difficulty, but we are committed to helping our readers develop their problem-solving skills.

Patience, attention to details, and a willingness to work hard in an organized manner are all needed to be a successful student of interest theory. Along with a strong foundation in algebra, these qualities are the most important prerequisite. Recognizing this, we have written this book so that an able student who has not yet studied calculus can follow along. There are a few places where calculus is used, but an alternate path is given for those yet to learn calculus.

Examples

A careful description of the theory is important; however, it is just as important that there be many examples presented in detail. The text includes more than 240 examples. Each was written, then revised, with the questions: "Have we given an explanation that is clear to a reader who was not previously adept at working such a problem?" and "Have we described why the path we have taken is a natural and good one?" We believe that complicated examples should be preceded by simple ones, and that it is often helpful to present more than one way to obtain the solution.

Problems

Of course, no matter how helpful our examples may be, it is important that a student tackles many problems. Like all mathematical endeavors, mastering interest theory requires a willingness to practice and to learn from mistakes. Mathematical Interest Theory includes more than 430 problems. Just as is the case with our examples, these range from the straightforward to the challenging. Some problems may be completed quickly, while others involve significant analysis followed by considerable calculation. There is an appendix in which numerical (and other short answers) are given. Detailed solutions to the odd-numbered problems are left for the author-prepared student solution manual. An instructor manual, again written by the authors, includes solutions to all the problems.

Special features

Most of the problems involve computation, and naturally we wish to provide our readers with information on how to effectively utilize technology. We have elected to include considerable discussion and key-by-key examples for our readers who have access to Texas Instruments BAII Plus or BAII Plus Professional calculators. Chapter 0 introduces the reader to these calculators. The calculator discussion is typeset in a different font from the rest of the text, making the book easy to use, whether or not a reader wants to consider the instructions for these calculators.

It is easy for students of interest theory to feel overwhelmed by the number of formulas. We therefore have placed the equations that are especially important in boxes. Readers may decrease the number of equations they need to memorize if they note which ones can quickly be derived from another formula they have already learned. Statements placed in boxes are also especially important.

We hope that our book will be considered by teachers wishing to offer a class with a significant writing component, but not a course on writing proofs. The end-of-chapter writing problems should be very useful to instructors of such courses.

Coverage

We present a classic introduction to interest theory in Chapters 1 through 5. However, we introduce yield rates at an earlier stage than is usual, then revisit them as we introduce new financial settings. Students often have difficulty with yield rates, and we believe that introducing them early and using them regularly goes a long way to help students with a topic that many find difficult.

Bonds are corporate and government loans, and have their own language and conventions. These are discussed in Chapter 6. If a company wishes to raise money without borrowing money, it may do so by issuing equity shares or stocks. Stocks are considered in Chapter 7.

The financial horizon has changed markedly over the last three-and-a-half decades. It is now important for actuaries and others working in financial services, not to mention those following the news headlines, to understand financial derivatives such as options and futures. In the aftermath of Hurricane Katrina, gas futures made the headlines. Chapter 8 introduces in detail modern financial concepts such as arbitrage, derivatives, options, futures, and swaps, and it does so without requiring the reader to know probability theory.

Risk management is important in a world where interest rates may change in a manner that is difficult to foresee. In Chapter 9 we introduce the reader to asset-liability matching and to immunization.

There is more material in this book than is apt to be covered in a one-semester undergraduate course. For many, the syllabus will likely be determined based on the syllabus for the joint SOA/CAS exam that includes interest theory. As of Spring 2006, this would mean covering Chapters 1–3, Chapter 4 (perhaps a bit quickly and excluding Section 4.4), Chapter 5, Chapter 6 through Section 6.9, Chapter 7 (omitting Section 7.2), Sections 8.2–8.3, and Chapter 9 through Section 9.4. There will be SOA/CAS syllabus changes taking effect in the spring of 2007, and at that point instructors will likely wish to consider including more of Chapter 8; of course, this depends on the content of other courses taken by actuarial students.

If a teacher's focus is not on covering the interest theory needed on actuarial exams, the choice of material to include will depend on individual preferences and the backgrounds of the students. However, we suggest all readers study Chapters 1–3, except possibly for sections requiring calculus.

Financial transactions

This book includes accounts of numerous financial arrangements. For the most part, these are fictional examples, designed to illustrate and reinforce general principles. We made up the names of participants and financial institutions and any overlap with the names of actual individuals or businesses is accidental. You will notice that the names have diverse ethnic roots. We hope that no gender or nationality appears too often as a borrower or unwise investor.

We have already mentioned that we are mathematicians rather than financial advisors. However, at this point we are duty-bound to give you one piece of financial advice. *Always* make sure you understand all details of any financial arrangement *before* you agree to it. Ask questions. Do not rely on everything working as it does in this book.

Acknowledgments

We offer our heartfelt thanks to our students. Their questions, comments, and even their mistakes have all been helpful to us as we wrote this book. Without our students, there would have been no book. We are thankful to those who class-tested large portions of the text and pointed out errors, omissions, or passages requiring clarification. Their enthusiasm for our project was very welcome. A special thanks to Carl Gillette, Karen Kimberly, and Gagan Nanda, who worked through the examples and problems as accuracy checkers!

We are grateful to those who reviewed our book at various stages in its production: Leonard Asimov, Louis Friedler, Lorenzo Garlappi, Victor Goodman, Matt Hassett, Bryan Hearsey, Krzysztof Ostaszewski, Steve Paris, Lucinda Zmarzly, and especially Warren Luckner and Susan Staples.

Finally, we thank our families. While we like language, words cannot express our gratitude. Each day, they provided us the support to work hard and also the needed breaks from doing so. They reminded us what really is important!

Contacting the authors

Unfortunately, there are undoubtedly still some mistakes. If you note errors of any sort, we would appreciate you informing us of them. Other comments and suggestions for future editions are also welcome. Contact us at `lvaaler@math.utexas.edu`.

James W. Daniel and Leslie Jane Federer Vaaler

C H A P T E R 0

An introduction to the Texas Instruments BA II Plus

0.1 CHOOSING A CALCULATOR

There are many modestly priced financial calculators on the market today. Among these is the Texas Instruments BA II Plus, and we have chosen to integrate instructions regarding its effective use for solving problems. Our choice of calculator for the illustrations is based on the fact that the BA II Plus calculator is very well suited for the tasks we pose, is among the calculators allowed for the SOA/CAS examinations including Financial Mathematics (SOA Exam FM, CAS Exam 2), and Texas Instruments representatives have informed us that they expect to continue selling the BA II Plus calculator for some time to come. You may also follow the instructions using the BA II Plus Professional calculator that was released in 2004. This is an enhanced version of the BA II Plus calculator, and has the same keys. Occasionally we will make notes specific to the BA II Plus Professional. Although the less expensive Texas Instruments BA-35 calculator performs well for many of the problems addressed, there are some problems (notably those relating to uneven cashflows) where it is deficient. We caution you not to follow our calculator instructions with the BA-35 calculator, since the correct operation is not always what you might guess. For example, the sign with which certain payments need to be entered is not always the same as it would be for the BA II Plus.

For those readers who choose to follow along using a calculator other than the BA II Plus calculator, for instance one of the Hewlett Packard financial calculators, perhaps our instructions will still be of some service.

0.2 FONT CONVENTION

Those parts of the text that are aimed at readers using the BA II Plus calculator will be given in sans serif (or **boldface sans serif**) type. The sans serif font is illustrated in Figure (0.1.1).

Focus on the third row of buttons on the BA II Plus Calculator. From left to right, the symbols are $\boxed{\text{N}}$, $\boxed{\text{I/Y}}$, $\boxed{\text{PV}}$, $\boxed{\text{PMT}}$, and $\boxed{\text{FV}}$.

FIGURE (0.1.1)

Readers who wish to concentrate on the theoretical flow, and not worry about how to perform calculations on the BA II Plus calculator, may ignore all discussion given in the sans serif font just displayed in Figure (0.1.1).

0.3 BA II PLUS BASICS

We will introduce you to the basic features of the BA II Plus calculator only to the extent necessary in order that our directions are clear. For instance, we presume familiarity with standard arithmetic operations of the calculator and how to change a newly entered number on the display if you have entered it incorrectly.

It is important that we explain to you the "calculation methods" available on the BA II Plus and BA II Plus Professional calculators; that is, how the calculator interprets a string of keystrokes. The default setting for the calculation method is the so-called **chain method**; arithmetic operations are done in the order you enter them, unless parentheses indicate instructions to the contrary. So, if you key $\boxed{4}$ $\boxed{+}$ $\boxed{2}$ $\boxed{\times}$ $\boxed{5}$ $\boxed{=}$ the calculator will compute $(4 + 2) \times 5 = 30$, and if you press $\boxed{2}$ $\boxed{\times}$ $\boxed{5}$ $\boxed{y^x}$ $\boxed{3}$ $\boxed{=}$ the calculator will compute $(2 \times 5)^3 = 1,000$. If instead, you need to compute $2 \times 5^3 = 250$, this may be accomplished by pressing the sequence of keys $\boxed{2}$ $\boxed{\times}$ $\boxed{(}$ $\boxed{5}$ $\boxed{y^x}$ $\boxed{3}$ $\boxed{)}$ $\boxed{=}$ or $\boxed{5}$ $\boxed{y^x}$ $\boxed{3}$ $\boxed{\times}$ $\boxed{2}$ $\boxed{=}$. The chain calculation method (**Chn**) is commonly used on financial calculators, and

---

our calculator instructions assume that you have your calculator set for the chain calculation method.

Remember, the chain calculation method is the default setting, and you must take action if you prefer your calculator to use the algebraic operating system (**AOS**) implemented by many scientific calculators.

To change the calculation method, you press
$\boxed{\text{2ND}}$ $\boxed{\text{FORMAT}}$ $\boxed{\uparrow}$ $\boxed{\text{2ND}}$ $\boxed{\text{SET}}$ $\boxed{\text{2ND}}$ $\boxed{\text{QUIT}}$.

If your calculator is set for **AOS**, keying $\boxed{4}\,\boxed{+}\,\boxed{2}\,\boxed{\times}\,\boxed{5}\,\boxed{=}$ results in the calculator computing $4 + (2 \times 5) = 14$, while pushing the sequence $\boxed{2}\,\boxed{\times}\,\boxed{5}\,\boxed{y^x}\,\boxed{3}\,\boxed{=}$
causes the calculator to find $2 \times 5^3 = 250$.

> **Warning: If your calculator is set with the algebraic operating system, adjustments to our key-by-key instructions are needed in some cases.**

Even if you are accustomed to using a scientific calculator, we encourage you to have your calculator set for the chain calculation method and to use parentheses to achieve the order of operation you desire.

There are many times when you need to raise a number to a rational power. For example, you might wish to calculate $(.96)^{\frac{4}{3}}$. This can be done efficiently by pushing the nine-key sequence

$$\boxed{\bullet}\,\boxed{9}\,\boxed{6}\,\boxed{y^x}\,\boxed{4}\,\boxed{y^x}\,\boxed{3}\,\boxed{1/x}\,\boxed{=}\,,$$

or by pressing the nine-key sequence

$$\boxed{\bullet}\,\boxed{9}\,\boxed{6}\,\boxed{y^x}\,\boxed{(}\,\boxed{4}\,\boxed{\div}\,\boxed{3}\,\boxed{)}\,\boxed{=}\,.$$

By either method, you should obtain that $.96^{\frac{4}{3}} \approx .947025437$.

The keys on the BA II Plus have their primary function printed on the key. Most of the keys also have a secondary function and it is printed above the key, in pale yellow on the BA II Plus calculator and in charcoal gray on the BA II Plus Professional calculator. To access a given secondary function, you first press the key $\boxed{\textbf{2ND}}$ and then the key whose secondary function you desire. For example, if you wish to reset your calculator to the factory default settings (as is required of candidates on the SOA/CAS exams), you will need to begin by pushing $\boxed{\textbf{2ND}}$ followed by the key $\boxed{+/-}$. Even though RESET is the second function on the key $\boxed{+/-}$, our directions to the reader will be to push $\boxed{\textbf{2ND}}\,\boxed{\textbf{RESET}}\,\boxed{\textbf{ENTER}}$ $\boxed{\textbf{2ND}}\,\boxed{\textbf{QUIT}}$. Our intention is to provide instructions that focus on the function we desire to access rather than the function written directly on the key pushed.

> To **reset** your calculator, use
> $\boxed{\textbf{2ND}}\,\boxed{\textbf{RESET}}\,\boxed{\textbf{ENTER}}\,\boxed{\textbf{2ND}}\,\boxed{\textbf{QUIT}}$.

If you reset the BA II Plus calculator you clear all ten memories and all worksheet data. The default settings include using angles measured in degrees rather than radians as the arguments of trigonometric functions.[1] More importantly, the default format setting is **2-formatting**. The BA II Plus calculator can

---

[1] Trigonometric functions and their inverses are rarely used in this subject, but see Problems

display at most ten digits (although it stores thirteen-digit numbers), and for α in $\{0, 1, 2, \dots, 9\}$, we say that we have α-**formatting** if, while not exceeding the ten digit limit, the calculator attempts to display up to α digits to the right of the decimal point. (So with 4-formatting the ten digit number 12345.54321 is displayed as 12345.5432 if it is the result of a calculation, while the twelve-digit number 123456789.543 is displayed as 123456789.5. Rounding is used, so with 4-formatting the ten digit number 12345.56786 is displayed as 12345.5679 if it is the result of a calculation, as is 12345.56785; but a resulting 12345.5678499 will be displayed as 12345.5678.) The authors most often prefer to keep their calculators with **9-formatting** so that an answer is displayed with maximal possible accuracy.

To obtain α-formatting, use

2ND | FORMAT | α | ENTER | 2ND | QUIT .

Note that you enter α when the calculator screen reads "DEC =   "
as it awaits your numerical input. Allowable inputs are in the interval
$(-.5, 9.5)$, but you might as well enter an integer from $\{0, 1, 2, \dots, 9\}$
as the calculator rounds the entered number to an integer.

One situation in which you may not want to have 9-formatting is if you are calculating a number of dollars to change hands, particularly if you then use that dollar amount for further computation. To find the correct number of cents and have it stored in the calculator for further calculation, you may use both the FORMAT key and the ROUND key. This technique is demonstrated in the solution to our first (and very elementary) example.

EXAMPLE 0.3.1

Problem: Mrs. Juanitez is a savvy shopper. At a drugstore, she finds a binder on sale for $14.97 plus 8.25% sales tax. She has a mail-in coupon for a $7.00 rebate on the binder, and she charges it on her credit card that refunds 5% of the charged amount to her. The store advertises that the binder is "$7.97 after mail-in rebate." Calculate how much the binder will actually cost her assuming that it costs 42 cents to mail in the rebate request (postage and the cost of the envelope).

Solution   The drugstore purchase generates a credit card charge of $16.21, since $1.0825 \times \$14.97 = \$16.205025$. As stated, Mrs. Juanitez spends $.42 to mail her $7.00 rebate request. She receives the $7.00 rebate and Mrs. Juanitez's credit card company gives her a rebate of $.81 since $.05 \times \$16.21 = \$.8105$. Therefore,

---

(0.3.2) and (1.12.10). The inverse of a trigonometric function is called by keying INV prior to pressing the key sequence for the function; for example INV 2ND TAN accesses $\tan^{-1}$.

Mrs. Juanitez's total cost for the binder is $16.21 + $.42 − $7.00 − $.81 = $8.82.

To perform this simple calculation on the BA II Plus calculator, first format to two decimal places (if necessary) by pressing

| 2ND | FORMAT | 2 | ENTER | 2ND | QUIT |.

Then press

| 1 | • | 0 | 8 | 2 | 5 | × | 1 | 4 | • | 9 | 7 | = | 2ND | ROUND | STO | 0 |
| + | • | 4 | 2 | = | − | 7 | = | − | (| RCL | 0 | × | • | 0 | 5 |) | = |.

Your display should now read 8.82, so Mrs. Juanitez's cost was $8.82. Note that had you pressed the above sequence of keys except that you omitted | 2ND | ROUND |, your answer would have been $8.81. This is because stored information carries more decimal values than those in the display.   ■

The use of 2-formatting and the | ROUND | key was appropriate in Example (0.1.2) because all rounding was to the **nearest** integral number of cents. However, you should be careful because sometimes a problem requires you to round **up** or round **down**.

Note 0.3.2 on rounding:

Usually, you wait until the end of the problem to round. The exception is that if money changes hand (for instance, you close a savings account or make a mortgage payment), you must round to an integral number of cents. Whether you should round to the nearest penny, round up, or round down depends on the particular situation. For instance, if you are asked for the smallest deposit you can make today at 3% so as to be able to withdraw $243.25 one year from now and $K deposited today will grow to $K(1.03)$, you **round up** $\frac{$243.25}{1.03} \approx 236.1650485 to $236.17. We have used the notation $\approx$ to mean "approximately equal to," and this will be used throughout the text. On the other hand, if you are asked how much you may withdraw in one year if you deposit $236.17 now to your new account, you compute $236.17(1.03) = 243.2551, and since the balance to the **nearest** penny is $243.26, you may withdraw $243.26 unless you have a particularly stingy bank that makes you round down to $243.25. (To determine whether the bank can round down, you might have to read the fine print or look at applicable state law, but we will not concern ourselves with this.)

As was the case in Example (0.1.2), our calculator instructions will sometimes include storing a partial result in one of the BA II Plus calculator's ten memories. It is a matter of taste (and habit) which memory is selected. Therefore, although in Example (0.1.2) we specified storage in memory 0, in our directions specific memory numbers will not be given. It is helpful to develop routines as to where you store certain types of intermediate results, but it would be presumptuous of us to impose a particular pattern.

> To **store a number** displayed on your calculator in memory **m**
> use $\boxed{\text{STO}}\ \boxed{\text{m}}$.

> To **recall a number stored in memory m** use $\boxed{\text{RCL}}\ \boxed{\text{m}}$.

A useful feature of the BA II Plus calculator is that you may replace the number stored in register m by its sum with the displayed value by pushing $\boxed{\text{STO}}\ \boxed{+}\ \boxed{\text{m}}$. Likewise, you may replace the entry in register m by its value minus the displayed value by pushing $\boxed{\text{STO}}\ \boxed{-}\ \boxed{\text{m}}$, by its value multiplied by the displayed value by pushing $\boxed{\text{STO}}\ \boxed{\times}\ \boxed{\text{m}}$, and by its value divided by the displayed value by pushing $\boxed{\text{STO}}\ \boxed{\div}\ \boxed{\text{m}}$. Pressing $\boxed{\text{STO}}\ \boxed{\text{y}^{\text{x}}}\ \boxed{\text{m}}$ will result in the value in memory m being raised to the number that was just displayed. If you wish to view the entries in all ten memory registers, it is most efficient to open the **Memory worksheet** by pressing $\boxed{\text{2ND}}\ \boxed{\text{MEM}}$. The calculator display will then show "M0 =   ". The number following the equal sign gives the number stored in memory register 0, or as many digits of it as the formatting requires. Go ahead and push $\boxed{\downarrow}$, and the display will include "M1 =   " and the number stored in memory register 1, again as dictated by formatting. Repeatedly pushing $\boxed{\downarrow}$ allows you to cycle through the memory registers. Should you wish to reverse the direction of your cycling, push $\boxed{\uparrow}$ rather than $\boxed{\downarrow}$.

The **memory worksheet** is one of twelve "worksheets" included in the BA II Plus calculator. Five of these we discuss in detail and three others receive at least passing treatment. In general, a worksheet contains registers for storing a set of variables, and the worksheets each have a set of built-in formulas relating how the entries in the worksheet's registers should be related (although in the case of the memory worksheet, the set of relationships is empty). For instance, the **Interest Conversion worksheet** has variables NOM, EFF, and C/Y, and if j, i, and m are their respective entries, the formula $1 + i = (1 + \frac{j}{m})^m$ should hold. The **TVM worksheet** [discussed in Section (3.2)] has registers that may be filled using the keys $\boxed{\text{N}}$, $\boxed{\text{I/Y}}$, $\boxed{\text{PV}}$, $\boxed{\text{PMT}}$, and $\boxed{\text{FV}}$ highlighted in Figure (0.1.1), but usually worksheet registers are accessed using the $\boxed{\text{2ND}}$ key. For instance, the **Interest Conversion worksheet** is accessed by keying $\boxed{\text{2ND}}\ \boxed{\text{ICONV}}$.

Except for the **TVM worksheet**, in each worksheet you cycle through a set of variables using the key $\boxed{\downarrow}$ or the key $\boxed{\uparrow}$. If you wish to change the value a variable is set equal to, key the new value and then press $\boxed{\text{ENTER}}$. Once you have finished your work on a particular worksheet, at least for the time being, you should leave it by pushing 2ND QUIT . Just as cookbook authors most often neglect to tell the cook to turn off the oven, even if they have given preheating instructions, many examples end with a displayed worksheet computation being discussed, and you are left to quit the worksheet on your own. Ovens left on too long present fire hazards, and calculators left in a worksheet mode have their own hazards!

> To exit a worksheet on your calculator, use 2ND QUIT .

The BA II Plus calculator has many other features that you may enjoy using. You are encouraged to browse through the instruction manual produced by Texas Instruments. In particular you might appreciate accessing your last answer by keying 2ND ANS or storing a number and an operation for use in repetitive calculations by keying 2ND K . You may quickly multiply a displayed entry by .01 by pushing % , a useful feature since we frequently obtain a rate as a percent and wish to change to its decimal representation.

0.4   PROBLEMS, CHAPTER 0

(0.0) Chapter 0 writing problems:

(1) We have introduced you to the chain calculation method and indicated that the text will give instructions for those who have selected this method. In the text, we gave a couple of examples where the AOS method would produce different results. Give three other examples where the two operating methods produce different results. In each instance, explain how parentheses might be added to the string so that the two methods would agree.

(2) Write a paragraph discussing the advantages and disadvantages of each of the two calculation methods, Chn and AOS, available on the BA II Plus calculator.

(0.3) BA II Plus calculator Basics

(1) Set your calculator with 9-formatting and the chain calculation method, calculate

$$(3.264)(1.0825) + (4.67)(1.065),$$

and round your answer to the nearest hundredth. How does this compare to the result you see displayed if you have 2-formatting and calculate $(3.264)(1.0825) + (4.67)(1.065)$? How do these answers compare to the result you obtain if you first round the products $(3.264)(1.0825)$ and

$(4.67)(1.065)$ to the nearest hundredth with 2-formatting and then add them? Explain how to obtain the sum of the two rounded products on your BA II Plus calculator.

(2) The BA II Plus calculator does not have a key with π on it. Find the first nine digits of pi by computing $2\sin^{-1}1$. Here the argument of the inverse sine function is given in radians! (If you do not see "RAD" on your calculator display, you will need to change from degrees to radians by keying 2ND FORMAT ↓ 2ND SET 2ND QUIT.) What kind of formatting is needed to obtain the first nine digits of pi? Explain how you can obtain the first eleven digits of pi with your BA II Plus calculator.

(3) Using the chain calculation method, compute the following to the nearest ten thousandth. (a) $500(1.04)^5 + 800(1.03)^4$ (b) $\frac{\ln(32,546)}{\ln(1.05)}$ (c) $e^{1.05} - 1$.

(4) With your calculator set to use the chain calculation method, consider the following sequences of keystrokes. In each case, describe what they accomplish if the calculator has just been cleared.

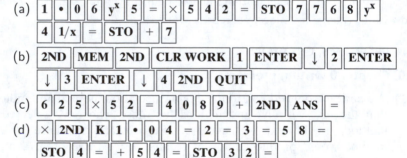

CHAPTER 1

The growth of money

1.1 INTRODUCTION

Throughout this book, the growth of money due to investment will be discussed. In our first section (Section (1.2)) we take a brief look at the rationale behind deposits increasing in value, and we end the chapter with Section (1.14) describing how inflation can erode that growth. In Section (1.3), we establish units of time and of money and the fundamental **amount functions**. Linear **simple interest** amount functions are discussed in Section (1.4), while in Section (1.5) we discuss the **usual compound interest accumulation function**. It is important to read that section carefully so that you understand why growth by compound interest is so prevalent. You are probably familiar with loans where interest is applied at the end of each time period, but the concept of loans **made at a discount** where interest is due at the beginning of the loan may be new to you. This is discussed in Section (1.6), and **discount rates** are introduced. The fundamental concept of **equivalence of rates** is also established. In Section (1.7) we introduce **discount functions**, which are reciprocals of amount functions. Discount functions are useful for determining the value of a promised future payment, and linear discount functions are discussed in Section (1.8). In Section (1.9) we discover that what we might reasonably call "compound discount" is **equivalent** to compound interest. The interest rates and discount rates introduced prior to Section (1.9) are called **effective** rates. **Nominal** rates are explained in Section (1.10). These are rates that do not give rise to the expected growth for periods differing from the so-called compounding period. As we look at nominal rates compounded more and more frequently, we are led to consider the **force of interest**. The force of interest is a measure of how strongly interest is working to increase investments at a given instant.**force of interest** In Section (1.11) we find that compound interest gives rise to a constant force of interest. For those with a background including calculus, Section (1.12) gives a general discussion of the force of interest. If you do not know calculus, skip Section (1.12), but Section (1.13) was written for you.

1.2   WHAT IS INTEREST ?

People frequently participate in financial transactions. These are varied and at times quite complicated. However, it is fair to say that when lenders invest their money, they usually do so with the expectation (or at least the hope) of financial gain. If an investment amount $\$K$ grows to an amount $\$S$, then the difference $\$S - \$K$ is **interest**. The interest may be thought of as a rent paid by the borrower for the use of the $\$K$.

You probably take it for granted that you will be charged interest if you borrow money from a bank, and that you will receive interest if you lend money to a bank by opening a savings account. Nowadays, except for loans to family members or friends, interest-free loans are a rarity. In Western society the charging of interest is generally accepted business practice, but this was not always true. A powerful moral argument suggests you help your neighbor if he is in need. In the Middle Ages, the Catholic church viewed the charging of interest as sinful, and this view is still held in much of the Islamic world.

Part of the economic rationale for the charging of interest is based on the economic productivity of capital. This is sometimes referred to as the **investment opportunities theory**. If a farmer borrows seed and then harvests the crop he grew from it, he may return the quantity of seed he borrowed plus a bit more as interest. The loan allowed productivity, and the farmer shares his gain. Likewise, if you borrow money to run a successful business, the borrowed money allows you to produce more money and you should assign some of that gain to the lender. (A problem occurs for the borrower if his venture does not regenerate at least the original loan amount plus the interest due.)

Another justification for the charging of interest is the **time preference theory**. Generally, people prefer to have money now rather than the same amount of money at some later date. After all, if you have the money now, you have a choice as to whether you use it now or save it for the future. If you lend it, you no longer have the option of immediately using your money. Interest compensates a lender for this loss of choice.

One of the most widely accepted excuses for interest being charged is that a lender should be compensated for the possibility that the borrower defaults and capital is lost.

In the real world, investments have an element of risk and investors sometimes lose money. Some investments, such as deposits insured by the FDIC [1] or Treasury securities backed by the United States government, have low risk. Risk-free investments earn a positive amount of interest. In Section (1.14) we briefly discuss the risk due to inflation, but until Chapters 8 and 9, we rarely consider questions of risk. Therefore, *we assume that a nonnegative amount, and usually a positive amount, of interest is paid.* Of course in real life, issues of risk are of great importance!

1.3   ACCUMULATION AND AMOUNT FUNCTIONS

We begin by introducing units of money. Most commonly we will think of these as dollars, and hence we will use a dollar sign to indicate our units of money. However, what follows is equally valid for other units of money such as euros, yen, or gold pieces. Occasionally, it might be more useful to have our basic unit of money be thousands of dollars or millions of dollars.

A common financial transaction is for one party to lend another party a lump sum of money. The lender views this as an investment since interest is expected in addition to the return of the loan amount. The amount K of money the investor loans is called the **principal**. If we turn this around to the borrower's perspective, the amount of money borrowed is the principal.

Let us introduce units of time. The units of time are most often years, but sometimes it is more convenient to use some other unit such as months or days or quarters of a year. We need to choose a time to call time 0 and it is natural to make that the time of our initial financial transaction. Thus, we suppose that K is invested at time $t = 0$.

We can now define a real-valued function $A_K(t)$ with domain $\{t | t \geqslant 0\}$ by insisting that $A_K(t)$ equals the balance at time t. The function $A_K(t)$ is called the **amount function** for principal K. It is standard to write $a(t)$ for $A_1(t)$ where $A_1(t)$ is the amount function for principal $1. The function $a(t)$

---

[1] The Federal Deposit Insurance Corporation (FDIC) is a federal government agency, that was created in 1933 in response to the many bank failures of the 1920s and 1930s. It oversees more than 5000 member banks and insures the deposits of individual investors at these banks. The amount insured per investor at each bank is limited to $100,000, but deposits held with different ownership (e.g., single with spouse as survivor, joint) may be insured individually, effectively allowing a depositor more insurance. The total amount of insured deposits surpasses three trillion dollars. FDIC does *not* insure investments other than deposits, even if they are offered by a member bank. The FDIC is funded by premiums that banks pay and by interest they earn.

is called the **accumulation function**.

> **FACT (1.3.1):**
> **$\$K$ invested at time 0 grows to $\$A_K(t)$ at time t, and $\$1$ invested at time 0 grows to $\$a(t)$ at time t.**

Typically $A_K(t) = Ka(t)$. This equality means that the amount your investment is worth at time t is proportional to the amount $\$K$ you deposited at time 0. However, this relationship need not hold. For instance, you may encounter tiered investment accounts in which the rate of interest you earn depends on the interval in which your balance lies. (See Example 1.5.5.) We will assume that $A_K(t) = Ka(t)$ unless a tiered growth structure is specifically indicated.

We note that $a(0) = 1$ and $A_K(0) = K$. The functions $a(t)$ and $A_K(t)$ are most often nondecreasing functions. However, if a fund loses money over an interval, the associated accumulation and amount functions will decrease. If we have continuous accrual of interest, the amount function is continuous. On the other hand, if interest is only paid at the end of each interest period, say at the end of each quarter of a year, the associated accumulation and amount functions are step functions.

EXAMPLE 1.3.2

Problem: An investment of $\$1,000$ grows by a constant amount of $\$250$ each year for five years. What does the graph of the amount function $A_{1,000}(t)$ look like if interest is paid continuously using the linear relationship $A_{1,000}(t) = \$1,000 + \$250t$? How about if interest is only paid at the end of each year?

Solution   If interest is earned continuously using the given linear relationship, the graph of $A_{1,000}(t)$ is a line segment with slope 250.

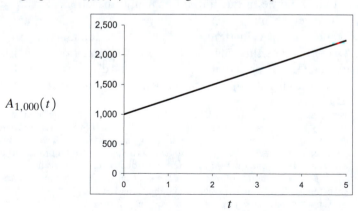

If interest is only paid at the end of each year, the graph of $A_{1,000}(t)$ is as follows:

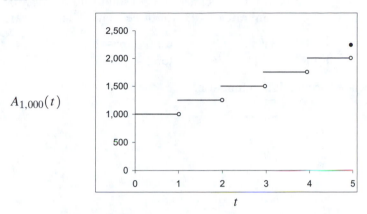

$A_{1,000}(t)$

■

EXAMPLE 1.3.3

Problem: Suppose that time is measured in years and an investment fund grows according to $a(t) = (1.2)^t$ for $0 \leqslant t \leqslant 5$. Then the investment fund grows at a constant rate of 20% per year. (In Section 1.5, we will call $a(t)$ a compound interest accumulation function with annual effective interest rate $i = .2$.) Graph the accumulation function $a(t)$.

Solution

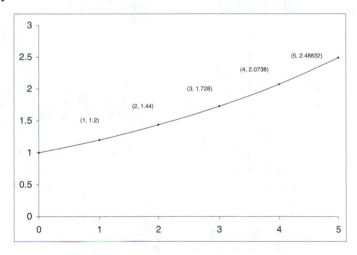

■

EXAMPLE 1.3.4

Problem: The Risky Investment Fund declines at a constant rate of 30% for two years, then grows at a constant rate of 40% for three years. That is to say,

$$a(t) = \begin{cases} (1 - .3)^t & \text{for } 0 \leqslant t \leqslant 2 \\ (1 - .3)^2(1 + .4)^{t-2} & \text{for } 2 \leqslant t \leqslant 5. \end{cases}$$

Graph the associated accumulation function.

Solution

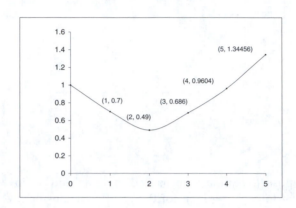

If $t_2 > t_1 \geq 0$, then $\$(A_K(t_2) - A_K(t_1))$ gives the **amount of interest** earned on an investment of $\$K$ (made at $t = 0$) between time t_1 and time t_2. We define the **effective interest rate for the interval** $[t_1, t_2]$ to be

(1.3.5)
$$i_{[t_1,t_2]} = \frac{a(t_2) - a(t_1)}{a(t_1)}.$$

Note that whenever $A_K(t) = Ka(t)$, we also have

(1.3.6)
$$i_{[t_1,t_2]} = \frac{A_K(t_2) - A_K(t_1)}{A_K(t_1)}.$$

An investor of $\$K$ at time 0 is concerned with $A_K(t)$ and hence when $A_K(t) \neq Ka(t)$, he still regards the right-hand side of (1.3.6) as being the rate of interest for the interval $[t_1, t_2]$, even if it is not equal to the left-hand side of (1.3.6). (Of course this only happens if $A_K(t) \neq Ka(t)$.)

If n is a positive integer, the interval $[n - 1, n]$ is called the **n-th time period** and we agree to write i_n for $i_{[n-1,n]}$. Thus,

(1.3.7)
$$i_n = \frac{a(n) - a(n - 1)}{a(n - 1)} \quad \text{and} \quad a(n) = a(n - 1)(1 + i_n).$$

In particular,

(1.3.8) $$i_1 = \frac{a(1) - a(0)}{a(0)} = \frac{a(1) - 1}{1} = a(1) - 1.$$

Note that i_n represents the interest rate earned by an investor during the n-th period in which the investment is governed by the accumulation function $a(t)$.

1.4   SIMPLE INTEREST / LINEAR ACCUMULATION FUNCTIONS

In the world of day-to-day financial transactions, you don't hear talk of accumulation and amount functions. However, investments and loans are usually made according to carefully spelled out rules, and many of these algorithms are easily translated into the language of amount functions. Among the simplest amount functions which occur in real life are linear functions, and we next consider how these occur.

Consider two parties negotiating a loan. They might agree that for each $1 borrowed, at the time of repayment the borrower will pay $.05T$ interest where T is the number of years until repayment. (They will also presumably agree upon allowable values of T.) Then the amount owed at time t per dollar borrowed is $a(t) = 1 + (.05)t$, a linear accumulation function with y-intercept 1 and slope .05.

Focus now on an arbitrary linear accumulation function $a(t)$. If $a(t)$ were linear, since $a(0) = 1$, the accumulation function $a(t)$ would satisfy $a(t) = 1 + st$ for some constant s. Then $a(1) = s + 1$. But by (1.3.8), $s = i_1$ and $a(t) = 1 + i_1 t$.

$A_K(t) = K(1 + st)$ is called the **amount function for K invested by simple interest at rate s. The function** $a(t) = 1 + st$ **is the simple interest accumulation function at rate** s.

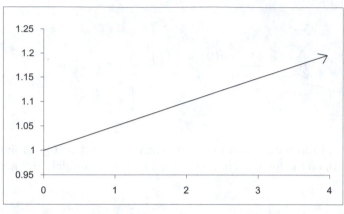

Graph of $a(t) = 1 + st$ **for** $s = .05$

We next consider a couple of problems involving simple interest.

EXAMPLE 1.4.1

Problem: Tonya loans Renu $1,600. Renu promises that in return, she will pay Tonya $2,000 at the end of four years. To what rate of simple interest does this correspond?

Solution   The equation $2,000 = \$A_{1,600}(4) = \$1,600(1 + 4s)$ is equivalent to $s = \frac{1}{4}(\frac{2,000}{1,600} - 1) = .0625$, so the loan corresponds to a simple interest rate of 6.25%.   ∎

EXAMPLE 1.4.2

Problem: Antonio loans his brother Bob $2,400 for three years at 5% simple interest. The brothers agree that if Bob wishes to repay the loan early, he may do so, and the repayment amount will still be based on 5% simple interest. Find the amount Bob would be required to pay if he makes his repayment at the end of three years. What if the repayment is after two years or after one year? Calculate i_1, i_2, and i_3 if the loan lasts the full three years.

Solution   If Bob repays the loan after three years, he must pay Antonio $2,400(1 + 3(.05)) = \$2,760$. If the repayment comes at the end of two years, the repayment amount is $2,400(1 + 2(.05)) = \$2,640$, while if it comes after one year, the amount due is $2,400(1 + 1(.05)) = \$2,520$. We therefore have the following annual effective interest rates.

$$i_1 = \frac{\$2{,}520 - \$2{,}400}{\$2{,}400} = 5\%,$$

$$i_2 = \frac{\$2{,}640 - \$2{,}520}{\$2{,}520} \approx 4.76\%,$$

and

$$i_3 = \frac{\$2{,}760 - \$2{,}640}{\$2{,}640} \approx 4.55\%.$$

Note that the annual effective interest rates are decreasing. Each year there is less incentive for Bob to repay the loan early since the rate of interest he is paying is lower. ∎

When the growth of money is governed by simple interest at rate s,

(1.4.3)
$$i_n = \frac{a(n) - a(n-1)}{a(n-1)} = \frac{(1 + sn) - (1 + s(n-1))}{1 + s(n-1)}$$
$$= \frac{s}{1 + s(n-1)}.$$

Hence $\{i_n\}$ is a decreasing sequence (and, for those of you familiar with calculus, $\{i_n\}$ converges to 0). **In part because $\{i_n\}$ is a decreasing sequence, simple interest is rarely used for loans of long duration.**

In Examples (1.4.1) and (1.4.2), the loans were for an integral number of years and we counted the length of time in the simplest possible manner. However, when simple interest is used, there are a number of different methods used for measuring the time of the loan. We now briefly mention some of the more common rules.

With the **exact simple interest** method, the term D of the loan is first measured in days and then divided by the number of days in a year (usually 365) to yield the length of the loan in years. Exact simple interest is sometimes referred to as the **"actual/actual"** method, the first "actual" for the number of days, the second "actual" for dividing by 365. To calculate D, it is important to know the number of days in each month. January, March, May, July, August, October, and December each have 31 days while April, June, September, and November each have 30. February has 28 days unless it occurs in a leap year, in which case it has 29 days. You are likely familiar with the fact that leap years only occur in years divisible by 4. A more obscure fact is that if a year is divisible by 100, it must also be divisible by 400 in order to be a leap year. Hence 1988, 1956, and 2000 were leap years, while 1987, 1953, and 1900 were not leap years.

EXAMPLE 1.4.3   **Exact simple interest actual/actual**

Problem: Brad borrows $5,000 from Julio on October 14, 1998 at 8% exact simple interest and agrees to repay the loan on May 7, 1999. What is the amount of Brad's required May repayment?

Solution   The duration of the loan in days is

$$\underbrace{(31 - 14)}_{\text{Oct.}} + \underbrace{30}_{\text{Nov.}} + \underbrace{31}_{\text{Dec.}} + \underbrace{31}_{\text{Jan.}} + \underbrace{28}_{\text{Feb.}} + \underbrace{31}_{\text{Mar.}} + \underbrace{30}_{\text{Apr.}} + \underbrace{7}_{\text{May}} = 205$$

and therefore Brad must repay Julio $5,000$(1 + .08(\frac{205}{365})) \approx$ $5,224.66. ■

With the **ordinary simple interest** method, you pretend each month has 30 days and hence a year has 360 days.

If the loan is from day d_1 of month m_1 of year y_1 to day d_2 of month m_2 of year y_2, then you compute the number of days using the formula

(1.4.5) $\boxed{d = 360(y_2 - y_1) + 30(m_2 - m_1) + (d_2 - d_1).}$

Ordinary simple interest is sometimes referred to as the **"30/360"** rule, the "30" for the number of days in a month, the "360" for the number of days in a year.

EXAMPLE 1.4.6   **Ordinary simple interest 30/360**

Problem: Suppose that the loan of Example (1.4.4) is made at 8% ordinary simple interest instead of 8% exact simple interest. What is the amount of Brad's required May repayment?

Solution   According to the method of ordinary simple interest, the duration of the loan in days is calculated to be

$$360(1999 - 1998) + 30(5 - 10) + (7 - 14) = 203.$$

Therefore Brad must repay Julio $5,000$(1 + .08(\frac{203}{360})) \approx$ $5,225.56. ■

The **Banker's rule** is a hybrid of the above methods that is usually more advantageous to the lender. As in the exact simple interest method, count the actual number of days but imagine that a year has 360 days. The Banker's rule is sometimes referred to as the **"actual/360"** method.

EXAMPLE 1.4.7   Banker's rule actual/360

Problem: Suppose that the loan of Example (1.4.4) is made by 8% simple interest computed using the Banker's rule instead of 8% exact simple interest. What is the amount of Brad's required May repayment?

Solution   The term of the loan is calculated as in Example (1.4.4) and we use 360 for the number of days in a year. Therefore, Brad must repay Julio $\$5,000(1 + .08(\frac{205}{360})) \approx \$5,227.78.$ ∎

The duration of the loan, be it an actual count of the number of days or an estimate based on the assumption that all months have thirty days, may be quickly determined using the BA II Plus calculator's **Date worksheet** so long as the loan takes place during the years 1950–2049. (In fact, in the event that you are interested in an interval other than during this one hundred year span, the worksheet is still quite helpful [see Problem (1.4.7)]. We next illustrate the use of the **Date worksheet** with the loan interval of Examples (1.4.4) and (1.4.6), namely the period from October 14, 1998 until May 7, 1999.

Press $\boxed{\text{2ND}}$ $\boxed{\text{DATE}}$ to open the date worksheet. Next push the keys $\boxed{\text{2ND}}$ $\boxed{\text{CLR WORK}}$ if you need to check whether the calculator is formatted to accept dates in U.S. order (Month Day Year) or European order (Day Month Year). If it is ready to accept U.S. order, the display will read "DT1 = 12 - 31 - 1990", and otherwise the display will show "DT1 = 31 - 12 - 1990", indicating European formatting. The default formatting is for the U.S. ordering, and assuming this is in place, you should push $\boxed{1}$ $\boxed{0}$ $\boxed{\bullet}$ $\boxed{1}$ $\boxed{4}$ $\boxed{9}$ $\boxed{8}$ $\boxed{\text{ENTER}}$ to enter a starting date of October 14, 1998. (Should your display have read "DT1 = 31 - 12 - 1990", indicating European formatting, you should enter October 14, 1998 by keying $\boxed{1}$ $\boxed{4}$ $\boxed{\bullet}$ $\boxed{1}$ $\boxed{0}$ $\boxed{9}$ $\boxed{8}$ $\boxed{\text{ENTER}}$.) Push $\boxed{\downarrow}$ and the display will show "DT2", indicating that the worksheet is ready to accept the loan completion date of May 7, 1999. Enter this date by keying $\boxed{0}$ $\boxed{5}$ $\boxed{\bullet}$ $\boxed{0}$ $\boxed{7}$ $\boxed{9}$ $\boxed{9}$ $\boxed{\text{ENTER}}$ or $\boxed{0}$ $\boxed{7}$ $\boxed{\bullet}$ $\boxed{0}$ $\boxed{5}$ $\boxed{9}$ $\boxed{9}$ $\boxed{\text{ENTER}}$ depending on whether you have U.S. or European formatting. (The BA II Plus calculator accepts calendar dates from January 1, 1950 through December 31, 2049. Regardless of which formatting is in effect, the year is entered by keying the last two-digits after the month and day have been entered. The month and day are each recorded by entering a two digit number, and they are separated by entering a decimal point.)

Now that the loan commencement and termination dates have been entered, you should press $\boxed{\downarrow}$ $\boxed{\downarrow}$. This will result in the display reading either "ACT" or "360." The first of these tells you that the calculator is prepared to make an exact calculation of the days the loan lasted, while the latter alerts you that it will estimate the loan duration using thirty-day months. Should you wish to have the other basis for the calculation of the loan duration, push $\boxed{\textbf{2ND}}\ \boxed{\textbf{SET}}\ \boxed{\uparrow}\ \boxed{\textbf{CPT}}$. On the other hand, if you are satisfied with the indicated calculation method, just push $\boxed{\uparrow}\ \boxed{\textbf{CPT}}$. Now read the display. If you used the "ACT" calculation, your screen should show "DBD = 205", while the "360" method will produce "DBD = 203".

1.5 COMPOUND INTEREST (THE USUAL CASE!)

Although simple interest is easy to compute, practical applications of this method are limited. We now explain why this is so. Suppose you invest at a bank where savings accounts earn simple interest at a rate i. As noted in Section (1.4), the effective interest rate in the n-th year is a decreasing function of n. Consequently, you would do well to go into the bank, close your account, and then instantly reopen it. But this would be inconvenient for you and for the bank. Therefore, it is sensible for the bank to design an account that grows in a manner where there is no advantage or disadvantage to closing an account and then instantly reopening it. In particular, we want the effective interest rate for the n-th period i_n to be independent of n.

Define

(1.5.1)
$$\boxed{i = i_1 = a(1) - 1.}$$

CLAIM 1.5.2:

If an accumulation function $a(t)$ has the associated periodic interest rates i_n all equal to the constant i, then the accumulation function must satisfy $a(k) = (1 + i)^k$ for all nonnegative integers k.

Claim (1.5.2) is a statement about the nonnegative integers. The Principle of Mathematical Induction is a valuable technique for proving facts about the nonnegative integers (or about any infinite set of consecutive integers with a smallest element), and we shall use this method to establish Claim (1.5.2). Mathematical Induction requires you to establish that your claim is true for the smallest integer in your set of consecutive integers (in this case for $k = 0$) and that the claim being true for a given integer k forces its validity for that integer's successor $k + 1$.[2]

Proof: (An induction argument) For a given k, refer to the equation $a(k) = (1 + i)^k$ as equation k. Our task is to establish equation k for all nonnegative

---

[2]The following commonly used analogy involves dominoes or other thin blocks. It may help you visualize why Mathematical Induction should be allowed. Suppose you place dominoes close together with their faces parallel to one another. Do not use glue! If the block at one end falls toward the next block, the blocks will all fall.

integers k. Note that equation 0 is true since $(1 + i)^0 = 1$ and for any accumulation function, $a(0) = 1$. Having established this first equation, our method is to show that if equation k is true, equation $(k + 1)$ must also be true. For this, note that we now have two assumptions, namely (1) $i_n = i$ for all n — in particular $i_{k+1} = i$, and (2) $a(k) = (1 + i)^k$. From these it follows that

$$i = i_{k+1} = \frac{a(k + 1) - a(k)}{a(k)} = \frac{a(k + 1) - (1 + i)^k}{(1 + i)^k}.$$

But this is equivalent to

$$a(k + 1) = i(1 + i)^k + (1 + i)^k = (1 + i)(1 + i)^k = (1 + i)^{k+1}.$$

So, we have verified equation $(k + 1)$. ■

The situation for nonintegral investment periods is more subtle. In terms of accumulation functions, the condition that there should never be an advantage or disadvantage to closing and immediately reopening one's account means that

(1.5.3) $a(s + t) = a(s)a(t)$ for all positive real numbers t and s.

In calculus-based Problem (1.5.11), we help the student show that assuming $a(t)$ is differentiable at all $t > 0$ and is differentiable from the right at $t = 0$, then equation (1.5.3) forces $a(t) = (1 + i)^t$ for all nonnegative t, not just for integral t. In practice however, banks do not always use this formula when t is nonintegral. Some banks pay compound interest for an integral number of years followed by simple interest for the final portion of a year. For instance, they may set $a(3.25) = a(3)(1 + .25i) = (1 + i)^3(1 + .25i)$ rather than using $a(3.25) = (1 + i)^{3.25}$.

Unless otherwise stated, we will use $a(t) = (1 + i)^t$ for all $t \geq 0$ and will call this the **compound interest accumulation function** at interest rate i.

For accounts governed by the compound interest accumulation function at interest rate i, money earns interest at the constant interest rate i. As interest is paid, it is reinvested and also earns interest at rate i.

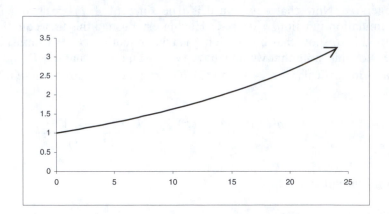

Graph of $a(t) = (1 + i)^t$ **for** $i = .05$

You might wonder how simple interest and compound interest compare. Here is a graph showing both kinds of accumulation with annual rates of 5%.

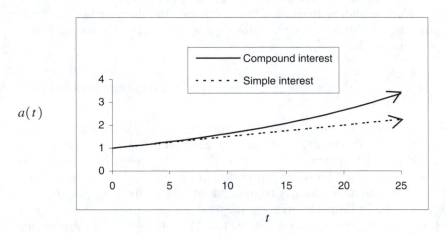

You can easily note that if you go far enough to the right (large t), the compound interest accumulation function lies above the simple interest accumulation function. The scale on our graph is probably not large enough for you to see that the simple interest function is larger up until time 1, at which point the compound interest function begins dominating.

EXAMPLE 1.5.4 Compound interest from deposit to withdrawal

Problem: Fernando deposits $12,000 in an account at Victory Bank where accounts grow according to the compound interest accumulation function $a(t) = (1.05)^t$ for all nonnegative t. He makes no further deposits or withdrawals until he closes the account six and a half years after he opened it. How much money does Fernando receive when he closes the account?

Solution   Fernando receives the account balance which is $12,000(1.05)^{6.5} = $16,478.27.   ∎

In the next example, and again later in the book, we use the **greatest integer function**. This function is defined by

$$\lfloor t \rfloor = \text{the largest integer which does not exceed } t.$$

You may say "the greatest integer in t" for $\lfloor t \rfloor$. Computer scientists refer to the function $\lfloor t \rfloor$ as the **floor function**.

EXAMPLE 1.5.5   Compound interest with simple interest for fractional parts

Problem: Sarafina also had $12,000 to deposit. She invested her money at Simpler Bank where accounts grow according to the accumulation function

$$a(t) = (1.05)^{\lfloor t \rfloor}(1 + .05(t - \lfloor t \rfloor)).$$

She also closes her balance after six and a half years. Compare her closing balance to the closing balance Fernando received in Example (1.5.4).

Solution   Sarafina's closing balance is

$$
\begin{aligned}
\$12,000a(6.5) &= \$12,000(1.05)^{\lfloor 6.5 \rfloor}(1 + .05(6.5 - \lfloor 6.5 \rfloor)) \\
&= \$12,000(1.05)^6(1 + .05(.5)) \approx \$16,483.18.
\end{aligned}
$$

Her balance is close to Fernando's but is slightly higher.   ∎

EXAMPLE 1.5.6   Tiered investment account

Problem: Patriot Bank offers a "bracketed account" with compound interest at an annual effective interest rate of 2% on balances of less than $2,000, 3% on balances of at least $2,000 but less than $5,000, and 4% on balances of at least $5,000. Moises opens an account with $1,800. Determine the amount function $A(t) = A_{1,800}(t)$ and show that the *function* $A_{1,800}(t)$ is different from the *function* $800a(t)$, although their values agree for t in $[0, 5.32]$.

Solution   Define t_1 to be the length of time it takes for Moises' balance to grow to $2,000 and t_2 to be the time it takes to grow from $2,000 to $5,000. Then $1,800(1.02)^{t_1} = $2,000 and $2,000(1.03)^{t_2} = $5,000. It follows that

$$t_1 = \frac{\ln(2,000/1,800)}{\ln(1.02)} \approx 5.320532174,$$

and

$$t_2 = \frac{\ln(5{,}000/2{,}000)}{\ln(1.03)} \approx 30.99891276.$$

Then

$$A(t) = \begin{cases} \$1{,}800(1.02)^t & \text{if } 0 \leqslant t \leqslant t_1 \approx 5.3205; \\ \$2{,}000(1.03)^{t-t_1} & \text{if } 5.3205 \approx t_1 \leqslant t \leqslant t_1 + t_2 \approx 36.319; \\ \$5{,}000(1.04)^{t-t_1-t_2} & \text{if } 36.319 \approx t_1 + t_2 \leqslant t. \end{cases}$$

On the other hand, $a(t) = \$1(1.02)^t$ for $0 \leqslant t \leqslant t_3$ where $(1.02)^{t_3} = \$2{,}000$, that is for

$$t_3 = \frac{\ln(2{,}000)}{\ln(1.02)} \approx 383.8330311.$$

So, $A_{1{,}800} = 1{,}800a(t)$ only on the interval $[0, t_1]$. Beyond that interval, for any argument t, $A_{1{,}800}(t)$ exceeds $1{,}800a(t)$. For instance, $A_{1{,}800}(10) = \$2{,}000(1.03)^{10-t_1} \approx 2{,}340.62$, and $1{,}800a(10) = 1{,}800(1.02)^{10} \approx 2{,}194.19$. ∎

Commonly, the interest rate applied will vary from bank statement to bank statement. It may be determined in a specified manner from a monetary index (for example, the Federal Funds rate or the prime rate), or the bank may be free to offer whatever rate it wants. In either case, supply and demand play a major role, and this in turn is influenced by the government's fiscal and monetary policies. **Fiscal policy** refers to the government's decisions concerning spending and taxation. If the government spends more, this introduces money into the economy and should eventually increase the amount of money available to consumers. This tends to drive interest rates down. On the other hand, taxation decreases the amount of money available for individuals and companies, so interest rates may rise due to increased demand for loans. **Monetary policy** refers to the regulation of the money supply and interest rates by a central bank. In the United States, the Federal Reserve has direct control of the rate it charges banks for "overnight" borrowing. The Federal Reserve may also strongly influence the **Federal Funds Rate** charged for interbank "overnight" loans by competing (or not competing) for bank's money with U.S. Treasury securities. The Federal Reserve has a less direct influence on the **prime rate**, the rate that a bank charges to its "best" customers. The level of these rates tends to impact the interest rates available to all borrowers.

Savings accounts are generally governed by compound interest, but the rate at which compound interest is paid is subject to change.

EXAMPLE 1.5.7   Varying interest rates

Problem: Arisha deposits $8,000 in a savings account at Victory Bank. For the first three years the money is on deposit, the annual effective rate of compound interest paid is 5%, for the next two years it is 5.5%, and for the following four years it is 6%. If Arisha closes her account after nine years, what will her balance be?

Solution   If Arisha closes her account after nine years, her balance will be $8,000(1.05)^3(1.055)^2(1.06)^4 \approx $13,013.26. ■

It is not uncommon for investors to wish to determine a rate of compound interest that would provide them with a certain amount of money at a given later date.

EXAMPLE 1.5.8   Unknown rate

Problem: Pedro wishes to have $12,000 in three years, so he can buy his father's car. Pedro has $9,800 to invest in a three year certificate of deposit. [3] What annual effective rate of interest must the CD earn so that it will have a redemption value of $12,000? When we speak of a CD, we will refer to an investment with a fixed term and fixed rate.

Solution   Denote the required effective interest rate by i. Then $12,000 = $A_{9,800}(3) = $9,800(1 + i)^3$. Solving for i, we find $i = \left(\frac{12,000}{9,800}\right)^{\frac{1}{3}} - 1 \approx$ .069838912. A rate of 7% would produce $9,800(1.07)^3 \approx $12,005.42. ■

1.6   INTEREST IN ADVANCE / THE EFFECTIVE DISCOUNT RATE

When you rent an apartment, usually you are required to pay rent for each month at the beginning of the month. In other words, you pay the rent before you have the use of the apartment. We said (Section (1.1)) that interest may be thought of as a rent for the use of the investor's money. It is therefore not surprising that there are financial arrangements in which the interest must be paid by the borrower before the borrowed money becomes available.

---

[3] A **certificate of deposit** (CD) requires the investor to deposit money to the issuing bank or savings and loan for a fixed term. Should the investor decide to withdraw deposited funds before the end of the term, there is usually a substantial penalty — perhaps one quarter's interest payment — but withdrawals of interest are usually allowed. **Liquid CDs** may allow one or more partial withdrawals without penalty before the CD matures. With a traditional CD, the interest rate is fixed at the time the account is opened. However, the CD market has expanded to include market-linked CDs and CDs for which the investor may request one-or-more adjustments to the interest rate should interest rates go up. Most often, additional funds may not be added to a CD once the account has been opened; however, some CDs allow the customer to make a limited number of additional deposits.

When money is borrowed with interest due before the money is released, we describe the relationship using discount rates. If an investor lends $\$K$ for one basic period at a **discount rate** D, then the borrower will have to pay $\$KD$ in order to receive the use of $\$K$. Therefore, instead of having the use of an extra $\$K$, the borrower only has the use of an extra $\$K - \$KD = (1 - D)\$K$. The quantity $\$KD$ is called the **amount of discount** for the loan.

Note that in Section (1.3), we defined the amount of interest on an interval, and now we have defined the amount of discount on an interval. In any cashflow with a beginning and ending balance and no withdrawals or deposits, the amount of interest and the amount of discount are the same; they are both equal to the change in the balance.

EXAMPLE 1.6.1

Problem: Chan borrows $1,000 at a discount rate of 7%. How much extra money does he have the use of?

Solution In order to get the $1,000, he must first pay $(.07)\$1,000 = \$70$. Chan therefore has the use of an extra $\$1,000 - \$70 = \$930$. ∎

In our example, the discount rate is

$$7\% = \frac{\$70}{\$1,000} = \frac{\$1,000 - \$930}{\$1,000}.$$

In other words, the discount rate for the loan period may be obtained by first calculating the difference between the stated amount of the loan and the amount of extra money actually available, then dividing this difference by the stated amount of the loan.

We wish to define an effective discount rate analogous to the effective interest rate [defined by (1.3.5)]. Suppose that the loan period is the interval $[t_1, t_2]$ from time t_1 to t_2. Also suppose that at time t_1 the borrower will have the use of an extra $\$a(t_1)$. At time t_2 this debt will have grown to $\$a(t_2)$. Therefore, the amount of discount for the interval $[t_1, t_2]$ is $\$a(t_2) - \$a(t_1)$, and the discount per dollar borrowed at a discount is $\$\left(\frac{a(t_2)-a(t_1)}{a(t_2)}\right)$. We define the for the interval $[t_1, t_2]$ to be

(1.6.2)
$$d_{[t_1,t_2]} = \frac{a(t_2) - a(t_1)}{a(t_2)}.$$

Comparing definition (1.6.2) with definition (1.3.5), we see that the definitions for $i_{[t_1,t_2]}$ and $d_{[t_1,t_2]}$ have the same numerators but different denominators. To compute the interest rate $i_{[t_1,t_2]}$, your denominator is the accumulated amount $a(t_1)$ at the **beginning** of the interval $[t_1, t_2]$. To compute the discount rate $d_{[t_1,t_2]}$, you divide by the accumulated amount $a(t_2)$ at the **end** of the interval $[t_1, t_2]$.

If, as is commonly the case, $A_K(t) = Ka(t)$, then

(1.6.3)
$$d_{[t_1,t_2]} = \frac{A_K(t_2) - A_K(t_1)}{A_K(t_2)}.$$

Recall that if n is a positive integer, the interval $[n - 1, n]$ is called the **n-th time period**. We agree to write d_n for $d_{[n-1,n]}$. Thus

(1.6.4)
$$d_n = \frac{a(n) - a(n - 1)}{a(n)} \quad \text{and} \quad a(n - 1) = a(n)(1 - d_n).$$

Compare this definition with the definition of i_n given in Equation (1.3.7).

EXAMPLE 1.6.5   Computing interest and discount rates; Solution includes important calculator information on using stored intermediate results

Problem: Suppose that the growth of money is governed by the accumulation function $a(t) = (1.05)^{\frac{t}{2}}(1 + .025t)$. Find d_4 and i_4.

Solution   Note that $a(4) = (1.05)^2(1.1) = 1.21275$ and $a(3) = (1.05)^{\frac{3}{2}}(1.075)$ ≈ 1.156624568. Therefore $d_4 = \frac{a(4)-a(3)}{a(4)} \approx .046279474$ and $i_4 = \frac{a(4)-a(3)}{a(3)} \approx$ .048525195.

Note that in our computations, we will always use the stored values resulting from previous calculations and these may include more places of accuracy than we have reported; we usually only report the displayed value resulting from 9-formatting. For example, the calculator stores 1.156624567708 for $a(3)$ rather than the announced 1.156624568. However, if you use the number 1.156624568 (displayed for $a(3)$ if you have 9-formatting) instead of the stored value 1.156624567708, your calculated value for i_4 will be approximately .048525194.   ∎

As was the case in Example (1.6.5), usually $i_{[t_1,t_2]}$ and $d_{[t_1,t_2]}$ are not equal. However, they are clearly related. With an eye to pursuing this relationship, we define what it means for a rate of interest and a rate of discount to be **equivalent**.

> **IMPORTANT DEFINITION 1.6.6**
> A rate of interest and a rate of discount are said to be **equivalent** for an interval $[t_1, t_2]$ if for each \$1 invested at time t_1, the two rates produce the same accumulated value at time t_2. More generally, two different methods of specifying an investment's growth (over a given time period) are called **equivalent** if they correspond to the same accumulation function.

Focus on a loan lasting from time t_1 to time t_2. If the loan is for an amount \$$L$ and the interest rate is $i_{[t_1,t_2]}$, then we must repay \$$L(1 + i_{[t_1,t_2]})$. On the other hand, if the loan is made at a discount with discount rate $d_{[t_1,t_2]}$ and the repayment amount is \$$L(1 + i_{[t_1,t_2]})$, then the borrower walked away at the beginning of the loan period with \$$L(1 + i_{[t_1,t_2]})(1 - d_{[t_1,t_2]})$ of the lender's money. Consequently, on the interval $[t_1, t_2]$, an interest rate of $i_{[t_1,t_2]}$ is **equivalent to** a discount rate of $d_{[t_1,t_2]}$ precisely when \$$L = $L(1 + i_{[t_1,t_2]})(1 - d_{[t_1,t_2]})$ for all loan amounts \$$L$. It follows that the rates are **equivalent** if and only if

(1.6.7)
$$1 = (1 + i_{[t_1,t_2]})(1 - d_{[t_1,t_2]}).$$

Equation (1.6.7) may also be algebraically derived using Equations (1.3.5) and (1.6.2), but this demonstration is less instructive from an interest theory point of view.

Equation (1.6.7) is equivalent to

$$0 = i_{[t_1,t_2]} - d_{[t_1,t_2]} - i_{[t_1,t_2]}d_{[t_1,t_2]}$$

which in turn gives rise to

(1.6.8)
$$i_{[t_1,t_2]} = \frac{d_{[t_1,t_2]}}{1 - d_{[t_1,t_2]}}$$

and

(1.6.9)
$$d_{[t_1,t_2]} = \frac{i_{[t_1,t_2]}}{1 + i_{[t_1,t_2]}}.$$

If we have a positive interest rate $i_{[t_1,t_2]}$, it follows from equation (1.6.9) that the discount rate $d_{[t_1,t_2]}$ is less than $i_{[t_1,t_2]}$.

Recalling (1.6.4) and that we write d_n for $d_{[n-1,n]}$, (1.6.7), (1.6.8), and (1.6.9) respectively, give us the equations

(1.6.10)
$$(1 + i_n)(1 - d_n) = 1,$$

(1.6.11)
$$i_n = \frac{d_n}{1 - d_n},$$

and

(1.6.12)
$$d_n = \frac{i_n}{1 + i_n}.$$

EXAMPLE 1.6.13   Discount rates and compound interest

Problem: Suppose that an account is governed by compound interest at an annual effective interest rate of 8%. Find an expression for d_n, the discount rate for the n-th year.

Solution   Since the account is governed by compound interest at an annual effective rate of 8%, $i_n = .08$ for all positive integers n. Therefore (1.6.12) yields $d_n = \frac{.08}{1+.08} = \frac{.08}{1.08} = \frac{2}{27}$, a constant. We will return to constant d_n in Section (1.9).   ∎

1.7   DISCOUNT FUNCTIONS / THE TIME VALUE OF MONEY

In a world where interest may be earned, an amount $\$K$ present on hand is worth more than a payment of $\$K$ t years in the future. This is because you could invest the $\$K$ today and after t years it would have grown to a larger amount $\$A_K(t)$. In particular, $1 invested now will grow to $\$a(t)$ in t years and, assuming the growth of money is proportional to the amount invested, $\frac{\$1}{a(t)}$ will grow to $1 in t years. This leads us to define the **discount function** $v(t)$ by

(1.7.1)
$$v(t) = \frac{1}{a(t)}.$$

The value $\$v(t_0)$ is the amount of money you must invest at time 0 to have $1 after t_0 years.

EXAMPLE 1.7.2

Problem: Suppose that the growth of money for the next five years is governed by simple interest at 5%. How much money should you invest now in order that you have a balance of $23,000 three years from now?

Solution   Note that the discount function is $v(t) = \frac{1}{a(t)} = \frac{1}{1+(.05)t}$. Therefore $v(3) = \frac{1}{1+(.05)3} = 1/1.15$. If we wish to have $23,000 three years from now, we should invest $23,000v(3) = \$23,000(1/1.15) = \$20,000$.   ∎

The next example considers a more subtle question since you wish to invest at a time which is not 0. The reason that this is potentially trickier is that the accumulation function is defined as giving the new value of 1, deposited *at time zero*, t years after it was deposited.

EXAMPLE 1.7.3

Problem: Again, suppose that the growth of money for the next five years is governed by the linear accumulation function $a(t) = 1 + .05t$ so that

$$v(t) = \frac{1}{a(t)} = \frac{1}{1 + (.05)t}.$$

If you wish to invest money two years from now so as to have \$23,000 five years from now, how much money should you invest?

Solution   Let's begin by focusing on the desired \$23,000 five years from now. Were you to invest money *now* so that it would grow to this, you would need to invest $23,000v(5) = \$23,000(1/1.25) = \$18,400$. But \$18,400 would grow after two years to $\$18,400a(2) = \$18,400(1 + 2(.05)) = \$20,240$. Hence you should invest \$20,240 at time two to achieve \$23,000 at time five.

It may be puzzling why the relevant accumulation function is not one that begins at the time ($t = 2$) when you invest. This mystery is resolved by the realization that the accumulation function was specified without regard to your particular investment behavior. Your involvement began after the accumulation function governing the growth of investments was set.   ■

The answer to the problem posed in Example (1.7.2) is \$20,000 while the answer to the problem asked in Example (1.7.3) is \$20,240. The lesson is that in general, the accumulation of money over time depends not only on the length of the time interval but also on where in time the interval lies.

If we review the solution to Example (1.7.3), we see that the answer was obtained by computing $\$23,000v(5)a(2) = \$23,000\frac{a(2)}{a(5)} = \$23,000\frac{v(5)}{v(2)}$. This method of solution applies more generally and gives us the following result.

IMPORTANT FACT 1.7.4

If we wish to invest money t_1 years from now in order to have \$S t_2 years from now, we should invest $\$Sv(t_2)a(t_1) = \$S\frac{a(t_1)}{a(t_2)} = \$S\frac{v(t_2)}{v(t_1)}$.

Next concentrate on the compound interest accumulation function $a(t) = (1 + i)^t$. We define the **discount factor**

(1.7.5)
$$\boxed{v = \frac{1}{1 + i}.}$$

Then
$$v(t) = \frac{1}{a(t)} = \frac{1}{(1 + i)^t} = v^t,$$

and
$$S\frac{v(t_2)}{v(t_1)} = S\frac{v^{t_2}}{v^{t_1}} = Sv^{t_2 - t_1}.$$

So, for compound interest, the amount you need to invest in order to have $\$S$ at a later time depends only on the amount of time until you need the $\$S$.

We next introduce **present value**, a concept that is very useful if you wish to make a comparison of investment alternatives. If the growth of money is proportional to the amount of money invested, and if we have an accumulation function $a(t)$ with associated discount function $v(t) = \frac{1}{a(t)}$, $\$Lv(t_0)$ invested at time 0 will grow to $\$L$ at time t_0. (This is true because $(\$Lv(t_0))a(t_0) = \$L(1/a(t_0))a(t_0) = \$L.$) We therefore define the **present value** *with respect to* $a(t)$ *of* $\$L$ **to be received at time** t_0 to be $\$Lv(t_0)$. We denote this present value by $PV_{a(t)}(\$L$ at $t_0)$.

(1.7.6)
$$\boxed{PV_{a(t)}(\$L \text{ at } t_0) = \$Lv(t_0).}$$

When there is a clear choice of accumulation function, we drop the subscript $a(t)$ in $PV_{a(t)}(\$L$ at $t_0)$.

EXAMPLE 1.7.7

Problem: What is the present value of $\$2,000$ to be paid in four years assuming money grows by compound interest at an annual effective interest rate of 6%? What if money grows by compound interest at an annual effective interest rate of 3%?

Solution In the two scenarios we have compound accumulation functions $a(t) = (1.06)^t$ and $a(t) = (1.03)^t$ respectively. We calculate that

$$PV_{(1.06)^t}(\$2,000 \text{ at } 4) = \$2,000(1.06)^{-4} \approx \$1,584.19$$

and
$$PV_{(1.03)^t}(\$2,000 \text{ at } 4) = \$2,000(1.03)^{-4} \approx \$1,776.97.$$

When we have the lower interest rate 3%, it takes more money invested now to produce $\$2,000$ four years from now. This is why $PV_{(1.03)^t}(\$2,000$ at $4) > PV_{(1.06)^t}(\$2,000$ at $4)$. ∎

Suppose that the growth of money is viewed as being governed by an accumulation function $a(t)$. Then the **net present value** or **NPV** of a sequence of investment returns R_0, R_1, R_2, ..., R_n received at times 0, t_1, t_2, ..., t_n is defined to be the sum

$$\sum_{k=0}^{n} R_k \, v(t_k).$$

Note that a return R_k is positive if the investor receives money and negative if the investor pays out money. If you have a savings account, you are the investor and from your perspective, withdrawals are positive cashflows and deposits are negative cashflows.

EXAMPLE 1.7.8

Problem: Compare the net present values of two certificates of deposit available to Helga, one at Bank Alpha with purchase price $1,000 and redemption value $1,250 at the end of two years, the other at Bank Beta with purchase price $1,000 and redemption value $1,300 at the end of three years, first using the compound interest accumulation function $a(t) = (1.05)^t$ and next using the compound interest accumulation function $a(t) = (1.03)^t$.

Solution   Under compound interest at 5%, the Bank Alpha certificate of deposit has NPV equal to $-\$1,000 + \$1,250(1.05)^{-2} \approx \$133.79$ and the Bank Beta certificate has NPV equal to $-\$1,000 + \$1,300(1.05)^{-3} \approx \$122.99$. So, using the compound interest accumulation function $a(t) = (1.05)^t$, the certificate of deposit at Bank Alpha has a higher NPV. On the other hand, if we repeat our calculation using the compound interest accumulation function $a(t) = (1.03)^t$, Bank Alpha's certificate has NPV $-\$1,000 + \$1,250(1.03)^{-2} \approx \$178.24$ which is lower than the NPV $-\$1,000 + \$1,300(1.03)^{-3} \approx \$189.68$ of Bank Beta's certificate.

Note: Using the accumulation function $a(t) = (1.05)^t$, at Bank Alpha we have a present value

$$PV_{(1.05)^t}(\$1,250 \text{ at } 2) = \$1,250(1.05)^{-2} \approx \$1,133.79,$$

while at Bank Beta we have a present value

$$PV_{(1.05)^t}(\$1,300 \text{ at } 3) = \$1,300(1.05)^{-3} \approx \$1,122.99.$$

We might look at the values $1,133.79 and $1,122.99 as follows. If Helga had a 5% savings account, $1,133.79 deposited at time 0 would grow to $(\$1,133.79)(1.05)^2 = \$1,250$ at time 2. Similarly, $1,122.99 deposited at time 0 would grow to $1,300 at time 3. Thus having the opportunity to invest in the two-year Bank Alpha CD is comparable to being given an extra $133.79 at time 0 to invest in a 5% savings account. The opportunity to invest in the three year Bank Beta CD is comparable to being given an extra $122.99 at time 0 to invest in a 5% savings account for three years. Of course $133.79 and $122.99 are the net present values we found using the accumulation function $a(t) = (1.05)^t$.   ∎

Example (1.7.8) involved a comparison of investments using net present values. Another way of comparing investments is by seeing what annual effective rate of interest each of the investments corresponds to.

EXAMPLE 1.7.9

Problem: As in Example (1.7.8), Helga has $1,000 to invest. She has a choice of two investments. Bank Alpha offers a two-year certificate of deposit in which her $1,000 would grow to $1,250. Bank Beta's three-year certificate of deposit would allow her $1,000 to grow to $1,300. Find the annual effective rate of compound interest to which each of these investments correspond.

Solution We first consider the certificate of deposit offered by Bank Alpha. If Helga opens this account, the $1,000 she deposits at time $t = 0$ grows to $1,250 at $t = 2$. Accordingly, the applicable annual effective rate of compound interest is i_α where

$$\$1,250 = \$1,000(1 + i_\alpha)^2.$$

Therefore,

$$i_\alpha = \left(\frac{\$1,250}{\$1,000}\right)^{\frac{1}{2}} - 1 \approx .118033989 \approx 11.8\%.$$

In contrast, the $1,000 she deposits at time $t = 0$ in a Bank Beta certificate of deposit grows to $1,300 at $t = 3$. So, the applicable annual effective rate of compound interest is i_β where

$$\$1,300 = \$1,000(1 + i_\beta)^3.$$

Therefore,

$$i_\beta = \left(\frac{\$1,300}{\$1,000}\right)^{\frac{1}{3}} - 1 \approx .091392883 \approx 9.1\%.$$

∎

In Example (1.7.9), the higher interest rate at Bank Alpha suggests that Helga might prefer banking there. However, she would have to reinvest her money two years from now, and expectations of what rates will be available at that point must be considered. We will look at problems with reinvestment rates later in the book.

We end this section with a more difficult net present value problem, followed by a discussion of special BA II Plus calculator capabilities for solving certain net present value problems.

EXAMPLE 1.7.10

Problem: Project 1 requires an investment of $10,000 at time 0 and an additional investment of $5,000 at $t = 1$. It returns $2,000 at $t = 3$ and $6,000 at times $t = 4$, $t = 5$, and $t = 6$. Project 2 requires an investment of $6,000 at $t = 0$ and returns $3,500 at $t = 1$ and $5,000 at an unknown time. The net present values of the two projects are equal when calculated using compound interest at 4%. Find the unknown time for the second return of Project 2.

Solution   The net present value of Project 1 is

$$-\$10,000 - \$5,000(1.04)^{-1} + \$2,000(1.04)^{-3} + \$6,000[(1.04)^{-4} + (1.04)^{-5} + (1.04)^{-6}] \approx \$1,772.575351.$$

Project 2 has net present value

$$-\$6,000 + \$3,500(1.04)^{-1} + \$5,000(1.04)^{-T}$$

where T is the time of the unknown $5,000 return. Therefore,

$$-\$6,000 + \$3,500(1.04)^{-1} + \$5,000(1.04)^{-T} \approx \$1,772.575351.$$

This is equivalent to

$$\$5,000(1.04)^{-T} \approx \$1,772.575351 + \$6,000 - \$3,500(1.04)^{-1} \approx \$4,407.190735,$$

so

$$T \approx \frac{\ln\left(\dfrac{5,000}{4,407.190735}\right)}{\ln 1.04} \approx 3.217709596.$$

(To obtain this value, we have followed through the calculations with the stored values, thereby using more significant figures than our equations record. If you just worked with the equations, you would get 3.217709594.) Project 2 has its $5,000 return after approximately 3.217698947 years or, based on a 365-day year, 1,174.46 days. ∎

Net present values may also be calculated using the **Cash Flow worksheet** and the associated **NPV subworksheet** of the BA II Plus calculator. These frequently used worksheets are the only two, except for the **TVM worksheet**, that are accessed without using the $\boxed{\text{2ND}}$ button. Observe that the second row of the calculator looks like

$$\boxed{\textbf{2ND}} \quad \boxed{\textbf{CF}} \quad \boxed{\textbf{NPV}} \quad \boxed{\textbf{IRR}} \quad \boxed{\rightarrow}.$$

Here, CF stands for cashflow, NPV stands for net present value, and IRR stands for internal rate of return. Internal rate of return will be introduced in Section (2.3) and the $\boxed{\textbf{IRR}}$ key will only be used in conjunction with the **Cash Flow worksheet**.

To open the **Cash Flow worksheet** of the BA II Plus calculator, push $\boxed{\textbf{CF}}$. At this point it is advisable to clear the worksheet by pushing

$$\boxed{\textbf{2ND}} \quad \boxed{\textbf{CLR WORK}}.$$

Your display should now include "CFo = 0". The register CFo is designed to hold any cashflow made at time 0. The remaining registers in the cashflow worksheet of the BA II Plus calculator are

C01=0, F01=0, C02=0, F02=0, C03=0, ..., C24=0, F24=0,

while the BA II Plus Professional contains registers numbered through C32=0 and F32=0. The letters C in the above sequence stand for "contribution" while the letters F signifies "frequency."

If the **Cash Flow worksheet** is open, repeatedly pushing $\boxed{\downarrow}$ will cause your calculator to cycle through the <u>filled</u> registers of the worksheet; so, if all the registers are filled, on a standard BA II Plus calculator, you will need to press $\boxed{\downarrow}$ forty-nine times to cycle through all the registers.

With a goal of filling these registers in a useful manner, first choose a time to denote as time $t = 0$. This will often be the time a financial relationship is originated. Enter any cashflow that occurs at time $t = 0$ in the CFo register by having the display show CFo and then keying the desired cashflow amount at time 0, followed by $\boxed{\textbf{ENTER}}$. Remember, you will need to enter all your cashflows with a consistent viewpoint relative to their signs.

Next choose an increment of time between successive cashflows. Here it is important to note that you may wish to enter 0 for some of your cashflows. The increment should in general be chosen to be the longest time interval that allows you to include all of your non-zero cashflows. So, if you have cashflows at three months, nine months, and twelve months after your $t = 0$ payment, go ahead and choose three months as your basic interval. (In this case your second contribution will be 0 since there is no cashflow at six months.) On the other hand, if contributions occur at three months and five months, the best you can do is to select one month as your time interval. Had you chosen three months, you would have skipped over the payment at the end of five months.

Furthermore, neither two months nor one-and-a-half months would have been a suitable selection.[4]

The cashflows are now entered successively by showing the next available contribution register, keying a numerical amount, and then remembering to depress $\boxed{\textbf{ENTER}}$ $\boxed{\downarrow}$. The just depressed $\boxed{\downarrow}$ keystroke moves you from a contribution register to the frequency register indexed by the same number. It will be showing a frequency of 1, but if there are consecutive payments that are for an identical amount (perhaps 0), save yourself time and registers by keying in an appropriate frequency greater than 1, then depressing $\boxed{\textbf{ENTER}}$ $\boxed{\downarrow}$.

EXAMPLE 1.7.11   Detailed instructions on the use of the Cash Flow worksheet and NPV subworksheet

Problem: Suppose that Ivy receives payments of $300 three months from now and again at the end of seven months. Furthermore, she receives $1,500 at the end of each month starting nine months from now and continuing through two years from now. That is to say, she receives a payment of $1,500 ten months from now and these payments continue monthly, the last payment occurring twenty-four months from now. Find the net present value of these payments if the monthly interest rate is $(1.04)^{\frac{1}{12}} - 1$, a rate that we will learn in Section (1.10) is equivalent to the annual effective interest rate 4%.

Solution   Take the basic time interval to be a month and let the present be $t = 0$. With these choices, consider the appropriate settings for the cashflow registers.

- CFo $= 0$ since there is no payment at $t = 0$.
- C01 $= 0$ and F01 $= 2$ since there are no payments at the ends of the first two months, that is to say at time $t = 1$ and $t = 2$.
- C02 $= 300$ and F02 $= 1$ since there is a payment of $300 at the end of the next month ($t = 3$), and it is not repeated the following month.
- C03 $= 0$ and F03 $= 3$ since there is no payment at the end of the next three months, that is to say at time $t = 4$, $t = 5$, and $t = 6$.
- C04 $= 300$ and F04 $= 1$ since there is a payment of $300 at the end of the next month ($t = 7$), and it is not repeated the following month.
- C05 $= 0$ and F05$=2$ since following the payment at $t = 7$, there are two successive months with no payment, namely $t = 8$ and $t = 9$.
- C06 $= 1,500$ and F06 $= 15$ since there are 15 successive months ($t = 10, 11, \ldots, 24$) at the end of each of which there is a payment of $1,500.

---

[4]If you know a little number theory, you might realize that the fact that the desired time increment is one month results from 3 and 5 having 1 as their greatest common divisor. In contrast, in our example where we used a three-month increment, there were non-zero cashflows at three, nine, and twelve months and the greatest common divisor of 3, 6, and 12 is 3.

Graphically, we may show this as

$$
\begin{array}{ccccccc}
0 & 1\ 2 & 3 & 4\ 5\ 6 & 7 & 8\ 9 & 10\ \ 11\ \ 12\ \ \ldots\ \ 24\ . \\
\end{array}
$$

CFo=0	C01=0 F01=2	C02=300 F02=1	C03=0 F03=3	C04=300 F04=1	C05=0 F05=2	C06=1,500 F06=15

Note that in this schematic, the contributions C01 C06 give you the amount of the cashflow at each of the times indicated above them, while F01 F06 each indicate the number of different times shown above them.

The sequence of keystrokes that accomplishes the entry of the desired entries for CFo through F06 is as follows. Push $\boxed{\text{CF}}$ $\boxed{\text{2ND}}$ $\boxed{\text{CLR WORK}}$ to open the **Cash Flow worksheet** and set the entries of all the registers equal to 0. Key $\boxed{\downarrow}$ $\boxed{\downarrow}$ to move past CF0 and C01 to F01. Next push $\boxed{2}$ $\boxed{\text{ENTER}}$ $\boxed{\downarrow}$. This enters 2 as the value of the F01 register and moves you on to the C02 register. You may take care of the C02 and F02 entries by pushing $\boxed{3}$ $\boxed{0}$ $\boxed{0}$ $\boxed{\text{ENTER}}$ $\boxed{\downarrow}$ $\boxed{\downarrow}$. Here you don't have to enter the F02 entry because when you enter a nonzero entry into a C-register, it automatically changes the contents of the corresponding F-register to 1. Next push $\boxed{\downarrow}$ $\boxed{3}$ $\boxed{\text{ENTER}}$ $\boxed{\downarrow}$ to fill the C03 and F03 registers with 0 and 3, respectively, and move to the registers indexed by 4. Keying $\boxed{3}$ $\boxed{0}$ $\boxed{0}$ $\boxed{\text{ENTER}}$ $\boxed{\downarrow}$ $\boxed{\downarrow}$ takes care of these and moves you to index 5. (Once again you took advantage of the default frequency of 1 being entered for a nonzero contribution.) The values for the registers indexed by 5 are correctly assigned by depressing $\boxed{\downarrow}$ $\boxed{2}$ $\boxed{\text{ENTER}}$ and you push $\boxed{\downarrow}$ to move to the registers indexed by 6. Key $\boxed{1}$ $\boxed{5}$ $\boxed{0}$ $\boxed{0}$ $\boxed{\text{ENTER}}$ $\boxed{\downarrow}$ $\boxed{1}$ $\boxed{5}$ $\boxed{\text{ENTER}}$ to fill these as indicated above.

You now are ready to compute the desired net present value. This requires you to open the **NPV subworksheet** by depressing $\boxed{\text{NPV}}$. The subworksheet will show "I =" and you need to enter the interest rate (as a percent) per month. You were given a monthly interest rate of $(1.04)^{\frac{1}{12}} - 1$. Enter this by pushing

$$\boxed{1}\ \boxed{\bullet}\ \boxed{0}\ \boxed{4}\ \boxed{y^x}\ \boxed{1}\ \boxed{2}\ \boxed{1/x}\ \boxed{=}\ \boxed{-}\ \boxed{1}\ \boxed{=}\ \boxed{\times}\ \boxed{1}\ \boxed{0}\ \boxed{0}\ \boxed{=}\ \boxed{\text{ENTER}}\ .$$

Now, push $\boxed{\downarrow}$ $\boxed{\text{CPT}}$ to move to the NPV register and compute (and enter) the net present value \$21,876.34572 corresponding to the above payments. ∎

In our next example, we practice the skills introduced in Example (1.7.11) by recalculating the net present value of Example (1.7.10). The only new feature of this example is that we have a cashflow at $t = 0$.

EXAMPLE 1.7.12   Cash Flow worksheet and NPV subworksheet

Problem: As in Example (1.7.10), Project 1 requires an investment of $10,000 at time 0 and an additional investment of $5,000 at $t = 1$. It returns $2,000 at $t = 3$ and $6,000 at times $t = 4$, $t = 5$, and $t = 6$. Use the **Cash Flow worksheet** and **NPV subworksheet** to find the net present value of this project if the effective interest rate per basic time period is 4%.

Solution   The desired entries are given schematically [as in Example (1.7.11)] by

0	1	2	3	4 5 6 .
CFo= $-10,000$	C01= $-5,000$	C02=0	C03=2,000	C04=6,000
	F01=1	F02=1	F03=1	F04=3

These may be achieved by keying

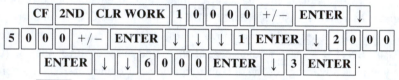

Next push NPV. The calculator should now display "I =   ". Enter the interest rate of 4% by pushing 4 **ENTER**, and subsequently depress $\downarrow$ CPT. Then 1,772.575351 is shown, and this is the net present value (in dollars). ∎

On occasion you wish to revise the holdings of the cashflow registers without having to clear the worksheet and start anew. We next explain how this may be done efficiently. The procedure, which is described in (1) below, replaces a previously-entered register value with a new one; it works for any BA II Plus calculator register that allows entered (as opposed to computed) values. You may have already discovered this with the **Date worksheet**.

(1) To change the amount entered as the k-th cashflow, display the current value of that cashflow, and then key the desired value and push ENTER. Similarly, to change the frequency of the k-th cashflow, view the current value stored as the frequency, then key the desired frequency and depress ENTER.

(2) To delete the k-th cashflow and its accompanying frequency, display the current holding of that register and key 2ND DEL. This will result in the previous k-th cashflow being removed. Any subsequent cashflows will be moved, along with their accompanying frequencies, to the registers with labels one less than those they previously occupied. For instance, if you delete the 5th cashflow, the old 6th cashflow (if any) becomes the 5th cashflow, the 7th cashflow (if any) becomes the 6th cashflow, etc.

(3) To insert a forgotten cashflow as the k-th cashflow and move any subsequent cashflows to registers indexed by one more than those they previously occupied, first display the label for the k-th register along with its current entry. Then push 2ND INS followed by the desired new numerical value of the k-th entry and ENTER. The frequency of this new entry will be entered as 1. If this is not as desired, change it according to the instructions given in (1). Naturally you may only insert a new cashflow if all your cashflow registers have not already been filled. (The display will include "↑ ↓ DEL INS" when inserting is possible. If all the registers are full, the "INS" will be omitted.)

1.8 SIMPLE DISCOUNT

In Section (1.4) we considered two parties negotiating a loan with a fixed amount of interest per basic time period for each \$1 borrowed. Suppose that we again consider two parties negotiating a loan, but this time they agree on a fixed amount of **discount** D per basic time period for each \$1 borrowed. Then, if the loan period is $[0, t]$ and K is the loan amount, the borrower receives $\$K - \$KtD = K(1 - tD)$. In particular, the borrower receives \$1 if $K = (1 - tD)^{-1}$. It follows that $a(t) = (1 - tD)^{-1} = \frac{1}{1-tD}$. Then the discount function $v(t) = \frac{1}{a(t)} = 1 - tD$ is linear.

Momentarily, we will further examine the situation where the discount function $v(t)$ is linear. We caution you that this is different from the accumulation function $a(t)$ being linear. When $a(t)$ is linear, say $a(t) = 1 + st$, the discount function $v(t) = (1 + st)^{-1}$ is a decreasing function that is asymptotic to the t-axis, hence is **not** linear.

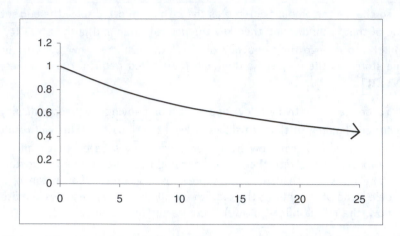

$$\textbf{Graph of } v(t) = (1 + st)^{-1} \textbf{ for } s = .05$$

If $v(t)$ is linear, $v(t) = mt + b$, we say that we have **simple discount**. In fact, since $v(0) = \frac{1}{a(0)} = \frac{1}{1} = 1$, the discount function must then satisfy $v(t) = mt + 1$. To find m, we combine two observations. First note that $m + 1 = v(1) = \frac{1}{a(1)} = \frac{1}{i_1+1}$. Secondly, according to Equation (1.6.10), $\frac{1}{i_1+1} = 1 - d_1$. Therefore, $m + 1 = 1 - d_1$, $m = -d_1$, and

$$v(t) = -d_1 t + 1.$$

So the **simple discount accumulation function** is $a(t) = \frac{1}{1-d_1 t}$.

$A_K(t) = \frac{K}{(1-dt)}$ is called the **amount function for \$K invested by simple discount at rate** d.

$a(t) = \frac{1}{1-dt}$ is called the **simple discount accumulation function at rate** d.

Then $a(t)$ is an increasing function that is asymptotic to the line $t = \frac{1}{d_1}$. Therefore, it only makes sense to talk about simple discount on the interval $[0, \frac{1}{d_1})$.

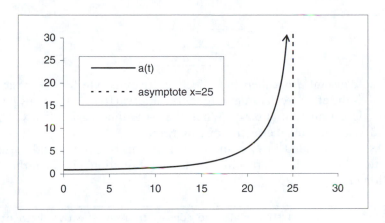

Graph of $a(t) = (1 - d_1 t)^{-1}$ for $d_1 = .04$

We next want to see how simple discount might arise in real life. Let us suppose that we have a borrowing relationship with interest paid in advance and that this loan is to extend for more than one period. If all the interest is required to be paid at the beginning, the interest per period is d, and the loan lasts for a length of time t, then the interest on \$1 is dt. We therefore get $1 - dt$ and are expected to repay 1 at time t. Hence, $v(t) = 1 - dt$, and we have simple discount.

The scenario of the previous paragraph (which resulted in accumulation being governed by the simple discount accumulation function) is analogous to the situation in which you rent an apartment for several months (or years) and are expected to pay for the whole rental in advance. This is an unlikely rental arrangement except for short periods.

1.9   COMPOUND DISCOUNT

In Section (1.5) we considered what happens when the interest rate $i_n = i_{[n-1,n]}$ is independent of n. We found that this forces us to have the compound interest accumulation function $a(t) = (1 + i)^t$. Now let us look at the consequences of stipulating that the discount function $d_n = d_{[n-1,n]}$ is a constant d.

If d_n is a constant d, then recalling (1.6.11), we see that $i_n = \dfrac{d_n}{1-d_n} = \dfrac{d}{1-d}$ is also constant. As usual we call this constant i. We then have

(1.9.1)
$$i = \frac{d}{1 - d},$$

and therefore

(1.9.2)
$$i = \frac{1}{1 - d} - 1.$$

The constant d is called the effective discount rate for the basic time period. It is the effective discount rate for the interval $[t - 1, t]$ for any positive integer t [see Equation (1.6.2)]. When the basic time period is a year, it is referred to as the **annual effective discount rate**.

As demonstrated in example (1.9.3), if one is given d and wishes to compute i, it may be more efficient to use Equation (1.9.2) rather than Equation (1.9.1).

> On the BA II Plus calculator, to convert from an effective discount rate d to an effective interest rate i, enter d (NOT as a percent), then push
>
> $$\boxed{+/-} \quad \boxed{+} \quad \boxed{1} \quad \boxed{=} \quad \boxed{1/x} \quad \boxed{-} \quad \boxed{1} \quad \boxed{=}$$
>
> to obtain i (again NOT as a percent).

EXAMPLE 1.9.3 Finding an effective interest rate equivalent to a given discount rate

Problem: Given that the annual effective discount rate is 4.386286%, compute the annual effective interest rate i as a percent.

Solution 1 Use formula (1.9.2) to obtain i. On the BA II Plus calculator, push

$$\boxed{\cdot}\,\boxed{0}\,\boxed{4}\,\boxed{3}\,\boxed{8}\,\boxed{6}\,\boxed{2}\,\boxed{8}\,\boxed{6}\,\boxed{+/-}\,\boxed{+}\,\boxed{1}\,\boxed{=}\,\boxed{1/x}\,\boxed{-}\,\boxed{1}\,\boxed{=}$$

to display .045875072. Then $i = 4.5875072\%$.

Solution 2 Use formula (1.9.1) to obtain i. On the BA II Plus calculator, push

$$\boxed{\cdot}\,\boxed{0}\,\boxed{4}\,\boxed{3}\,\boxed{8}\,\boxed{6}\,\boxed{2}\,\boxed{8}\,\boxed{6}\,\boxed{STO}\,\boxed{\alpha}\,\boxed{+/-}\,\boxed{+}\,\boxed{1}\,\boxed{=}\,\boxed{1/x}\,\boxed{\times}\,\boxed{RCL}\,\boxed{\alpha}\,\boxed{=},$$

thereby displaying .045875072. Then $i = 4.5875072\%$. ■

Once it is known that i and d are constant, equation (1.6.12) gives

(1.9.4)
$$d = \frac{i}{1 + i}.$$

That is, the amount of discount is the ratio of the interest i for the period divided by the amount 1 will have grown to at the end of the period. Two consequences of Equation (1.9.4) are

(1.9.5) $d = iv,$

and

(1.9.6) $d = 1 - \dfrac{1}{1 + i}.$

Note that by the definition of v as the reciprocal of $1 + i$ [Equation (1.7.5)], it is immediate from (1.9.6) that

(1.9.7) $d = 1 - v$ and $v = 1 - d.$

A verbal interpretation of Equation (1.9.5) is that d, the amount of discount on 1, is equal to the amount of interest on 1, discounted for one year by multiplying by the discount factor v.

Equations (1.9.4) and (1.9.6) give formulas for calculating the effective discount rate from the effective interest rate. Just as Equation (1.9.2) is often more efficient than (1.9.1) for changing rates, many times Equation (1.9.6) leads more quickly to the interest rate than Equation (1.9.4). The reader might observe this by converting from $i = .03263529$ to $d \approx .031603888$.

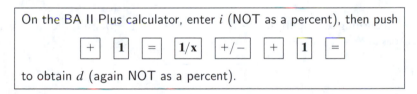

On the BA II Plus calculator, enter i (NOT as a percent), then push

$\boxed{+}$ $\boxed{1}$ $\boxed{=}$ $\boxed{1/\text{x}}$ $\boxed{+/-}$ $\boxed{+}$ $\boxed{1}$ $\boxed{=}$

to obtain d (again NOT as a percent).

EXAMPLE 1.9.8   Comparing interest and discount rates

Problem: Cassandra needs to borrow money to pay her tuition. She has a choice of borrowing at an annual effective interest rate of 5.1% or at an annual effective discount rate of 4.9%. Which rate should she choose?

Solution 1   Using Equation (1.9.1), an annual effective discount rate of 4.9% is equivalent to an annual effective interest rate of $\dfrac{.049}{1-.049} \approx .0515$. Since $.0515 > .051 = 5.1\%$, and it benefits the borrower to have a lower effective interest rate, Cassandra should borrow at the 5.1% interest rate.

Solution 2   According to Equation (1.9.4), the annual effective interest rate $i = 5.1\% = .051$ corresponds to an annual effective discount rate of $\dfrac{.051}{1.051} \approx .0485$. Since $.0485 < .049 = 4.9\%$, Cassandra should borrow at the 5.1% interest rate. This is because it assists the borrower of a loan to have the loan governed by a lower effective discount rate.   ■

EXAMPLE 1.9.9   **Equation (1.9.7) is useful here.**

Problem: Radhika is guaranteed a payment of \$5,000 in exactly four years. She needs \$4,500 now in order to pay her tuition bill. The best loan Radhika qualifies for has a discount rate of 4.9% and requires repayment in exactly four years. Radhika can borrow the full \$4,500, in which case she will owe $\$4,500(1 - d)^{-4} = \$4,500(.951)^{-4} \approx \$5,501.62$ at the end of four years. She can repay this with her guaranteed \$5,000 and an additional \$501.62 that she will have to raise at the end of four years. Alternatively, Radhika may be able to sell her guaranteed \$5,000 payment and use the proceeds from the sale to cover all or part of her tuition payment. How much should she be willing to sell her \$5,000 payment for?

Solution   The present value of \$5,000 four years from now with respect to $a(t) = (1 + i)^{t} = (1 - d)^{-t} = (1 - .049)^{-t}$ is $\$5,000v(4) = \$5,000v^{4} = \$5,000(1 - d)^{4} = \$5,000(1 - .049)^{4} \approx \$4,089.705844$. Radhika should be willing to sell her \$5,000 payment if she is offered more than \$4,089.70. For example, if she can sell it for \$4,200, then Radhika will only have to borrow \$4,500 $-$ \$4,200 $=$ \$300 now.   In four years she would need to repay $\$300(1 + i)^{4} = \$300(1 - d)^{-4} = \$300(.951)^{-4} \approx \$366.77$. Thus the amount she needs to raise after four years would be reduced from \$501.62 to \$366.77.   ■

According to Equation (1.9.1), $(1 - d)^{-1} = 1 + i$, and since a constant i_n forces $a(t) = (1 + i)^{t}$,

(1.9.10)
$$a(t) = (1 + i)^{t} = (1 - d)^{-t}.$$

Equation (1.9.10) tells us that compound discount at a discount rate d is equivalent to compound interest at an interest rate $i = \dfrac{d}{1-d}$.

> The accumulation function $a(t) = (1 - d)^{-t}$ is called the **compound discount accumulation function** at discount rate d. It is equal to the compound interest accumulation function $a(t) = (1 + i)^{t}$ if i is the effective interest rate equivalent to d.

When the accumulation function is $a(t) = (1 - d)^{-t}$, if at time 0 you borrow \$$K$ at a discount and agree to repay it at time n, you walk away with $\$K(1 - d)^{n}$ of the lender's money. The amount of discount for the loan is $K - K(1 - d)^{n}$.

Another way of looking at the situation where you have constant d_n is the following. At time 0 you ask the lender to allow you to borrow \$$K$. Since the lender charges a discount rate of d per basic period, he initially views this inquiry as a request that you use $K(1 - d)$ of his money from time 0 to time 1 and then you repay K. However, you inform him that you wish to have a loan of longer duration. Consider first the case where this longer loan period is two basic time periods. The lender then thinks of your loan as a sequence of two loans, each for one basic time period, and you have promised to pay him K at time $t = 2$. Since you will be paying the lender K at time $t = 2$, the lender understands that you are owed $K(1 - d)$ at the beginning of the second period, that is to say at time $t = 1$. Rather than actually paying you this $K(1 - d)$, he will forgive you $K(1 - d)$ that you would otherwise be required to pay him. This latter $K(1 - d)$ is the amount you would owe at $t = 1$ if the lender loaned you $K(1 - d)^2$ at $t = 0$. Thus, if you borrow K from $t = 0$ to $t = 2$ at a discount rate d, you will receive $K(1 - d)^2$ at $t = 0$. The situation for a two-period loan is represented in Figure (1.9.11).

get $K(1 - d)^2$ pay $K(1 - d)$
get $K(1 - d)$ pay K

| 0 | 1 | 2 |

FIGURE (1.9.11)

More generally, if the loan lasts k basic time periods, the lender thinks of your loan as a sequence of k loans, each lasting one basic time period. As is suggested by Figure (1.9.12), the amount you receive at $t = 0$ is $K(1 - d)^k$.

get $K(1 - d)^k$ pay $K(1 - d)^{k-1}$
get $K(1 - d)^{k-1}$

$\cdots$

pay $K(1 - d)^2$
get $K(1 - d)^2$ pay $K(1 - d)$
get $K(1 - d)$ pay K

| 0 | 1 | $\cdots$ | $k - 2$ | $k - 1$ | k |

FIGURE (1.9.12)

EXAMPLE 1.9.13

Problem: Ezra has the opportunity to borrow money according to compound discount at an annual effective discount rate of 8%. He would like to borrow money in order to completely pay for a $3,000 used piano. Ezra wishes to repay the loan with a payment in exactly five years. How much money must he ask to borrow at 8% discount?

Solution   Let $K denote the amount Ezra requests to borrow for five years at 8% discount. Then the amount he receives is $K(1 - .08)^5 = \$K(.92)^5$. In order for this amount to be $3,000, it is necessary that $K = 3,000(.92)^{-5} \approx$ $4,551.79. Since $4,551.79(.92)^5 \approx \$3,000.000686$, if Ezra borrows $4,551.79 for five years at 8% discount, he receives exactly $3,000.   ∎

1.10   NOMINAL RATES OF INTEREST AND DISCOUNT

Suppose that we have an investment governed by compound interest. This means that if i is the applicable effective interest rate for the investment, then the growth of the money is governed by the compound accumulation function $a(t) = (1 + i)^t = 1 + [(1 + i)^t - 1]$. Therefore, 1 invested for a period of length T earns $(1 + i)^T - 1$ interest. The quantity $[(1 + i)^T - 1]$ may be thought of as the effective interest rate for a period of length T. In particular, the effective interest rate for a period of length $T = \frac{1}{m}$ is $[(1 + i)^{\frac{1}{m}} - 1]$.

Banks commonly credit interest more than once per year, say m times per year. They advertise a **nominal** (annual) **interest rate** of $i^{(m)}$ **convertible** or **compounded** or **payable m times per year**.[5] The word "nominal" means "in name only," and this is indeed the case. The bank pays interest at a rate of $\frac{i^{(m)}}{m}$ per m-th of a year. But we just observed that the rate for such an interval is $[(1 + i)^{\frac{1}{m}} - 1]$. We therefore have the following important statement.

IMPORTANT FACT 1.10.1: If an account is governed by a nominal interest rate of $i^{(m)}$ payable m times per year, the bank pays interest at a rate of $\frac{i^{(m)}}{m} = [(1 + i)^{\frac{1}{m}} - 1]$ per m-th of a year.

Observe that it follows from the equation of Fact (1.10.1) that

(1.10.2)
$$i^{(m)} = m[(1 + i)^{\frac{1}{m}} - 1],$$

and

(1.10.3)
$$i = \left(1 + \frac{i^{(m)}}{m}\right)^m - 1.$$

---

[5]You may see **APR**, which is an acronym for "annual percentage rate," used to indicate a nominal rate.

There is a nice way to visualize why Equation (1.10.3) must hold. Money grows by a factor of $1 + \frac{i^{(m)}}{m}$ each m-th of a year, and it therefore changes by a factor of $\left(1 + \frac{i^{(m)}}{m}\right)^m$ each year. This implies that $1 + i = \left(1 + \frac{i^{(m)}}{m}\right)^m$, and this equality is essentially Equation (1.10.3).

Banks are required to report the rate i as the "annual percentage yield" or **APY**. It follows from the binomial theorem and equation (1.10.3), that if m is an integer greater than 1 and $i^{(m)} > 0$, then $i > i^{(m)}$. This is as it should be since if compounding takes place more frequently, interest earns interest, and the stated rate $i^{(m)}$ need not be as large as i.

EXAMPLE 1.10.4

Problem: National Bank advertises a savings account paying 4% nominal interest compounded quarterly. What is the annual effective yield **APY** for this account?

Solution To say that the account pays 4% interest compounded quarterly means that the interest rate per quarter is $\frac{4\%}{4} = 1\%$. Therefore the annual effective yield is $(1.01)^4 - 1 = 4.060401\%$.   ∎

EXAMPLE 1.10.5   Varying rates

 Problem: Sandra inherits $10,000. She deposits it in a five-year certificate of deposit paying 6% nominal interest compounded monthly and the interest remains on deposit. At the end of the five years, Sandra decides to renew her CD for another five years at the then current nominal interest rate of 7.5% compounded quarterly. Again, interest is left to accrue. At the time her second CD matures, what is her investment worth?

Solution The interest rate on the first five-year CD was 6% convertible monthly and there are $5 \times 12 = 60$ months in five years. Therefore, at the time her first CD matured, it was worth $10,000(1 + \frac{.06}{12})^{60}$. [Each month the interest rate is $\frac{.06}{12}$, so each month the balance grows by a factor of $(1 + \frac{.06}{12})$. The original balance was $10,000, so after sixty months it is $10,000(1 + \frac{.06}{12})^{60}$.] So, Sandra's initial deposit to open her second five-year CD was $10,000(1 + \frac{.06}{12})^{60}$. The interest rate for the five-year reinvestment was 7.5% nominal convertible quarterly, and there are $5 \times 4 = 20$ quarters in five years. It follows that when Sandra's second CD matures, it has a value of $10,000(1 + \frac{.06}{12})^{60}(1 + \frac{.075}{4})^{20} \approx \$19,557.63$.

Note: At the time that the money was transferred to the second CD, the balance was $10,000(1 + \frac{.06}{12})^{60} \approx \$13,488.50153$. At that point, the bank might have rounded it to the nearest cent. In this case her final balance would be $\$13,488.50(1 + \frac{.075}{4})^{20} \approx \$19,557.62394 \approx \$19,557.62$, a penny less than it would have been without the rounding.   ■

EXAMPLE 1.10.6   Unknown rate

Problem: Vladimir deposits $50,000 in a three-year certificate of deposit for which interest is compounded quarterly and is left to accrue. At the end of the three years, the balance in the CD is $63,786.11. What is the nominal annual interest rate $i^{(4)}$ convertible quarterly?

Solution   The three years that the money is on deposit consists of $3 \times 4 = 12$ quarters, and the quarterly interest rate is $\frac{i^{(4)}}{4}$. Therefore $\$50,000(1 + \frac{i^{(4)}}{4})^{12} = \$63,786.11$ and $i^{(4)} = 4[(\frac{\$63,786.11}{\$50,000})^{\frac{1}{12}} - 1] \approx .082000004 \approx .082 = 8.2\%$.   ■

EXAMPLE 1.10.7 $i^{(m)}$ for m not an integer

Problem: Jolene invests money in a fund for which interest is paid once every two years. The effective rate per two-year period is 14%. Find the nominal interest rate convertible biennially[6] and the annual effective interest rate governing the fund.

Solution   We are asked to find $i^{(\frac{1}{2})}$ and also i. $14\% = \frac{i^{(\frac{1}{2})}}{\frac{1}{2}}$ so $i^{(\frac{1}{2})} = 7\%$ and $1.14 = (1 + i)^2$ so $i = (1.14)^{\frac{1}{2}} - 1 \approx .067707825 \approx 6.77\%$.   ■

Just as banks might advertise nominal interest rates when the interest period is other than a year, you might be presented with nominal discount rates when the discount period is other than yearly. More specifically, you might be offered a **nominal discount rate** $d^{(m)}$ **convertible or compounded or payable m times for year**. Then the year is divided into m subintervals of equal length, and the discount rate per each subinterval is $\frac{d^{(m)}}{m}$. It follows that

$$(1.10.8) \qquad 1 - d = \left(1 - \frac{d^{(m)}}{m}\right)^m \qquad \text{and} \qquad d = 1 - \left(1 - \frac{d^{(m)}}{m}\right)^m.$$

---

[6]**Biennially** means once every two years. **Semiannually** means twice a year. The word biannually is sometimes used as a synonym for biennially and othertimes as a synonym for semiannually. We will not use biannual.

Therefore,

(1.10.9)
$$d^{(m)} = m[1 - (1 - d)^{\frac{1}{m}}].$$

Just as we derived formula (1.6.7), we can show that

(1.10.10)
$$\left(1 - \frac{d^{(m)}}{m}\right)\left(1 + \frac{i^{(m)}}{m}\right) = 1.$$

This is equivalent to

(1.10.11)
$$0 = \frac{i^{(m)}}{m} - \frac{d^{(m)}}{m} - \frac{i^{(m)}}{m}\frac{d^{(m)}}{m},$$

from which one easily obtains

(1.10.12)
$$\frac{i^{(m)}}{m} = \frac{\dfrac{d^{(m)}}{m}}{1 - \dfrac{d^{(m)}}{m}} \qquad \text{and} \qquad i^{(m)} = \frac{d^{(m)}}{1 - \dfrac{d^{(m)}}{m}}.$$

From (1.10.11) one also derives

(1.10.13)
$$\frac{d^{(m)}}{m} = \frac{\dfrac{i^{(m)}}{m}}{1 + \dfrac{i^{(m)}}{m}} \qquad \text{and} \qquad d^{(m)} = \frac{i^{(m)}}{1 + \dfrac{i^{(m)}}{m}}.$$

It is also worth noting that if n and p are integers, since Equations (1.10.3) and (1.10.8) hold for any integer m,

(1.10.14)
$$\left(1 + \frac{i^{(n)}}{n}\right)^n = 1 + i = (1 - d)^{-1} = \left(1 - \frac{d^{(p)}}{p}\right)^{-p}.$$

This equation is useful for converting between a nominal interest rate and a nominal discount rate, perhaps having different compounding frequencies. We shall illustrate this in Example (1.10.17).

EXAMPLE 1.10.15

Problem: Trust Bank offers a savings account with a nominal discount rate of 4.8% payable monthly. Find the annual effective rate of interest i and the annual effective rate of discount d.

Solution Note that $\frac{d^{(12)}}{12} = \frac{4.8\%}{12} = .4\% = .004$. It follows that $1 - d = (1 - \frac{d^{(12)}}{12})^{12} = (1 - .004)^{12} = (.996)^{12}$ and $d = 1 - (.996)^{12} \approx .046957954 \approx 4.7\%$. Therefore, $1 + i = (1 - d)^{-1} = (.996)^{-12}$ and $i = (.996)^{-12} - 1 \approx 4.927165\%$. ∎

We observe that, in Example (1.10.15), $i > d^{(12)} > d$. In general, $d^{(m)} > d$ for $m > 1$ since with more frequent discounting, discount is discounted and the stated rate $d^{(m)}$ must be larger than d. Moreover, recalling (1.10.13), we see that

$$d^{(m)} = \frac{i^{(m)}}{1 + \frac{i^{(m)}}{m}} < i^{(m)}$$

provided $i^{(m)} > 0$. Also, as noted following Equation (1.10.3), $i > i^{(m)}$ when $i^{(m)} > 0$. Finally, it follows from (1.10.3) that $i^{(m)} > 0$ if and only if $i > 0$. We therefore have

---

IMPORTANT FACT (1.10.16): If $i > 0$ and $m > 1$, then

$$i > i^{(m)} > d^{(m)} > d.$$

Sometimes one is presented with a problem whose solution might be accomplished most easily by converting between an interest rate (possibly nominal) and a discount rate (again possibly nominal) or vise versa.

EXAMPLE 1.10.17

Problem: Given that $d^{(4)} = 4.6\%$, find $i^{(12)}$.

Solution   Equation (1.10.14) tells us that $\left(1 + \frac{i^{(12)}}{12}\right)^{12} = \left(1 - \frac{d^{(4)}}{4}\right)^{-4}$.

Consequently, $1 + \frac{i^{(12)}}{12} = \left(1 - \frac{d^{(4)}}{4}\right)^{-\frac{1}{3}}$ and $i^{(12)} = 12\left[\left(1 - \frac{d^{(4)}}{4}\right)^{-\frac{1}{3}} - 1\right]$.

If $d^{(4)} = 4.6\%$, then this gives us $i^{(12)} = 4.6355852\%$.   ∎

The BA II Plus calculator has an **Interest Conversion worksheet** that can be used to change from an effective interest rate to an equivalent nominal interest rate or from a nominal interest rate to an equivalent effective interest rate. To open this worksheet, push 2ND ICONV . At this point the display will show "NOM =   ".

Suppose that the Interest Conversion worksheet is open and displays "NOM =   ". To find an effective interest rate equivalent to a given nominal interest rate $i^{(m)}$, push calculator buttons to display the numerical value of $i^{(m)}$ as a percent and then push ENTER ↑ , at which time the display will show "C/Y =   ". Push calculator buttons to display the numerical value of m and then push ENTER ↑ , at which time the display will show "EFF =   ". Push CPT . The equivalent effective interest rate is then displayed as a percent.

EXAMPLE 1.10.18

Problem: Use the BA II Plus calculator **Interest Conversion worksheet** to find an annual effective interest rate equivalent to a nominal rate of interest of 6% convertible quarterly.

Solution   Push
2ND ICONV 6 ENTER ↑ 4 ENTER ↑ CPT .
Displayed is now the desired effective rate as a percent, namely 6.136355063%. ∎

If you have an effective interest rate and desire an equivalent nominal rate, proceed by pushing 2ND ICONV to open the worksheet, then push ↓ so that "EFF =   " is displayed.

Suppose that the **Interest Conversion worksheet** is open and displays "EFF =   ". To find a nominal interest rate convertible m times per period that is equivalent to an effective interest rate i for the period, push calculator buttons to display the numerical value of i as a percent and then push ENTER ↓ , at which time the display will show "C/Y =   ". Push calculator buttons to display the numerical value of m and then push ENTER ↓ , at which time the display will show "NOM =   ". Push CPT . The equivalent nominal interest rate convertible m times per year is then displayed as a percent.

EXAMPLE 1.10.19

Problem: Use the BA II Plus calculator **Interest Conversion worksheet** to find a nominal interest rate convertible bimonthly[7] that is equivalent to an annual effective interest rate of 3%.

---

[7]We will always use **bimonthly** to mean once every two months. When we wish to indicate twice a month, we use **semimonthly**.

Solution   Push | 2ND | | ICONV | | ↓ | | 3 | | ENTER | | ↓ | | 6 | | ENTER | | ↓ | | CPT |.
Displayed now is the desired nominal rate as a percent, namely 2.963173219%.   ∎

The **Interest Conversion worksheet** may also be used for discount rates, more specifically to change from an effective discount rate to an equivalent nominal discount rate or from a nominal discount rate to an equivalent effective discount rate. To figure out how this should be accomplished, note that Equation (1.10.8) can be rewritten as

$$(1.10.20) \qquad d = 1 - \left(1 + \frac{(-d^{(m)})}{m}\right)^m,$$

and equation (1.10.9) may be rewritten as

$$(1.10.21) \qquad d^{(m)} = -m\left[\left((1 + (-d))^{\frac{1}{m}} - 1\right)\right].$$

Comparing Equation (1.10.20) with Equation (1.10.3) and Equation (1.10.21) with Equation (1.10.2), one sees that an equivalent rate may be found for discount rates just as for interest rates except for the need for minus signs just after a discount rate is entered and just before the equivalent rate is found.

Suppose that **the Interest Conversion worksheet** is open and displays "NOM =   ".   To find an effective discount rate equivalent to a given nominal discount rate $d^{(m)}$, push calculator buttons to display the numerical value of $d^{(m)}$ as a percent and then push | +/− | | ENTER | | ↑ |, at which time the display will show "C/Y =   ". Push calculator buttons to display the numerical value of m and then push | ENTER | | ↑ |, at which time the display will show "EFF =   ". Push | CPT | | +/− |. The equivalent effective discount rate is then displayed as a percent.

EXAMPLE 1.10.22

Problem: Use the BA II Plus calculator **Interest Conversion worksheet** to find an annual effective discount rate equivalent to a nominal rate of discount of 4% convertible semiannually.

Solution   Push

| 2ND | | ICONV | | 4 | | +/− | | ENTER | | ↑ | | 2 |
| ENTER | | ↑ | | CPT | | +/− |.

Displayed is now the desired effective rate as a percent, namely 3.96%.   ■

> Suppose that the **Interest Conversion worksheet** is open and displays "EFF = ". To find a nominal discount rate convertible m times per period that is equivalent to an effective discount rate d for the period, push calculator buttons to display the numerical value of d <u>as a percent</u> and then push $\boxed{+/-}$ $\boxed{\textbf{ENTER}}$ $\boxed{\downarrow}$, at which time the display will show "C/Y = ". Push calculator buttons to display the numerical value of m and then push $\boxed{\textbf{ENTER}}$ $\boxed{\downarrow}$, at which time the display will show "NOM = ". Push $\boxed{\textbf{CPT}}$ $\boxed{+/-}$. The equivalent nominal discount rate convertible m times per year is then displayed <u>as a percent</u>.

EXAMPLE 1.10.23

Problem: Use the BA II Plus calculator **Interest Conversion worksheet** to find a nominal discount rate convertible monthly that is equivalent to an annual effective discount rate of 7%.

Solution   Push

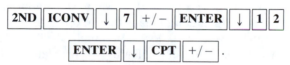

$$\boxed{\textbf{ENTER}} \; \boxed{\downarrow} \; \boxed{\textbf{CPT}} \; \boxed{+/-}.$$

Displayed is now the desired nominal rate as a percent, namely approximately 7.235169679%.   ■

Sometimes one wishes to convert between two nominal rates. If one of these is a discount rate and the other an interest rate, it is easiest to convert without using the special **Interest Conversion worksheet** of the BA II Plus calculator. If both the rates are interest rates or both are discount rates, the Interest Conversion worksheet may be used as may the equations of this section.

EXAMPLE 1.10.24

Problem: Given that $i^{(5)} = 3.2\%$, find $i^{(7)}$.

Solution 1   (**Interest Conversion worksheet**)   Key $\boxed{\textbf{2ND}}$ $\boxed{\textbf{ICONV}}$ to open the Interest Conversion worksheet. Follow this by pressing the sequence of keys $\boxed{3}$ $\boxed{\bullet}$ $\boxed{2}$ $\boxed{\textbf{ENTER}}$ $\boxed{\uparrow}$ $\boxed{5}$ $\boxed{\textbf{ENTER}}$ $\boxed{\uparrow}$ $\boxed{\textbf{CPT}}$ to obtain the intermediate result

"EFF= 3.241222984".   Next push $\boxed{\downarrow}\ \boxed{7}\ \boxed{\textbf{ENTER}}\ \boxed{\downarrow}\ \boxed{\textbf{CPT}}$.   The display should show "NOM = 3.197082281".

Solution 2   According to Equation (1.10.3), $\left(1 + \dfrac{i^{(5)}}{5}\right)^5 = \left(1 + \dfrac{i^{(7)}}{7}\right)^7$.

So, $i^{(7)} = 7\left[\left(1 + \dfrac{i^{(5)}}{5}\right)^{\frac{5}{7}} - 1\right]$. When $i^{(5)} = .032$, this gives $i^{(7)} \approx$ 3.197082281%. (You may obtain this by the twenty-three key sequence $\boxed{\bullet}\ \boxed{0}\ \boxed{3}$ $\boxed{2}\ \boxed{\div}\ \boxed{5}\ \boxed{=}\ \boxed{+}\ \boxed{1}\ \boxed{=}\ \boxed{y^x}\ \boxed{5}\ \boxed{=}\ \boxed{y^x}\ \boxed{7}\ \boxed{1/x}\ \boxed{=}\ \boxed{-}\ \boxed{1}\ \boxed{=}\ \boxed{\times}\ \boxed{7}\ \boxed{=}$, where some of the $\boxed{=}$ keys are not necessary.) ∎

1.11   A FRIENDLY COMPETITION (CONSTANT FORCE OF INTEREST)

(calculus needed here)

Let us assume that we have compound interest at an annual effective rate of $i > 0$. Further suppose that m and p are positive integers, with $p > m$ and that the nominal rates $i^{(m)}$ and $i^{(p)}$ are each equivalent to the effective rate i. Then $i^{(p)} < i^{(m)}$ because when we compute interest p times per year, there is more compounding of interest than when we compute it m times per year. This additional compounding means that in order to produce the same effective interest rate i, we need a lower nominal rate $i^{(p)}$. The inequality $i^{(p)} < i^{(m)}$ can also be derived using calculus [see problem (1.11.4)].

As m increases, $i^{(m)}$ decreases. The following argument uses calculus to show that $i^{(m)}$ gets closer and closer to the number $\delta = \ln(1 + i)$ as m grows without bound. Recalling equality (1.10.2)

$$i^{(m)} = m((1 + i)^{\frac{1}{m}} - 1),$$

and l'Hospital's rule, we may find the limit of the sequence $\{i^{(m)}\}$.

$$\lim_{m \to \infty} i^{(m)} = \lim_{m \to \infty} m((1 + i)^{\frac{1}{m}} - 1)$$

$$= \left(\lim_{m \to \infty} \frac{((1 + i)^{\frac{1}{m}} - 1)}{\frac{1}{m}}\right) = \lim_{m \to \infty} \frac{(1 + i)^{\frac{1}{m}} \ln(1 + i)(-m^{-2})}{-m^{-2}}$$

$$= \lim_{m \to \infty} (1 + i)^{\frac{1}{m}} \ln(1 + i) = \lim_{m \to \infty} \ln(1 + i) = \ln(1 + i).$$

We use the letter δ to denote this limit, and refer to it as the **force of interest**.

We have shown that

(1.11.1)
$$\delta = \lim_{m \to \infty} i^{(m)} = \ln(1 + i).$$

Equivalently,

(1.11.2)
$$i = e^{\delta} - 1 \quad \text{and} \quad e^{\delta} = 1 + i.$$

Since we have compound interest at an effective rate i, $a(t) = (1 + i)^t$ and (1.11.2) yields

(1.11.3)
$$a(t) = e^{\delta t}.$$

This limiting process is rather abstract. We wish to visualize how it may occur in real life. Imagine that there are two banks competing hard for depositors. Bank A advertises that it pays 5% annual effective interest. Bank B decides to offer a better deal, namely 5% nominal interest convertible quarterly. This is equivalent to an annual effective interest rate of $(1 + \frac{.05}{4})^4 - 1 \approx 5.0945\%$. Not to be bested, Bank A changes its accounts to pay 5% nominal interest convertible monthly. This is equivalent to an annual effective interest rate of $(1 + \frac{.05}{12})^{12} - 1 \approx 5.1162\%$. Bank B continues the competition by offering 5% nominal interest convertible daily. In a nonleap year, this is equivalent to an annual effective interest rate of $(1 + \frac{.05}{365})^{365} - 1 \approx 5.1267\%$. Finally, in order not to be surpassed in this competition, at least at a 5% rate of interest, Bank A offers 5% nominal interest convertible continuously. Recalling (1.11.2), the equivalent annual effective interest rate is $i = e^{.05} - 1 \approx 5.1271\%$.

We note that the effective interest rates corresponding to nominal rates of 5% convertible daily and 5% convertible continuously are very close. Continuous compounding is sometimes used to approximate daily compounding.

The discussion in this section has involved nominal interest rates. However, δ may also be realized as a limit of nominal discount rates. To see this, note that (1.11.1) implies that $\lim_{m \to \infty} \left(1 + \frac{i^{(m)}}{m}\right) = 1$. Consequently, again recalling (1.11.1) and (1.10.13),

(1.11.4)
$$\lim_{m \to \infty} d^{(m)} = \lim_{m \to \infty} \left(\frac{i^{(m)}}{1 + \frac{i^{(m)}}{m}} \right) = \lim_{m \to \infty} i^{(m)} = \delta.$$

Since $\{i^{(m)}\}$ is decreasing and $\{d^{(m)}\}$ is increasing, we deduce from the limits $\lim_{m \to \infty} i^{(m)} = \delta$ and $\lim_{m \to \infty} d^{(m)} = \delta$ that

$$i^{(m)} > \delta > d^{(m)}.$$

We can thus extend the ordering (1.10.16) to

IMPORTANT FACT (1.11.5): If $i > 0$ and $m > 1$, then

$$i > i^{(m)} > \delta > d^{(m)} > d.$$

In this section we have assumed that we had a compound interest accumulation function. Before moving on to consider the force of interest when you have a general accumulation function, we consider two examples.

EXAMPLE 1.11.6

Problem: Estelle deposits $12,500 in a four-year certificate of deposit at Community Trust Bank. If the annual percentage yield (APY) is 4.235%, find the nominal interest rate δ convertible continuously.

Solution We are given that $i = 4.235\%$. So, Equation (1.11.1) gives us $\delta = \ln(1.04235) \approx .041477779$. As expected [see Fact (1.11.5)], $\delta < i$. ∎

The expression "money grows at $r\%$ compounded continuously" means that the force of interest δ is numerically equal to $r\%$.

EXAMPLE 1.11.7

Problem: Money at Swift National Bank grows at 3.8% compounded continuously. Rafael Ortiz closed his savings account at Swift exactly three years after he opened it, and he received his balance which was $24,812. Mr. Ortiz made a $6,500 deposit exactly one year after he opened the account, but he made no other deposits or withdrawals except for an unspecified opening deposit X. Find X and his balance exactly two years after he opened the account?

Solution We are given that $\delta = 3.8\%$ so $a(t) = e^{.038t}$. At the time the account was closed, the $6,500 deposit has grown to $6,500\frac{a(3)}{a(1)} = \$6,500e^{.076}$, and the initial deposit X has accumulated to $Xe^{.038\times3} = Xe^{.114}$. So, $\$24,812 \approx \$6,500e^{.076} + Xe^{.114}$, and $X = e^{-.114}(\$24,812 - \$6,500e^{.076}) \approx \$15,881.07029$. X must be an integral number of cents so $X = \$15,881.07$. His balance at the end of two years was $\$15,881.07a(2) + 6,500\frac{a(2)}{a(1)} = \$15,881.07e^{.076} + 6,500e^{.038} \approx \$23,886.83$. This balance may also be found by computing $\$24,812\frac{a(2)}{a(3)} = \$24,812e^{-.038} \approx \$23,886.83$. ∎

1.12   FORCE OF INTEREST

(calculus needed here)

Suppose that you want to measure how well an investment is doing near a particular point in time, say time t. The interest rate for the unit interval $[t, t + 1]$ is $\frac{a(t+1)-a(t)}{a(t)}$. However, if the interest rate varies a lot, this interest rate may not give a good idea of what is happening at the instant t: that is, it may not be helpful for determining how money grows on very short intervals containing t. Suppose that instead of looking at $[t, t + 1]$, you let m denote a positive integer and look at the interval $[t, t + \frac{1}{m}]$. Then $\frac{a(t+\frac{1}{m})-a(t)}{a(t)}$ is the interest rate for your interval of length $\frac{1}{m}$, hence corresponds to the nominal interest rate $\left(\frac{a(t+\frac{1}{m})-a(t)}{a(t)}\right) \Big/ \frac{1}{m} = m\left(\frac{a(t+\frac{1}{m})-a(t)}{a(t)}\right)$. As the integer m under consideration increases, this nominal interest rate convertible m times per basic period represents a better and better approximation to the interest rate at time t. As m tends to infinity, $\frac{1}{m}$ tends to 0 and

$$m\left(\frac{a(t + \frac{1}{m}) - a(t)}{a(t)}\right) = \frac{\left(\frac{a(t+\frac{1}{m})-a(t)}{\frac{1}{m}}\right)}{a(t)}$$

tend to $\frac{a'(t)}{a(t)}$. Thus $\frac{a'(t)}{a(t)}$ can be thought of as a nominal interest rate convertible continuously that describes the performance of the investment at the instant t. We call the ratio $\frac{a'(t)}{a(t)}$ the **force of interest** δ_t at time t.

(1.12.1)
$$\boxed{\delta_t = \frac{a'(t)}{a(t)}.}$$

Note that if money grows by compound interest, you have the force of interest δ from Section (1.11) as well as the newly defined force of interest δ_t. This is acceptable since, as we will verify in Example (1.12.4), when you have a compound interest accumulation function, $\delta_t = \delta$ for all t.

EXAMPLE 1.12.2 δ_t for simple interest

Problem: Suppose that you have simple interest at a rate r. Find the force of interest δ_t.

Solution   The accumulation function is $a(t) = 1 + rt$. Therefore, $\delta_t = \frac{a'(t)}{a(t)} = \frac{r}{1+rt}$.   ∎

EXAMPLE 1.12.3 δ_t for simple discount

Problem: Suppose that you have simple discount at a rate d. Find the force of interest δ_t.

Solution   The accumulation function is $a(t) = (1 - dt)^{-1}$. So, $\delta_t = \frac{a'(t)}{a(t)} = \frac{-(1-dt)^{-2}(-d)}{(1-dt)^{-1}} = \frac{d}{1-dt}$.   ∎

While the force of interest may vary with t (as it does in our examples with simple interest or simple discount), the next example shows that under compound interest, δ_t is equal to the constant δ of Section (1.11).

EXAMPLE 1.12.4 δ_t for compound interest

Problem: Suppose that you have compound interest at a rate i. Find the force of interest δ_t as a function of i.

Solution   The accumulation function is $a(t) = (1 + i)^t$. Therefore,

$$\delta_t = \frac{a'(t)}{a(t)} = \frac{(1 + i)^t \ln(1 + i)}{(1 + i)^t} = \ln(1 + i).$$

Recalling (1.11.1),

$$\delta_t = \ln(1 + i) = \lim_{m \to \infty} i^{(m)} = \delta.$$

∎

The definition (1.12.1) of the force of interest δ_t tells you how to obtain δ_t from the accumulation function $a(t)$. Observe that the ratio $\frac{a'(t)}{a(t)}$ is equal to the derivative $\frac{d}{dt}(\ln a(t))$. Therefore,

(1.12.5)
$$\boxed{\delta_t = \frac{d}{dt}(\ln a(t)).}$$

This gives us a second formula with which to calculate the force of interest δ_t from a given accumulation function $a(t)$.

EXAMPLE 1.12.6   **Finding δ_t from $a(t)$ using (1.12.5)**

Problem: Suppose $a(t) = (1.05)^{\frac{t}{2}}(1.04)^{\frac{t^2}{3}}(1.03)^{\frac{t^3}{6}}$. Find δ_t.

Solution   According to (1.12.5),

$$\delta_t = \frac{d}{dt}(\ln a(t)) = \frac{d}{dt}(\ln((1.05)^{\frac{t}{2}}(1.04)^{\frac{t^2}{3}}(1.03)^{\frac{t^3}{6}}))$$

$$= \frac{d}{dt}(\frac{t}{2}\ln(1.05) + \frac{t^2}{3}\ln(1.04) + \frac{t^3}{6}\ln(1.03))$$

$$= \frac{1}{2}\ln(1.05) + \frac{2t}{3}\ln(1.04) + \frac{t^2}{2}\ln(1.03)$$

$$= \ln((1.05)^{\frac{1}{2}}(1.04)^{\frac{2t}{3}}(1.03)^{\frac{t^2}{2}}).$$

■

We next learn how the accumulation function $a(t)$ may be found if the force of interest function δ_t is known.

It follows immediately from the definition of the force of interest at time r that if $r \in [0, t]$, then $\delta_r = \frac{a'(r)}{a(r)} = \frac{d}{dr}\ln a(r)$. (Note that $r \in [0, t]$ is a standard notation denoting that r is in the set $[0, t]$, and more generally $s \in S$ means that s is an element of the set S. We will henceforth use the symbol $\in$ without comment.) Integrating over the interval $[0, t]$ you obtain

$$\int_0^t \delta_r dr = \int_0^t \frac{d}{dr}\ln(a(r))dr.$$

Now by the Fundamental Theorem of Calculus, $\int_0^t \frac{d}{dr}\ln a(r)dr = \ln a(t) - \ln a(0)$. Since $a(0) = 1$ and $\ln 1 = 0$, this gives us

$$\int_0^t \delta_r dr = \ln a(t).$$

Consequently,

(1.12.7)

$$\boxed{a(t) = e^{\int_0^t \delta_r dr}.}$$

EXAMPLE 1.12.8   **Finding $a(t)$ from δ_t**

Problem: Suppose $\delta_t = \frac{3}{1-3t}$. Find the corresponding accumulation function $a(t)$.

Solution   Note that

$$\int_0^t \delta_r dr = \int_0^t \frac{3}{1-3r} dr = -\ln(1-3r)\Big|_0^t$$

$$= -\ln(1-3t) + \ln 1 = -\ln(1-3t) = \ln((1-3t)^{-1}).$$

Therefore

$$a(t) = e^{\int_0^t \delta_r dr} = e^{\ln((1-3t)^{-1})} = (1-3t)^{-1}.$$

This is the accumulation function for simple discount at a rate of 300% per basic period. ∎

1.13   NOTE FOR THOSE WHO SKIPPED SECTIONS (1.11) AND (1.12)

Imagine that there are two banks competing hard for depositors. Bank A advertises that it pays 5% annual effective interest. Bank B decides to offer a better deal, namely 5% nominal interest convertible quarterly. The reason that this is a better deal is that there is more frequent compounding, and therefore more earning of interest by previous interest. The nominal interest rate of 5% convertible quarterly is equivalent to an annual effective interest rate of $(1 + \frac{.05}{4})^4 - 1 \approx 5.0945\%$. Not to be bested, Bank A changes its accounts to pay 5% nominal interest convertible monthly. This is equivalent to an annual effective interest rate of $(1 + \frac{.05}{12})^{12} - 1 \approx 5.1162\%$. Bank B continues the competition by offering 5% nominal interest convertible daily. In a nonleap year, this is equivalent to an annual effective interest rate of $(1 + \frac{.05}{365})^{365} - 1 \approx 5.1267\%$. Finally, in order not to be surpassed in this competition, at least at a 5% rate of interest, Bank A declares that it offers 5% nominal interest **convertible continuously**. That is, they say they will compound interest constantly, and this is the limiting case of compounding more and more frequently. Calculus is the mathematics of limits, so it takes calculus to analyze to what annual effective rate of interest this continuous compounding is equivalent. In fact, it is equivalent to an annual effective interest rate of $i = e^{.05} - 1 \approx 5.1271\%$.

More generally,

A nominal interest rate of δ convertible continuously is equivalent to an annual effective rate of $i = e^{\delta} - 1$. When these equivalent rates govern the growth of money, $a(t) = (1+i)^t = e^{\delta t}$ and $v = \frac{1}{1+i} = e^{-\delta}$.

The constant δ is called the **force of interest**.

Alternatively, we may start with an annual effective interest rate and look for an equivalent force of interest (nominal rate convertible continuously).

> An annual effective interest rate of i is equivalent to a nominal rate of interest δ convertible continuously where $\delta = \ln(1 + i)$.

EXAMPLE 1.13.1

Problem: Swift Bank promises 3.5% interest compounded continuously. If Ken deposits $3,000 in Swift Bank, what will his balance be four years later?

Solution   At Swift Bank, $\delta = .035$, $a(t) = e^{\delta t} = e^{.035t}$, and Ken's balance after four years is $\$3,000e^{(.035)4} \approx \$3,450.82$.   ∎

1.14   INFLATION

In Section (1.2) we introduced units of money, saying that we would usually take these to be dollars. But what exactly does one mean by a dollar? A dollar at the beginning of the year 2000 was worth — that is, had the same purchasing power as — only about 89 cents in January 1, 1995 dollars and about seventy five cents in January 1, 1990 dollars. So, if a loaf of bread cost 75 cents in 1990, the price of a comparable loaf in the year 2000 might be about one dollar. Put another way, it would take approximately $\frac{100}{75} \approx 1.33$ dollars in 2000 to buy what you could buy for one dollar ten years earlier. This loss of purchasing power per dollar (or other monetary unit) is called **inflation**.

Inflation is formidable to measure because it gives the increase over time of a vague economic function, the price level function $p(t)$. Focus on an interval of time $[t_1, t_2]$. Analogously to how the effective interest rate $i_{[t_1,t_2]}$ was defined [see (1.3.5)], the inflation rate should be defined by

$$(1.14.1) \qquad\qquad r_{[t_1,t_2]} = \frac{p(t_2) - p(t_1)}{p(t_1)}.$$

However, it is unclear how best to define $p(t)$. Usually the price of some well-defined "basket of goods" is taken to define $p(t)$, say the "basket of goods" used to calculate a national **Consumer Price Index**[8]. But, for a particular investor, a price index of a specialized segment of the economy might be more relevant.

To analyze the impact of inflation on investments, consider the following illustration. Suppose that Marcel has $\$D$ now and that he could currently purchase u units of some good with his money. That is, each dollar could be

---

[8]In the United States, the Consumer Price Index is calculated by the Bureau of Labor Statistics.

used to purchase $\frac{u}{D}$ units of the good. If the inflation rate over one unit of time is r, then one unit of time later, the cost of the u units would be $(1 + r)\$D$ and one dollar would buy $\frac{u}{(1+r)\$D}$ units. If Marcel invests the $\$D$ for one unit of time at an effective interest rate i, then it will have grown to $(1 + i)\$D$, which would then be able to buy $\left[(1 + i)\$D\right]\frac{u}{(1+r)\$D}$ units. So, Marcel's purchasing power has changed from u units to $\frac{1+i}{1+r}u$ units. Since this occurred over a time interval of one unit, Marcel's **inflation adjusted** (or **real**) interest rate j on his investment satisfies

(1.14.2)
$$1 + j = \frac{1 + i}{1 + r},$$

and this is equivalent to

(1.14.3)
$$j = \frac{i - r}{1 + r}.$$

When inflation is expected, investors will require higher interest rates because they are interested in actual growth in their buying power, not just a growth in the number of dollars that they possess. Borrowers who anticipate a depreciation in the value of a dollar should be willing to agree to repay more future dollars than they would otherwise [see Problem (1.14.4)]. The fact that interest rates reflect predicted inflation rather than actual inflation is a complication to those studying the correlation between interest rates and inflation rates.

In practice, of course, one cannot possibly know what the inflation rate will be in the future, and this makes investment decisions difficult. For example, suppose Marcel intends to only make investments for which the inflation-adjusted interest rate is at least the number j' and that he is presented with an investment opportunity with nonadjusted interest rate i. Does this investment meet Marcel's criterion? The answer to this question depends on the inflation rate for the period of the investment. That is to say, the assessment is contingent upon what happens in the future, and the correctness of the response is therefore uncertain. If Marcel *believes* that the inflation rate for this period will be r', then by (1.14.3), he *believes* that his inflation-adjusted interest rate will be $\frac{i-r'}{1+r'}$. Assuming Marcel acts based on his belief, he will only invest if his anticipated real interest rate is at least j', that is if

(1.14.4)
$$\frac{i - r'}{1 + r'} \geq j'.$$

Inequality (1.14.4) is equivalent to

(1.14.5)
$$i \geq j' + r' + r'j'.$$

EXAMPLE 1.14.6   Anticipated inflation

Problem: Tamilla wishes to invest $10,000 for one year provided that she can anticipate a 4% growth in her buying power. She forecasts that the inflation rate for the upcoming year will be 3%.

(a) What is the lowest rate at which she would be willing to make a loan?

(b) What rate of growth in her buying power does she look forward to if she is able to loan out her money for exactly one year at an annual effective interest rate of 8%?

(c) What was her actual growth in buying power if the inflation rate for the year was 3.5% and she was able to loan out her money for exactly one year at an annual effective interest rate of 8%?

Solution

(a) Tamilla's inflation prediction amounts to setting $r' = .03$, and her desired growth rate puts $j' = .04$. According to inequality (1.14.5), the lowest rate at which she should make a loan is $j' + r' + r'j' = .04 + .03 + (.03)(.04) = .0712 = 7.12\%$.

(b) If Tamilla is able to loan out her money at the higher rate of 8%, she foresees that her buying power will grow at the rate

$$\frac{i - r'}{1 + r'} = \frac{.08 - .03}{1.03} = \frac{5}{103} \approx .048543689 \approx 4.85\%.$$

(c) If Tamilla underestimated the rate of inflation and it actually grew at 3.5%, her actual growth in buying power was at the rate

$$\frac{i - r}{1 + r} = \frac{.08 - .035}{1.035} = \frac{45}{1035} \approx .043478261 \approx 4.35\%.$$

■

We note that the answer to question (c) of Example (1.14.1) was found using Equation (1.14.3). Had we ignored the denominator $1 + r$ of $\frac{i-r}{1+r}$, $i - r$ would have given us a good approximation to Tamilla's inflation-adjusted rate of interest, namely 4.5% rather than the true rate that was slightly below 4.35%. More generally, in times of low inflation, using the numerator $i - r$ as an estimate to j will not lead to great inaccuracies. *However*, when there is high inflation, it is critical that one include the denominator for a meaningful analysis.

EXAMPLE 1.14.7   Illustrating the importance of the denominator of Equation (1.14.3) in economies with high inflation.

Problem: Gustavo noted that the annual inflation rate was very high, some **months** in excess of 20%. He therefore usually converted his salary, paid in Brazilian Cruzados, to a more stable currency. However, one January 1st, Gustavo loaned his sister 20,000 Cruzados for one year at an effective interest rate of 620%. If the annual rate of inflation for the year was 600%, what was Gustavo's real interest rate on this family loan?

Solution   According to (1.14.3), the real monthly rate of interest is $\frac{6.20-6}{1+6} \approx$ .028571429. Thus, Gustavo had a modest return of about 2.86%, not the large 20% gain that would result if one ignored the denominator. ∎

Since i is only equal to the real interest rate j if $r = 0$ [see Problem (1.14.2)], the interest rate i may be referred to as a **nominal** interest rate. Beware that this is a different use of the term "nominal interest rate" than was introduced in Section (1.10). In this book, except when explicitly stated to the contrary, we will ignore inflation. In real life problems, the stated effective interest rate should always be replaced by the real interest rate j if it is known, or by the anticipated real interest rate j' if you are dealing with growth in the future. If the actual inflation rate significantly exceeds the anticipated one, the real interest rate is quite likely to be negative. However, $1 + j$ is equal to the quotient $\frac{1+i}{1+r}$ and is hence positive except if there is depreciation exceeding 100% for the period under consideration. In periods where the actual rate of inflation is de-accelerating, overestimates of the inflation rate are common, and these result in real interest rates that are unexpectedly high.

1.15   PROBLEMS, CHAPTER 1

(1.0) Chapter 1 writing problems

(1) Call a local bank. Learn what accounts and rates are available with an initial deposit of $1,000

    (a) if you want access to your money at any time.

    (b) if you are willing to invest your money for a fixed term. (Ask about Certificates of Deposit.)

(2) Interest is a rent for the use of money. Learn about rates charged for the rental of another commodity, e.g., vacuum cleaners, house, car. Then write a clear paragraph describing the rental situation. Be sure to include the item to be rented, the approximate value of the item at the

beginning of the rental period, the length of the rental period, the price paid for the rental, and estimates of the depreciation (or appreciation) and maintenance costs during the rental period.

(3) In western cultures, the charging of interest is now well accepted as fair business practice. This was not always the case. Learn about the history of interest, and then write an essay describing what you have learned. You might include Aristotle's views on interest, the views of various religious groups (over the ages) regarding interest, and Henry VIII of England's contribution to the acceptance of interest.

(4) Learn what "usury" means. Write an essay in which you either support or argue against laws limiting the rates of interest that lenders may charge.

(5) Learn the word meaning "interest" in several foreign languages. Discover the etymologies of these words, and try to think of English words with the same roots. Write a clear paragraph (or paragraphs) describing what you have discovered.

(6) [following Section (1.14)] Inflation rates vary over time and from country to country. Pick a country and describe its inflation history. Discuss political and social developments that accompanied periods of remarkable inflation or deflation.

(7) The United States Treasury issues savings bonds for debt financing. These include EE bond and I bonds. Learn about how each of these are purchased and make returns to the investor. Write a short prospectus for each type of bonds. Include a paragraph discussing their relative advantages and disadvantages.

(1.3) Accumulation and amount functions

(1) Given that $A_K(t) = \frac{1,000}{100-t}$ for $0 \leqslant t < 100$, find K and $a(20)$.

(2) If you invest \$2,000 at time 0 and $a(t) = 1 + .04t$, how much will you have at time 5?

(3) Suppose that an account is governed by a quadratic accumulation function $a(t) = \alpha t^2 + .01t + \beta$ and the interest rate i_1 for the first year is 2%. Find α, β, and the interest rate for the fourth year i_4.

(4) Suppose $a(t) = \alpha t^2 + \beta t + \gamma$. If \$100 invested at time 0 accumulates to \$152 at time 4 and \$200 invested at time 0 accumulates to \$240 at time 2, find the accumulated value at time 8 of \$1,600 invested at time 6. [HINT: First find the unknowns α and β. Then determine how much money X that you would need to deposit at time 0 in order to have an accumulation of \$1,600 at time 6. The desired answer is the accumulated value of X at time 8.]

(5) It is known that for each positive integer k, the amount of interest earned by an investor in the k-th period is k. Find the amount of interest earned by the investor from time 0 to time n, n a fixed positive integer.

(6) It is known that for each positive integer k, the amount of interest earned by an investor in the k-th period is 2^k. Find the amount of interest earned by the investor from time 0 to time n, n a fixed positive integer.

(7) Let $A_K(t) = 3t^2 + 2t + 800$. Show that the sequence of interest rates $\{i_n\}$ is decreasing for $n \geqslant 17$.

(8) Prof. Oops reports to his class three facts: (a) $a(t) = \alpha t^2 + \beta t + 1$; (b) $1,000 invested at time 0 accumulates to $1,200 at time 2; (c) $1,000 invested at time 0 accumulates to $10,000 at time 4. Explain why all three of these statements cannot simultaneously be true. Give an example of an accumulation function such that facts (b) and (c) hold.

(1.4) Simple interest

(1) How much interest is earned in the fourth year if $1,000 is invested under simple interest at an annual rate of 5%? What is the balance at at the end of the fourth year?

(2) In how many years will $500 accumulate to $800 at 6% simple interest?

(3) The monthly simple interest rate is .5%. What is the yearly simple interest rate?

(4) Find the yearly simple interest rate so that $1,000 invested at time 0 will grow to $1,700 in eight years.

(5) At a particular rate of simple interest, $1,200 invested at time $t = 0$ will accumulate to $1,320 in T years. Find the accumulated value of $500 invested at the same rate of simple interest and again at $t = 0$, but this time for $2T$ years.

(6) A loan is made at time 0 at simple interest at a rate of 5%.

   (a) In which period is this equivalent to an effective rate of $\frac{1}{23}$?
   (b) What is the effective interest rate for the interval $[4,6]$?

(7) [BA II Plus Calculator] Use the Date worksheet and what you know about leap years to calculate the number of days Albert Einstein lived. His date of birth was March 14, 1979, and he died on April 18, 1955. [HINT: The calculator can count the days between March 14, 1979 and December 31, 2049. It can also count the days between January 1, 1950 and April 18, 1955.]

(1.5) Compound interest

(1) Alice invests $2,200. Her investment grows according to compound interest at an annual effective interest rate of 4% for T years, at which time it has accumulated to $8,000. Find T.

(2) Elliott received an inheritance from his Aunt Ruth when she died on his fifth birthday. On his eighteenth birthday, the inheritance has grown to $32,168. If the money has been growing by compound interest at an annual effective interest rate of 6.2%, find the amount of money Aunt Ruth left to Elliott.

(3) Horatio invests money in an account earning compound interest at an unknown annual effective interest rate i. His money doubles in nine years. Find i.

(4) How much interest is earned in the fourth year by $1,000 invested under compound interest at an annual effective interest rate of 5%?

(5) At a certain rate of compound interest, money will double in α years, money will triple in β years, and money will increase tenfold in γ years. At this same rate of compound interest, $5 will increase to $12 in n years. Find integers a, b, and c so that $n = a\alpha + b\beta + c\gamma$.

(6) Sean deposits $826 in a savings account that earns interest at Increasing Rates Bank. For the first three years the money is on deposit, the annual effective interest rate is 3%. For the next two years the annual effective interest rate is 4%, and for the following five years the annual effective interest rate is 5%. What is Sean's balance at the end of ten years?

(7) For a fourteen-year investment, what level annual effective rate of interest gives the same accumulation as an annual effective rate of interest of 5% for eight years followed by a monthly effective rate of interest .6% for six years?

(8) Suppose you invest $2,500 in a fund earning 10% simple interest annaually. After two years you have the option of moving your money to an account that pays compound interest at an annual effective rate of 7%. Should you move your money to the compound interest account

   (a) if you wish to liquidate in five more years?
   (b) if you are confident your money will stay on deposit for a total of ten years?

(9) In 1963, an investor opened a savings account with K earning simple interest at an annual rate of 2.5%. Four years later, the investor closed the account and invested the accumulated amount in a savings account earning 5% compound interest. Determine the number of years (since 1963) necessary for the balance to reach $3K$.

(10) On March 1, 1993, Mr. Hernandez deposited $4,200 into an account that used a 4% annual effective interest rate when the balance was under $5,000 and a 5.5% annual effective interest rate when the balance is at least $5,000. Mr. Hernandez withdrew $1,000 on March 1, 1999. If there were no other deposits or withdrawals, find Mr. Hernandez's account balance on March 1, 2003.

(11) In terms of accumulation functions, the condition that there should never be an advantage or disadvantage to closing and immediately reopening one's account means that for all positive real numbers s and t, $a(s + t) = a(s)a(t)$. Assume that $a(t)$ is differentiable for all $t \geqslant 0$ and differentiable from the right at $t = 0$. These conditions on derivatives amount to the accumulation function being continuous and not having any sudden changes in direction.

(a) Use the definition of the derivative to show that
$a'(s) = a(s) \lim_{h \to 0} \frac{a(h)-1}{h}$.

(b) Show that $a'(s) = a(s)a'(0)$ where $a'(0)$ is a right-handed derivative.

(c) Note that $\frac{d}{ds} \ln a(s) = \frac{a'(s)}{a(s)}$. Use (b) to show that
$$\int_0^t \frac{d}{ds} \ln a(s)ds = a'(0)t.$$

(d) Deduce $\ln a(t) = \ln a(t) - \ln a(0) = a'(0)t$ from (c).

(e) Show that $a'(0) = \ln(1 + i)$.

(f) Show that $a(t) = (1 + i)^t$.

(1.6) Effective discount rates/ Interest in advance

(1) Antonio borrows $3,000 for one year at an annual discount rate of 8%. How much extra money does he have the use of?

(2) Grace borrows X for one year at a discount rate of 6%. She has use of an extra $2,400. Find X.

(3) Jonathan borrows $1,450 for one year at a discount rate of D. He has the use of an extra $1,320. Find D and the annual interest rate that this is equivalent to.

(4) The amount of interest earned on K for one year is $256. The amount of discount paid on a one year loan "for K," transacted on a discounted basis at a discount rate that is equivalent to the interest rate of the first transaction, is $236. Find K.

(5) A savings account earns compound interest at an annual effective interest rate i. Given that $i_{[2,4.5]} = 20\%$, find $d_{[1,3]}$.

(1.7) Discount functions/ The time value of money

(1) Suppose money grows according to the simple interest accumulation function $a(t) = 1 + .05t$. How much money would you need to invest at time 3 in order to have $3,200 at time 8?

(2) Find the time value at $t = 6$ of $4,850 to be paid at time 12 if $a(t) = (1 - .04t)^{-1}$.

(3) Frances Morgan purchased a house for $156,000 on July 31, 2002. If real estate prices rose at a compound rate of 6.5% annually, how much was the home Frances bought worth on July 31, 1998?

(4) A payment of X two years from now along with a payment of $2X$ four years from now repays a debt of $6,000 at 6.5% annual effective compound interest. Find X.

(5) What is the present value of $5,000 due in ten years assuming money grows according to compound interest and the annual effective rate of interest is 4% for the first three years, 5% for the next two years, and 5.5% for the final five years?

(6) Show that if the growth of money is governed by compound interest at an annual effective interest rate $i > 0$, then the sum of the current value of a payment of K made n periods ago and a payment of K to be made n periods from now is greater than $2K$. More generally, what must be true about the operative accumulation function $a(t)$ in order that the stated conclusion holds?

(7) You have two options to repay a loan. You can repay $6,000 now and $5,940 in one year, or you can repay $12,000 in 6 months. Find the annual effective interest rate(s) i at which both options have the same present value.

(8) Two projects have equal net present values when calculated using a 6% annual effective interest rate. Project 1 requires an investment of $20,000 immediately and will return $8,000 at the end of one year and $15,000 at the end of two years. Project 2 requires investments of $10,000 immediately and $X in two years. It will return $3,000 at the end of one year and $14,000 at the end of three years. Find the difference in the net present values of the two projects if they are calculated using a 5% annual effective interest rate.

(1.8) Simple discount

(1) If money grows according to simple discount at an annual rate of 5%, what is the value at time 4 of $3,460 to be paid at time 9?

(2) Sylvia invests her money in an account earning interest based on simple discount at a 2% annual rate. What is her effective interest rate in the fifth year?

(3) Suppose you can invest $1,000 in a fund earning simple discount at an annual rate of 8% or in a fund earning simple interest at an annual rate of 12%. How long must you invest your money in order for the simple discount account to be preferable?

(4) On July 1, 1990, John invested $300 in an account that earned 8% simple interest. On July 1, 1993 he closed this account and deposited the liquidated funds in a new account earning q% simple discount. On July 1, 1998, John had a balance of $520 in the simple discount account. How much interest did he earn between July 1, 1993 and July 1, 1994?

(5) Suppose you invest $300 in a fund earning simple interest at 6%. Three years later you withdraw the investment (principal and interest) and invest it in another fund earning 8% simple discount.

　(a) How much time (including the three years in the simple interest account) will be required for the original $300 to accumulate to $650?

　(b) At what annual effective rate of compound interest would $300 accumulate to $650 in the same amount of time?

(1.9) Compound discount

(1) A savings account starts with $1,000 and has a level annual effective discount rate of 6.4%. Find the accumulated value at the end of five years.

(2) Latisha wishes to obtain $4,000 to pay her college tuition now. She qualifies for a loan with a level annual effective discount rate of 3.5%

　(a) How much will she have to repay if the loan term is six years?

　(b) What is the annual effective interest rate of Latisha's loan?

(3) The annual effective interest rate on Mustafa's loan is 6.8%. What is the equivalent effective quarterly discount rate on the loan?

(4) An account is governed by compound interest. The interest for three years on $480 is $52. Find the amount of discount for two years on $1000.

(5) An account is governed by compound interest. The discount for three years on $2,120 is $250. Find the amount of interest for two years on $380.

(6) An account with amount an initial amount B earns compound interest at an annual effective interest rate i. The interest in the third year is $426 and the discount in the seventh year is $812. Find i.

(7) The amount of interest on X for two years is $320. The amount of discount on X for one year is $148. Find the annual effective interest rate i and the value of X.

(1.10) Nominal rates of interest and discount

(1) Suppose we have compound interest and $d^{(4)} = 8\%$. Find equivalent rates d, $d^{(3)}$, i, and $i^{(6)}$.

(2) The annual effective interest rate on Rogelio's loan is 6.6%. What is the equivalent nominal discount rate convertible monthly on the loan? What is the effective monthly discount rate?

(3) Suppose we have compound interest and an effective monthly interest rate of 0.5%. Find equivalent rates $i^{(12)}$, i, and d.

(4) Find the accumulated value of $2,480 at the end of twelve years if the nominal interest rate was 2% convertible monthly for the first three years, the nominal rate of discount was 3% convertible semiannually for the next two years, the nominal rate of interest was 4.2% convertible once every two years for the next four years, and the annual effective rate of discount was .058 for the last three years.

(5) Given equivalent rates $i^{(m)} = .0469936613$ and $d^{(m)} = .046773854$, find m.

(6) Given that

$$1 - \frac{d^{(n)}}{n} = \frac{1 + \frac{i^{(7)}}{7}}{1 + \frac{i^{(6)}}{6}},$$

find n.

(7) Let m be a positive real number. Suppose interest is paid once every m years at a nominal interest rate $i^{\left(\frac{1}{m}\right)}$. This means that the borrower pays interest at an effective rate of $\frac{i^{\left(\frac{1}{m}\right)}}{\frac{1}{m}} = mi^{\left(\frac{1}{m}\right)}$ per m year period.

(a) Find an expression for $i^{\left(\frac{1}{m}\right)}$ in terms of i.

(b) If $i^{\left(\frac{2}{3}\right)} = .06$, find i.

(c) Define $d^{\left(\frac{1}{m}\right)}$ to be the nominal discount rate payable once every m years. This means that the borrower pays discount at an effective rate of $\frac{d^{\left(\frac{1}{m}\right)}}{\frac{1}{m}} = md^{\left(\frac{1}{m}\right)}$ per m year period. Find a formula that gives $d^{\left(\frac{1}{m}\right)}$ in terms of $i^{\left(\frac{1}{m}\right)}$, and a formula that gives $d^{\left(\frac{1}{m}\right)}$ in terms of d.

(1.11) A friendly competition (Constant force of interest)

(1) Suppose $d^{(4)} = 3.2\%$. Find δ.

(2) Given that $\delta = .04$, find the accumulated value of $300 five years after it is deposited.

(3) You have a choice of depositing your money in account A which has an annual effective interest rate of 5.2%, account B which has an effective monthly rate of .44%, or account C that is governed by force of interest $\delta = .0516$. Which account should you choose? Which account would give you the lowest accumulation?

(4) Let $i(x) = x[(1 + i)^{\frac{1}{x}} - 1]$, $x > 1$. Note that $i(m) = i^{(m)}$.

(a) Find the derivative $i'(x)$ and show that the condition that $i(x)$ is decreasing is equivalent to $(1 + i)^{\frac{1}{x}}\left(1 - x^{-1}\ln(1 + i)\right) < 1$.

(b) Show that the condition $i(x)$ is decreasing is equivalent to the equation

$$\ln\left[(1 + i)^{\frac{1}{x}}(1 - x^{-1}\ln(1 + i))\right] < \ln 1 = 0.$$

(c) Let $z = \frac{1}{x}\ln(1 + i)$. Show that $i(x)$ is decreasing is equivalent to $z + \ln(1 - z) < 0$ and therefore to $1 - z < e^{-z}$.

(d) Show that $1 - z < e^{-z}$ follows from Taylor's theorem with remainder.

(e) Explain how this problem allows us to conclude that if $p > m > 1$, then $i^{(p)} < i^{(m)}$.

(1.12) Force of interest

(1) Given that the force of interest is $\delta_t = .05 + .006t$, find the accumulated value after three years of an investment of $300 made at

(a) time 0.

(b) time 4.

(2) Given that the force of interest is $\delta_t = \frac{2t}{1+t^2}$, find the effective rate of discount for the sixth year.

(3) Given that the force of interest is $\delta_t = \frac{t^2}{1+t^3}$, find the present value at time 0 of $300 to be paid at time $t = 4$.

(4) Given that $a(t) = e^{.03t+.002t^2}$, find δ_2.

(5) Given that $a(t) = (1 + .02)^t(1 + .03t)(1 - .05t)^{-1}$, find δ_3.

(6) Fund 1 accumulates with a discount rate of 2.4% convertible monthly. Fund 2 accumulates with a force of interest $\delta_t = \frac{t}{12}$ for all $t \geqslant 0$. At time 0, $100 is deposited in each fund. Determine all later times at which the two funds have equal holdings, assuming that there are no further contributions to either fund.

(7) As in Problem (1.5.8), suppose you invest $2,500 in a fund earning 10% simple interest. Further suppose that you have the option at any time of closing this account and opening an account earning compound interest at an annual effective interest rate of 7%. At what instant should you do so in order to maximize your accumulation at the end of five years? How about if you wish to maximize the accumulation at the end of ten years?

(8) (a) Fund I grows according to simple interest at rate r. Find the force of interest $\delta_t^{(I)}$ acting on fund I at time t.

  (b) Fund D grows according to simple discount at rate s. Find the force of interest $\delta_t^{(D)}$ acting on fund D at time t.

  (c) Suppose $r > s$. Find all t such that $\delta_t^{(I)} = \delta_t^{(D)}$.

(9) Fund A has a balance of $600 at time $t = 0$ and its growth is determined by a force of interest $\delta_t^{(A)} = \frac{.08}{1+.08t}$. Fund B has a balance of $300 at time $t = 0$, and its growth is determined by a force of interest $\delta_t^{(B)} = .01t$. Fund C has amount function $A^{(C)}(t) = A^{(A)}(t) + 2A^{(B)}(t)$ where $A^{(A)}(t)$ is the amount function giving the growth of fund A and $A^{(B)}(t)$ is the amount function giving the growth of fund B. The force of interest for fund C is $\delta_t^{(C)}$. Find $\delta_4^{(C)}$.

(10) Mr. Valdez has $10,000 to invest at time $t = 0$, and three possible ways to invest it. Investment account I is governed by compound interest with an annual effective discount rate of 3%. Investment account II has force of interest equal to $\frac{.04}{1+.05t^2}$. Investment account III is governed by the accumulation function $a^{III}(t) = (1 - .005t^2)^{-1}$. Mr. Valdez can transfer his money between the three investments at any time. What is the maximum amount he can accumulate at time $t = 5$? [HINT: At all times, Mr. Valdez wishes to have his money in the account that has the greatest force of interest at that moment. Therefore, begin by determining the force of interest function for each of the investment accounts. Next decide for which time interval Mr. Valdez should have his money in each of the accounts. Assume that he accordingly moves his money to maximize his return. You will then need the accumulation functions for the accounts in order to determine Mr. Valdez's balance at $t = 5$. Remember to use Important Fact (1.7.4).]

(1.13) Note for those who skipped Section (1.11) and (1.12)

(1) What rate of annual effective interest does John earn if his money is invested in a bank that credits interest continuously at a rate of 3.75%?

(2) Andrea is upset because her bank does not compound interest continuously. Instead, they use a nominal discount rate of 4% compounded quarterly. To what rate of continuous compounding is that equivalent?

(1.14) Inflation

(1) Inflation is forecast to be at an annual rate of 3% for the next year.

  (a) What will the real rate of interest be if the forecast holds true, and the stated effective rate for the year is 4.2%?

  (b) What will the real rate be if the actual rate of inflation is 4.6%?

(2) Show that the nominal interest rate i and the real interest rate j are equal if and only if the inflation rate is zero.

(3) The nominal rate of discount is 3% convertible quarterly. The inflation-adjusted (effective) rate of interest is 1.24%. Find the rate of inflation.

(4) We have noted that the effective rate i should be replaced by the inflation-adjusted rate $j = \frac{1+i}{1+r} - 1$ in real life calculations. With notation as in Section (1.7), find Y such that

$$PV_{(1+j)^t}(\$X \text{ at time } n) = PV_{(1+i)^t}(\$Y \text{ at time } n).$$

(5) Money is invested in a savings account with a nominal interest rate of 2.4% convertible monthly for three years. The rate of inflation is 1.5% for the first year, 2.8% for the second year, and 3.4% for the third year. Find the percentage of purchasing power lost during the time the money is invested; that is, find p so that if you could purchase exactly u units at the time the money was invested, three years later you could purchase $u(1 - .01p)$.

(6) Suppose that there is compound interest at an annual inflation adjusted rate of 1.8% and that the annual rate of inflation is 2.3%. Find the force of interest corresponding to the stated rate of interest and the inflation-adjusted force of interest corresponding to the real rate of interest. In general, what can you say about the difference between the force of interest and the inflation-adjusted force of interest?

Chapter 1 review problems

(1) Find the accumulated value of $6,208 at the end of eight years if the nominal rate of discount is 2.3% convertible quarterly for the first two years, the nominal rate of interest is 3% convertible monthly for the

next year, the annual effective rate of discount is 4.2% for the next three years, and the force of interest is .046 for the last two years.

(2) Suppose that $d^{(2)} - d = .00107584$. Find $i^{(3)} - i$.

(3) Assume that an investment is governed by compound interest at a level interest rate. The present value of $\$K$ payable at the end of two years is $\$1,039.98$. If the force of interest is cut in half (resulting in a change in the level interest rate), the present value of $\$K$ payable at the end of two years is $\$1,060.78$. What is the present value of $\$K$ payable at the end of two years if the annual effective discount rate is cut in half?

(4) (a) Express $\frac{d}{d\delta}v$ as a function of d.

  (b) Express $\left(\frac{d}{dv}\delta\right)\left(\frac{d}{di}d\right)$ as a function of d.

(5) A borrower will have two options for repaying a loan. The *first option* is to make a payment of $\$6,000$ on December 1, 2003 and a payment of $\$4,000$ on December 1, 2004. The *second option* is to make a single payment of $\$12,000$ N months after December 1, 2003. Assuming that the two options have the same value on December 1, 2003, if the interest rate is an annual effective rate of 5%, determine N.

(6) On June 3, 1977, Alan borrowed $\$3,000$ from Chan and gave Chan a promissory note at an annual rate of simple interest of 10% and a maturity date of May 15, 1978. Exact simple interest is used to compute the repayment amount on this note. Javier purchased the promissory note from Chan on December 20, 1977 based on simple discount at an annual rate of 12%, with time measured using the "actual/actual" method. Determine Javier's purchase price, and the equivalent annual effective interest rates earned by Chan and by Javier. (In Chapter 2, such rates are called annual yield rates.)

(7) Suppose that an investment is governed by an accumulation function

$$a(t) = \begin{cases} at^2 + bt + c, & \text{for } 0 \leqslant t \leqslant 6 \\ (at^2 + bt + c)(1 + .05(t - 6)), & \text{for } t > 6. \end{cases}$$

Further suppose that $i_3 = 50/1{,}088$ and $d_4 = 54/1{,}192$.

  (a) Determine the constants a, b, and c.
  (b) Find the value at $t = 3$ of $\$1,000$ to be paid at $t = 8$.

(8) Shanon deposits $\$K$ into an account that earns compound interest at an annual effective interest rate i. She makes no further deposits or withdrawals. Her interest in the fifth year is $\$175.37$, and the discount for the second year is $\$153.59$. Find i.

(9) Suppose that $\delta_t = \frac{4}{t-1}$ for $2 \leqslant t \leqslant 8$. For $2 \leqslant n \leqslant 7$, let $f(n) = i_{n+1} + 1$. Write a simple formula for $f(n)$.

CHAPTER 2

Equations of value and yield rates

2.1   INTRODUCTION

A financial transaction may be as simple as making a single deposit C, and the quantity being studied may be $A_C(t)$, the accumulation t time periods in the future. However, commonly there is a series of deposits and withdrawals before one focuses on the accumulation. For instance, a young couple may be saving to make a down payment on a home. At the end of each month the couple makes a deposit into a savings account using any available money. Occasionally, the couple has a month with many expenses and a withdrawal must be made from the savings account. Thus there is a contribution at the end of each month and while most of the contributions are positive, sometimes one is negative or conceivably zero. The couple might have a target date of two years from the opening of the savings account for the purchase of the house, and the accumulation at that time (the balance of the account) determines the amount available for a down payment. Alternatively, the couple might have a predictable contribution schedule, and the question is at what point will the balance reach $20,000, the minimum the couple wants to have saved before becoming homeowners. To lay down the mathematical foundations for the study of such questions, in Sections (2.2) and (2.3) we introduce **equations of**

value. *These are essential to the discussion in the remainder of the text.*

A problem in Interest Theory may often be solved using one or more equations of value. Often you must solve for unknowns. You have already seen problems where there is an unknown rate of interest, and in Section (2.4) we define the **(dollar-weighted) yield rate(s)** to be the unknown rate(s) of interest. Questions of existence and uniqueness are discussed. Reinvestments have yield rate ramifications, which are first studied in Section (2.5). Sometimes an exact solution to a yield rate problem is difficult or even impossible to obtain. A method for finding an approximate solution is introduced in Section (2.6). Finally, a different sort of "yield rate," the so-called **time-weighted yield rate**, is discussed in Section (2.7). This new "yield rate" is a measure of how well a fund is managed.

2.2 EQUATIONS OF VALUE FOR INVESTMENTS INVOLVING A SINGLE DEPOSIT MADE UNDER COMPOUND INTEREST

The simplest of interest theory problems involves a single investment C for an interval of time of length T under an amount function $A_C(t)$. If we have compound interest at an annual effective rate i and our basic units of time are years, then $A_C(t) = C(1 + i)^t$. The value of the investment at time T is then $A_C(T) = C(1 + i)^T$. If any three of the four basic quantities C, T, i, and $A_C(T)$ are known, we can use the equation

$$(2.2.1) \qquad\qquad A_C(T) = C(1 + i)^T$$

to solve for the fourth unknown quantity. Equation (2.2.1) is a time T equation of value for the investment under consideration. More generally, a time T equation of value equates two different expressions, each giving the value of some cashflow or sequence of cashflows at time T.

EXAMPLE 2.2.2   Unknown principal

Problem: Aiko's aunt opens an account for her niece. The account earns compound interest at an annual effective interest rate of 8%. The aunt notifies Aiko of the gift but not of the amount deposited. Five years after the account was opened, Aiko closes the account and receives $3,673.32. What was the amount C that Aiko's aunt deposited?

Solution   We are given that the final account balance, which was the balance after five years, was $3,673.32. On the other hand, the compound interest accumulation function (defined for an effective interest rate of 8%) has value $C(1 + .08)^5$ after five years. Therefore, our time 5 equation of value is $3,673.32 = C(1 + .08)^5$, and we find $C = \$3,673.32(1 + .08)^{-5} \approx$ $2,499.999869. Her aunt must have deposited an integral number of cents, so her deposit was for $2,500. ∎

EXAMPLE 2.2.3   Unknown length of investment

Problem: Chen opens an account with $1,000. The account earns interest at an annual effective rate of 4%. Chen learns that the balance is currently $1,342. How long did it take to earn the $342 of interest?

Solution Let T denote the length of time. Then our time T equation of value is $\$1,342 = \$1,000(1.04)^T$. Hence $T = \frac{\ln(1,342/1,000)}{\ln(1.04)} \approx 7.500145074$. Thus the money has been invested for approximately $7\frac{1}{2}$ years.   ∎

EXAMPLE 2.2.4   Doubling your money / "The rule of seventy-two"

Problem: Antonio plans to open an account. He notes that if the account has an annual effective interest rate of i, then the length of time T required to double his money satisfies the time T equation of value $2 = (1 + i)^T$, whence $T = \frac{\ln(2)}{\ln(1+i)}$. If Antonio knows i and has a calculator handy, T may thus be quickly found. However, Antonio would like to make approximate calculations even when no calculator is available. Being mathematically inclined, he notes that $\frac{\ln(2)}{\ln(1+i)} = \left(\frac{\ln(2)}{i}\right)\left(\frac{i}{\ln(1+i)}\right)$, the function $f(i) = \frac{i}{\ln(1+i)}$ grows very slowly for i positive and close to zero, and $\ln(2)f(.08) = \frac{(\ln(2)(.08))}{\ln(1.08)} \approx .720517467$. Use Antonio's observations to derive a rule for approximating T, then check how well this rule works when $i = 4\%$ and when $i = 9\%$.

Solution $T = \frac{\ln(2)}{\ln(1+i)} = \left(\frac{\ln(2)}{i}\right)\left(\frac{i}{\ln(1+i)}\right) = \left(\frac{\ln(2)}{i}\right)f(i) \approx \left(\frac{\ln(2)}{i}\right)f(.08) = \frac{\ln(2)f(.08)}{i} \approx \frac{.72}{i}$. Thus if i is given as a percent, you can approximate T by dividing 72 by that percentage. For example, if $i = 4\%$, then $T \approx \frac{72}{4} = 18$ and if $i = 9\%$, then $T \approx \frac{72}{9} = 8$. In fact, the formula $T = \frac{\ln(2)}{\ln(1+i)}$ yields that the values are approximately 17.67298769 and 8.043231727 so these are excellent approximations!   ∎

We note that in Example (2.2.4) we used $f(.08)$ to approximate $f(i)$. This is a standard choice and yields the number 72 which has many divisors. Had we approximated $f(i)$ by 1,[1] we would have had a rule of 69, had we approximated $f(i)$ by $f(.12)$ we would have had a rule of 73, while using $f(.2)$ to approximate $f(i)$ would have given us a rule of 76. The rule of seventy-two has commonly been used.

---

[1] $f(0)$ is undefined but if you apply l'Hospital's rule from calculus, you may show that $\lim_{i \to 0} f(i) = 1.$, So approximating $f(i)$ by 1 corresponds to using an interest rate of 0%.

EXAMPLE 2.2.5   Unknown interest rate

Problem: Adrianna deposits $400 in an account for her daughter that grows by compound interest. After five years the balance has grown to $463.71. Assume there is a constant annual effective interest rate i earned by the account, and find that rate.

Solution $400(1 + i)^5 = \$463.71$. Consequently, $i = (463.71/400)^{\frac{1}{5}} - 1 \approx .030000164 \approx 3\%$.   ∎

2.3   EQUATIONS OF VALUE FOR INVESTMENTS WITH MULTIPLE CONTRIBUTIONS

In Section (2.2) we considered investments under compound interest with a single contribution of principal. More generally, an interest theory problem involves a sequence of contributions $\{C_{t_k}\}$, the investment amount at time t_k being C_{t_k}. We have a negative contribution C_{t_k} if there is a withdrawal at time t_k. The growth of the contributions is governed by an amount function. The sequence of investments is liquidated at time T, producing a balance of B. If we choose a common date τ at which to value all the contributions, since the total value of the contributions must equal the value of the liquidated amount, we obtain an equation, called the **time τ equation of value** for the investment. If the growth of money is governed by an accumulation function $a(t)$, then by Important Fact (1.7.4), the time τ value of C_{t_k} contributed at time t_k is $C_{t_k}\frac{a(\tau)}{a(t_k)}$ and the time τ value of B at time T is $B\frac{a(\tau)}{a(T)}$. We hence have the time τ equation of value

$$(2.3.1)_\tau \qquad \boxed{\sum_k C_{t_k}\frac{a(\tau)}{a(t_k)} = B\frac{a(\tau)}{a(T)}} \qquad \text{(time } \tau \text{ equation of value)};$$

note that τ is a number, and if you replace τ with another number τ', you get another equation that we will reference as $(2.3.1)_{\tau'}$.

In particular, since $a(0) = 1$ and the discount function $v(t)$ is defined by $v(t) = \frac{1}{a(t)}$, the time 0 equation of value is

$$(2.3.2) \qquad \boxed{\sum_k C_{t_k}v(t_k) = Bv(T)} \qquad \text{(time 0 equation of value)}.$$

The time T equation of value is

$$(2.3.3) \qquad \boxed{\sum_k C_{t_k}\frac{a(T)}{a(t_k)} = B} \qquad \text{(time } T \text{ equation of value)}.$$

Equation $(2.3.1)_\tau$ and Equation $(2.3.1)_{\tau'}$ are equivalent for all choices of τ and τ' since equation $(2.3.1)_{\tau'}$ may be obtained from $(2.3.1)_\tau$ by multiplying by $\frac{a(\tau')}{a(\tau)}$. Therefore, if you solve for an unknown using a time τ equation of value, the result is independent of the choice of τ.

Before turning to some examples involving multiple contributions, we wish to explain how equation (2.2.1) is an equation of type (2.3.3). With notation as in Section (2.2), let $B = A_C(T)$, the accumulated value of our single contribution at the end of our investment period. Then, as we assumed in Section (2.2) that $a(t) = (1 + i)^t$, (2.2.1) may be rewritten as $B = Ca(T)$. Alternatively, since $a(0) = 1$, (2.2.1) may be thought to be of the form $B = C\frac{a(T)}{a(0)}$. This is (2.3.3) for the single contribution time t_1.

The examples we are about to consider are fairly simple, and you could probably follow them without drawing **timelines**. However, as the financial situations you consider become more complicated, you will likely find it helpful to construct timelines. Therefore, to familiarize you with this illustrative technique, we include timelines for each of the examples in this section.

EXAMPLE 2.3.4   Unknown payment

Problem: John borrows \$1,000. The loan is governed by compound interest at an annual effective interest rate of 10%. John repays \$600 at the end of one year and plans to complete repayment of the loan with a payment of P at the end of the second year. Find P.

Solution   From John's perspective, the financial arrangement may be represented by the following timeline.

PAYMENT:	$-\$1,000$	$\$600$	P
TIME:	0	1	2

To determine P consider a time 2 equation of value

$$\$1,000(1.1)^2 = \$600(1.1) + P.$$

Solving for P, we find $P = \$1,000(1.1)^2 - \$600(1.1) = \$550$. We note that P could also have been found by noting that John's original debt was \$1,000. After one year this debt has grown to $\$1,000(1.1) = \$1,100$ but John made a \$600 payment, reducing outstanding debt to \$500. At the end of another year, this debt has grown to $\$500(1.1) = \$550$.

We note that the problem may also be done starting with a different (but equivalent) equation of value, for instance a time 0 equation of value; but the solution is simpler if you begin as we did. ∎

EXAMPLE 2.3.5   Unknown rate of interest and the quadratic formula

Problem: Caitlin opens a savings account with a deposit of $5000. She deposits $3000 a year later and $2,000 a year after that. The account grows by compound interest at a constant annual effective rate i. Just after Caitlin's $2,000 deposit, her balance is $11,000. Find the effective rate of interest i.

Solution   Consider a timeline for Caitlin's two-year investment.

PAYMENT:	$-\$5,000$	$-\$3,000$	$-\$2,000$
TIME:	0	1	2
BALANCE:			$\$11,000$

To determine i, note that the time 2 equation of value is

$$\$5,000(1 + i)^2 + \$3,000(1 + i) + \$2,000 = \$11,000.$$

This is equivalent to the equation $5(1 + i)^2 + 3(1 + i) - 9 = 0$. Letting $x = 1 + i$, we have $5x^2 + 3x - 9 = 0$ which is a quadratic equation. Recall that the quadratic formula tells us that the solutions to the quadratic equation $ax^2 + bx + c = 0$ are $x = \frac{-b + \sqrt{b^2 - 4ac}}{2a}$ and $x = \frac{-b - \sqrt{b^2 - 4ac}}{2a}$. Therefore, the quadratic formula and the fact that $x = 1 + i$ is positive tell us that $x = \frac{-3 + \sqrt{189}}{10} \approx 1.074772708$, and $i = x - 1 \approx 7.4772708\%$.

BA II Plus calculator solution   Use the **Cash Flow worksheet** to enter the deposit of $5,000 at time 0 and of $3,000 at time 1. This, along with advancing to the contribution register C02, is accomplished by pushing

| CF | 2ND | CLR WORK | 5 | 0 | 0 | 0 |

| ENTER | ↓ | 3 | 0 | 0 | 0 | ENTER | ↓ | ↓ |.

Rather than entering $2,000 at time 2, we enter the negative of the balance at time 2 just prior to the $2,000 deposit, namely $-\$9,000 = -(\$11,000 - \$2,000)$. We do so because had we made a withdrawal of $9,000 at time 2, neither Caitlin nor the savings institution would owe the other party any money. The entering of the $9,000 imaginary withdrawal is accomplished by continuing with our **Cash Flow worksheet** and pushing | 9 | 0 | 0 | 0 | +/− | ENTER |. Now push | IRR | CPT | and the display will show | IRR | =7.477270849.   ∎

  If we alter the above problem so that we know the balance at time 3 rather than at time 2, then we can still easily write down an equation of value, but we can not easily solve this equation in closed form for $1 + i$. However, for those equipped with the BA II Plus calculator, the **Cash Flow worksheet** may still be used. Alternatively, a graphing calculator can be used to find a good approximation to $1 + i$. Moreover, as we now demonstrate, we can obtain the rate i to any desired degree of accuracy by the "**guess and check method**" or, for those familiar with Calculus, by **Newton's method**.

EXAMPLE 2.3.6   Unknown rate of interest and an equation of value that has no easy algebraic solution

Problem: Daphne opens a savings account with a deposit of $5,000. She deposits $3,000 a year later and $2,000 a year after that. The account grows by compound interest at a constant annual effective rate i. Three years after Daphne opened the account, it has a balance of $11,000. Find the effective rate of interest i.

Timeline Let's represent Daphne's three-year investment experience by a timeline.

PAYMENT:	−$5,000	−$3,000	−$2,000	
TIME:	0	1	2	3
BALANCE:				$11,000

BA II Plus calculator solution   Use the **Cash Flow worksheet** to enter the deposits of $5,000 at time 0, $3,000 at time 1, and $2,000 at time 2. This, along with advancing to the register C03, is done by pushing

| CF | 2ND | CLR WORK | 5 | 0 | 0 | 0 | ENTER | ↓ | 3 | 0 | 0 | 0 |

| ENTER | ↓ | ↓ | 2 | 0 | 0 | 0 | ENTER | ↓ | ↓ | .

A withdrawal of $11,000 at time 3 would result in neither Daphne nor the savings institution owing the other party any money so we continue with our **Cash Flow worksheet** by pushing | 1 | 1 | 0 | 0 | 0 | +/− | ENTER |. Now push

| IRR | CPT | and the display will show | IRR | = 4.207651054.

Guess and Check Solution   To determine i, note that the time 3 equation of value is

$$\$5,000(1 + i)^3 + \$3,000(1 + i)^2 + \$2,000(1 + i) = \$11,000.$$

Comparing this to the scenario in Example (2.3.5), we see that it has taken an extra year for our account balance to grow to $11,000. Hence, the annual effective interest governing Daphne's account must be lower than the 7.48% rate we found in (2.3.5). Let $P(i) = \$5,000(1 + i)^3 + \$3,000(1 + i)^2 + \$2,000(1 + i)$. The polynomial $P(i)$ has only positive coefficients and therefore must be an increasing function of i. We then seek i so that $P(i) = \$11,000$. Note that $P(.03) = \$10,706.335 < \$11,000$, so $.03 < i$, and $P(.04) = \$10,949.12 < \$11,000$. so .04 is also lower than the desired rate but fairly close. Further observe that $P(.042) \approx \$10,998.12$. Thus i is just slightly more than 4.2%. In fact $P(.0421) \approx \$11,000.58$. Consequently, i is between 4.2% and 4.21%. Noting that $P(.04205) \approx \$10,999.35$, we see that i is in fact between 4.205% and 4.21%. Hence, to the nearest hundredth of a percent it is 4.21%. Of course this procedure can be continued to get i to any desired degree of accuracy.

Newton's Method Solution (calculus required) Recall that Newton's method is an iterative method used to find a solution of an equation of the form $f(x) = 0$. An initial approximation x_0 to the solution is specified and then one forms a sequence of approximations using the formula

$$x_{n+1} = x_n - \frac{f(x_n)}{f'(x_n)}.$$

The equation we wish to solve here is

$$5,000(1 + x)^3 + 3,000(1 + x)^2 + 2,000(1 + x) = 11,000,$$

so we set

$$f(x) = 5,000(1 + x)^3 + 3,000(1 + x)^2 + 2,000(1 + x) - 11,000.$$

Note that this polynomial has derivative

$$f'(x) = 15,000(1 + x)^2 + 6,000(1 + x) + 2,000.$$

Just as in our "guess and check" solution, our initial approximation is .03. That is, we set $x_0 = .03$. We then calculate that

$$x_1 = x_0 - \frac{f(x_0)}{f'(x_0)} \approx .03 - \frac{-293.665}{24,093.5} \approx .042188557.$$

Continuing,

$$x_2 = x_1 - \frac{f(x_1)}{f'(x_1)} \approx .042188557 - \frac{2.75000276}{24,545.48617} \approx .04207652,$$

$$x_3 = x_2 - \frac{f(x_2)}{f'(x_2)} \approx .04207652 - \frac{.00023389}{24,541.31123} \approx .042076511,$$

and

$$x_4 = x_3 - \frac{f(x_3)}{f'(x_3)} \approx .04076511 - \frac{.00000001}{24,541.31087} \approx .042076511.$$

Note that the approximations x_3 and x_4 are equal. Therefore, $i \approx 4.2076511\%$.

∎

The "guess and check" method and Newton's method proceed much better if you make a good initial guess. In general, we may not know as desirable a starting point for our analysis as we did in Example (2.3.6). In that case, just go ahead and guess. Suppose that in Example (2.3.6), we had started by guessing that the effective rate of interest was 8%. In the "guess and check" method, we would then have computed $P(.08)$. Since $P(.08) = \$11,957.76$, we would have known that we needed a considerably lower rate. Noting that $P(0) = \$10,000$, we would next have looked for a rate approximately halfway between 8% and 0% (perhaps slightly closer to 8% since $P(.08)$ is closer to the target $11,000 than $P(0)$ is). In Newton's method, an initial guess of $x_0 = .08$ would have lead to $x_1 \approx .043129042$ and we would have again quickly obtained the approximation $i \approx 4.2076511\%$. However, a sufficiently poor initial estimate in Newton's method may lead to a sequence of estimates that is nonconvergent .

Problems involving an unknown time period may also be solved using an equation of value and the "guess and check method" or Newton's method if needed.

EXAMPLE 2.3.7

Problem: Franklin borrows $1,000 and the loan is governed by compound interest at an annual effective interest rate of 10%. Franklin agrees to repay the loan by two equally spaced payments of $525. When should he make these payments?

Algebraic solution The situation may be exhibited by a timeline.

PAYMENT:	$-\$1,000$	$\$525$	$\$525$
TIME:	0	T	$2T$
BALANCE:			$\$0$

We let the time of the loan be $t = 0$. Then, considering a time 0 equation of value, we note that the repayments should take place at times T and $2T$ where

$$\$1,000 = \$525v^T + \$525v^{2T}.$$

Letting $x = v^T$, we can rewrite this equation as $\$1,000 = \$525x + \$525x^2$. Since $x > 0$, the quadratic formula yields

$$x = \frac{-525 + \sqrt{(525)^2 + 2,100,000}}{1,050} \approx .967910728.$$

Then, taking natural logarithms of each side of the equation $x = v^T$, we find $T = \frac{\ln(x)}{\ln(v)} \approx \frac{\ln(.967910728)}{-\ln(1.1)} \approx .342202894$. Consequently, Franklin should make payments at times .342202894 and .684405787.

BA II Plus calculator solution Push [CF] [2ND] [CLR WORK] to open and clear the Cash Flow worksheet. Then push

[1] [0] [0] [0] [ENTER] [↓] [5] [2] [5] [+/−] [ENTER] [↓] [2] [ENTER] .

At this point you have entered CFo $= 1{,}000$, C01 $= -525$, and F01 $= 2$. Now push [IRR] [CPT] to obtain IRR $= 3.315313209$. Thus the length of time T until the first payment satisfies $(1.1)^T \approx 1.03315313209$. Equivalently, $T \approx \dfrac{\ln(1.03315313209)}{\ln(1.1)} \approx .342202894$.   ∎

In the solution of Example (2.3.7), had we not thought to use the quadratic formula, we could have proceeded by setting $F(T) = \$525v^T + \$525v^{2T} = \$525(1.1)^{-T} + \$525(1.1)^{-2T}$. Then $F(T)$ is a decreasing function of T, $F(0) = \$1{,}050$, and $F(.5) \approx \$977.84$ so T is clearly between 0 and .5. We get better and better estimates of T using the "guess and check" method. Newton's method can also be used to find T [see Problem (2.3.12)].

We end this section by considering a method, the so-called **method of equated time** for finding approximate solutions to the following unknown time problem. The method of equated time is sometimes useful for finding an initial value in Newton's method, the "guess and check method," or another iterative method.

PROBLEM (2.3.8) An investor makes a sequence of contributions to an account governed by compound interest. There is a contribution of amount C_{t_k} at time t_k, $k = 1, 2, ..., n$. C_{t_k} is positive if we have a deposit and negative if we have a withdrawal. Find T so that a single payment of $C = \sum_{k=1}^{n} C_{t_k}$ at time T has the same value at $t = 0$ as the sequence of n contributions of our original scenario.

An exact solution to Problem (2.3.8) may be obtained using the time 0 equation of value

$$Cv^T = \sum_{k=1}^{n} C_{t_k} v^{t_k}.$$

This equation is equivalent to

$$v^T = \frac{\sum_{k=1}^{n} C_{t_k} v^{t_k}}{C}$$

and hence to

$$T \ln v = \ln\left(\frac{\sum_{k=1}^{n} C_{t_k} v^{t_k}}{C}\right).$$

Therefore,

(2.3.9)
$$T = \frac{\ln(\frac{\sum_{k=1}^{n} C_{t_k} v^{t_k}}{C})}{\ln v}.$$

An *approximate* solution to Problem (2.3.8) is given by the simpler expression

(2.3.10)
$$\overline{T} = \frac{\sum_{k=1}^{n} C_{t_k} t_k}{C} = \sum_{k=1}^{n} (\frac{C_{t_k}}{C}) t_k.$$

This weighted average of payment times is called the **method of equated time approximation** to the solution to Problem (2.3.8). If all the contributions are positive, it can be shown [see Problems (2.3.7) and (2.3.8)] that

(2.3.11)
$$\overline{T} \geqslant T.$$

EXAMPLE 2.3.12

Problem: A loan is negotiated with the lender agreeing to accept $12,000 after five years and again after 10 years, then $30,000 after fifteen years in full repayment of the loan. The borrower wishes to renegotiate the loan so that these three payments are replaced by a single payment of $54,000. It is agreed that the new payment should have the same present value as the old sequence of payments if the present values are calculated using compound interest at an annual effective interest rate of 4.5%. When is the exact time T this payment should be made, and what is the method of equated time approximation $\overline{T}$?

Solution This time there are two timelines to consider, namely one for the originally negotiated repayment plan and another for the proposed payment plan. The second of these includes an unknown time T, and each timeline includes an unknown payment of $-\$L$ at time 0, $\$L$ being the loan amount.

ORIGINAL

PAYMENT:	$-L$	$12,000	$12,000	$30,000
TIME:	0	5	10	15
BALANCE:				$0

RENEGOTIATED

PAYMENT:	$-L$	$54,000
TIME:		T
BALANCE:		$0

A time 0 equation of value giving T is

$$\$12,000(1.045)^{-5} + \$12,000(1.045)^{-10} + \$30,000(1.045)^{-15}$$
$$= \$54,000(1.045)^{-T}.$$

Equivalently,

$$T = \frac{\ln\left(\left(12,000(1.045)^{-5} + 12,000(1.045)^{-10} + 30,000(1.045)^{-15}\right)\Big/54,000\right)}{-\ln(1.045)}.$$

Calculating we find $T \approx 11.28621406$.
   Next we wish to find $\overline{T}$. By (2.3.10),

$$\overline{T} = \left(\frac{12,000}{54,000}\right)5 + \left(\frac{12,000}{54,000}\right)10 + \left(\frac{30,000}{54,000}\right)15 = \frac{630,000}{54,000} = 11\frac{2}{3}.$$

The values T and $\overline{T}$ are close. Moreover, as promised by (2.3.11), T does not exceed $\overline{T}$.   ∎

2.4   INVESTMENT RETURN

In Examples (2.2.5), (2.3.5), and (2.3.6), we determined an unknown interest rate. The rates we found are known as **yield rates** for the investments. More generally, consider the time τ equation of value (under compound interest at the rate i)

(2.4.1) $$\sum_k C_{t_k}(1 + i)^{\tau - t_k} = B(1 + i)^{\tau - T}.$$

   A rate of interest which satisfies (2.4.1) is called an (annual) **yield rate** or **internal rate of return** for the investment giving rise to (2.4.1). Yield rates are sometimes called **dollar-weighted yield rates** to distinguish them from time-weighted yield rates that are introduced in Section (2.7). Think of a yield rate as an interest rate on savings and loans that would result in the contributions C_{t_k} accumulating to B at time T [as shown by setting $\tau = T$ in Equation (2.4.1)].

Yield rates are a measure of how attractive a particular financial transaction may be. A lender wishes to have a high yield rate while a borrower searches for a low yield rate. However, as we will consider in Section (2.5) and have already noted in our discussion of Example (1.7.8), there are complications posed for the lender unless repayments may be invested at a rate equal to the original interest rate. Furthermore, if the signs of the contributions C_{t_k} fluctuate, there may not be a consistent "borrower" and "lender."

We shall see that Equation (2.4.1) does not always have a solution and when it does, that solution need not be unique. However, in (2.4.7) we discuss hypotheses under which uniqueness is guaranteed, and these hypotheses are frequently satisfied in real life applications.

EXAMPLE 2.4.2   A unique yield rate

Problem: Gautam invested $1,000 on March 1, 1998 and $600 on March 1, 2000. In return he received $600 on March 1, 1999 and $1,265 on March 1, 2001. Show that $i = 10\%$ is the unique yield rate.

Solution   A March 1, 2001 equation of value describing Gautam's investment is

$$\$1{,}000(1 + i)^3 - \$600(1 + i)^2 + \$600(1 + i) - \$1{,}265 = \$0.$$

Let $p(x) = 1{,}000x^3 - 600x^2 + 600x - 1{,}265$. Then i is a yield rate if and only if $1 + i$ is a real root of $p(x) = 0$. Note that $p(1.1) = 0$. Therefore, $i = .1 = 10\%$ is a yield rate, and $x - 1.1$ divides the polynomial $p(x)$. In fact, $p(x)$ has the factorization $p(x) = (x - 1.1)(1{,}000x^2 + 500x + 1{,}150)$. The quadratic formula shows that $1{,}000x^2 + 500x + 1{,}150 = 0$ has no real roots, and consequently 1.1 is the only real root of $p(x) = 0$. So, $.1 = 10\%$ is the only yield rate. We note that if a graphing calculator is available, you might use it to quickly locate this rate. ∎

EXAMPLE 2.4.3   No yield rate

Problem: Ace Manufacturing agrees to pay $100,000 immediately and again in exactly two years in return for a loan of $180,000 one year from now (to be used to replace a piece of machinery). Their CEO is asked what yield rate is associated with this transaction, but he is unable to answer the question. Why must this be the case?

Solution   A time 2 equation of value describing this financial arrangement is

$$\$100{,}000(1 + i)^2 - \$180{,}000(1 + i) + \$100{,}000 = \$0.$$

Equivalently, $10(1 + i)^2 - 18(1 + i) + 10 = 0$. Thus, by the quadratic equation, $1 + i = \frac{18 \pm \sqrt{18^2 - 400}}{20}$. Since $18^2 - 400 < 0$, this leaves no real solutions to the yield equation. ∎

EXAMPLE 2.4.4   Undefined yield

Problem: Banker Johnson is always on the lookout for opportunities to make money without risking any of his own funds. Johnson is able to borrow $10,000 for one year at an annual effective interest rate of 4%, then loan out the $10,000 for one year at an annual effective rate of 6%. What is Johnson's yield rate on this transaction?

Solution   Banker Johnson must pay .04($10,000) = $400 of interest for the money he borrows. However, he receives interest of .06($10,000) = $600 on the $10,000 he loans out. He thus makes $200 = $600 − $400 on the transaction without tying up any of his own money. No finite yield rate describes this situation. It might be tempting to say that the yield rate is infinite. However, we refrain from that because this would not give us a way to distinguish Johnson's situation from the even more favorable situation in which he is able to loan out the $10,000 at a rate higher than 6%. ∎

EXAMPLE 2.4.5   Multiple yield rates

Problem: Alice and Afshan are friends. Afshan agrees to give Alice $1,000 today and $1,550 two years from now if Alice will give her $2,500 in one year. What is the yield rate for this transaction?

Solution   A time two equation of value for the given situation is

$$\$1,000(1 + i)^2 - \$2,500(1 + i) + \$1,550 = 0.$$

This is quadratic in $(1 + i)$ and the quadratic formula tells us that

$$1 + i = \frac{2,500 \pm \sqrt{6,250,000 - 6,200,000}}{2,000} = \frac{2,500 \pm \sqrt{50,000}}{2,000} = \frac{25 \pm \sqrt{5}}{20}.$$

Therefore, $i = \frac{5+\sqrt{5}}{20} \approx .361803399$ or $i = \frac{5-\sqrt{5}}{20} \approx .138196601$. We have two distinct positive interest rates. ∎

EXAMPLE 2.4.6   Three distinct yield rates

Problem: Parties A and B agree that A will pay B $1,000,000 at $t = 0$ and $3,471,437 at $t = 2$. In return, B will pay A $3,228,000 at $t = 1$ and $1,243,757 at $t = 3$. Find the possible yield rates, estimating each to at least the nearest 10,000th of a percent.

Solution   For those equipped with a BA II Plus calculator, the rate closest to zero may easily be found. (In general, **if the cashflow worksheet gives you a yield rate, it is the one closest to zero.**) The correct sequence of buttons to produce the approximate yield rate of 4.184306654% is

CF	2ND	CLR WORK	1	0	0	0	0	0	0

	ENTER	↓	3	2	2	8	0	0	0	+/−

	ENTER	↓	↓	3	4	7	1	4	3	7

	ENTER	↓	↓	1	2	4	3	7	5	7	+/−

	ENTER	IRR	CPT	.

For those not using the BA-II Plus, this first yield may be found using either Newton's method or "guess and check," each of which might be facilitated by a good graph of the function

$$f(x) = 1,000,000(1 + x)^3 - 3,228,000(1 + x)^2 + 3,471,437(1 + x) - 1,243,757,$$

perhaps obtained on a graphing calculator.

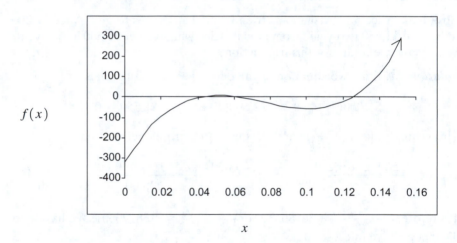

The graph is helpful because the yield rates are the roots of this polynomial, and for either Newton's method or the "guess and check" method, an initial approximation is needed. A graph shows that as well as being a root near 4%, there is a root close to 6% and another close to 12%. Here we use the "guess and check" method to find the root close to 6% and calculus-based Newton's method to find the root near 12%.

The root near 6%: We successively calculate $f(.06) = 1.42$, $f(.061) = .25$, $f(.0611) = .127951$, $f(.0612) = .005008$, $f(.06121) = -.007335$, $f(.061205) = -.001163$, $f(.061204) = .000072$, and $f(.0612041) = -.000052$. So, the yield rate is between 6.1204% and 6.12041%.

The root near 12%: Note that

$$f'(x) = 3,000,000(1 + x)^2 - 6,456,000(1 + x) + 3,471,437$$

and set $x_0 = .12$. Then $x_1 = x_0 - \dfrac{f(x_0)}{f'(x_0)} = .12 - \dfrac{-22.76}{3917} \approx .125810569$,

$x_2 = x_1 - \dfrac{f(x_1)}{f'(x_1)} \approx .125810569 - \dfrac{4.652858}{5552.278446} \approx .124972561, x_3 = x_2 - \dfrac{f(x_2)}{f'(x_2)} \approx$

$.124972561 - \dfrac{.104354}{5,303.93506} \approx .124952886$, and $x_4 = x_3 - \dfrac{f(x_3)}{f'(x_3)} \approx .124952886 -$

$\dfrac{.000054}{5,298.155063} \approx .124952876$. So, the rate is approximately equal to 12.49529%.

Using the **BA II Plus** calculator for these methods: In the "guess and check" method and also in Newton's method, one is repeatedly evaluating the polynomial f. It is helpful to store the coefficients of f in successive memories and also to reserve a register for one plus the argument at which the function is evaluated. Likewise, in Newton's method, use successive memories for the coefficients of the derivative f'. As one calculates f or f', store the partial results in an available register, remembering that the BA II Plus calculator allows you to add a displayed number to the value stored in register **m** by pushing STO + m . ∎

EXAMPLE 2.4.7   Three party example

Problem: Brian, Filemon, and Harold are friends. Brian will pay Filemon $1,000 now. Filemon will pay Brian $300 and Harold $800 in exactly one year. Finally, Harold will pay Brian $900 two years from now. What is Brian's annual yield for this three-way transaction spanning two years?

Solution   Brian's time 2 equation of value is $\$1,000(1 + i)^2 - \$300(1 + i) - \$900 = \$100[10(1 + i)^2 - 3(1 + i) - 9]$. The quadratic equation then tells us that $1 + i = \dfrac{3 \pm \sqrt{9 + 360}}{20}$. Since Brian invests $1,000 and receives $1,200 > $1,000, he has a positive yield rate, namely $i = \dfrac{3 + \sqrt{9 + 360}}{20} - 1 \approx .11046836$. ∎

Note that the three-way financial transaction of Example (2.4.7) can be viewed in terms of two-party loans. We could think of it as the following package of loans.

(1) Brian loans Filemon $1,000 at $t = 0$ and receives $1,100 from Filemon at time $t = 1$ in full repayment of the loan.

(2) Brian loans Harold $800 at $t = 1$ and receives $900 from Harold at time $t = 2$ in full repayment of the loan.

Note that the interest rate paid by Filemon is $\dfrac{\$1,100}{\$1,000} - 1 = .1$ and the interest rate paid by Harold is $\frac{\$900}{\$800} - 1 = .125$. Brian's yield rate is in between these two rates.

When computing Brian's yield rate in Example (2.4.7), we were not concerned with the source of the money coming in or the recipient of outgoing money. Rather, we took a "**bottom line approach**." We were only concerned with the times and amounts of all contributions (positive or negative) by Brian. This approach is fundamental to the successful determination of yield rates.

> When computing the yield rate received by an investor, take a "**bottom line approach**." Consider only the times and amounts of all contributions (positive or negative) by the investor.

Our next task is to analyze conditions that assure that there is a unique yield rate $i > -1$. We continue with our previous notation and suppose that i is a yield rate. That is, we suppose that we have the equation of value

$$\sum_{k=1}^{n} C_{t_k}(1 + i)^{T-t_k} = B.$$

Define

$$(2.4.8) \qquad B_{t_k}(i) = \sum_{q=1}^{k} C_{t_q}(1 + i)^{t_k-t_q}.$$

Then $B_{t_k}(i)$ represents the outstanding balance at time t_k, just after the contribution C_{t_k}, provided that money grows by compound interest at a rate i. An important observation is that

$$B_{t_k}(i) = B_{t_{k-1}}(i)(1 + i) + C_{t_k} \qquad \text{for } k = 2, 3, ..., n.$$

We now are ready to expain a set of conditions that forces there to be exactly one yield rate that is also greater than -1. The proof of Important fact (2.4.9) is discussed at the end of this section, and may be skipped without hindering your ability to understand the remainder of the book.

> **IMPORTANT FACT (2.4.9):** If the contributions take place at times $t_1 < t_2 < ... < t_{n-1} < t_n$ and there is a yield rate $i > -1$ such that the first $(n - 1)$ consecutive balances $B_{t_1}(i)$, $B_{t_2}(i), ..., B_{t_{n-1}}(i)$ are all positive or are all negative, then i is the unique yield rate that is greater than -1.

The condition of (2.4.9) is the condition that one party remains the lender throughout the period of the financial transaction. Among the situations in which this is the case is when all deposits take place before any withdrawals. When this happens, the existence of a unique yield rate i for which $i > -1$ may also easily be seen using Descartes' rule of signs. Note that in Example (2.4.5), at time 0 Afshan is the lender, but at time 1 Afshan becomes the borrower. Therefore, (2.4.9) does not apply, and it was demonstrated that there are two distinct positive interest rates in Example (2.4.5).

Proof of Important fact (2.4.9):

Assume that the hypotheses of (2.4.9) are satisfied.

Note that $B_{t_1}(i) = C_{t_1}$, and we will assume that $B_{t_1}(i) > 0$. If this is not the case, interchanging the roles of borrower and lender changes the signs of all the contributions and makes it true, as well as changing the signs of the other balances so $B_{t_k}(i) > 0$ for $k = 1, 2, \dots, n - 1$.

Now suppose that there is another interest rate j with $j > i$ and

$$\sum_{k=1}^{n} C_{t_k}(1 + j)^{T - t_k} = B.$$

Then j is also a yield rate. Just as we did for the rate i, we define

$$B_{t_k}(j) = \sum_{q=1}^{k} C_{t_q}(1 + j)^{t_k - t_q},$$

and observe that

$$B_{t_k}(j) = B_{t_{k-1}}(j)(1 + j) + C_{t_k} \qquad \text{for } k = 2, 3, \dots, n.$$

We now note that $B_{t_1}(i) = C_{t_1} = B_{t_1}(j)$. But then $B_{t_2}(i) = B_{t_1}(i)(1 + i) + C_{t_2}$, $B_{t_2}(j) = B_{t_1}(j)(1 + j) + C_{t_2}$, and since $B_{t_1}(i) > 0$, the inequality $j > i$ forces $B_{t_2}(j) > B_{t_2}(i)$. Since $B_{t_3}(i) = B_{t_2}(i)(1 + i) + C_{t_3}$, $B_{t_3}(j) = B_{t_2}(j)(1 + j) + C_{t_3}$, $j > i$, and $B_{t_2}(i) > 0$, this gives us $B_{t_3}(j) > B_{t_3}(i)$. Continuing inductively, we arrive at $B_{t_n}(j) > B_{t_n}(i)$. However, provided that $1 + i > 0$, the inequalities $j > i$ and $B_{t_n}(j) > B_{t_n}(i)$ combine to force

$$B = B_{t_n}(j)(1 + j)^{T - t_n} > B_{t_n}(i)(1 + i)^{T - t_n} = B.$$

With this contradiction, we see that there is no second yield rate j larger than the yield rate i.

Similarly [see Problem (2.4.7)], we can show that there is no smaller yield rate.

2.5   REINVESTMENT CONSIDERATIONS

In Example (2.4.7), we saw a three-party financial transaction. The yield rate calculated by the lending party, Brian, was not equal to the interest rate paid by either of the other parties. When we have an investor who reinvests money that is paid to him, that investor may be involved with multiple parties and his yield rate will not in general be equal to the interest rate of his initial borrower. In this section we look at two yield rate calculations, each involving reinvestments. We will look at further important examples in Section (3.10).

EXAMPLE 2.5.1

Problem: Jose loans Martin $12,000. Martin repays the loan by paying $5,000 at the end of two years and $10,000 at the end of four years. The money received at $t = 2$ is immediately reinvested at an annual effective interest rate of 2.4%. Find Martin's rate of interest and Jose's annual yield rate.

Solution Martin's time 4 equation of value is $12,000(1+i)^4 - \$5,000(1 + i)^2 - \$10,000 = \$0$. Let $x = (1 + i)^2$. Then Martin's equation of value is equivalent to $12x^2 - 5x - 10 = 0$. From the quadratic formula and the fact that x is non-negative, we deduce that $x = \frac{5+\sqrt{505}}{24} \approx 1.144675211$ and Martin's yield rate is $i = \sqrt{x} - 1 \approx .069894953$. (Another technique is to use the BA II Plus calculator Cash Flow worksheet with CFo $= 12,000$, C01 $= 0$, F01 $= 1$, C02 $= -5,000$, F02 $= 1$, C03 $= 0$, F03 $= 1$, C04 $= -10,000$, and F04 $= 1$. Pushing IRR CPT then gives the rate.)

To calculate Jose's yield rate, we remember the "bottom line " approach of Section (2.4). Jose contributes $12,000 at $t = 0$. The next time he has any of this $12,000 available to him is at time $t = 4$ since it is stipulated that the $5,000 that comes in at $t = 2$ is immediately invested at an annual effective rate of 2.4%. At time $t = 4$ Jose has available the proceeds of this account, namely $5,000(1.024)^2 = \$5,242.88$ and the $10,000 he is paid at time 4. Thus, the equation we must solve to find Jose's annual yield rate i_J is

$$\$12,000(1 + i_J)^4 = \$5,242.88 + \$10,000.$$

It follows that $i_J = (\frac{15,242.88}{12,000})^{\frac{1}{4}} - 1 \approx .061625755$.

Jose's yield rate is lower than Martin's interest since Jose reinvested the $5,000 at an interest rate lower than Martin's rate.   ■

EXAMPLE 2.5.2

Problem: Julie pays Chan $1,000. In return, he pays her $300 at the end of 1,2, 3, and 4 years. Each time Julie receives a payment, she reinvests it at 3%. She closes the 3% account at the end of 4 years. Find the annual effective interest rate paid by Chan and the annual yield earned by Julie.

Solution   Chan's time 0 equation of value is

$$\$1,000 - \$300v - \$300v^2 - \$300v^3 - \$300v^4 = \$0.$$

There is only one change of signs in the sequence of coefficients $-1,000$, 300, 300, 300, 300 so, by Descartes rule of signs, there is a unique interest rate i which makes this equation true. Chan's rate may be found to be about 7.713847295% using the BA II Plus calculator Cash Flow worksheet with CFo = 1,000, C01 = -300, and F01 = 4. Alternatively, the rate may be estimated using the "guess and check method" or Newton's method.

Julie pays $1,000 at time 0 and the next time she has any money available is at time 4 when she gets the accumulated balance in her 3% savings account. This amount is $\$300(1.03)^3 + \$300(1.03)^2 + \$300(1.03) + \$300 \approx \$1255.09$. Therefore, her yield rate is i_J where

$$\$1,000(1 + i_J)^4 = 1,255.09.$$

Consequently, $i_J = (\frac{1,255.09}{1,000})^{\frac{1}{4}} - 1 \approx 5.8446028\%$. Julie's rate is lower than Chan's since she reinvested money in a savings account paying interest at 3%, a rate lower than Chan's.   ∎

2.6   APPROXIMATE DOLLAR-WEIGHTED YIELD RATES

The yield rates we have defined are solutions to equations of value. Sometimes, these equations may be solved easily by algebraic methods. Other times, the "guess and check" or other iterative method may reasonably be used to arrive at a good approximation(s) to the yield rate(s). Or, perhaps there is a single rate and the timing of the payments is such that the BA II Plus calculator Cash Flow worksheet applies . However, many times these methods are cumbersome, and one might wish for an elementary method to obtain an approximation to the yield rate. The yield rate is sometimes referred to as a dollar-weighted or money-weighted yield rate, and we will call an approximation *by the methods of this section* an **approximate dollar-weighted yield rate.**

Suppose that we wish to study an investment fund over some finite time interval. We choose new units of time so that in our new units, the interval of our investment has length one. For example, if we wish to study an investment lasting from July 1, 1994 to January 1, 1997, then we let our new unit of time be a period of length two-and-a-half years. Furthermore, we let the starting time of the investment be denoted by time 0 and the ending point be denoted by time 1.

Denote the amount of money in the fund at the beginning ($t = 0$) by A and the amount of money in the fund at the end ($t = 1$) by B. For each time t with $0 < t < 1$,[2] let C_t denote the contribution to the fund at time t. By assumption (and following real life), there are only finitely many times t at which C_t is nonzero, and we may have a mixture of positive C_t's (deposits) and negative C_t's (withdrawals). Define the **net contributions** to the fund by

$$C = \sum_{t \in (0,1)} C_t$$

and let I denote the interest earned by the fund from $t = 0$ to $t = 1$. Then,

(2.6.1) $B = A + C + I.$

If j denotes the interest rate for the period $[0,1]$ of the investment, then

(2.6.2) $I = Aj + \sum_{t \in (0,1)} C_t[(1 + j)^{1-t} - 1].$

This last term, giving the interest on the contributions C_t as determined by compound interest, may be hard to calculate if there are a large number of contributions. We therefore *approximate* it by supposing that on each contribution C_t we earn **simple interest** at rate j from the time of deposit until $t = 1$.

(2.6.3) $C_t[(1 + j)^{1-t} - 1] \approx C_t j(1 - t)$ for $t \in (0,1)$.

Combining (2.6.2) and (2.6.3), we obtain the important approximation

(2.6.4) $I \approx Aj + \sum_{t \in (0,1)} C_t j(1 - t).$

Equivalently, we have the following approximation for j

(2.6.5) $$j \approx \frac{I}{A + \sum_{t \in (0,1)} C_t(1 - t)}.$$

The denominator in (2.6.5) should be thought of as an average amount of money invested. Approximation (2.6.5) tends to be fairly good if the C_t's are relatively small compared to A.

If we approximate each $t \in (0,1)$ for which $C_t \neq 0$ by a constant k, then (2.6.5) gives us the new approximation

(2.6.6) $$j \approx \frac{I}{A + \sum_{t \in (0,1)} C_t(1 - k)} = \frac{I}{A + C(1 - k)}.$$

_____

[2]That t lies in this interval is denoted $t \in (0,1)$.

Note that approximation (2.6.6) only requires the total contribution C, not all the individual contributions and their times. Recalling (2.6.1), it may be rewritten as

$$(2.6.7) \quad j \approx \frac{I}{A + (B - A - I)(1 - k)} = \frac{I}{kA + (1 - k)B - (1 - k)I}.$$

A good k to use would be the dollar-weighted average $\sum_{t\in(0,1)} \frac{C_t}{C} t$. However, its precise value may take too long to calculate, so perhaps a quick estimate of it will suffice. If one uses $k = \frac{1}{2}$, then one obtains

$$(2.6.8) \quad \boxed{j \approx \frac{I}{\frac{1}{2}A + (1 - \frac{1}{2})B - (1 - \frac{1}{2})I} = \frac{2I}{A + B - I}.}$$

EXAMPLE 2.6.9   One year investment period

Problem: On January 1, 1990, Martin's investment account had a balance of $10,210. He deposited $4,000 on March 1, 1990, withdrew $3,000 on June 1, 1990, and deposited $1,000 on December 1, 1990. At the end of 1990, Martin's balance was $12,982. Find Martin's approximate dollar-weighted yield for the year 1990 using Equation (2.6.5) and again using Equation (2.6.8).

Solution

DATE	CONTRIBUTION	BALANCE
Jan. 1, 1990	$0	$A = \$10,210$
Mar. 1, 1990	$C_{\frac{2}{12}} = \$4,000$	?
June 1, 1990	$C_{\frac{5}{12}} = -\$3,000$	?
Dec. 1, 1990	$C_{\frac{11}{12}} = \$1,000$	?
Dec. 31, 1990	$0	$B = \$12,982$

Since we are considering an investment over a period of exactly one year, our unit of time is a year. With the notation of this section, $A = \$10,210$, $B = \$12,982$, $C_{\frac{2}{12}} = \$4,000$, $C_{\frac{5}{12}} = -\$3,000$, and $C_{\frac{11}{12}} = \$1,000$. Then $C = \$4,000 - \$3,000 + \$1,000 = \$2,000$ and $I = B - A - C = \$12,982 - \$10,210 - \$2,000 = \$772$. Equation (2.6.5) gives us

$$j \approx \frac{\$772}{\$10,210 + \$4,000(1 - \frac{2}{12}) - \$3,000(1 - \frac{5}{12}) + \$1,000(1 - \frac{11}{12})} \approx .065.$$

On the other hand, Equation (2.6.8) gives us

$$j \approx \frac{2(\$772)}{\$10,210 + \$12,982 - \$772} \approx .069.$$

■

We note that the exact dollar-weighted yield rate j of Example (2.6.9) satisfies the time 1 equation of value

$$\$12,982 = \$10,210(1 + j) + \$4,000(1 + j)^{\frac{10}{12}} - \$3,000(1 + j)^{\frac{7}{12}}$$
$$+ \$1,000(1 + j)^{\frac{1}{12}}.$$

We let

$$p(j) = \$10,210(1 + j) + \$4,000(1 + j)^{\frac{10}{12}} - \$3,000(1 + j)^{\frac{7}{12}}$$
$$+ \$1,000(1 + j)^{\frac{1}{12}},$$

the left-hand side of this time one equation of value. Then $p(.064985) \approx$ $\$12,982.00004$, and therefore the exact dollar-weighted yield j is extremely close to 6.4985%. (In this example the BA II Plus calculator Cash Flow worksheet could be used to find the yield rate $(1 + j)^{\frac{1}{12}} - 1$ for a 12-th of a year and then to find j.)

EXAMPLE 2.6.10 Investment period not equal to one year

Problem: On January 1, 1995, Siobhan's investment account had a balance of $8,412. She deposited $1,000 on January 1, 1996 and withdrew $600 on January 1, 1997. Her balance on January 1, 1998 is $9,620. What was Siobhan's approximate annual dollar-weighted yield for the three-year period from January 1, 1995 to January 1, 1998, obtained using (2.6.5)?

Solution

DATE	CONTRIBUTION	BALANCE
Jan. 1, 1995	$0	$8,412
Jan. 1, 1996	$1,000	?
Jan 1, 1997	−$600	?
Jan 1, 1998	$0	$9,620

Note that $A = \$8,412$, $B = \$9,620$, $C_{\frac{1}{3}} = \$1,000$, and $C_{\frac{2}{3}} = -\$600$. It follows that $C = \$1,000 - \$600 = \$400$ and $I = \$9,620 - \$8,412 - \$400 = \$808$.

Therefore, we have an approximate *three year* yield

$$j \approx \frac{808}{\$8{,}412 + \$1{,}000(\frac{2}{3}) - \$600(\frac{1}{3})} \approx .091004655.$$

It follows that the approximate annual dollar-weighted yield for the three-year period is $(1 + .09100465)^{\frac{1}{3}} - 1 \approx 2.9458557\%$. We note that the BA II Plus calculator **Cash Flow worksheet** could be used to find the yield rate in this example. Clear the worksheet and enter CFo = 8,412, C01=1,000, F01=1, C02=−600, F02 = 1, C03=−9,620, F03=1. Then $\boxed{\text{IRR}}\ \boxed{\text{CPT}}$ gives a yield rate of 2.947128765% which is close to the above approximation.   ■

The next example is designed so that the **Cash Flow worksheet** (of the BA II Plus calculator or of the BA II Plus Professional calculator) can not be used to calculate the yield rate. (The reason for this is that one is limited to one initial Cash Flow and then at most twenty-four others, each made at the end of up to 9,999 <u>successive</u> periods.) However, it still might be used to estimate the rate by shifting all the deposits forward one month.

EXAMPLE 2.6.11   Investment period not equal to one year

Problem: On January 1, 1995, Saul's investment account had a balance of $7,688. He deposited $100 on February 1, 1995 and then again at the end of every two months (that is on April 1, 1995, June 1, 1995, ... , December 1, 1997). His balance on January 1, 1998 is $10,830. What was Saul's approximate annual dollar-weighted yield for the three-year period, calculated using (2.6.8)?

Solution   Note that $A = \$7{,}688$, $B = \$10{,}830$, and since there are eighteen $100 deposits and no withdrawals $C = \$1{,}800$. Therefore, $I = \$10{,}830 - \$7{,}688 - \$1{,}800 = \$1{,}342$ and using (2.6.8) we have the approximate *three-year* yield

$$j \approx \frac{2 \times 1{,}342}{7{,}688 + 10{,}830 - 1{,}342} \approx .156264555.$$

It follows that the approximate annual dollar-weighted yield for the three-year period is $(1 + .156264555)^{\frac{1}{3}} - 1 \approx 4.959\%$.

To approximate the rate using (2.6.5) would be quite time-consuming. However, in Chapter 3 we will come back to this problem [see Example (3.2.23)] and see a good approximation to the rate.

Let's take a moment to see why this problem cannot be solved using the BA II Plus calculator. We are given the balance on January 1, 1995 and enter this as our initial cashflow CFo. Then, we have a payment one month later (of $100) which we enter as C01. This establishes that the time period being used in the **Cash Flow worksheet** is one month. Thus, we would wish to alternately enter cashflows of $0 (for the months there were no deposits) and $100 (for the months when Saul deposited $100). But then, to span the three-year time period from January 1, 1995 to January 1, 1998 would require registers to enter 36 payments. The problem is that the BA II Plus calculator has 24 cashflow registers, and the BA II Plus Professional has 32 registers, rather than 36 appropriate registers.

Since we can't solve the given problem, we look at the similar problem where the deposits occurred every two months starting with March 1, 1995. The **Cash Flow worksheet** can then be used with CF0 $= 7{,}688$, C01 $= 100$, F01 $= 17$, C02 $= -10{,}830 + 100 = -10{,}730$, F02$=1$. Then $\boxed{\text{IRR}}\ \boxed{\text{CPT}}$ gives a two-month yield of $J \approx .81628332\%$ and this corresponds to an annual yield of $(1 + J)^6 - 1 \approx 4.998742257\%$. Our shift of each of the deposits forward one month (from the original problem) results in this rate being a bit higher than the yield rate for the given problem. ∎

We end this section with an example showing the usefulness of approximation (2.6.6).

EXAMPLE 2.6.12   The average date for contributions

Problem: An investment fund had a balance of $320,000 on January 1, 2002 and a balance of $374,000 one year later. The amount of interest earned during the year was $14,000, and the annual yield rate on the fund was 4%. Estimate the (dollar-weighted) average date of contributions to the account.

Solution   We are given that $A = \$320{,}000$, $B = \$374{,}000$, and $I = \$14{,}000$. So, $C = B - A - I = \$40{,}000$, and if k denotes the average (dollar-weighted) time of the contributions, approximation (2.6.6) gives

$$.04 \approx \frac{\$14{,}000}{\$320{,}000 + \$40{,}000(1 - k)}.$$

Equivalently, $k \approx \frac{1}{4}$. The average date of contributions to the account is a quarter of the way through the year, that is to say on April 1. ∎

2.7   FUND PERFORMANCE

As in Section (2.6), let us consider an investment fund over a given interval of time. However, rather than considering the annual dollar-weighted yield which is a measure of how the investment actually grew, we define a new "yield rate," called the annual **time-weighted yield rate** that quantifies *how well the fund was managed*. The dollar-weighted yield rate and time-weighted yield rates are often close, but this is not always the case. For example, if the investor makes large deposits right before the fund does particularly well and makes large withdrawals just before the fund does poorly, the investor's dollar-weighted yield will be considerably higher than the time-weighted yield associated with the fund. This is the case considered in Example (2.7.5).

As in Section (2.6), introduce units so that the given investment period has length one unit. Let B_t denote the account balance at time t, just before any time t contributions. (In the notation of Section (2.6), $B_0 = A$ and $B_1 = B$.) Let C_t denote the contribution at time t, and assume that there are only finitely many t's in the interval $(0,1)$ for which $C_t \neq 0$, namely $t_1, t_2, ..., t_r$ where $0 < t_1 < t_2 < ... < t_r \leq 1$. Let $t_{r+1} = 1$. The balance then progresses as is indicated in the following diagram:

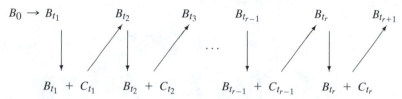

The vertical arrows arise due to contributions to the investment fund, and the nonvertical arrows reflect the growth (or shrinkage) of the investment fund over time. In the case of a non-vertical arrow to B_{t_k}, the balance grows by multiplying by $1 + j_k$ where

(2.7.1)
$$1 + j_k = \begin{cases} \dfrac{B_{t_1}}{B_0} & \text{if } k = 1 \\ \dfrac{B_{t_k}}{B_{t_{k-1}} + C_{t_{k-1}}} & \text{if } k = 2, 3, ..., r + 1. \end{cases}$$

Think of j_k as the yield rate for the interval $[t_{k-1}, t_k]$. The **time-weighted yield** j_{tw} for the investment period is defined so that the sum $1 + j_{tw}$ is equal to the product $(1 + j_1)(1 + j_2) ... (1 + j_{r+1})$; standard mathematical notation for this is $\prod_{k=1}^{r+1}(1 + j_k)$. Thus,

(2.7.2)
$$j_{tw} = \left[\prod_{k=1}^{r+1}(1 + j_k) \right] - 1.$$

The motivation for the definition of j_{tw} is as follows. The growth of money over the interval $[0,1]$ is given by its successive growth over the subintervals $[t_{k-1}, t_k]$, and on the subinterval $[t_{k-1}, t_k]$ the growth of money is governed by multiplication by $1 + j_k$.

Note that j_{tw} is a "yield rate" for the interval of interest. If the interval lasts T years, the **annual time-weighted yield rate** i_{tw} is

(2.7.3)
$$i_{tw} = (1 + j_{tw})^{\frac{1}{T}} - 1 = \left[\prod_{k=1}^{r+1}(1 + j_k) \right]^{\frac{1}{T}} - 1.$$

EXAMPLE 2.7.4   Investment period not equal to one year

Problem: Mohammed had $20,000 in his investment account on August 15, 1999. On August 15, 2000 his balance was $21,200 and he deposited an additional $5,000, giving him a new balance of $26,200. On August 15, 2001, Mohammed's account had a balance of $27,300. Assuming that there are no other contributions to the account, find the annual time-weighted yield and note that it is very close to the annual dollar-weighted yield.

Solution   The balance grows as follows:

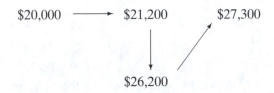

Therefore $i_{tw} = \left[\left(\dfrac{\$21{,}200}{\$20{,}000} \right) \left(\dfrac{\$27{,}300}{\$26{,}200} \right) \right]^{\frac{1}{2}} - 1 \approx .050953765.$

On the other hand, the dollar-weighted yield i satisfies the equation of value

$$\$20{,}000(1 + i)^2 + \$5{,}000(1 + i) = \$27{,}300.$$

This equation may be solved using the quadratic equation, and $i = .05$ is the only positive yield rate. ∎

EXAMPLE 2.7.5   Time-weighted yield less than dollar-weighted yield

Problem: Astute Mr. Haywood notices that although the "Tomorrow Fund" has an excellent performance history, it performed less well when the price of gasoline experienced a sharp rise. He decides to invest in the fund but, to the extent possible, move his money away from the fund during periods when he anticipates a sharp increase in gasoline prices. On January 1, Mr. Haywood deposits $100,000 in the fund. On March 1 his balance is $102,000, and he withdraws $50,000. On May 1 his balance is $52,500 and he deposits $50,000. At the end of the year Mr. Haywood's fund balance is $111,000. Find the time-weighted yield for the "Tomorrow Fund" for the year, and show that this is lower than Mr. Haywood's dollar-weighted yield.

Solution   The balance grows as follows:

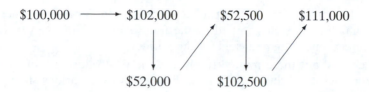

$$\$100{,}000 \longrightarrow \$102{,}000 \qquad \$52{,}500 \qquad \$111{,}000$$

$$\$52{,}000 \qquad \$102{,}500$$

Therefore $i_{tw} = j_{tw} = (\frac{\$102,000}{\$100,000})(\frac{\$52,500}{\$52,000})(\frac{\$111,000}{\$102,500}) - 1 \approx .115206379$. The dollar-weighted yield i satisfies

$$\$111{,}000 = \$100{,}000(1 + i) - \$50{,}000(1 + i)^{\frac{10}{12}} + \$50{,}000(1 + i)^{\frac{8}{12}}.$$

Let $q(i) = \$100{,}000(1 + i) - \$50{,}000(1 + i)^{\frac{10}{12}} + \$50{,}000(1 + i)^{\frac{8}{12}}$. Then $q(i_{tw})$ $\approx \$110{,}534.53 < 111{,}000$. Therefore, Mr. Haywood's dollar-weighted yield is higher than i_{tw}. In fact, using the "guess and check" method or Newton's method, one may determine that Mr. Haywood has a yield rate of approximately .1203. More easily, the BA II Plus calculator **Cash Flow worksheet** may be used to find a monthly yield rate of .951201439 is equivalent to an annual yield rate of 12.03092019%.

The reason that Mr. Haywood's dollar-weighted yield exceeds the Tomorrow Fund's time-weighted yield is that he timed his deposits and withdrawals well. Had they been poorly timed, his dollar-weighted yield would have been worse than the Fund's time-weighted yield. ∎

2.8   PROBLEMS, CHAPTER 2

(2.0) Chapter 2 writing problems

(1) [following Section (2.3)] Consider the equation

$$\$2{,}000(1 + i)^4 - \$300(1 + i)^2 + \$600(1 + i) = \$2{,}925.$$

Describe a financial situation for which this is the associated time 4 equation of value. Give details explaining the sources of any deposits and the reasons for any withdrawals.

(2) [following Section (2.3)] Consider the equation

$$\sum_{k=1}^{35} \$2{,}000(1 + i)^{35-k} = \sum_{m=1}^{30} \$10{,}000 v^m.$$

Describe a financial situation for which this is the associated time 35 equation of value.

(3) [following Section (2.7)] Write an advertisement for an investment fund, giving annual yields earned by the fund over each of five years and an annual time-weighted yield for the period.

(4) Learn about three mutual funds and the portfolio focus of each fund. For each fund, note the annual yield rates over various time periods. Include the period from the date of date of inception to a recent date and a recent five-year period. Comment on similarities and differences in the performance of the funds. Speculate on the causes of any performance discrepancies.

(2.2) Equations of value for investments involving a single deposit made under compound interest

(1) Mr. Lopez deposits $\$K$ in an account paying 4% annual effective discount. The balance at the end of three years is $982. Find K.

(2) Marianne deposits $2,000 in a five-year certificate of deposit. At maturity the balance is $2,580.64. Find the annual effective rate of interest governing the account.

(3) Suzanne remembers that her only deposit into her savings account was a $1,800 deposit. She knows that the account has had a constant nominal interest rate of 3.2% convertible monthly and that the balance is now $1,965.35. How long ago did Suzanne make her deposit?

(4) Use the rule of seventy-two to approximate the length of time it takes money to double at an annual effective interest rate of 5% and then at an annual effective rate of 10%. Then find the exact time it takes for money to double at each of these interest rates.

(5) Derive a "rule of n" to approximate the length of time it takes for money to triple. As in the derivation of the "rule of seventy-two," your rule should be derived to give an especially good estimate when the annual effective interest rate is 8%. After you have stated your rule, compare the approximations it gives for annual effective interest rates of 4% and 12% with the true values at these rates.

(2.3) Equations of value for investments with multiple contributions

(1) Sidney borrows $12,000. The loan is governed by compound interest and the annual effective rate of discount is 6%. Sidney repays $4,000 at the end of one year, X at the end of two years, and $3,000 at the end of three years in order to exactly pay off the loan. Find X.

(2) Rafael opens a savings account with a deposit of $1,500. He deposits $500 one year later and $1,000 a year after that. Just after Rafael's deposit of $1,000, the balance in his account is $3,078. Find the annual effective interest rate governing the account.

(3) Esteban borrows $20,000, and the loan is governed by compound interest at an annual effective interest rate of 6%. Esteban agrees to repay the loan by making a payment of $10,000 at the end of T years and a payment of $12,000 at the end of $2T$ years. Find T.

(4) Shakari opens a savings account with a deposit of $3,500. She deposits $500 six months later and $800 nine months after opening the account. The balance in Shakari's account one year after she opened it is $5,012. Assuming that the account grows by compound interest at a constant annual effective interest rate i, find i.

(5) A loan is negotiated with the lender agreeing to accept $8,000 after one year, $9,000 after two years, and $20,000 after four years in full repayment of the loan. The loan is renegotiated so that the borrower makes a single payment of $37,000 at time T and this results in the same total present value of payments when calculated using an annual effective rate of 5%. Estimate T using the method of equated time. Also find T exactly.

(6) Anne and Frank Smith each borrow $12,000 from their father. Anne and Mr. Smith have agreed that she will repay her loan in full by paying $6000 in two years and $8000 in four years. Frank prefers to make one lump payment of $15,000 to fully repay his loan. When should he make that payment so that he and his sister will each have the same effective interest rate?

(7) Let $b_1, b_2, \ldots, b_n$ be positive real numbers. Set $A = \left(\sum_{k=1}^{n} b_k\right)/n$, the arithmetic mean of the numbers, and $G = \left(\prod_{k=1}^{n} b_k\right)^{\frac{1}{n}}$, the geometric mean of the numbers. The objective of this problem is to establish that $A \geqslant G$ and that $A > G$ whenever $b_1, b_2, \ldots, b_n$ are not all equal.

   (a) Write the point-slope equation for the tangent line to $y = \ln x$ at $(A, \ln A)$.

   (b) Use (a) and concavity to show that $\ln x \leqslant A^{-1}(x - A) + \ln A$ for all positive x. Moreover, show that equality holds if and only if $x = A$.

   (c) Use (b) to prove that $\ln G \leqslant \frac{1}{n}\sum_{k=1}^{n} A^{-1}(b_k - A) + \ln A$ and that this is a strict inequality unless all the b_k's are equal.

   (d) Show that $\frac{i}{n}\sum_{k=1}^{n} A^{-1}(b_k - A) + \ln A = \ln A$.

   (e) Conclude that $G \leqslant A$ and $G < A$ unless all the b_k's are equal.

(8) Let C_{t_k} denote the contribution in cents made at distinct times t_k, $k = 1, 2, \ldots, n$. Suppose that these are all positive so that we have deposits, but no withdrawals. Then the C_{t_k}'s are positive integers. As in (2.3.9), let $T = \ln(\frac{\sum_{k=1}^{n} C_{t_k} v^{t_k}}{C})/\ln v$. As in (2.3.10), let $\overline{T} = \sum_{k=1}^{n}(\frac{C_{t_k}}{C})t_k$. The objective of this problem is to use the result of Problem (2.3.7) to establish that $\overline{T} \geq T$ and that this is a strict inequality unless $n = 1$.

(a) Consider C_{t_k} quantities each equal to v^{t_k}, $k = 1, 2, \ldots n$. Then in all we are considering $C = C_{t_1} + C_{t_2} + \cdots C_{t_n}$ quantities. Use the result of Problem [2.3.7(e)] to show that

$$\frac{C_{t_1} v^{t_1} + C_{t_2} v^{t_2} + \cdots + C_{t_n} v^{t_n}}{C} \geq v^{\overline{T}}$$

and that this is a strict inequality unless $n = 1$.

(b) Use (a) to show that the present value of the deposits is at least as large as the present value given by the method of equated time and that it is strictly larger unless $n = 1$.

(c) Show that $\overline{T} \geq T$ with strict inequality unless $n = 1$.

(9) Suppose that you pay $1,000 at time 0, get $4,000 at time 1, and pay $2,000 at time 2. Let $C_0 = \$1,000$, $C_1 = -\$4,000$, and $C_2 = \$2,000$. Set $C = C_0 + C_1 + C_2 = \$1,000 - \$4,000 + \$2,000 = -\$1,000$. Find T such that getting an inflow of $-C$ at time T has the same present value as the above sequence of financial transactions, assuming that the growth of money is governed by compound interest at $i = 1\%$. Show that T is greater than the weighted average $\overline{T} = \frac{C_0}{C}0 + \frac{C_1}{C}1 + \frac{C_2}{C}2$. [This shows that Inequality (2.3.11) need not hold if you have a negative contribution.]

(10) [recommended for those with a BA II Plus calculator]
Dax borrows $300,000 and the loan is governed by compound interest at an annual effective interest rate of 4.75%. Dax agrees to repay the loan by ten equally spaced payments, the first four of which are for $25,000 and the next six of which are for $40,000. When should he make the first payment?

(11) Find the amount to be paid at the end of eight years that is equivalent to a payment of $400 now and a payment of $300 at the end of four years

(a) if 6% simple interest is earned from the date each payment is made and use a comparison date of right now.

(b) if 6% simple interest is earned from the date each payment is made and use a comparison date of eight years from now.

(c) Explain why the fact you get different answers in parts (a) and (b) does not contradict the fact that equations of value at different times are equivalent equations.

(d) Repeat parts (a) and (b) except replace "simple interest" with "compound interest."

(12) [calculus needed] Use Newton's method to solve the problem of Example (2.3.7). More specifically, set $f(t) = 525(1.1)^{-2t} + 525(1.1)^{-t} - 1,000$, make an initial guess T_1 for a root T, and find a sequence of approximations $\{T_1\}$ to T that allow you to obtain T to the nearest hundredth of a percent. Why might $T_1 = .3$ be a reasonable initial guess?

(2.4) Investment return

(1) Payments of $3,000 now and $8,000 eight years from now are equivalent to a payment of $10,000 four years from now at either rate i or rate j. Find $|i - j|$. Explain why the yield rate is not unique in this case.

(2) Success, Inc. enters into a financial arrangement in which it agrees to pay $100,000 today and $101,000 two years from now in exchange for $200,000 one year from now. Show that there is no yield rate that can be assigned to this two-year transaction.

(3) Sigmund, Inc. agrees to pay $150,000 today and $40,000 four years from today in return for $210,000 two years from today. What is the yield rate for this four-year financial arrangement?

(4) Firms A, B, C, and D enter into a financial arrangement. Money flush firm A will pay expanding firms B and C each $1,000,000 today. B will pay D $2,200,000 three years from today. C will pay B $800,000 two years from today and D $350,000 two years from today. Finally, D will pay A $3,200,000 six years from today. Calculate the yield rate or interest rate, to the nearest hundredth of a percent, that each firm experiences over the period of their involvement (6 years for A, 3 years for B, 2 years for C, and 4 years for D).

(5) Sulsa invests $8,572.80 at $t = 0$ and $28,500 at $t = 2$. In return, she receives $27,074 at $t = 1$ and $10,000 at $t = 3$. Write down a time 0 equation of value and verify that it is satisfied for $v = .94$, $v = .95$, and $v = .96$. Find the corresponding three yield rates.

(6) Pedro invests $100,000 at $t = 0$ and $60,000 at $t = 2$. In return he gets $60,000 at $t = 1$ and $126,500 at $t = 3$. Write down a time 3 equation of value describing Pedro's investment. Explain why there is a unique yield rate and find it.

(7) Show that if $-1 < j \leqslant i$, $\sum_{k=1}^{n} C_{t_k}(1 + i)^{T-t_k} = \sum_{k=1}^{n} C_{t_k}(1 + j)^{T-t_k}$, and $B_{t_1}(i)$, $B_{t_2}(i)$, ..., $B_{t_{n-1}}(i)$ are all positive [with $B_{t_k}(i)$ as given in (2.4.8)], then $j = i$.

(8) [recommended for those with a BA II Plus calculator]
On January 1, Ezequiel opens an account at Friendly Bank. His opening deposit is for $50 and he makes deposits at the end of each quarter for four years, then makes no more deposits. He closes the account exactly seven years after he opens it and receives $3423.28. Find his annual yield rate for this seven-year period if his quarterly deposits were $60 in the first year, $75 in the second year, $50 in the third year, and were successively $300, $450, $800, and $240 in the fourth year.

(9) [recommended for those with a BA II Plus calculator]
A loan of $20,000 is to be repaid by thirty-three end-of-month payments. The first payment is $400 and then each payment is $25 more than the previous payment. Find the annual yield rate correct to the nearest hundredth of a percent. [HINT: The **Cash Flow worksheet** only accepts twenty-four payments, or thirty-two if you have a BA II Plus Professional calculator. If you are working with the BA II Plus calculator, suppose that the payments beyond the twenty-fourth, which you do not have registers to accommodate, are made along with the twenty-fourth. Now use the "guess and check" method, obtaining your first estimate by using $\boxed{\text{IRR}}\ \boxed{\text{CPT}}$. This is a challenging problem, especially for those of you with only 24 registers. However, when performing the successive calculations required by the "guess and check" method, you may make judicious use of the **NPV subworksheet** to decrease your work.]

(2.5) Reinvestment considerations

(1) Angela loans Kathy $8,000. Kathy repays the loan by paying $6,000 at the end of one and a half years and $4,000 at the end of three years. The money received at $t = 1\frac{1}{2}$ is immediately reinvested at an annual effective interest rate of 6%. Find Kathy's annual effective rate of interest and Angela's annual yield.

(2) Kurt loans Randy $14,000. Randy repays the loan by paying Kurt $4,000 at the end of one year and $6,000 at the end of two years and as well as at the end of three years. The money received at $t = 1$ and at $t = 2$ is immediately reinvested at an annual effective interest rate of 3%. Correct to the nearest tenth of a percent, find Randy's annual effective interest rate and Kurt's annual yield.

(3) On January 15, 2000, Enterprise A loans $6,000 to Enterprise B and $17,000 to Enterprise C. Enterprise B repays Enterprise A $7,000 on January 15, 2002 and this money is reinvested at a 5% annual effective

rate. Enterprise C repays Enterprise A $22,500 on January 15, 2004. What is the annual yield received by Enterprise A over the four-year interval. Compare it to the annual effective interest rates paid by Enterprises B and C.

(2.6) Approximate dollar-weighted yield rates

(1) Sandra invests $10,832 in the Wise Investment Fund. Three months later her balance has grown to $11,902 and she deposits $2,000. Two months later her fund holdings are $14,308 and she withdraws $7,000. Two years after her initial investment, she learns that her holdings are worth $12,566.

   (a) Write an equation of value involving the exact dollar-weighted annual yield i over the two-year period.
   (b) Compute the approximate dollar-weighted annual yield over the investment period using (2.6.5) and again using (2.6.8).

(2) On February 1, Arshak's investment account has a balance of $19,800. He deposited $1,200 on April 1 and $2,600 on May 1. He withdrew $8,400 on July 1. On November 1, Arshak's balance was $14,820. Find Arshak's approximate dollar-weighted annual yield for this nine-month period using (2.6.5).

(3) Franklin's investment fund had a balance of $290,000 on January 1, 1995 and a balance of $448,000 two years later. The amount of interest earned during the two years was $34,000, and the annual yield rate on the fund was 5.4%. Estimate the (dollar-weighted) average date of contributions to the account.

(4) The investment balance of a firm is $5,000,000 at the beginning of a two-year period and $7,000,000 at the end. The firm makes a single contribution during the two-year interval of $1,200,000. What is the difference between the approximate annual dollar-weighted yield earned by the firm if the contribution is made after 6 months as opposed to it being made after one year?

(2.7) Fund performance

(1) On January 1, 1988, Antonio invests $9,400 in an investment fund. On January 1, 1989 his balance is $10,600 and he deposits $2,400. On July 1, 1989 his balance is $14,400 and he withdraws $1,000. On January 1, 1992 his balance is P. Express his annual time-weighted yield as a function of P.

(2) Arthur buys $2,000 worth of stock. Six months later, the value of the stock has risen to $2,200 and Arthur buys another $1,000 worth of stock.

After another eight months, Arthur's holdings are worth $2,700 and he sells off $800 of them. Ten months later, Arthur finds that his stock has a value of $2,100.

(a) Compute the annual time-weighted yield rate of the stock over the two-year period.

(b) Compute the annual dollar-weighted yield for Arthur over the two-year period.

(c) Should the answer in part (a) or part (b) be larger? Why?

(3) Bright Future Investment Fund has a balance of $1,205,000 on January 1. On May 1, the balance is $1,230,000. Immediately after this balance is noted, $800,000 is added to the fund. If there are no further contributions to the fund for the year and the time-weighted annual yield for the fund is 16%, what is the fund balance at the end of the year?

Chapter 2 review problems

(1) Sohail makes an initial investment of $20,000. In return, he receives $4,000 at the end of one year and another $18,000 at the end of three years.

(a) Assuming that the investment is made at simple interest at rate r, write down an equation of value for the investment and find r.

(b) Assuming that the investment is made at compound interest at effective interest rate i, write down an equation of value for the investment and justify the statement that there is a unique yield rate. Use the "guess and check method" to estimate i to the nearest hundredth of a percent.

(c) Starting with the same initial guess for i that you used in (b), check you answer using Newton's method.

(d) (recommended for those with a BA II Plus calculator) Use the Cash Flow worksheet to find i to the nearest millionth.

(2) (a) Suppose now that investments are governed by compound interest at an effective interest rate $i \geqslant 0$. By how much does the sum of the time n value of $\$K$ paid at time 0 and the time n value of $\$K$ paid at time $2n$ exceed $\$2K$? Express your answer as a function $g(n)$ and show that $g(n) > 0$ if $n > 0$.

(b) Suppose that investments are governed by the simple interest accumulation function $a(t) = 1 + rt, r \geqslant 0$. Does the sum of the time n value of $\$K$ paid at time 0 and the time n value of $\$K$ paid at time $2n$ exceed $\$2K$ for all r and n? Justify your answer.

(3) Elyse invests $16,312 at $t = 0$. In return, she gets $8,000 at $t = 1$ and $10,000 at $t = 2$. Half of the time 1 payment, she reinvests at an annual effective interest rate of 5%. What is her annual yield rate for the two-year period?

(4) Sports Manufacturing needs capital for expansion. It borrows $1,000,000 from Venture Bank for three years at 6% nominal interest convertible quarterly, and $500,000 for five years from a private investor at a 5% effective discount rate. At the end of two years, Sports Manufacturing makes a $200,000 three-year loan to its supplier of titanium (for baseball bats) at 7% annual effective interest. What annual internal rate of return should Sports Manufacturing report for the combined cashflows over the five-year period?

(5) Abiyote invested $24,500 on January 1, 1994 in the Utopia Fund. On May 1, 1995, his balance was $28,212 and he withdrew $10,000. On December 1, 1995, his balance was $15,892, and he deposited $8,000. On January 1, 1997 his balance was $30,309.

   (a) Find the annual time-weighted yield for the Utopia Fund for the three-year period from January 1, 1994 until January 1, 1997.

   (b) Find an approximate annual dollar-weighted yield received by Abiyote for the three-year period from January 1, 1994 until January 1, 1997 using (2.6.5)

   (c) (recommended for those with a BA II Plus calculator) Find the dollar-weighted yield received by Abiyote for the three-year period from January 1, 1994 until January 1, 1997, correct to the nearest millionth of a percent.

   (d) Compare the time-weighted yield experienced by the Utopia Fund and the dollar-weighted yield received by Abiyote from his investment in the Utopia Fund. Discuss why the inequality between them is in the direction it is.

(6) Xiang and Dmitry are friends. They agree that Xiang will pay Dmitry $800 immediately and another $200 at the end of three years. In return, Dmitry will pay Xiang K in exactly one year and again at the end of exactly two years.

   (a) Find K if the transaction is based on compound interest at a nominal discount rate of 6% convertible monthly.

   (b) If $K = 600$, is there a unique positive yield rate for the transaction? Justify your answer.

CHAPTER 3

Annuities (annuities certain)

3.1   INTRODUCTION

Common financial obligations include the repayment of a loan by a series of equally spaced payments. Deposits to a retirement fund are also often made at regular intervals, as are interest payments on an investment. An **annuity** is a series of payments made at specified intervals for a fixed or contingent period. When the payments are equally spaced and form an orderly sequence, for instance a constant, arithmetic, or geometric sequence, and accumulation is by compound interest at a level interest rate, the value of the sequence at a given time may be most easily found using algebraic methods. Understanding the summation of geometric series is essential and is discussed in Section (3.2).

The Latin root of the word annuity is *annus* meaning year. The original use of the word annuity was a sequence of payments with the remittances at yearly intervals. However, the current usage allows other payment intervals, and the interval is called the **payment period**. It is convenient to have coincident payment periods and interest periods, and in this chapter (by finding an equivalent interest rate if necessary) this will be the case. So, in addition to considering payment periods of one year, we will consider periods of, for example, a month or a quarter. If payments occur at the end of each payment period, the annuity is called an **annuity-immediate**. Annuities-immediate with level payments are the subject of Section (3.2). If payments occur at the beginning of each payment period, the annuity is an **annuity-due**. In Section (3.3) we consider annuities-due with level payments.

When the length of time that the payments will be made is predetermined and no risk is involved, the annuity is called an **annuity-certain**. In contrast, a **contingent annuity** has payments for an uncertain term. If that term is the lifetime of the recipient, it is called a **life annuity**. In this text we only consider annuities-certain, and these will be referred to as "annuities." An annuity that has an infinite term is called a **perpetuity**. Level perpetuities are discussed in Section (3.4).

Suppose we have an annuity beginning at time T and lasting for n periods. In Sections (3.2) and (3.3), we concentrate on the value of the annuity at times T and $T + n$. In Section (3.5) we look at its value on an arbitrary date, and the important technique of introducing imaginary payments is introduced.

Imagine that a loan is to be repaid by an annuity lasting n periods with level payments, except perhaps for a slightly reduced final payment. Examples of such loans include most mortgages. In Section (3.6) we consider the intermediate outstanding loan balances. For a home mortgage, these are important if the homeowners wish to sell their home or refinance before the end of the loan.

Nonlevel payments are also common in annuities, and such annuities are called **nonlevel annuities**. If the sequence of payments has a pattern, computing the annuity's value at a given date may be facilitated if one makes clever use of level annuities. Examples of how to do this are presented in Section (3.7). Annuities with payments forming a geometric sequence are the topic of Section (3.8), and those whose payments form an arithmetic sequence are looked at in Section (3.9).

In Chapter 2 we introduced the important concept of the yield rate. In Section (3.10) we compute the yield rate for a number of financial transactions that involve annuities.

The methods of this chapter include the introduction of a number of **annuity symbols**. These are defined to give the value of a particular annuity at the beginning of the term of the annuity or at the end of the term of the annuity, and this term is an integral number of interest conversion periods. For each of the annuity symbols that we introduce, we derive an algebraic expression assuming growth is governed by compound interest. These algebraic expressions make sense even when the variable giving the term of the annuity does not take on an integer value. In Section (3.11) we use this fact to define annuity symbols with nonintegral terms.

The basic definitions and ideas of Sections (3.2)–(3.7) apply to investments governed by arbitrary accumulation functions. However, our examples prior to Section (3.12) all involve compound interest accumulation functions with a level interest rate. In Section (3.12) we consider examples concerning annuities where the growth of money is not governed by an accumulation function of the form $a(t) = (1 + i)^t$. Our examples include growth by simple interest and growth by nonlevel rates of compound interest.

Following Section (3.12), it is natural to introduce the investment year method that is used by investment funds that wish to consider the year of investment as well as the current time in determining the rate of interest to be applied. This is done in Section (3.13).

3.2 ANNUITIES - IMMEDIATE

An annuity lasting n interest periods with payments at the **end** of each interest period is called an **annuity-immediate**. Students are often confused by the term annuity-immediate as they view "immediate" as meaning "occurring without delay." However, when we speak of an annuity-immediate, the "immediate" signifies "of or near the future time." Just as the immediate successor to the monarchy must wait until the present king or queen steps down or dies, the owner (the annuitant) of an annuity-immediate must wait until the end of one period to receive his or her first annuity payment. This is in contrast to the annuitant of an annuity-due who is due to receive a payment as soon as the annuity begins. We introduce annuities-due in Section (3.3).

By a **basic annuity-immediate**, we mean an annuity-immediate with level payments, each of which is equal to 1. Thanks to the distributive law, the value at a given time of a level annuity-immediate with payments of Q is just Q times the value of the basic annuity-immediate.

Provided that the growth of money is governed by an increasing accumulation function, receiving a basic annuity-immediate lasting n periods is less advantageous than receiving $\$n$ immediately. The present value of the basic annuity immediate lasting n periods equals the sum of the present value of the n end-of-period payments, namely

$$v(1) + v(2) + v(3) + \cdots + v(n).$$

When we have compound interest at an effective rate of i per interest period, $v(t) = (1 + i)^{-t} = v^t$, and this sum is the geometric series $v + v^2 + v^3 + \cdots + v^n$.

EXAMPLE 3.2.1

Problem: Marguerite receives an allowance of $\$20$ at the end of each month. Her four-year old brother Stevie receives an allowance of $\$1$ at month's end.

Find the value on January 1 of each of their allowances for the three-month period consisting of January, February, and March, using a monthly effective interest rate of .5%.

Solution   Stevie receives $1 at the end of each of the three months. The January 1 value of his end of January payment is $\$1/1.005 \approx \$.99502$, of his end of February payment is $\$1/(1.005)^2 \approx \$.99007$, and of his end of March payment is $\$1/(1.005)^3 \approx \$.98515$. The sum of these is $\$1/1.005 + \$1/(1.005)^2 + \$1/(1.005)^3 \approx \$2.97$. Each month big sister Marguerite receives $20, so the January 1 value of her three allowance payments is $\$20/1.005 + \$20/(1.005)^2 + \$20/(1.005)^3 = 20(\$1/1.005 + \$1/(1.005)^2 + \$1/(1.005)^3) \approx \$59.40$. Note that $\$59.40 = 20 \times \$2.97$.   ∎

We next recall that there is a nice algebraic expression for the sum of a geometric series. Specifically,

(3.2.2)
$$c + cr + cr^2 + \cdots + cr^{n-1} = \frac{c(1 - r^n)}{1 - r} \quad \text{for} \quad r \neq 1.$$

This is easy to derive. Let $S = c + cr + cr^2 + \cdots + cr^{n-1}$, then $rS = cr + cr^2 + cr^3 \cdots + cr^n$, and $(1 - r)S = S - rS = (c + cr + cr^2 + \cdots + cr^{n-1}) - (cr + cr^2 + cr^3 \cdots + cr^n) = c - cr^n = c(1 - r^n)$. Solving for S, the left-hand side of (3.2.2), we obtain $\frac{c(1-r^n)}{1-r}$.

When $c = r = v$, since $v(1 + i) = 1$, formula (3.2.2) gives

(3.2.3) $$v + v^2 + v^3 + \cdots + v^n = \frac{v(1 - v^n)}{1 - v} = \frac{(1 + i)v(1 - v^n)}{(1 + i)(1 - v)} = \frac{1 - v^n}{i}.$$

The **present value of the basic annuity-immediate lasting** n **interest periods** is used frequently. If the accumulation function to be used is clear, the present value is denoted by the **annuity symbol** $a_{\overline{n}|}$. If a compound interest accumulation function is operative, and the effective interest rate for the payment period is i, then you often write $a_{\overline{n}|i}$ instead of just $a_{\overline{n}|}$. This allows you to emphasize the effective interest rate used. The annuity symbol $a_{\overline{n}|i}$ is read "a angle n at interest rate i." Equation (3.2.3) may now be rewritten

(3.2.4)
$$a_{\overline{n}|i} = v + v^2 + v^3 + \cdots + v^n = \frac{1 - v^n}{i}.$$

IMPORTANT FACT (3.2.5): With an annuity-immediate, you must wait one period to start receiving payments. The present value of a basic annuity-immediate lasting n periods is denoted by $a_{\overline{n}|}$. It gives the value of the annuity **one period before the first payment.**

It is also useful to have a symbol giving the accumulated value of a basic annuity-immediate lasting n periods at the time of its final payment. The annuity symbol $s_{\overline{n}|}$ or $s_{\overline{n}|i}$ denotes this accumulated value. Of course the latter is only used when the accumulation is governed by the compound interest accumulation function $a(t) = (1 + i)^t$. Read the symbol $s_{\overline{n}|i}$ as "s angle n at interest rate i."

IMPORTANT FACT (3.2.6): With an annuity-immediate lasting n periods, the final payment is made at time n. The accumulated value of a basic annuity-immediate lasting n periods **at the time of the last payment** is denoted $s_{\overline{n}|}$.

Note that $s_{\overline{n}|}$ and $a_{\overline{n}|}$ each measure the value of the same sequence of payments, but $s_{\overline{n}|}$ measures the value n interest periods later.

PAYMENT:		1	2	1	$\cdots$	1
TIME:	0	1	2	3		n
VALUE:	$a_{\overline{n}|}$					$s_{\overline{n}|}$

FIGURE (3.2.7)

When $a(n) = (1 + i)^n$, we have the important equalities

(3.2.8)
$$s_{\overline{n}|i} = (1 + i)^n a_{\overline{n}|i} \quad \text{and} \quad a_{\overline{n}|i} = v^n s_{\overline{n}|i}.$$

More generally,

(3.2.9)
$$s_{\overline{n}|} = a(n) a_{\overline{n}|} \quad \text{and} \quad a_{\overline{n}|} = v(n) s_{\overline{n}|}.$$

It follows from (3.2.4) and (3.2.8), or directly from the definition of $s_{\overline{n}|i}$, that

(3.2.10)
$$s_{\overline{n}|i} = (1 + i)^{n-1} + (1 + i)^{n-2} + \cdots + 1 = \frac{(1 + i)^n - 1}{i}.$$

Here $(1 + i)^{n-k}$ is the accumulated value at time n of the k-th payment.

EXAMPLE 3.2.11   An unknown present value

Problem: Candace wishes to buy a new car. Her budget allows her to make monthly payments of $500, and she has saved $1,800 for a down payment. Candace qualifies for a 36-month auto loan which has a nominal interest rate of 4.8% convertible monthly. How expensive a car can Candace buy?

Solution   The monthly interest rate is $\frac{i^{(12)}}{12} = 4.8\%/12 = .4\% = .004$, and Candace is prepared to make 36 successive end-of-month payments of $500. The present value of her payments must equal the balance due on the car after the down payment. Therefore, she can take out a loan for $500a_{\overline{36}|.004} = \$500(\frac{1-(1.004)^{-36}}{.004}) \approx \$16,732.94$. Since she has $1,800 for a down payment, she can buy a car costing $16,732.94 + \$1,800 = \$18,532.94$. ∎

EXAMPLE 3.2.12   An unknown contribution

Problem: Seth wishes to borrow $4,400 so he can pay his tuition. He qualifies for a two-year loan with a monthly effective interest rate of .25% and level payments. Find the amount of Seth's monthly payments under this loan.

Solution   The loan lasts twenty-four months. Therefore, if Q is the payment amount, the time 0 equation of value stating that his payments have present value $4,400 is $4,400 = Qa_{\overline{24}|.0025}$. So, $Q = \frac{\$4,400}{a_{\overline{24}|.0025}} = \frac{(\$4,400)(.0025)}{1-(1.0025)^{-24}} \approx$ $189.1173327. Since Seth's monthly payments must be an integral number of cents, they are $189.12. These payments will in fact repay a loan of $4,400.06 since $189.12a_{\overline{24}|.0025} \approx \$4,400.062057114$. (The display will just show "PV = 4,400.062057," but if you subtract 4,400.06, you obtain .002057114.) Thus the last payment may be reduced by $.062057114(1.0025)^{24} \approx .065889578 \approx .07$. The last payment should therefore be $189.05. ∎

We note that in the solution to Example (3.2.12), we used the following basic fact.

> **IMPORTANT FACT 3.2.13:** If a loan of amount L is to be repaid by n level end-of-period payments of Q and the effective interest rate for the payment period is i, then $Q = \frac{L}{a_{\overline{n}|i}}$.

We also note the following useful algorithm.

ALGORITHM 3.2.14:

Suppose that a loan of amount L is to be repaid by n end-of-period payments, the effective interest rate for the payment period is i, and it is stipulated that the payments be level except that the last payment be slightly reduced if that is necessary so that the borrower does not repay more than the loan amount (rounded to the nearest penny).

(1) Compute $\dfrac{L}{a_{\overline{n}|i}}$ and round this to the nearest penny. Call the result Q_1.

(2) Compute $Q_1 a_{\overline{n}|i}$ and round to the nearest penny. If this is L, then a slightly reduced final payment is not needed and the payments are all equal to Q_1.

(3) Otherwise, $\dfrac{L}{a_{\overline{n}|i}}$ is not an integral number of cents and we **round-up** to the next cent. Call this Q_2. The first $n-1$ payments are Q_2.

(4) Compute $E = Q_2 a_{\overline{n}|i} - L$. This is the amount beyond the loan amount that level payments of Q_2 would pay off.

(5) The final irregular payment is $R = Q_2 - E(1+i)^n$.

Henceforth, **use Algorithm (3.2.14)** [or the equivalent BA II Plus calculator algorithm (3.2.19)] **to find the payments of an annuity set up to have level payments** *unless an alternate method is specified.*

EXAMPLE 3.2.15   An unknown contribution

Problem: Mr. Liang wishes to accumulate $200,000 in a college fund for his newborn daughter. He wishes the money to be available to her on her eighteenth birthday, and he is prepared to make level contributions on her first through eighteenth birthdays. The fund has an annual effective discount rate of 5%. How large will Mr. Liang's contributions need to be?

Solution The annual interest rate earned by the fund is $i = \frac{.05}{1-.05} = \frac{1}{19}$. Denote the amount of Mr. Liang's contributions by Q. Then, we need $\$200,000 = Qs_{\overline{18}|\frac{1}{19}} = Q\left(\frac{(1+\frac{1}{19})^{18}-1}{\frac{1}{19}}\right) \approx 28.8331196Q$. Therefore $Q \approx \frac{\$200,000}{28.8331196} \approx \$6,936.47$. In fact, these payments will produce $\$200,000.07$, so Mr. Liang could reduce his last payment by seven cents. ■

The basic principle used in Example (3.2.15) was

> **IMPORTANT FACT 3.2.16:** If an amount B is to be accumulated by n level end-of-period payments of Q and the effective interest rate for the payment period is i, then $Q = \frac{B}{s_{\overline{n}|i}}$.

EXAMPLE 3.2.17   An unknown number of contributions

Problem: Hyun wishes to contribute $\$100$ at the end of each month to his savings account until the account has a balance of at least $\$3,000$, at which time he will buy his grandmother's piano. Hyun's savings account has an annual interest rate of 5.4% convertible monthly. How many months it will take Hyun to accumulate the needed money, and how much money will he then have saved?

Solution The monthly interest rate is 5.4%/12 = .45% = .0045, and Hyun makes $\$100$ end-of-month payments. We wish to find the smallest integer n such that $\$100s_{\overline{n}|.0045} \geq \$3,000$. But $\$100s_{\overline{n}|.0045} = \$100\left(\frac{(1.0045)^n-1}{.0045}\right)$, so what we need is the smallest integer n with $\frac{(1.0045)^n-1}{.0045} \geq 30$. This inequality is equivalent to $(1.0045)^n \geq 1 + 30(.0045) = 1.135$. Taking natural logarithms, we need $n \geq \frac{\ln 1.135}{\ln 1.0045} \approx 28.2$. So, $n = 29$ and Hyun's balance is $\$100s_{\overline{29}|.0045} = \$100\left(\frac{(1.0045)^{29}-1}{.0045}\right) \approx \$3,090.32$.

Another method would have been to make a guess as to the number of payments and calculate the balance resulting from that guess. Had we found a balance less than $\$3,000$, we would have increased our guess and tried again. Had we guessed 30 or more, the balance would have been over $\$3,200$ and we would have decreased our guess. ■

The fundamental relation between $a_{\overline{n}|i}$ and $s_{\overline{n}|i}$ is given by (3.2.8). There is a second connection that we now derive using (3.2.10) and (3.2.4). First note that

$$i + \frac{1}{s_{\overline{n}|i}} = i + \frac{i}{(1+i)^n - 1} = \frac{i[(1+i)^n - 1] + i}{(1+i)^n - 1} = \frac{i(1+i)^n}{(1+i)^n - 1}.$$

Multiplying this last expression by $\frac{v^n}{v^n} = 1$ and using $v(1+i) = 1$, we find

$$i + \frac{1}{s_{\overline{n}|i}} = \frac{i}{1 - v^n} = \frac{1}{a_{\overline{n}|i}}.$$

We have derived the relation

(3.2.18)
$$\boxed{i + \frac{1}{s_{\overline{n}|i}} = \frac{1}{a_{\overline{n}|i}}.}$$

It is instructive to give a second demonstration of Equation (3.2.18) that involves interest theory interpretations of the terms involved. This latter argument will be important in Chapter 5 when we study sinking fund loans. Consider a loan of amount 1, to be repaid at the end of n periods, and let i denote the effective interest rate per period. The loan may be repaid by n level end-of-period payments. Recalling (3.2.12), these level payments will each be $\frac{1}{a_{\overline{n}|i}}$. An alternate way to repay the loan would be as follows. At the end of each period, pay the lender the interest for that period, namely i. Also, make a deposit to a savings account (paying interest at an effective rate i per payment period) so that at time n the balance in the savings account will be 1. According to (3.2.16), the deposit amounts should be $\frac{1}{s_{\overline{n}|i}}$. Finally, at time n, remove the balance 1 from the savings account, and give it to the lender. Each of these repayment scenarios have level costs to the borrower, utilize the interest rate i, and result in the loan being exactly repaid at time n. The level payments under the two schemes must be equal. This equality is (3.2.18).

The BA II Plus calculator has special buttons

$$\boxed{\text{N}} \quad \boxed{\text{I/Y}} \quad \boxed{\text{PV}} \quad \boxed{\text{PMT}} \quad \boxed{\text{FV}}$$

that, along with $\boxed{\text{CPT}}$, are useful for solving level annuity problems that have one unknown. These special keys appear on the third row of the BA II Plus calculator, and together they govern the so-called **Time - Value - Money** (or **TVM) worksheet**. All five of the registers governed by the **TVM** buttons may simultaneously be set to zero by pushing $\boxed{\text{2ND}}$ $\boxed{\text{CLR TVM}}$. This also happens when you press $\boxed{\textbf{2ND}}$ $\boxed{\textbf{RESET}}$ but, as discussed in Chapter 0, the latter has additional consequences.

Recall that an annuity-immediate has end-of-period payments. In order for the **TVM worksheet** of the BA II Plus calculator to handle a problem involving a level annuity-immediate, it must be set to interpret payments as coming at the end of the period. This is the factory default setting, but if you have not just reset your calculator, you should confirm that your calculator is so designated by glancing at the display window. If the display contains "BGN" at the top of the right-hand side, the calculator is not as desired, and you should remedy the situation by pushing $\boxed{\textbf{2ND}}$ $\boxed{\textbf{BGN}}$ $\boxed{\textbf{2ND}}$ $\boxed{\textbf{SET}}$ $\boxed{\textbf{2ND}}$ $\boxed{\textbf{QUIT}}$. The calculator will be said to be in "**END mode**" when "BGN " is not displayed.

The registers **PMT**, **PV**, and **FV** accept dollar amounts, and the labels are abbreviations for **payment**, **present value**, and **future value**. One uses $\boxed{\textbf{PMT}}$ to enter the dollar amount of each of the level annuity payments. This amount (and a cashflow amount in general) should be entered as positive if it is received (an **inflow**) and negative if it is paid out (an **outflow**). Of course whether a payment is an inflow or an outflow depends on which party is being considered, but the important thing here is to take a consistent viewpoint, thereby minimizing the number of error messages and incorrect answers.

> Choose a perspective, and then enter inflows
> as positive and outflows as negative.

If the annuity is to repay a loan , from the borrower's perspective , the annuity payments will be out payments, while if the annuity is to accumulate money in a savings account, from the investor's perspective, the annuity payments will be out payments. Sticking with a consistent viewpoint, in the first case there will be an inflow (the loan) at the beginning of the annuity and this should be entered in the register $\boxed{\textbf{PV}}$ while zero should be entered in the $\boxed{\textbf{FV}}$ register. In the latter case, the accumulated amount at the end of the annuity period (the time of the last payment) needs to go in $\boxed{\textbf{FV}}$ as an inflow (one imagines the investor closing the account and withdrawing his money), while zero should be put in the $\boxed{\textbf{PV}}$. In these simple situations the entry in either $\boxed{\textbf{PV}}$ or $\boxed{\textbf{FV}}$ will be zero, but later there will be many examples where one enters nonzero numbers in both.

Registers **I/Y, P/Y,** and **C/Y** are all filled using the $\boxed{\textbf{I/Y}}$ key, but **P/Y** and **C/Y** first require $\boxed{\textbf{2ND}}$ to be pushed to access their worksheet. Note that **I/Y** is a shorthand for "interest rate per year" and is designed to store a nominal interest rate, **P/Y** stands for "payments per year," and **C/Y** indicates the number of interest conversion periods per year. The "per year" designation is somewhat confusing in that the user may choose a basic time period other than a year when choosing entries for registers **I/Y, P/Y,** and **C/Y**, for instance, a month if you have an <u>effective</u> monthly interest rate and monthly payments. It is essential that you consider the same basic period when filling all three of these registers!

Use $\boxed{\text{I/Y}}$ to enter a nominal interest rate convertible m times per period <u>as a percentage</u>. A convenient choice for m depends on what sort of an interest rate you know. When an effective interest rate <u>per payment period</u> is handy, enter that rate <u>as a percentage</u>. In this case, you will wish the $\boxed{\text{P/Y}}$ and $\boxed{\text{C/Y}}$ registers to each hold the value 1. (You may prefer to keep the number 1 in the $\boxed{\text{P/Y}}$ and $\boxed{\text{C/Y}}$ registers, and convert to an effective interest rate per payment period if one is not provided. This choice makes your calculator seem more similar to the Texas Instruments BA-35 calculator and is the factory default setting on the BA II Plus Professional calculator. The factory default setting on the BA II Plus calculator for the $\boxed{\text{P/Y}}$ register and also for the $\boxed{\text{C/Y}}$ register is 12, which is perfect if you have monthly payments and a nominal interest rate convertible monthly.)

If you have a nominal interest rate convertible m times per period, k payments for that same period, and do not wish to convert to an effective interest rate per payment period, set the $\boxed{\text{P/Y}}$ and $\boxed{\text{C/Y}}$ registers to m and k respectively by first pushing $\boxed{\text{2ND}}$ $\boxed{\text{P/Y}}$, then entering m, pushing $\boxed{\downarrow}$, entering k, and finally pushing $\boxed{\text{2ND}}$ $\boxed{\text{QUIT}}$. For example, if you wish to have **P/Y** $= 6$ and **C/Y**$= 4$, depress

$$\boxed{\text{2ND}}\ \boxed{\text{P/Y}}\ \boxed{6}\ \boxed{\text{ENTER}}\ \boxed{\downarrow}\ \boxed{4}\ \boxed{\text{ENTER}}\ \boxed{\text{2ND}}\ \boxed{\text{QUIT}}.$$

If $m = k$, you can skip pushing $\boxed{\downarrow}$ and entering k, since when a value is entered in $\boxed{\text{P/Y}}$, it is also automatically entered in $\boxed{\text{C/Y}}$. So, if you desire to have **P/Y** $= 6$ and **C/Y**$= 6$, just push $\boxed{\text{2ND}}\ \boxed{\text{P/Y}}\ \boxed{6}\ \boxed{\text{2ND}}\ \boxed{\text{QUIT}}$. Again note that the factory default entries for the registers $\boxed{\text{P/Y}}$ and $\boxed{\text{C/Y}}$ on the BA II Plus calculator (but not on the BA Plus Professional) are both 12. So, if you wish to store **P/Y** $= 12$ and **C/Y** $= 12$ on the BA II Plus calculator, you may accomplish this by simply keying $\boxed{\text{2ND}}\ \boxed{\text{P/Y}}\ \boxed{\text{2ND}}\ \boxed{\text{CLR WORK}}$. (A similar shortcut is available on the BA II Plus Professional if you want to enter 1 into the P/Y and C/Y registers.)

The register **N** is designed to hold the number q of payments that the annuity has. This may be done directly by entering the number q and then depressing $\boxed{\text{N}}$. (Alternatively, if you have an annuity lasting n periods with m

payments per period and m has been entered in the **P/Y** register, you may enter n and then push $\boxed{\textbf{2ND}}\,\boxed{\times\textbf{P/Y}}\,\boxed{\textbf{N}}$ to correctly fill the **N** register.)

The **TVM** worksheet is designed so that if values are entered in any order into the four of the registers and the key $\boxed{\textbf{CPT}}$ is depressed, followed by the fifth **TVM key**, the value of the fifth **TVM** variable is displayed and simultaneously entered into the fifth **TVM** register. Of course this is only possible when consistent values are stored in the first four registers. If incompatible values are found, "ERROR 5" is displayed. (A common cause of this error message appearing is for an inflow or outflow to be entered with the incorrect sign.) If the unknown is the interest rate, there may be a delay of a number of seconds before a display appears. This is because the calculation requires an iterative method. It is possible that the calculator will fail to find a solution, even if one exists, if its limit of time for performing iterations has expired. In this case, "ERROR 7" will be displayed. Note that the **TVM** worksheet covers a situation where there is one borrower until the end of the annuity term, hence a unique yield rate.

Utilizing the **Cash Flow worksheet** is an alternate way of letting the BA II Plus calculator help solve annuity-immediate problems if the unknown is the present value of the sequence of annuity payments or the interest rate. (On the BA II Plus Professional calculator, the **NPV subworksheet** is more extensive, and the **Cash Flow worksheet** can also be used to compute unknown future values.) In the case of an unknown present value, the amount of the level payments needs to be entered as C01 and the number of payments is entered as F01. Then push $\boxed{\textbf{NPV}}$. The display will show "I =" and you need to enter the effective interest rate per payment period. Do this by entering the numerical value of the rate as a percentage and then depressing $\boxed{\textbf{ENTER}}$. Now push $\boxed{\downarrow}$, or $\boxed{\uparrow}$ so that "NPV =" appears. Pushing $\boxed{\textbf{CPT}}$ will give you the previously unknown present value. If the interest rate per payment period is the unknown, enter the present value of the sequence of annuity payments as CFo, the negative of the amount of the individual level payments as C01, and the number of payments as F01. Now hit $\boxed{\textbf{IRR}}\,\boxed{\textbf{CPT}}$ to display a computed value of the effective interest rate per payment period as a percentage. Again, it is possible to receive ERROR 5 if you have entered inconsistent values or ERROR 7 if the iteration fails to produce a rate in the allotted time. If the interest rate required was other than the effective rate per payment period, make a suitable interest rate conversion.

Algorithm (3.2.14) may be efficiently carried out using the BA II Plus calculator. We refer to this equivalent algorithm as Algorithm (3.2.19).

ALGORITHM 3.2.19:
(Entries made from the lender's perspective)

Suppose that a loan of amount L is to be repaid by n end-of-period payments, the effective interest rate for the payment period is i, and it is stipulated that the payments be level except that the last payment be slightly reduced if that is necessary so that the borrower does not repay more than the loan amount (rounded to the nearest penny).

(1) Put your calculator in END mode with P/Y=C/Y=1. Then fill the N register with the number n, the I/Y register with the number i, the PV register with the number $-L$, and the FV register with 0. Push **CPT** **PMT**. Since you are working from the lender's perspective, the resulting computed payment will be positive. Round it to the nearest .01, so that in dollars it represents an integral number of cents, and enter this value into the PMT register. Call this value Q_1.

(2) Push **CPT** **PV**. If this is equal to $-L$ when rounded to the nearest .01, a slightly reduced final payment is not needed. All the payments are equal to the number Q_1 currently stored in the PMT register.

(3) Otherwise, you still need to find the payments. Let

$$Q_2 = \begin{cases} Q_1 & \text{if you rounded up to find } Q_1 \text{ in (2)} \\ Q_1 + .01 & \text{if you rounded down to find } Q_1 \text{ in (2).} \end{cases}$$

So, Q_2 is the number obtained in (1) by pushing **CPT** **PMT**, rounded up to the nearest .01. Thus, $\$Q_2$ represents an integral number of cents. Enter Q_2 into the PMT register. The first $n-1$ payments are $\$Q_2$.

(4) Reenter $-L$ into the PV register. Push **CPT** **FV**. This will be negative because you rounded up, increasing the lender's total contribution. The negative number F is the amount (in dollars) by which the borrower would overpay if all n payments were for Q_2.

(5) The final irregular payment is $\$R = \$Q_2 + \$F$.

We end this section with some examples illustrating the use of the BA II Plus calculator for the solution of problems involving an annuity-immediate, beginning with one illustrating Algorithm (3.2.19)

EXAMPLE 3.2.20   Example 3.2.12 revisited, an unknown present value

Problem: Seth wishes to borrow \$4,400 so he can pay his tuition. He qualifies for a two-year loan with a monthly effective interest rate of .25% and level payments. Find the amount of Seth's monthly payments under this loan using Algorithm (3.2.19).

Solution   In END mode with P/Y $= 1$ and C/Y$=1$, depress

| 2 | 4 | N | • | 2 | 5 | I/Y | 4 | 4 | 0 | 0 | +/− | PV | 0 | FV | CPT | PMT |.

The display should now show "PMT $= 189.1173327$". Key

| 1 | 8 | 9 | • | 1 | 2 | PMT |, and then | CPT | PV |.

We thereby obtain "PV $= -4,400.062057$". So, level payments of \$189.12 pay off a loan of \$4,400.06, which is 6 cents more than Seth wishes to borrow. Push | 4 | 4 | 0 | 0 | +/− | PV | to reenter the desired loan amount, then | CPT | FV |. The display reads "FV $= -.065889577$". Rounding to the nearest .01, the value displayed in the FV register is $-.07$. So, the final payment is \$189.12 $+ (-\$.07) = \$189.05.$ ∎

EXAMPLE 3.2.21   Example 3.2.11 revisited, An unknown present value

Problem: Candace wishes to buy a new car. Her budget allows her to make monthly payments of \$500, and she has saved \$1,800 for a down payment. Candace qualifies for a 36-month auto loan which has a nominal interest rate of 4.8% convertible monthly. Use the BA II Plus calculator **TVM worksheet** to determine how expensive a car Candace can buy. Check your answer using the BA II Plus calculator **Cash Flow worksheet**.

TVM worksheet solution 1   In END mode with P/Y $= 12$ and C/Y$=12$, depress

| 3 | 6 | N | 4 | • | 8 | I/Y | 5 | 0 | 0 | +/− | PMT | 0 |

| FV | CPT | PV | + | 1 | 8 | 0 | 0 | = |.

The calculator display then shows 18,532.93794, assuming the calculator was formatted for nine decimal places. Rounding to the nearest penny, Candace may buy a car for \$18,532.94.

TVM worksheet solution 2   In END mode with P/Y $= 1$ and C/Y$=1$, depress

| 3 | 6 | N | • | 4 | I/Y | 5 | 0 | 0 | +/− | PMT | 0 |

| FV | CPT | PV | + | 1 | 8 | 0 | 0 | = |.

The calculator again displays 18,532.93794, once again assuming the calculator was formatted for nine decimal places. Rounding to the nearest penny, Candace may buy a car for $18,532.94.

Cash Flow worksheet solution   Press

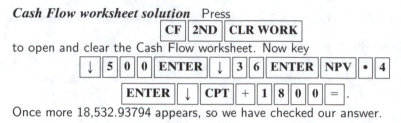

to open and clear the Cash Flow worksheet. Now key

Once more 18,532.93794 appears, so we have checked our answer.   ∎

EXAMPLE 3.2.22   Example 3.2.17 revisited. An unknown number of payments

Problem: Hyun wishes to contribute $100 at the end of each month to his savings account until the account has a balance of at least $3,000, at which time he will buy his grandmother's piano. Hyun's savings account has an annual interest rate of 5.4% convertible monthly. Use the **TVM worksheet** to determine how many months it will take Hyun to accumulate the needed money, and how much money he will then have saved.

Solution   With $P/Y = 12$ and $C/Y = 12$, push the following sequence of keys

to display 28.20385804. The smallest integer greater than this number is 29, so Hyun needs to make twenty-nine $100 deposits. Depressing

the calculator display shows 3090.32074, so Hyun will have saved $3,090.32.   ∎

Note that when we wished to answer the second question posed in Example (3.2.22), we only had to make an entry to one **TVM** register. The reason for this was that the other registers where we wished to hold data already held the desired entries. This is a typical savings of work.

EXAMPLE 3.2.23   Example 2.6.11 revisited. An unknown yield

Problem: On January 1, 1995, Saul's investment account had a balance of $7,688. He deposited $100 on February 1, 1995 and then again at the end of each two months (that is on April 1, 1995, June 1, 1995, ... , December 1, 1997). His balance on January 1, 1998 is $10,830. Approximate Saul's annual dollar-weighted yield, correct to the nearest hundredth of a percent.

Solution   We view this as a three-year investment beginning on January 1, 1995 and ending on January 1, 1998 by pretending that Saul deposits $7,688 on January 1, 1995 and liquidates the account on January 1, 1998 by making a withdrawal of $10,830. A January 1, 1995 equation of value for this investment is

$$(3.2.24) \qquad 0 = \$7,688 + (1 + J)^{\frac{1}{2}}\$100a_{\overline{18}|J} - \$10,830(1 + J)^{-18}$$

where J is the yield rate for a two-month period. We note that the interest rate that satisfies this equation is unique since the deposits all precede the withdrawal, but it eludes solution by algebraic means, the **TVM worksheet**, or the **Cash Flow worksheet** (since there are fewer than thirty-six contribution registers). However, there is a modified equation that may be solved by either the **TVM worksheet** or the **Cash Flow worksheet** and allows you to obtain a good approximation J' to J, an approximation that you may then modify using the "guess and check" method and (3.2.4). This modified equation

$$0 = \$7,688 + \$100a_{\overline{18}|J'} - \$10,830(1 + J')^{-18}$$

corresponds to Saul delaying each of his $100 deposits by one month. Since such delays would be advantageous to Saul, the rate J' is an overestimate of J. To use the **TVM worksheet** to find J' (which like J is unique), set your calculator in END mode with $P/Y = 1$ and $C/Y = 1$. Then key

| 7 | 6 | 8 | 8 | PV | 1 | 8 | N | 1 | 0 | 0 | PMT |

| 1 | 0 | 8 | 3 | 0 | +/− | FV | CPT | I/Y |

Your display should now read "I/Y = .816283332". This gives J' as a percent. Alternatively, you may obtain this rate using the **Cash Flow worksheet** and keying

| CF | 2ND | CLR WORK | 7 | 6 | 8 | 8 | ENTER | ↓ | 1 | 0 | 0 |

| ENTER | ↓ | 1 | 7 | ENTER | 1 | 0 | 0 |

| − | 1 | 0 | 8 | 3 | 0 | = | ENTER | IRR | CPT |

We note that this **Cash Flow worksheet** calculation was already mentioned in Example (2.6.11).

We now know that the two-month yield rate J is a little less than .00816283332. If you compute the right-hand side of (3.2.24) with J' substituted for J, you obtain approximately 6.79, a positive number. Calculating the right-hand side of (3.2.24) using .0081 for J, gives a negative number (approximately -2.79). So, the rate J is sandwiched between .0081 and .00816283332. You may narrow the interval that must contain J by further "guess and check." In fact, $.0081182 < J < .0081183$. Since $(1.0081182)^6 - 1 \approx .049708547$ and $(1.0081183)^6 - 1 \approx .049709168$, Saul's annual dollar-weighted yield is approximately 4.97%. ∎

3.3   ANNUITIES - DUE

An annuity lasting n interest periods with payments at the **beginning** of each interest period is called an **annuity-due**. By a **basic annuity-due**, we mean an annuity-due with level payments, each of which is equal to 1. The value at a given time of a level annuity-due with payments of Q is just Q times the value of the basic annuity-due.

Having fixed a choice of accumulation function $a(t)$, let $\ddot{a}_{\overline{n}|}$ denote the value at the time of the first payment of the basic annuity-due that lasts n periods, and write $\ddot{s}_{\overline{n}|}$ for the value at the end of the n-th payment period of this annuity. Note that time n is one period after the final payment of an annuity-due with duration n periods. If the chosen accumulation function is the compound interest accumulation function $a(t) = (1 + i)^t$, the symbol $\ddot{a}_{\overline{n}|i}$ is commonly used for the value at the time of the first payment, and the symbol $\ddot{s}_{\overline{n}|i}$ indicates the value at time n. The annuity symbol $\ddot{a}_{\overline{n}|i}$ is read "a double-dot, angle n, at interest rate i" or "a due, angle n, at interest rate i." Likewise, $\ddot{s}_{\overline{n}|i}$ is "s double-dot, angle n, at interest rate i" or "s due, angle n, at interest rate i."

PAYMENT:	1	1	1	1	$\cdots$		1	
TIME:	0	1	2	3	$\cdots$	$n-1$		n
VALUE:	$\ddot{a}_{\overline{n}|}$							$\ddot{s}_{\overline{n}|}$

FIGURE (3.3.1)

Analogously to (3.2.8) and (3.2.9), we have

(3.3.2)
$$\ddot{s}_{\overline{n}|i} = (1 + i)^n \ddot{a}_{\overline{n}|i} \quad \text{and} \quad \ddot{a}_{\overline{n}|i} = v^n \ddot{s}_{\overline{n}|i},$$

and more generally,

(3.3.3)
$$\ddot{s}_{\overline{n}|} = a(n)\ddot{a}_{\overline{n}|} \quad \text{and} \quad \ddot{a}_{\overline{n}|} = v(n)\ddot{s}_{\overline{n}|}.$$

> **IMPORTANT FACT (3.3.4):** With an annuity-due, you start receiving payments immediately. The value of a **basic annuity-due lasting n periods** at the time of the first payment is denoted by $\ddot{a}_{\overline{n}|}$. The last payment is made at the beginning of the n-th period, and $\ddot{s}_{\overline{n}|}$ gives the annuity's value one period after this last payment, that is to say at the end of the n-th period.

The annuity symbol $\ddot{a}_{\overline{n}|i}$ is given by a geometric series, the sum of which may be computed using (3.2.2). Recalling that $1 - v = d$ (Equation 1.9.7), we find

$$(3.3.5) \quad \ddot{a}_{\overline{n}|i} = 1 + v + v^2 + v^3 + \cdots + v^{n-1} = \frac{1(1 - v^n)}{1 - v} = \frac{1 - v^n}{d}.$$

A formula for the annuity symbol

$$\ddot{s}_{\overline{n}|i} = (1 + i)^n + (1 + i)^{n-1} + (1 + i)^{n-2} + \cdots + (1 + i)^1$$

can also be computed using the formula for summing a geometric series (3.2.2), but it is somewhat simpler to use (3.3.2) and (3.3.5). One arrives at

$$(3.3.6) \quad \begin{aligned} \ddot{s}_{\overline{n}|i} &= (1 + i)^n + (1 + i)^{n-1} + (1 + i)^{n-2} + \cdots + (1 + i)^1 \\ &= \frac{(1+i)^n - 1}{d}. \end{aligned}$$

Note that the formula for the present value $\ddot{a}_{\overline{n}|i}$ given in (3.3.5) is the same as the formula for the present value $a_{\overline{n}|i}$ presented in (3.2.4) except that in the formula for the annuity-due there is a d in the denominator, while the equation for the annuity-immediate has an i. There is a similar statement for the accumulated values. Perhaps you will find the following memory aid helpful. The word "immediate" begins with the letter "i" and the annuity symbols for a basic annuity-immediate have an "i" in the denominator. In contrast, "due" begins with "d" and the annuity symbols for a basic annuity-due have a "d" in the denominator.

The symbols $\ddot{a}_{\overline{n}|i}$ and $\ddot{s}_{\overline{n}|i}$ are closely related to the symbols $a_{\overline{n}|i}$ and $s_{\overline{n}|i}$. Since $\ddot{a}_{\overline{n}|i}$ measures the value of an n-payment basic annuity at the time of the first payment, and $a_{\overline{n}|i}$ measures the value of an n-payment basic annuity one period before the first payment,

$$(3.3.7) \quad \ddot{a}_{\overline{n}|i} = (1 + i)a_{\overline{n}|i}.$$

Similarly,

$$(3.3.8) \quad \ddot{s}_{\overline{n}|i} = (1 + i)s_{\overline{n}|i}.$$

There is a second type of relationship between the annuity-immediate symbols and the annuity-due symbols. The annuity symbol $\ddot{a}_{\overline{n}|i}$ measures the value at time 0 of payments of 1 made at times $0, 1, 2, \ldots, n - 1$, while the annuity symbol $a_{\overline{n-1}|i}$ measures the value at time 0 of payments of 1 made at times $1, 2, \ldots, n - 1$. So, these two symbols give the time 0 value of the same series of payments, except that $\ddot{a}_{\overline{n}|i}$ includes the value of an additional payment of 1 at time 0. Therefore

$$(3.3.9) \quad \ddot{a}_{\overline{n}|i} = a_{\overline{n-1}|i} + 1.$$

Similarly,

$$(3.3.10) \qquad\qquad \ddot{s}_{\overline{n}|i} + 1 = s_{\overline{n+1}|i},$$

since $s_{\overline{n+1}|i}$ measures the value at time n of payments of 1 made at times $1, 2, \ldots, n + 1$, while the annuity symbol $\ddot{s}_{\overline{n}|i}$ measures the value at time n of payments of 1 made at times $1, 2, \ldots, n$.

Given the relations (3.3.7)–(3.3.10) between the annuity-due and annuity-immediate symbols, it is clear that the annuity-due symbols are not strictly speaking essential. However, if you continue your actuarial studies, you will likely encounter heavy use of the annuity-due symbols. The following examples will hopefully convince you of their convenience and encourage you to master them. Moreover, calculators with annuity buttons are usually designed to handle the annuity-due symbols as well as the annuity-immediate symbols, and this adds to their usefulness.

The BA II Plus calculator handles problems with level annuities-due just like those with level annuities-immediate except now one wishes to have "BGN" showing on the display before making **TVM worksheet** computations. The state of having "BGN" displayed is called **BGN mode** ("begin mode") and if the calculator is in END mode, BGN mode is activated by pushing

$$\boxed{\text{2ND}}\ \boxed{\text{BGN}}\ \boxed{\text{2ND}}\ \boxed{\text{SET}}\ \boxed{\text{2ND}}\ \boxed{\text{QUIT}}.$$

This same sequence of keystrokes will also change the calculator's status from BGN mode to END mode.

EXAMPLE 3.3.11

Problem: During 1998, Omar had a dividend payment of $100 directly deposited to his savings account on the first day of each month. Find the accumulated value of these payments at the end of the year if his savings account had a nominal discount rate of 4.8% payable monthly.

Solution His account has a monthly discount rate of 4.8%/12 = .4% = .004. This is equivalent to a monthly interest rate of $\dfrac{.004}{1-.004} = \dfrac{4}{996}$. The accumulated value at the end of the year is therefore $\$100\ddot{s}_{\overline{12}|\frac{4}{996}} = \$100\left(\dfrac{(\frac{1000}{996})^{12}-1}{.004}\right) \approx$ $\$1,231.79$.

BA II Plus calculator solution With the calculator in BGN mode and P/Y = 1 and C/Y = 1, depress the following sequence of keys

$$\boxed{1}\ \boxed{2}\ \boxed{\text{N}}\ \boxed{4}\ \boxed{\div}\ \boxed{9}\ \boxed{9}\ \boxed{6}\ \boxed{=}\ \boxed{\times}\ \boxed{1}\ \boxed{0}\ \boxed{0}$$

$$\boxed{=}\ \boxed{\text{I/Y}}\ \boxed{0}\ \boxed{\text{PV}}\ \boxed{1}\ \boxed{0}\ \boxed{0}\ \boxed{+/-}\ \boxed{\text{PMT}}\ \boxed{\text{CPT}}\ \boxed{\text{FV}}.$$

The calculator display then shows 1,231.791249, and Omar's accumulated value is $1,231.79.   ∎

EXAMPLE 3.3.12

Problem: Dr. Hillary Street began making contributions to a new retirement account on her thirtieth birthday. She made a contribution of $4,000 at the beginning of each year through her sixty-fourth birthday. Starting at age sixty-five and continuing through her eightieth birthday, she made a level withdrawal on her birthday. Find the amount of these withdrawals if they completely exhaust the balance in her account, and the annual effective interest rate is 6% until she is sixty-five, then 5% thereafter.

Solution Let W denote the amount of the level withdrawals. A time diagram for the problem is as follows.

PAYMENT:	$4,000	$4,000	...	$4,000	$-W$	$-W$	...	$-W$
TIME:	30	31	...	64	65	66	...	80
RATE:								

$$6\% = .06 \qquad\qquad 5\% = .05$$

There are thirty-five deposits of $4,000 and the accumulated value of these deposits one period after the last of them, that is to say on her sixty-fifth birthday, is $4,000\ddot{s}_{\overline{35}|.06}$. There are also sixteen withdrawals of W and the value of these at time 65 (her sixty-fifth birthday) is $W\ddot{a}_{\overline{16}|.05}$. Since the withdrawals exactly exhaust the balance, we must therefore have $4,000\ddot{s}_{\overline{35}|.06} = W\ddot{a}_{\overline{16}|.05}$.

Thus, $W = \dfrac{\$4,000\ddot{s}_{\overline{35}|.06}}{\ddot{a}_{\overline{16}|.05}} \approx \dfrac{\$472,483.4667}{11.37965804} \approx \$41,520.01.$

BA II Plus calculator solution With the calculator in BGN mode and P/Y = 1 and C/Y= 1, depress the following sequence of keys

$$\boxed{3}\,\boxed{5}\,\boxed{N}\,\boxed{6}\,\boxed{I/Y}\,\boxed{0}\,\boxed{PV}\,\boxed{4}\,\boxed{0}\,\boxed{0}\,\boxed{0}\,\boxed{+/-}\,\boxed{PMT}\,\boxed{CPT}\,\boxed{FV}$$

to calculate Dr. Street's accumulated value at the time of her sixty-fifth birthday. With the accumulated value still displayed on the calculator, push

$$\boxed{PV}\,\boxed{1}\,\boxed{6}\,\boxed{N}\,\boxed{5}\,\boxed{I/Y}\,\boxed{0}\,\boxed{FV}\,\boxed{CPT}\,\boxed{PMT}.$$

The calculator now shows $-41,520.0057$. The retirement plan pays out to Dr. Street $41,520.01 each year. ∎

3.4 PERPETUITIES

A **perpetuity** is an annuity with an infinite term. That is to say, a perpetuity has payments lasting forever. The basic perpetuity-immediate has payments of 1 at the end of each period, and we denote its present value (calculated using an effective rate of i per payment period) by $a_{\overline{\infty}|i}$. The symbol $a_{\overline{\infty}|i}$ gives the present value (calculated using an effective rate of i per payment period) of the basic perpetuity-due, the perpetuity that has payments of 1 at the beginning of each period. More generally, for an arbitrary accumulation function, the symbol $a_{\overline{\infty}|}$ gives the value one period before the first payment of a perpetuity with payments of 1, and $\ddot{a}_{\overline{\infty}|}$ gives the value of such a perpetuity at the time of the first payment.

The annuity involved in the definition of $a_{\overline{\infty}|}$ has payments of 1 at times 1, 2, 3, ... The annuity involved in the definition of $\ddot{a}_{\overline{\infty}|}$ has all these payments and an additional payment of 1 at time 0. Therefore,

(3.4.1) $$\ddot{a}_{\overline{\infty}|} = 1 + a_{\overline{\infty}|}.$$

Suppose one deposits $\frac{1}{i}$ in a bank account, which pays out interest at the end of each period at an effective rate of $i > 0$ per period. Then, the balance at the beginning of each period is $\frac{1}{i}$, and hence the amount of each interest payment is $(\frac{1}{i})i = 1$. Therefore, it takes $\frac{1}{i}$ remaining on deposit forever (beginning at time 0) to create a basic level annuity-immediate. So,

(3.4.2) $$\boxed{a_{\overline{\infty}|i} = \frac{1}{i}.}$$

It follows from (1.9.4) that $1 + \frac{1}{i} = \frac{1+i}{i} = \frac{1}{d}$. Equations (3.4.1) and (3.4.2) therefore combine to yield

(3.4.3) $$\boxed{\ddot{a}_{\overline{\infty}|i} = \frac{1}{d}.}$$

Those who know calculus should note alternate derivations of (3.4.2) and (3.4.3). The annuity symbols $a_{\overline{\infty}|i}$ and $\ddot{a}_{\overline{\infty}|i}$ equal the sums of infinite geometric series. More precisely,

(3.4.4) $$a_{\overline{\infty}|i} = v + v^2 + v^3 + \cdots + v^n + \cdots \quad ,$$

and

(3.4.5) $$\ddot{a}_{\overline{\infty}|i} = 1 + v + v^2 + v^3 + \cdots + v^n + \cdots \quad .$$

To sum an infinite series, look at the limit of the partial sums. **Assuming $i > 0$,** $0 < v < \frac{1}{1+i}$, and therefore

$$a_{\overline{\infty}|i} = \lim_{n \to \infty} a_{\overline{n}|i} = \lim_{n \to \infty} \frac{1 - v^n}{i} = \frac{1}{i},$$

and

$$\ddot{a}_{\overline{\infty}|i} = \lim_{n \to \infty} \ddot{a}_{\overline{n}|i} = \lim_{n \to \infty} \frac{1 - v^n}{d} = \frac{1}{d}.$$

EXAMPLE 3.4.6

Problem: Gustave Larson has saved $20,000. On January 1, he purchases a perpetuity with annual end-of-year payments. The perpetuity price is based on an annual effective interest rate of 5%. What are the annual payments for Gustave's perpetuity?

Solution   The value on January 1 of a perpetuity with annual end-of-year payments of Q is $Q(\frac{1}{.05}) = 20Q$. Setting this value equal to $20,000, we find $Q = $1,000$. ∎

EXAMPLE 3.4.7

Problem: Norman wishes to leave an inheritance to three charities. The total inheritance is a series of level payments at the beginning of each year forever. He wishes charities A and B (which support medical research) to share the payments equally for ten years, and then all payments revert to charity C (that helps needy children). If the shares received by all three charities have the same value at the time of the bequest, find the annual effective interest rate i.

Solution   Since the shares have the same value at the time of the bequest, they have the same value at the time of the first January 1. Let P denote the amount paid out each January 1. Since charities A and B share the first ten payments equally, each receives a level annuity-due with payments of $\frac{P}{2}$. The value of charity A's share on the first January 1, the time of the first payment, is therefore $\frac{P}{2}\ddot{a}_{\overline{10}|i}$. Recalling (3.3.5), the value of charity A's share may also be expressed as $\frac{P}{2}\left(\frac{1-v^{10}}{d}\right)$. Charity C receives level payments of P beginning exactly ten years after the bequest. The value of charity C's share at the time of its first cash inflow is $\frac{P}{d}$, and the value on the first January 1 is $\frac{P}{d}v^{10}$. Equating the values of charity A's and charity C's shares on that first January 1, we have $\frac{P}{2}(\frac{1-v^{10}}{d}) = \frac{P}{d}v^{10}$. Equivalently, $1 - v^{10} = 2v^{10}$. Therefore, $v^{10} = \frac{1}{3}$ and $i = (1 + i) - 1 = v^{-1} - 1 = 3^{\frac{1}{10}} - 1 \approx .116123174$. ∎

3.5   DEFERRED ANNUITIES AND VALUES ON ANY DATE

In previous sections, we have introduced annuity symbols to measure the present value of an annuity or perpetuity at the time of the first payment or one period before the first payment. We also have symbols for the accumulated value of an annuity at the time of the last payment and at one period after the last payment. The symbols $a_{\overline{n}|i}$, $\ddot{a}_{\overline{n}|i}$, $s_{\overline{n}|i}$, and $\ddot{s}_{\overline{n}|i}$ each measure the value of n consecutive payments of 1, but they do so at different times. We next illustrate the times they give the value, supposing that the first payment is at time $k + 1$.

PAYMENT:		1	1	1	$\cdots$	1	
TIME:	k	$k+1$	$k+2$	$k+3$	$\cdots$	$k+n$	$k+n+1$
VALUE:	$a_{\overline{n}|i}$	$\ddot{a}_{\overline{n}|i}$				$s_{\overline{n}|i}$	$\ddot{s}_{\overline{n}|i}$

If we want the value of an annuity at some other time, we can multiply these symbols by an appropriate power of $1 + i$. For example, the value of the above annuity at time $(k + 5)$ is

$$a_{\overline{n}|i}(1 + i)^5 = \ddot{a}_{\overline{n}|i}(1 + i)^4 = s_{\overline{n}|i}(1 + i)^{5-n} = \ddot{s}_{\overline{n}|i}(1 + i)^{4-n}.$$

However, sometimes it is preferable to avoid introducing powers of $(1 + i)$. Assuming that the evaluation date is an integral number of payment periods from each payment date, an alternative is to represent the value as a sum of two annuity symbols or the difference between two annuity symbols. Exactly how this is done depends on whether the desired value is before, during, or after the term of the annuity. We look at an example in each of these cases.

EXAMPLE 3.5.1   The value before the term of the annuity

Problem: On her eighteenth birthday, Latisha receives an annuity that is to pay $5,000 on her twenty-fifth through thirty-ninth birthdays. Calculate the value of the annuity on her eighteenth birthday using an annual effective interest rate of 5%.

Solution If the annuity had payments on Latisha's nineteenth through thirty-ninth birthdays, its value on her eighteenth birthday would have been $5,000a_{\overline{21}|.05}$. But the first six of these payments (those boxed in the figure below) did not actually take place (and their value is $5,000a_{\overline{6}|.05}$), so the value of Latisha's annuity is $5,000a_{\overline{21}|.05}$ – $5,000a_{\overline{6}|.05} \approx \$64,105.76354$ – $\$25,378.46034 \approx \$38,727.3032$.

PAYMENT:									
		$\boxed{1}$	$\boxed{1}$	$\cdots$	$\boxed{1}$	1	1	$\cdots$	1
TIME:		18	19	20		24	25	26	39
TIME 18 VALUE:			$\$5,000a_{\overline{21}|.05}$		$-$	$\boxed{\$5,000a_{\overline{6}|.05}}$			

This difference of present values is very quick to calculate if your calculator has special buttons to calculate the value of an annuity symbol. (As a historical note, this would also be an attractive method if you depended on interest tables for the value of annuity symbols.) Otherwise, it is just as quick to calculate $(1.05)^{-6}\$5,000a_{\overline{15}|.05} \approx \$38,727.3032$. Yet another possibility is to express the value as a difference of two annuity symbols referring to annuities-due. If we add imaginary payments (boxed in the figure below) on Latisha's eighteenth through twenty-fourth birthdays, the value would be $\$5,000\ddot{a}_{\overline{22}|.05}$. These imaginary payments have value $\$5,000\ddot{a}_{\overline{7}|.05}$, so the value of the original annuity is $\$5,000\ddot{a}_{\overline{22}|.05} - \$5,000\ddot{a}_{\overline{7}|.05}$.

PAYMENT:		1	1	1	$\cdots$	1	1	1	$\cdots$	1
TIME:		18	19	20		24	25	26		39
TIME 18 VALUE:			$\$5,000\ddot{a}_{\overline{22}|.05}$		$-$	$\$5,000\ddot{a}_{\overline{7}|.05}$				

EXAMPLE 3.5.2   The value after the term of the annuity

Problem: On her eighteenth birthday, Latisha receives an annuity that is to pay $\$5,000$ on her twenty-fifth through thirty-ninth birthdays. Calculate the value of the annuity on her fiftieth birthday using an annual effective interest rate of 5%.

Solution   If the annuity had payments on Latisha's twenty-fifth through fiftieth birthdays, its value on her fiftieth birthday would have been $\$5,000s_{\overline{26}|.05}$. But the last eleven of these payments did not actually take place, so the value is $\$5,000s_{\overline{26}|.05} - \$5,000s_{\overline{11}|.05} \approx \$255,567.2688 - 71,033.93581 \approx \$184,533.3333$.

PAYMENT:		1	1	$\cdots$	1	1	1	$\cdots$	1
TIME:		24	25	26		39	40	41		50
TIME 50 VALUE:			$\$5,000s_{\overline{26}|.05}$		$-$	$\$5,000s_{\overline{11}|.05}$			

Alternatively, the value is $(1.05)^{11}\$5,000s_{\overline{15}|.05} \approx \$184,533.3333$. Again, a third possibility is to use symbols corresponding to annuities-due. Add in imaginary payments on Latisha's fortieth through forty-ninth birthdays, and find the value is $\$5,000\ddot{s}_{\overline{25}|.05} - \$5,000\ddot{s}_{\overline{10}|.05}$.

PAYMENT:		1	1	$\cdots$	1	1	1	$\cdots$	1
TIME:		25	26		39	40	41		49
TIME 50 VALUE:			$\$5,000\ddot{s}_{\overline{25}|.05}$		$-$	$\$5,000\ddot{s}_{\overline{10}|.05}$		

EXAMPLE 3.5.3   The value during the term of the annuity

Problem: On her eighteenth birthday, Latisha receives an annuity that is to pay \$5,000 on her twenty-fifth through thirty-ninth birthdays. Calculate the value of the annuity on her thirtieth birthday using an annual effective interest rate of 5%.

Solution   The value of the annuity on her thirtieth birthday is the sum of the value of the payments already made and the value of the payments yet to be made. On her thirtieth birthday she receives her sixth payment and she has nine yet to receive. Thus the value is $\$5,000s_{\overline{6}|.05} + \$5,000a_{\overline{9}|.05} \approx \$34,009.56406 + \$35,539.10838 \approx \$69,548.67244$. Had annuities-due been used, the value would have the expression $\$5,000\ddot{s}_{\overline{5}|.05} + \$5,000\ddot{a}_{\overline{10}|.05}$. You could also have calculated that the value is $(1.05)^6\$5,000a_{\overline{15}|.05} \approx \$69,548.67244$.   ∎

The annuity of Examples (3.5.1)–(3.5.3) is called a **deferred annuity** since there is a wait of more than one payment period for a payment. You can consider an annuity-immediate as an annuity-due deferred for one period. You may see the notation $_{w|n}a$ used to denote the value, w periods before the first payment, of an annuity that pays 1 each period for n periods. The notation $_{w|n}a$ gives the value, $w + 1$ periods before the first payment, of an annuity that pays 1 each period for n periods. With this notation $a_{\overline{n}|} = {}_{1|n}\ddot{a} = {}_{0|n}a$, and Example 3.5.1 asks for $_{7|15}\ddot{a} = {}_{6|15}a$. Note that the methods of this section can also be applied to perpetuities, and that in Example 3.4.7, charity C received a deferred-perpetuity, the value of which we could express as $P_{10|\infty}\ddot{a}$.

In our solutions to Examples (3.5.1) and (3.5.2), we used the important concept of introducing *imaginary payments*. As we did in the diagrams of these solutions, it is our convention that *imaginary payments will be indicated by putting them in boxes*. We note that one way of thinking of an imaginary payment is that at the instant the money is received, an offsetting outgoing payment is made. This viewpoint is seen in the first solution to the following example concerning an annuity with nonlevel payments.

EXAMPLE 3.5.4

Problem: Marlene received a gift of a thirty year annuity-immediate on the day she was born. The annuity pays \$1,000 on her first ten birthdays, \$2,000 on her next ten birthdays, and \$3,000 on the following ten birthdays. (i) If the nominal rate of interest is 5% convertible quarterly, find the value of the annuity on the day she was born. (ii) If the value of the annuity on the day she was born is \$25,000 when calculated using an annual effective rate of interest of j, find j correct to the nearest one-hundredth of a percent.

Solution   (i) Since the annuity payments are made annually, we want the annual effective rate of interest equivalent to $i^{(4)} = 5\%$. This annual effective rate is $i = (1 + \frac{.05}{4})^4 - 1 \approx .050945337$.

The $1,000 payments that Marlene receives at times 1,2,..., 10 may be thought of as a $3,000 payment, $2,000 dollars of which is imaginary, less two imaginary payments of $1,000. The $2,000 payments she receives at times 11,12,..., 20 may be viewed as a $3,000 payment, $1,000 dollars of which is imaginary, less an imaginary payment of $1,000. So, Marlene receives $3,000 payments at times $1, 2, \ldots, 30$ and makes $1,000 payments at times $1, 2, \ldots, 20$ and additional $1,000 payments at times $1, 2, \ldots, 10$. Therefore, the value of Marlene's annuity on the day she was born is $3,000a_{\overline{30}|i} - \$1,000a_{\overline{20}|i} - \$1,000a_{\overline{10}|i} \approx \$45,624.5245 - \$12,362.92174 - \$7,686.408379 \approx \$25,575.19$. Alternatively, the value is

$$\$1,000a_{\overline{10}|i} + (1 + i)^{-10}\$2,000a_{\overline{10}|i} + (1 + i)^{-20}\$3,000a_{\overline{10}|i}$$
$$= (1 + 2(1 + i)^{-10} + 3(1 + i)^{-20})\$1,000a_{\overline{10}|i}$$
$$\approx (3.327327032)(\$7,686.408379)$$
$$\approx \$25,575.19.$$

(ii) To find the rate j, we use the "guess and check method" [see Section (2.3)]. Since the rate i gives a present value that is greater than $25,000, $j > i \approx .0509$. As we guess various rates for j, we may either use $3,000a_{\overline{30}|j} - \$1,000a_{\overline{20}|j} - \$1,000a_{\overline{10}|j}$ or $(1 + 2(1 + j)^{-10} + 3(1 + j)^{-20})\$1,000a_{\overline{10}|j}$ to calculate the corresponding present value. *If you have a calculator with annuity buttons, using the first of these is more efficient.* We try the rate 5.2% and obtain a present value of about $25,183.48, then 5.25% and get $25,000.64. So, j is slightly greater than 5.25%. Since the rate 5.255% yields a present value of $24,982.45, $5.25\% < j < 5.255\%$, and correct to the nearest one hundredth, $j = 5.25\%$.

Cash Flow worksheet solution (i) Open and clear the **Cash Flow worksheet** by depressing $\boxed{\text{CF}}$ $\boxed{\text{2ND}}$ $\boxed{\text{CLR WORK}}$. Then push

$\boxed{\text{ENTER}}$ $\boxed{\downarrow}$ $\boxed{1}\boxed{0}\boxed{0}\boxed{0}$ $\boxed{\text{ENTER}}$ $\boxed{\downarrow}$ $\boxed{1}\boxed{0}$ $\boxed{\text{ENTER}}$ $\boxed{\downarrow}$ $\boxed{2}\boxed{0}\boxed{0}\boxed{0}$

$\boxed{\text{ENTER}}$ $\boxed{\downarrow}$ $\boxed{1}\boxed{0}$ $\boxed{\text{ENTER}}$ $\boxed{\downarrow}$ $\boxed{3}\boxed{0}\boxed{0}\boxed{0}$ $\boxed{\text{ENTER}}$ $\boxed{\downarrow}$ $\boxed{1}\boxed{0}$ $\boxed{\text{ENTER}}$

to set CFo = 0, C01 = 1,000, C02 = 2,000, C03 = 3,000, and F01 = F02 = F03 = 10. Continue by keying

$\boxed{\text{NPV}}$ $\boxed{\bullet}\boxed{0}\boxed{5}$ $\boxed{\div}$ $\boxed{4}$ $\boxed{=}$ $\boxed{+}$ $\boxed{1}$ $\boxed{=}$ $\boxed{y^x}$ $\boxed{4}$ $\boxed{=}$ $\boxed{-}$ $\boxed{1}$

$\boxed{=}$ $\boxed{\times}$ $\boxed{1}\boxed{0}\boxed{0}$ $\boxed{=}$ $\boxed{\text{ENTER}}$ $\boxed{\downarrow}$ $\boxed{\text{CPT}}$.

At this point the calculator displays the net present value 25,575.19437, and the answer to (i) is approximately $25,575.19.

(ii) To compute the desired annual effective interest rate, we continue with our calculator just as it is. Push $\boxed{\text{CF}}$ $\boxed{2}\boxed{5}\boxed{0}\boxed{0}\boxed{0}$ $\boxed{+/-}$ $\boxed{\text{ENTER}}$ to change the setting of CFo so that CFo $= -25{,}000$. Depressing $\boxed{\textbf{IRR}}\,\boxed{\textbf{CPT}}$ displays an approximate value for the desired interest rate as a percent, namely 5.250175513.

$\blacksquare$

3.6   OUTSTANDING LOAN BALANCES

A loan of amount L is to be repaid at the end of n time periods by payments at the end of each period. These payments are to be level except for a slightly reduced final payment if need be. We now consider the important question of how to find the loan balance at an intermediate date between the loan origination date and the date of the final payment. Once we have developed methods to do this, we apply them to problems concerning home mortgages and car loans.

The first method is called the **retrospective method**. The word "retrospective" means "looking back on or directed towards the past." In this method we note that had there been no payments, if the effective interest rate per payment period is i, then the loan balance at time k would be $L(1 + i)^k$. If OLB_k denotes the loan-balance at the end of k payment period, just after any time k payment, and Q denotes the amount of each of the first k payments, then OLB_k is equal to the accumulated value of the original loan amount $L(1 + i)^k$, less the accumulated value $Qs_{\overline{k}|i}$ of the payments that have been made.

$$(3.6.1) \qquad \boxed{\text{OLB}_k = L(1 + i)^k - Qs_{\overline{k}|i} \qquad \text{(Retrospective Method).}}$$

More generally, if the loan is made at time 0 and you desire the balance OLB_k just after the payment at time k, it is given by

$$(3.6.2) \qquad \boxed{\text{OLB}_k = La(k) - Qs_{\overline{k}|} \qquad \text{(Retrospective Method).}}$$

Note that the retrospective method does not require you to first calculate the amount of any irregular final payment. In fact, the only payments that must be level are the first k, and the remaining payments could be haphazard. Moreover, the retrospective method does not require you to know the total number of payments scheduled for the loan repayment.

The second method is referred to as the **prospective method**. "Prospective" means "potential, looking forward in time." In the prospective method we observe that OLB_k should equal the time k value just after the time k payment of the remaining payments, since those payments must pay off the outstanding balance. If Q denotes the amount of all but the last payment, the accumulation function is $a(t) = (1 + i)^t$, and as in (3.2.14) R denotes the amount of the last payment, then the value of the remaining payments is $Qa_{\overline{n-k-1}|}i + R(1 + i)^{n-k}$.

(3.6.3) $\boxed{OLB_k = Qa_{\overline{n-k-1}|i} + R(1 + i)^{n-k}}$ (Prospective Method).

If the payments are all equal (that is if $Q = R$), then we have the simpler formula

(3.6.4) $\boxed{OLB_k = Qa_{\overline{n-k}|i}.}$ (Prospective Method when all payments equal Q).

When all the payments are equal to a known amount Q, accumulation is at a level rate of compound interest i, and the total number of payments n is known, (3.6.4) is the most desirable formula to utilize. Formulas (3.6.3) and (3.6.4) do not require the initial loan balance L to be known.

EXAMPLE 3.6.5

Problem: Sasha is obligated to repay a loan made on the first of the month by paying $80 at the end of each of the next thirty months, including this month. The effective monthly rate on the loan is .4%. Find Sasha's loan balance immediately after her twelfth payment.

Solution   By the prospective method, the desired loan balance is

$$\$80a_{\overline{30-12}|.004.} = \$80\left(\frac{1-(1.004)^{-18}}{.004}\right) \approx \$1,386.71.$$ ∎

EXAMPLE 3.6.6

Problem: A loan of $20,000 is being repaid by payments of $2,500 at the end of each year and a final smaller payment one year after the last $2,500 payment. The annual effective interest rate on the loan is 8%. Find the outstanding loan balance just after the borrower has made payments totaling $15,000.

Solution   We want the outstanding balance OLB_6. According to the retrospective method, $OLB_6 = \$20,000(1.08)^6 - \$2,500s_{\overline{6}|.08} \approx \$31,737.48646 - \$18,339.82259 \approx \$13,397.66387$. (Note that the prospective method was unsuitable since we did not know the number of payments, nor the amount of the last payment.)   ∎

EXAMPLE 3.6.7   Home Mortgage

Problem: Mr. and Mrs. Harper purchase a home for $256,000. They make a down payment of $40,000 and finance the remainder of the purchase price with a thirty-year mortgage at an annual effective interest rate of 6.5%. The Harpers sell their home at the end of eight years, just after having made their ninety-sixth end-of-month mortgage payment. The sales price is $282,000, and the Harpers closing costs are 3% of the selling price. The outstanding loan balance is deducted from the amount the Harpers receive and sent to the lender. How large a check do the Harpers receive, and how much interest did they pay over the eight years?

Solution   The initial loan amount on the Harper's mortgage was $256,000 − $40,000 = $216,000. There are $30 \times 12 = 360$ monthly payments with a thirty-year mortgage. The monthly interest rate is $(1.065)^{\frac{1}{12}} - 1 = j \approx$ .005261694. Therefore, by algorithm (3.2.14), the amount of the first 359 monthly payments is $\frac{\$216,000}{a_{\overline{360}|j}}$ *rounded up* to the nearest penny.   This is calculated to be $1,338.96. Then, by the retrospective method, the outstanding loan balance at the time the Harpers sell the house is $216,000(1.065)^8 −$ $1,338.96 s_{\overline{96}|j} \approx \$357,479.065 - \$166,678.8219 \approx \$190,800.24$. The closing costs reduce the Harper's check amount by ($282,000)(.03) = $8,460. Putting this all together, the Harpers get a check for $282,000 − $190,800.24 − $8,460 = $82,739.76. The interest the Harpers paid is the difference in the total payments (in this case $96 \times \$1,338.96 = \$128,540.16$) and the change in the debt ($216,000 − $190,800.24 = $25,199.76). So, the amount of interest they paid is $128,540.16 − $25,199.76 = $103,340.40.   ∎

To calculate the interest paid in Example (3.6.7), we used the following important fact.

> **IMPORTANT FACT (3.6.8):** The amount of interest paid over a period may be calculated by first determining the difference between the outstanding balance at the beginning of the period and the outstanding balance at the end of the period. This is called the **amount of principal paid**. The amount of interest paid is obtained by subtracting the amount of principal paid from the total amount paid.

Formulas (3.6.1)–(3.6.4) apply when all of the payments are made as originally scheduled. In real life, this does not always happen. The next example illustrates how one can compute an outstanding loan balance if there have been missed payments.

EXAMPLE 3.6.9   Missed payments

Problem: Waswate took out a sixty-month car loan with an interest rate of 3% convertible monthly. To repay the loan, Waswate was to make monthly payments of $252.65. During the first three years of the loan, he made all of the payments as scheduled except that he neglected to pay the fourteenth and thirtieth. What is the outstanding balance at the end of three years, just after he has made the payment scheduled at that time?

Solution   To calculate the sought after outstanding loan balance, we imagine that Waswate made all thirty-six payments he was scheduled to make during the first three years and calculate what the balance would be by the prospective method.   To obtain the actual balance, we must add on the accumulated value at time 36 months of the two skipped payments.   We thus find that the outstanding balance at the end of three years is $252.65a_{\overline{60-36}|.0025}$ + $252.65(1.0025)^{36-14}$ + $252.65(1.0025)^{36-30} \approx$ $5,878.149739 + $266.9166657 +$256.463515 \approx$ $6401.53. (Note that the prospective method was used to calculate the outstanding loan balance assuming the two missed payments had been made, since this was simpler than applying the retrospective method. The original loan amount was not given.) ∎

We next consider an example in which we have level payments, none of which are missed, but we have the complication of an interest rate change.

EXAMPLE 3.6.10   Varying interest rate

Problem: Suppose that a loan is to be repaid by twenty end-of-quarter payments of $1,000. The interest rate for the first two years is 6% convertible quarterly, and for the last three years it is 8% convertible quarterly. Find the loan amount, the outstanding loan balance just after the sixth payment, and the outstanding loan balance immediately after the fifteenth payment.

Solution   The loan amount is

$$\$1,000a_{\overline{8}|.015} + (1.015)^{-8}\$1,000a_{\overline{12}|.02} \approx \$7,485.925 + \$9,387.848 \approx \$16,873.77.$$

For use in the following timeline, we set $L = 16.87377$, the loan amount expressed in thousands of dollars. In the timeline, due to space constraints, payments and balances are given in *thousands of dollars*. Time is given in *quarters*.

PAYMENT: 1   1 $\cdots$ 1   1   1   1   1 $\cdots$ 1   1   1   1   1   1

TIME: **0**   1   2 $\cdots$ **6**   7   8   9   10 $\cdots$ **15**   16   17   18   19   20

quarterly interest rate 1.5%   quarterly interest rate 2%

BALANCE: L $L(1.015)^6 - s_{\overline{6}|1.5\%}$ $a_{\overline{5}|2\%}$

time **6** balance (retrospective)   time **15** balance (perspective)

We prefer the retrospective method for computing the balance just after the sixth payment because the initial loan balance has already been calculated and, looking back from the time 6 to the time of the loan initiation, the quarterly interest rate was level at 1.5%. The outstanding loan balance just after the sixth payment is

$$OLB_6 = \$16{,}873.77(1.015)^6 - \$1{,}000s_{\overline{6}|.015} \approx \$18{,}450.510 - \$6{,}229.551 \approx \$12{,}220.96.$$

Note that had we not already calculated the loan amount, we might have preferred to use the prospective method

$$OLB_6 = \$1{,}000a_{\overline{2}|.015} + (1.015)^{-2}\$1{,}000a_{\overline{12}|.02}$$
$$\approx \$1{,}955.883 + \$10{,}265.079 \approx \$12{,}220.96.$$

In any case, the prospective method most easily yields the outstanding loan balance immediately after the fifteenth payment since the quarterly interest rate stays level at 2% from time 15 quarters through the end of the loan five quarters later. In fact,

$$OLB_{15} = \$1{,}000a_{\overline{5}|.02} \approx \$4{,}713.46.$$

■

Outstanding loan balances may also be calculated using the **Amortization worksheet** of the BA II Plus calculator. This worksheet will be discussed in greater depth in Section (5.2). For the time being, we note that the worksheet should only be opened after a compatible set of five values is stored in the five **TVM** registers. Here the last of these is usually entered by computing its value. The Amortization worksheet is opened by keying $\boxed{\text{2ND}}$ $\boxed{\text{AMORT}}$ and once it is opened "P1 = " appears. Pushing $\boxed{\downarrow}$ repeatedly causes the entries "P2 = ", "BAL = ", "PRN = ", and "INT =" to appear in that order. One more push of $\boxed{\downarrow}$ would cause "P1 = " to reappear and further depressing of $\boxed{\downarrow}$ would cause the entries to continue to cycle. (The key $\boxed{\uparrow}$ may be used to cycle in the opposite order.) Numbers for the values of P1 and P2 must be entered or left at their previous values, the default entries both being 1. Together the values of P1 and P2 give a range of payment numbers to focus on, the P1 entry being the first of these numbers and the P2 entry being the last. (You will get the error message "ERROR 2" if you have entered a larger value in the P2 register than is stored in the P1 register and then proceed to either the BAL or INT registers. Registers P1 and P2 will only accept positive integers, but should you attempt to enter a number that is at least .5, the calculator will enter the nearest integer, rounding up if there is a tie.) The BAL entry is an automatically calculated amount. It represents the balance just after the payment number stored as P2, and INT (again an automatically calculated entry) will give the total amount of interest paid in the designated payment range. [PRN will give the amount paid in the assigned payment range that is not interest, but we will leave further discussion of this to Section (5.2).]

The solutions to the problems posed in Examples (3.6.5) (3.6.7) may all be checked using the **Amortization worksheet**. For each we indicate how this can be done, assuming at all times that BA II Plus calculator is in END mode with P/Y = C/Y = 1.

Amortization worksheet solution to (3.6.5) Fill the five **TVM** worksheet entries by pushing

| 3 | 0 | N | • | 4 | I/Y | 0 | FV | 8 | 0 | +/− | PMT | CPT | PV |

Now push | 2ND | AMORT | ↓ | 1 | 2 | ENTER | ↓ |. At this point "BAL = 1,386.709088" will be displayed.

Amortization worksheet solution to (3.6.6) To use the **Amortization work-sheet** here, you first need to figure out the number of payments. Key

| 8 | I/Y | 2 | 0 | 0 | 0 | 0 | PV | 0 | FV | 2 | 5 | 0 | 0 | +/− | PMT | CPT | N |.

At this point "N = 13.27491459" is displayed. We therefore will need fourteen payments, the last being less than $2,500. Push | 1 | 4 | N | CPT | FV |. This cause the amount the last payment is reduced to be displayed and stored in the FV register. Now open the **Amortization worksheet** by depressing | 2ND | AMORT | and once it is opened push | ↓ | 6 | ENTER | to set P2 = 6. Then pushing | ↓ | displays "BAL = 13,397.66387".

Amortization worksheet solution to (3.6.7) We begin by working with the **TVM worksheet**. Push | 3 | 6 | 0 | N | 1 | • | 0 | 6 | 5 | y^x | 1 | 2 | 1/x | = | − | 1 | × | 1 | 0 | | 0 | = | I/Y | 2 | 1 | 6 | 0 | 0 | 0 | PV | 0 | FV | CPT | PMT | to obtain "PMT = −1,338.957713". We wish to round up the payment so push | 1 | 3 | 3 | 8 | • | 9 | 6 | | +/− | PMT |, then | CPT | FV | so that the **TVM worksheet** holds a set of consistent values. Keying the sequence $\boxed{\text{2ND}}$ $\boxed{\text{AMORT}}$ $\boxed{\downarrow}$ $\boxed{9}$ $\boxed{6}$ $\boxed{\text{ENTER}}$ $\boxed{\downarrow}$ gives you "BAL $= 190{,}800.243$". Pushing $\boxed{\downarrow}$ $\boxed{\downarrow}$ shows "INT $= -103{,}340.403$".

3.7   NONLEVEL ANNUITIES

In Example (3.5.4) we looked at an annuity that was not level (that is, the payments varied) but for which level annuities were used in the present value calculation. In this section we look at more examples of nonlevel annuities for which the payments form a pattern that is neither geometric [see Section (3.8)] nor arithmetic [see Section (3.9)].

EXAMPLE 3.7.1

Problem: On July 1, Dimas wins a lottery. His prize is a twenty-year annuity-due with payments of $1,000 each July 1st and payments of $2,000 each January 1. If the payments are all left to accumulate in a new account earning interest at an annual effective interest rate of 5%, what is the accumulated value exactly twenty years after Dimas won the lottery?

Solution   There are twenty annual (July 1) payments of $1,000, and the accumulation date we are interested in is one year after the last of these payments. Therefore, the accumulated value of the $1,000 payments is $1,000\ddot{s}_{\overline{20}|.05} \approx \$34{,}719.252$.

There are twenty annual (January 1) payments of $2,000 and the accumulated value of these at the time of the last payment is $2,000s_{\overline{20}|.05}$. The accumulation date we are interested in is six months later, at which point the $2,000 payments have accumulated to $(1.05)^{\frac{1}{2}}\$2,000s_{\overline{20}|.05} \approx \$67{,}765.041$.

Exactly twenty years after Dimas won the lottery, his prize payments have accumulated to about $\$34{,}719.252 + \$67{,}765.041 \approx \$102{,}484.29$.   ∎

EXAMPLE 3.7.2

Problem: An annuity lasts for twelve calendar years. At the end of each quarter, there is a payment. First quarter payments are $200, second quarter payments are $300, third quarter payments are $150, and fourth quarter payments are $400. Find the accumulated value of this annuity just after the last payment using a nominal quarterly interest rate of 7.2%.

Solution   The quarterly interest rate is $\frac{7.2\%}{4} = 1.8\%$. Therefore, at the end of each year the accumulated value of that year's payments are $Y = \$200(1.018)^3 + \$300(1.018)^2 + \$150(1.018) + \$400 \approx \$1{,}074.592766$.

The annual effective interest rate is $i = (1.018)^4 - 1 \approx 7.396743298\%$. So, the accumulated value of the annuity at the end of the twelfth year is the accumulated value of these accumulated values Y, namely $Ys_{\overline{12}|i} \approx \$19,677.64$. ∎

The **Cash Flow worksheet** of the BA II Plus calculator is ideal for solving many problems with nonlevel payments, but is ill-suited for direct use for the problems of Examples (3.7.1) and (3.7.2) because the payments cycle. (Alternating payments present a cycle of length two, while (3.7.2) has a cycle of length four.) The **Cash Flow worksheet** may be used indirectly if the cycle length is no more than twenty-four. For instance, in Problem (3.7.2) you can use the **Cash Flow worksheet** and $\boxed{\text{NPV}}$ to calculate the value of each year's payments at the beginning of each year. Then enter this amount in the **TVM** and use BGN mode to calculate the desired accumulated value.

Here is a loan problem with nonlevel payments for which the BA II Plus calculator's **Cash Flow worksheet** is a perfect tool.

EXAMPLE 3.7.3   Cash Flow worksheet

Problem: Dr. Sara Hamilton makes monthly contributions to her savings account which has a nominal discount rate of 5.4% convertible monthly. During her first year of practice, at the end of each month she deposits $600 and the following year the end-of-month deposits are level at $1,000. During Dr. Hamilton's third year, the sequence of her end-of-month contributions begins with $1,200, $1,400, −$5,000, $2,000, $2,200, and then ends with seven deposits of $1,600. Find the balance of Dr. Hamilton's savings account at the end of the three years.

Solution   Open and clear the **Cash Flow worksheet** by pushing $\boxed{\text{CF}}$ $\boxed{\text{2ND}}$ $\boxed{\text{CLR WORK}}$. Then push

$\downarrow$ $\boxed{6}$ $\boxed{0}$ $\boxed{0}$ $\boxed{\text{ENTER}}$ $\downarrow$ $\boxed{1}$ $\boxed{2}$ $\boxed{\text{ENTER}}$ $\downarrow$ $\boxed{1}$ $\boxed{0}$ $\boxed{0}$ $\boxed{0}$ $\boxed{\text{ENTER}}$

$\downarrow$ $\boxed{1}$ $\boxed{2}$ $\boxed{\text{ENTER}}$ $\downarrow$ $\boxed{1}$ $\boxed{2}$ $\boxed{0}$ $\boxed{0}$ $\boxed{\text{ENTER}}$ $\downarrow$ $\downarrow$ $\boxed{1}$ $\boxed{4}$ $\boxed{0}$ $\boxed{0}$ $\boxed{\text{ENTER}}$

$\downarrow$ $\downarrow$ $\boxed{5}$ $\boxed{0}$ $\boxed{0}$ $\boxed{0}$ $\boxed{+/-}$ $\boxed{\text{ENTER}}$ $\downarrow$ $\downarrow$ $\boxed{2}$ $\boxed{0}$ $\boxed{0}$ $\boxed{0}$ $\boxed{\text{ENTER}}$

$\downarrow$ $\downarrow$ $\boxed{2}$ $\boxed{2}$ $\boxed{0}$ $\boxed{0}$ $\boxed{\text{ENTER}}$ $\downarrow$ $\downarrow$ $\boxed{1}$ $\boxed{6}$ $\boxed{0}$ $\boxed{0}$ $\boxed{\text{ENTER}}$ $\downarrow$ $\boxed{7}$ $\boxed{\text{ENTER}}$.

At this point Dr. Hamilton's contributions are all properly stored in the **Cash Flow worksheet** so it is time to compute their net present value, remembering that the nominal rate .054 we are given is a **discount** rate convertible monthly. The computation of the net present value may be accomplished by depressing

$\boxed{\text{NPV}}$ $\boxed{\bullet}$ $\boxed{0}$ $\boxed{5}$ $\boxed{4}$ $\boxed{\div}$ $\boxed{1}$ $\boxed{2}$ $\boxed{=}$ $\boxed{+/-}$ $\boxed{+}$ $\boxed{1}$ $\boxed{=}$ $\boxed{1/x}$ $\boxed{-}$ $\boxed{1}$ $\boxed{=}$

$\boxed{\text{STO}}$ $\boxed{0}$ $\boxed{\times}$ $\boxed{1}$ $\boxed{0}$ $\boxed{0}$ $\boxed{=}$ $\boxed{\text{ENTER}}$ $\downarrow$ $\boxed{\text{CPT}}$.

Now the monthly effective interest rate is stored in register 0 in decimal notation and "NPV = 29,268.50864" is displayed. To move from this time 0 value to the desired time thirty-six months value, one may key

| STO | 1 | 2ND | QUIT | RCL | 0 | + | 1 | y^x | 3 | 6 | = | × | RCL | 1 | = |.

The time three years (thirty-six months) accumulation is now displayed in dollars. Rounding the displayed 34,428.26021, you see that Dr. Hamilton's account has a balance of $34,428.26 at the end of her third year of practice. Of course the choices of registers number 0 and 1 is not required here. Any two distinct registers would be fine. ■

3.8 ANNUITIES WITH PAYMENTS IN GEOMETRIC PROGRESSION

Problems concerning the value of an annuity whose payments form a geometric progression are best solved by noting that the value of the annuity is the sum of the values of the individual payments of the annuity, and these values also form a geometric progression. So Equation (3.2.2), which gives the sum of a geometric series, may be used to compute the desired value of the annuity.

EXAMPLE 3.8.1

Problem: On June 15, 1975, Roy purchased an annuity-immediate with annual payments for twenty-five years. The first payment was $800 and the payments increased by 3% each year. The purchase price was based on an annual effective interest rate of 7%. Find this price.

Solution The price is the June 15, 1975 value of the annuity payments. The k-th payment occurred k years after the purchase and had amount $\$800(1.03)^{k-1}$. So, the value of this payment on June 15, 1975 was $\$800(1.03)^{k-1}(1.07)^{-k}$. Therefore, the June 15 value of the annuity is

$$\sum_{k=1}^{25} \$800(1.03)^{k-1}(1.07)^{-k} = \sum_{k=1}^{25} \$800(1.07)^{-1}\left(\frac{1.03}{1.07}\right)^{k-1}$$

$$= \$800(1.07)^{-1}\left[1 + \left(\frac{1.03}{1.07}\right) + \left(\frac{1.03}{1.07}\right)^2 + \cdots + \left(\frac{1.03}{1.07}\right)^{24}\right]$$

$$= \$800(1.07)^{-1}\left(\frac{1 - \left(\frac{1.03}{1.07}\right)^{25}}{1 - \left(\frac{1.03}{1.07}\right)}\right) = \$800\left(\frac{1 - \left(\frac{1.03}{1.07}\right)^{25}}{1.07 - 1.03}\right)$$

$$\approx \$12,284.46.$$

■

In Example (3.8.1), the 3% rate meant that each payment was for an amount 1.03 times that of the previous payment. On the other hand, to compute the present value of a payment, we multiply by an extra $\frac{1}{1.07}$ beyond what we needed to compute the present value of the previous payment; this was because we need to bring back the value one more year. Hence, when looking at the present values of the payments, we found they formed a sequence where each term was $\frac{1.03}{1.07}$ times the previous term. Their sum therefore was a geometric series.

More generally, suppose that an annuity has payments where each payment is $1 + g$ times its predecessor. Let i denote the effective interest rate for the payment period. If the annuity is an annuity-immediate lasting for n payment periods with initial payment P and $i \neq g$, then the value of the annuity one period before the first payment is

$$\sum_{k=1}^{n} P(1 + g)^{k-1}(1 + i)^{-k} = \sum_{k=1}^{n} P(1 + i)^{-1}\left(\frac{1 + g}{1 + i}\right)^{k-1}$$

$$= P(1 + i)^{-1}\left[1 + \left(\frac{1 + g}{1 + i}\right) + \left(\frac{1 + g}{1 + i}\right)^2 + \cdots + \left(\frac{1 + g}{1 + i}\right)^{n-1}\right]$$

$$= P(1 + i)^{-1}\left(\frac{1 - \left(\frac{1+g}{1+i}\right)^n}{1 - \left(\frac{1+g}{1+i}\right)}\right) = P(1 + i)^{-1}\left(\frac{1 - \left(\frac{1+g}{1+i}\right)^n}{\frac{i-g}{1+i}}\right)$$

$$= P\left(\frac{1 - \left(\frac{1+g}{1+i}\right)^n}{i - g}\right).$$

Alternatively, if $j = \frac{i-g}{1+g}$, then $v_j = \frac{1}{1+j} = \frac{1}{1 + \frac{i-g}{1+g}} = \frac{1}{\frac{(1+g)+(i-g)}{1+g}} = \frac{1+g}{1+i}$, and the value of the annuity one period before the first payment is given by

$P(1 + i)^{-1}\left(\frac{1-v_j^n}{1-v_j}\right) = P(1 + i)^{-1}\ddot{a}_{\overline{n}|j}$. [In Example (3.8.1), $g = .03$, $i = .07$, and $j = \frac{.04}{1.03} \approx 3.883495146\%$.]

Note that if the situation is as described in the previous paragraph but $i = g$, then the calculation of the value of the annuity one period before the first payment is fine until we use formula (3.2.2) to sum the geometric series. But if $i = g$,

$$\left[1 + \left(\frac{1 + g}{1 + i}\right) + \left(\frac{1 + g}{1 + i}\right)^2 + \cdots + \left(\frac{1 + g}{1 + i}\right)^{n-1}\right] = \underbrace{1 + 1 + \cdots + 1}_{n \text{ times}} = n,$$

and therefore the value of the annuity one period before the first payment is $nP(1 + i)^{-1}$.

> **SUMMARY 3.8.2**: Suppose that an annuity has payments with each payment equal to $(1 + g)$ times its predecessor. Let i denote the effective interest rate for the payment period. If the annuity is an annuity-immediate lasting for n payment periods, the initial payment is P, and $i \neq g$, then the value of the annuity one payment period before the first payment is $P\left(\dfrac{1-\left(\frac{1+g}{1+i}\right)^n}{i-g}\right)$. This value is also equal to $P(1 + i)^{-1}\ddot{a}_{\overline{n}|j}$ where $j = \frac{i-g}{1+g}$. If $i = g$, the value is $nP(1 + i)^{-1}$.

Our next example involves a perpetuity rather than an annuity, and it has the further complication of a change in interest rate.

EXAMPLE 3.8.3

Problem: On January 1, 2002, Andrea inherits a perpetuity-immediate with annual payments. The first payment is $2,000 and after that the payments increase by 2% each year. Find the value of this perpetuity on January 1, 2002 if the annual effective rate of interest is 3% from January 1, 2002 through January 1, 2009 and 4% thereafter.

Solution Indicating the payment amounts in thousands of dollars, a time diagram for this problem is as follows:

PAYMENT		2	2(1.02)	$\cdots$	2(1.02)⁶	2(1.02)⁷	2(1.02)⁸	$\cdots$
TIME:	02	03	04	$\cdots$	09	10	11	$\cdots$
RATE:								

3% = .03   4% = .04

The January 1, 2002 value of the payments that occur in 2003 through 2009 is

$$\$2,000(1.03)^{-1}\left[1 + \left(\frac{1.02}{1.03}\right) + \left(\frac{1.02}{1.03}\right)^2 + \cdots + \left(\frac{1.02}{1.03}\right)^6\right]$$

$$= \$2,000(1.03)^{-1}\left(\frac{1 - \left(\frac{1.02}{1.03}\right)^7}{1 - \left(\frac{1.02}{1.03}\right)}\right)$$

$$= \$200,000\left[1 - \left(\frac{1.02}{1.03}\right)^7\right]$$

$$\approx \$13,202.68689.$$

The January 1, 2002 value of the deferred perpetuity that begins with the 2010 payment is

$$(1.03)^{-7}\$2{,}000(1.02)^7(1.04)^{-1}\left[1 + \left(\frac{1.02}{1.04}\right) + \left(\frac{1.02}{1.04}\right)^2 + \cdots\right]$$

$$= (1.03)^{-7}\$2{,}000(1.02)^7(1.04)^{-1}\left(\frac{1}{1 - \frac{1.02}{1.04}}\right)$$

$$\approx \$93{,}398.65656.$$

Thus Andrea's inheritance has a January 1, 2002 value of about $13,202.68688 +$93,398.65656 ≈ $106,601.34. ■

3.9   ANNUITIES WITH PAYMENTS IN ARITHMETIC PROGRESSION

Consider an annuity lasting n interest periods with a payment of $P + Q(j - 1)$ at the end of the j-th interest period. The first payment is P and the payments increase by the constant amount Q each interest period, hence form an "arithmetic progression." We introduce the annuity symbol $(I_{P,Q}a)_{\overline{n}|i}$ to denote the present value of this annuity, and the annuity symbol $(I_{P,Q}s)_{\overline{n}|i}$ denotes its accumulated value at the time of the last payment.

PAYMENT:		P	$P + Q$	$P + 2Q$	$\cdots$	$P + (n-1)Q$
TIME:	0	1	2	3	$\cdots$	n
VALUE:	$(I_{P,Q}a)_{\overline{n}|i}$					$(I_{P,Q}s)_{\overline{n}|i}$

FIGURE (3.9.1)

Our next task is to derive a concise formula for the accumulated value $(I_{P,Q}s)_{\overline{n}|i}$. The accumulated value $(I_{P,Q}s)_{\overline{n}|i}$ is obtained by adding the accumulated values of the individual payments. Therefore,

(3.9.2)
$$\begin{aligned}(I_{P,Q}s)_{\overline{n}|i} = {}&P(1 + i)^{n-1} + (P + Q)(1 + i)^{n-2}\\&+ (P + 2Q)(1 + i)^{n-3}\\&+ \cdots + (P + (n - 1)Q).\end{aligned}$$

Multiplying Equation (3.9.2) by $(1 + i)$, we have

$$\begin{aligned}(1 + i)(I_{P,Q}s)_{\overline{n}|i} = {}&P(1 + i)^n + (P + Q)(1 + i)^{n-1}\\&+ (P + 2Q)(1 + i)^{n-2}\\&+ \cdots + (P + (n - 1)Q)(1 + i).\end{aligned}$$

This may be combined with Equation (3.9.2) to obtain

$$i(I_{P,Q}s)_{\overline{n}|i} = (1 + i)(I_{P,Q}s)_{\overline{n}|} - (I_{P,Q}s)_{\overline{n}|}$$

$$= P(1 + i)^n + Q(1 + i)^{n-1} + Q(1 + i)^{n-2} + \cdots +$$
$$Q(1 + i) - (P + (n - 1)Q)$$

$$= P[(1 + i)^n - 1] + Q[(1 + i)^{n-1} + (1 + i)^{n-2} + \cdots +$$
$$(1 + i) + 1] - Qn$$

$$= P[(1 + i)^n - 1] + Q(s_{\overline{n}|i} - n).$$

So, dividing by i yields the equation

$$(I_{P,Q}s)_{\overline{n}|i} = P\left[\frac{(1 + i)^n - 1}{i}\right] + \frac{Q}{i}(s_{\overline{n}|i} - n).$$

Recalling Equation (3.2.10), this may be rewritten as

(3.9.3)
$$\boxed{(I_{P,Q}s)_{\overline{n}|i} = Ps_{\overline{n}|i} + \frac{Q}{i}(s_{\overline{n}|i} - n).}$$

This completes our task of finding a simple formula for $(I_{P,Q}s)_{\overline{n}|i}$. Moreover, if you multiply Equation (3.9.3) by v^n, you have the companion equation

(3.9.4)
$$\boxed{(I_{P,Q}a)_{\overline{n}|i} = Pa_{\overline{n}|i} + \frac{Q}{i}(a_{\overline{n}|i} - nv^n).}$$

When $P = Q = 1$, it is customary to write $(Ia)_{\overline{n}|i}$ for $(I_{P,Q}a)_{\overline{n}|i}$ and $(Is)_{\overline{n}|i}$ for $(I_{P,Q}s)_{\overline{n}|i}$. Then, according to (3.9.3),

$$(Is)_{\overline{n}|i} = (I_{1,1}s)_{\overline{n}|i} = s_{\overline{n}|i} + \frac{1}{i}(s_{\overline{n}|i} - n).$$

On the other hand,

$$s_{\overline{n}|i} + \frac{1}{i}(s_{\overline{n}|i} - n) = \frac{is_{\overline{n}|i} + s_{\overline{n}|i} - n}{i} = \frac{((1 + i)^n - 1) + s_{\overline{n}|i} - n}{i}$$

$$= \frac{s_{\overline{n+1}|i} - (n + 1)}{i}.$$

Recalling (3.3.10) $s_{\overline{n+1}|i} - 1 = \ddot{s}_{\overline{n}|i}$, we have

(3.9.5)
$$\boxed{(Is)_{\overline{n}|i} = \frac{s_{\overline{n+1}|i} - (n + 1)}{i} = \frac{\ddot{s}_{\overline{n}|i} - n}{i}.}$$

Multiplying (3.9.5) by v^n yields

(3.9.6)
$$(Ia)_{\overline{n}|i} = \frac{\ddot{a}_{\overline{n}|i} - nv^n}{i}.$$

There is also a special notation for $(I_{P,Q}a)_{\overline{n}|i}$ and for $(I_{P,Q}s)_{\overline{n}|i}$ when $P = n$ and $Q = -1$, that is, when the sequence of payments is $n, n - 1, n - 2,$ ..., 2, 1. We let $(Da)_{\overline{n}|i} = (I_{n,-1}a)_{\overline{n}|i}$ and $(Ds)_{\overline{n}|i} = (I_{n,-1}s)_{\overline{n}|i}$. The D reminds us that this is a decreasing annuity. Observe that (3.9.4) gives us

$$(Da)_{\overline{n}|i} = na_{\overline{n}|i} - \frac{1}{i}(a_{\overline{n}|i} - nv^n).$$

On the other hand,

$$na_{\overline{n}|i} - \frac{1}{i}(a_{\overline{n}|i} - nv^n)$$

$$= \frac{nia_{\overline{n}|i} - a_{\overline{n}|i} + nv^n}{i} = \frac{n(1 - v^n) - a_{\overline{n}|i} + nv^n}{i} = \frac{n - a_{\overline{n}|i}}{i}.$$

Therefore,

(3.9.7)
$$(Da)_{\overline{n}|i} = \frac{n - a_{\overline{n}|i}}{i}.$$

Of course, one can multiply this by $(1 + i)^n$ to obtain a formula for $(Ds)_{\overline{n}|i}$.

EXAMPLE 3.9.8

Problem: Susan receives an annuity for her eighteenth birthday. The annuity pays $2,000 on her nineteenth birthday and then has an annual payment on each of her birthdays through her thirty-fifth birthday. Each year the payments increase by $500. Find the value of the annuity on Susan's eighteenth birthday, assuming the annual effective rate of interest is 4.2%.

Solution   The desired value is $(I_{\$2,000,\$500}a)_{\overline{17}|.042}$. According to (3.9.4), this is equal to $\$2,000a_{\overline{17}|.042} + \frac{\$500}{.042}(a_{\overline{17}|.042} - 17(1.042)^{-17})$. Calculating, we find that this is approximately $23,958.24444 + $42,050.18436 \approx $66,008.43$. ∎

EXAMPLE 3.9.9

Problem: At the end of each year Mr. Dunn deposits $6,000 into an investment fund. The fund pays out interest each year at an annual effective interest rate of 6%. Mr. Dunn is only able to reinvest this interest at an annual effective

interest rate of 4%. Assuming Mr. Dunn continues to make his $6,000 deposits
for twenty years, that he always immediately deposits all the interest paid
out into the 4% account, and that he makes no withdrawals, what is the
accumulated value of Mr. Dunn's investments at the time of the last deposit?

Solution   Mr. Dunn makes twenty $6,000 annual payments into the invest-
ment fund.   Refer to the times he made these as times 1, 2, ..., 20. If
$k \in \{1, 2, ..., 20\}$, then during the year that begins at time k, Mr. Dunn
earns 6% interest on k $6,000 deposits.   Therefore, at the end of the
$(k + 1)$-st year (which begins at time k), he earns $k(.06)(\$6,000) = \$360k$.
These earnings are deposited into his 4% savings account.   Thus, during the nineteen-year period from time 1 to time 20, the deposits into
the 4% account form an annuity-immediate which is $360 times the one
whose accumulated value is denoted $(Is)_{\overline{19}|.04}$.   So, the balance in the

4% account at time 20 is $360(Is)_{\overline{19}|.04} = \$360 \left(\dfrac{\ddot{s}_{\overline{19}|.04} - 19}{.04} \right) \approx \$88,002.71$.

There is also $20(\$6,000) = \$120,000$ in the 6% fund, so at the time of
Mr. Dunn's last deposit (time 20), the accumulated value of his investments is
$88,002.71 + \$120,000 = \$208,002.71$.   ∎

Heretofore, the studied annuities with payments in arithmetic progression have all been annuities-immediate. However, we can also define annuity
symbols for annuities-due. Let $(I_{P,Q}\ddot{a})_{\overline{n}|i}$ denote the present value of an annu-
ity lasting n interest periods of an annuity with a payment of $P + Q(k - 1)$
at the beginning of the k-th interest period, and let $(I_{P,Q}\ddot{s})_{\overline{n}|i}$ denote its value
at the end of the n-th period.

PAYMENT:	P	$P + Q$	$P + 2Q$	$\cdots$	$P + (n-1)Q$	
TIME:	0	1	2	$\cdots$	$n-1$	n
VALUE:	$(I_{P,Q}\ddot{a})_{\overline{n}|i}$					$(I_{P,Q}\ddot{s})_{\overline{n}|i}$

FIGURE (3.9.10)

Comparing Figures (3.9.1) and (3.9.10) and recalling (3.9.3), we see that

$$(I_{P,Q}\ddot{s})_{\overline{n}|i} = (1 + i)(I_{P,Q}s)_{\overline{n}|i} = (1 + i)\left[Ps_{\overline{n}|i} + \frac{Q}{i}(s_{\overline{n}|i} - n) \right].$$

Since $(1 + i)s_{\overline{n}|i} = \ddot{s}_{\overline{n}|i}$ (3.3.8) and $d = \frac{i}{1+i}$ (1.9.4),

(3.9.11) $\boxed{(I_{P,Q}\ddot{s})_{\overline{n}|i} = P\ddot{s}_{\overline{n}|i} + \dfrac{Q}{d}(s_{\overline{n}|i} - n).}$

Multiplying (3.9.11) by v^n yields

$$(3.9.12) \qquad (I_{P,Q}\ddot{a})_{\overline{n}|i} = P\ddot{a}_{\overline{n}|i} + \frac{Q}{d}(a_{\overline{n}|i} - nv^n).$$

Set

$$(I\ddot{a})_{\overline{n}|i} = (I_{1,1}\ddot{a})_{\overline{n}|i}, \ (I\ddot{s})_{\overline{n}|i} = (I_{1,1}\ddot{s})_{\overline{n}|i}, \ (D\ddot{a})_{\overline{n}|i} = (I_{n,-1}\ddot{a})_{\overline{n}|i},$$
$$\text{and } (D\ddot{s})_{\overline{n}|i} = (I_{n,-1}\ddot{s})_{\overline{n}|i}.$$

If one multiplies equation (3.9.5) by $(1 + i)$, one finds

$$(3.9.13) \qquad (I\ddot{s})_{\overline{n}|i} = \frac{s_{\overline{n+1}|i} - (n + 1)}{d} = \frac{\ddot{s}_{\overline{n}|i} - n}{d}.$$

Similarly, from (3.9.7) one obtains

$$(3.9.14) \qquad (D\ddot{a})_{\overline{n}|i} = \frac{n - a_{\overline{n}|i}}{d}.$$

Note that $(Ia)_{\overline{n}|i} + (Da)_{\overline{n}|i} = (n + 1)a_{\overline{n}|i}$, and $(I\ddot{a})_{\overline{n}|i} + (D\ddot{a})_{\overline{n}|i} = (n + 1)\ddot{a}_{\overline{n}|i}$. These equalities give us alternate derivations of equations (3.9.7) and (3.9.14) [see Problem (3.9.8)].

EXAMPLE 3.9.15

Problem: Alfonso has an annuity that will pay $4,000 on October 30, 2010 and has annual payments through October 30, 2024. The payments increase by $800 each year. As soon as Alfonso receives a payment, he will deposit it in a savings account with a 4% annual effective interest rate, and he will not make any withdrawals. What will be the balance in Alfonso's 4% account on October 30, 2025?

Solution　The annuity has fifteen payments and we are looking for an accumulated value one period after the final payment. When $i = .04, d = \frac{.04}{1.04}$. The desired value is therefore $(I_{\$4,000,\$800}\ddot{s})_{\overline{15}|.04} = \$4,000\ddot{s}_{\overline{15}|.04} + \frac{\$800}{.04/1.04}(s_{\overline{15}|.04} - 15)$ $\approx \$83,298.12457 + \$20,800(20.02358764 - 15) \approx \$187,788.75.$ ■

　　Perpetuities with payments in arithmetic progression also deserve mention. In this case, we replace the n in our present value annuity symbols by ∞. For example, let $(I_{P,Q}a)_{\overline{\infty}|i}$ denote the present value of a perpetuity-immediate that has an initial payment of P and payments increasing by Q each interest period, and $(I\ddot{a})_{\overline{\infty}|i}$ signify the present value of an annuity-due with a payment of k in the k-th interest period.

We may take limits (*calculus!*) in (3.9.4) and (3.9.12). In each case, we will need $\lim_{n \to \infty} n v^n$. Using l'Hospital's rule, we find

$$\lim_{n \to \infty} n v^n = \lim_{n \to \infty} \frac{n}{(1+i)^n} = \lim_{n \to \infty} \frac{1}{(1+i)^n \ln n} = 0.$$

Recalling (3.4.2), from (3.9.4) we obtain,

(3.9.16) $$\boxed{(I_{P,Q}a)_{\overline{\infty}|i} = P a_{\overline{\infty}|i} + \frac{Q}{i} a_{\overline{\infty}|i} = \frac{P}{i} + \frac{Q}{i^2}.}$$

Likewise, referencing (3.4.2) and (3.4.3), (3.9.12) gives us

(3.9.17) $$\boxed{(I_{P,Q}\ddot{a})_{\overline{\infty}|i} = P \ddot{a}_{\overline{\infty}|i} + \frac{Q}{d} a_{\overline{\infty}|i} = \frac{P}{d} + \frac{Q}{id}.}$$

Equation (3.9.17) may also be derived by multiplying (3.9.16) by $(1+i)$. You may choose just to memorize (3.9.16), but be forewarned that the expression for $(I_{P,Q}\ddot{a})_{\overline{\infty}|i}$ is *not* just obtained by replacing the i in the expression for $(I_{P,Q}a)_{\overline{\infty}|i}$ by a d.

EXAMPLE 3.9.18

Problem: Rafael purchases a perpetuity-immediate. The perpetuity pays $1,000 at the end of each of the first eleven years and then has payments that increase by $180 each year. So, the payment at time 11 is $1,000, and the payment at time 12 is $1,180. The purchase price was determined using a 5% annual effective interest rate. Find Rafael's purchase price.

Solution Think of the perpetuity as a level annuity-immediate paying $1,000 for ten years, followed by a deferred perpetuity-immediate. The perpetuity is deferred for ten years, has an initial payment of $1,000, and has payments that increase by $180 each year. The ten-year annuity has a present value of $1,000 a_{\overline{10}|.05} \approx \$7,721.73$ and the deferred perpetuity has a present value

$$(1.05)^{-10}(I_{\$1,000,\$180}a)_{\overline{\infty}|.05} = (1.05)^{-10}\left(\frac{\$1,000}{.05} + \frac{\$180}{(.05)^2}\right)$$

$$= (1.05)^{-10}(\$20,000 + \$72,000) \approx \$56,480.02.$$

So, the purchase price is $\$7,721.73 + \$56,480.02 = \$64,201.75$.   ∎

This section contains formulas for computing present value of annuities with payments in arithmetic progression. In particular, we draw your attention to Equation (3.9.4), along with the more specialized Equation (3.9.6). We now wish to explain how, given these equations, you may use your BA II Plus calculator to efficiently calculate the quantitites $(Ia)_{\overline{n}|i}$, and $(I_{P,Q}a)_{\overline{n}|i}$.

According to equation (3.9.6), to calculate $(Ia)_{\overline{n}|i}$, you just need to calculate $\ddot{a}_{\overline{n}|i} - nv^n$ and then divide your answer by the interest rate i. This can be done in BGN mode by first entering n in the $\boxed{\text{N}}$ register and also in the $\boxed{\text{FV}}$ register, i as a percent in the $\boxed{\text{I/Y}}$ register, -1 in the $\boxed{\text{PMT}}$ register, and then keying $\boxed{\text{CPT}}\,\boxed{\text{PV}}$; next divide by i as a decimal rather than as a percent.

EXAMPLE 3.9.19

Problem:

Use the BA II Plus calculator to compute $(Ia)_{\overline{180}|.5\%}$.

Solution Put your calculator in BGN mode and key

$$\boxed{1}\,\boxed{8}\,\boxed{0}\,\boxed{\text{N}}\,\boxed{\text{FV}}\,\boxed{\bullet}\,\boxed{5}\,\boxed{\text{I/Y}}\,\boxed{1}\,\boxed{+/-}\,\boxed{\text{PMT}}\,\boxed{\text{CPT}}\,\boxed{\text{PV}}\,.$$

At this point you will have computed $\ddot{a}_{\overline{180}|.5\%} - 180(1.005)^{180}$, and the display should read "PV $= 45.74919544$". To complete the calculation, divide by $i = .005$ by pressing $\boxed{\div}\,\boxed{\bullet}\,\boxed{0}\,\boxed{0}\,\boxed{5}$. This gives the result \$9,149.839088. ∎

Let's now move on to the more general $(I_{P,Q}a)_{\overline{n}|i}$. You may rewrite equation (3.9.4) as

$$(I_{P,Q}a)_{\overline{n}|i} = \left(P + \frac{Q}{i}\right)a_{\overline{n}|i} - \left(\frac{Qi}{n}\right)v^n.$$

So, you may compute $(I_{P,Q}a)_{\overline{n}|i}$ by putting $-\left(P + \frac{Q}{i}\right)$ in the $\boxed{\text{PMT}}$ register, $\frac{Qi}{n}$ in the $\boxed{\text{FV}}$ register, i in the $\boxed{\text{I/Y}}$ register, n in the $\boxed{\text{N}}$ register, and then pressing $\boxed{\text{CPT}}\,\boxed{\text{PV}}$.

Here we have calculated the present value of an n-period annuity-immediate that pays P in the first period and whose payments increase by Q each subsequent period. Should you wish to find the accumulated value at the end of the period, since equation (3.9.3) is equivalent to

$$(I_{P,Q}a)_{\overline{n}|i} = \left(P + \frac{Q}{i}\right)s_{\overline{n}|i} - \left(\frac{Qi}{n}\right),$$

you may easily compute $\left(P + \frac{Q}{i}\right)s_{\overline{n}|i}$, and then subtract $\frac{Qi}{n}$. $(I_{P,Q}\ddot{a})_{\overline{n}|i}$ and $(I_{P,Q}\ddot{s})_{\overline{n}|i}$ are perhaps most easily obtained by first obtaining $(I_{P,Q}a)_{\overline{n}|i}$ and $(I_{P,Q}s)_{\overline{n}|i}$ respectively, and then mulitplying by $(1 + i)$.

3.10 YIELD RATE EXAMPLES INVOLVING ANNUITIES

In Chapter 2 we introduced the yield rate. We now look at several yield rate problems that involve annuities.

EXAMPLE 3.10.1

Problem: Melanie pays $24,000 and receives a fifteen-year annuity with end-of-year payments of $2,100. What is Melanie's annual yield on this investment?

Solution   The time 0 equation of value describing Melanie's investment is

$$\$24{,}000 = \$2{,}100 a_{\overline{15}|i}.$$

Recalling (3.2.4), we see that this equation is equivalent to the equation $24{,}000i - 2{,}100(1 - (1 + i)^{-15}) = 0$. So, the unknown yield rate is the root of the function $f(x) = 240x - 21(1 - (1 + x)^{-15})$. This root may be approximated by the "guess and check method" or by Newton's method. More easily, it may quickly be found in two different ways using financial buttons on the BA II Plus calculator. All four solutions are now presented.

Solution by the "guess and check" method   Melanie receives a single payment of $24,000 and her repayments total $15 \times \$2{,}100 = \$31{,}500$. Were she to have made one lump repayment of $31,500 at the end of eight years (the average of her repayment times), the yield rate would have been $(31{,}500/24{,}000)^{\frac{1}{8}} \approx 3.5\%$ so it is reasonable to start the "guess and check method" with an initial guess of 3.5%. We calculate that $f(.035) \approx -.065$. The guess is therefore close, but since f evaluated at this initial guess is negative, looking at the definition of f indicates we should raise our guess for i a little. Check that $f(.036) \approx -.0056$, $f(.0361) \approx .00056$, and $f(.03609) \approx -.00005$. So, the rate is between .03609 and .0361, probably closer to .03609. If need be, we can continue refining our guesses to get any needed degree of accuracy.

Solution by Newton's method   Once again, we begin with an initial guess $x_1 = .035$. Note that $f'(x) = 240 - 315(1 + x)^{-16}$, $x_2 = x_1 - \dfrac{f(x_1)}{f'(x_1)} \approx .035 - \dfrac{-.065297009}{58.33763781} \approx .036119295$, and $x_3 = x_2 - \dfrac{f(x_2)}{f'(x_2)} \approx .036119295 - \dfrac{.001748424}{61.45225324} \approx .036090843$. One may continue to find $x_4 = x_3 - \dfrac{f(x_3)}{f'(x_3)} \approx .036090825$, so .0360908 is a very good estimate. In fact $f(.0360908) \approx -.00000152$.

Solution using the TVM In End Mode with $C/Y = P/Y = 1$, key

$$\boxed{2}\,\boxed{4}\,\boxed{0}\,\boxed{0}\,\boxed{0}\,\boxed{+/-}\,\boxed{\textbf{PV}}\,\boxed{1}\,\boxed{5}\,\boxed{\textbf{N}}\,\boxed{2}\,\boxed{1}\,\boxed{0}\,\boxed{0}$$

$$\boxed{\textbf{PMT}}\,\boxed{0}\,\boxed{\textbf{FV}}\,\boxed{\textbf{CPT}}\,\boxed{\textbf{I/Y}}$$

to obtain $I/Y = 3.609082476$.

Solution using the Cash Flow worksheet   Push $\boxed{\textbf{CF}}\,\boxed{\textbf{2ND}}\,\boxed{\textbf{CLR WORK}}$ to open and clear the **Cash Flow worksheet**. Then press

$$\boxed{2}\,\boxed{4}\,\boxed{0}\,\boxed{0}\,\boxed{0}\,\boxed{+/-}\,\boxed{\textbf{ENTER}}\,\boxed{\downarrow}\,\boxed{2}\,\boxed{1}\,\boxed{0}\,\boxed{0}\,\boxed{\textbf{ENTER}}$$

$$\boxed{\downarrow}\,\boxed{1}\,\boxed{5}\,\boxed{\textbf{ENTER}}\,\boxed{\textbf{IRR}}\,\boxed{\textbf{CPT}}$$

to obtain IRR $= 3.609082476$.   ■

The next two examples involve reinvestment (partial or complete) of annuity payments received.

EXAMPLE 3.10.2

Problem: Gian Carlo invests \$58,000 and receives an annuity of \$7,000 at the end of each year for thirteen years. Each time Gian Carlo gets a \$7,000 payment, he immediately deposits \$4,000 in a savings account that earns 9%. Find the annual yield received by Gian Carlo.

Solution   Recall that one takes a "bottom line approach" when computing a yield rate. So, the first task is to determine the amounts and times of all expenditures and receipts by the investor. The only expenditure is the initial investment of \$58,000. We denote the time of this payment time 0. As for receipts, he receives \$3,000 = \$7,000 − \$4,000 at times 1, 2, ...13. He also receives the balance of the 9% savings account at time 13. This balance is $\$4,000 s_{\overline{13}|.09}$, or to the nearest penny \$91,813.54. We therefore have the time 0 equation of value

$$\$58,000 = \$3,000 a_{\overline{13}|i} + (\$91,813.54)v^{13}.$$

As in the previous example, the yield rate may be found using the "guess and check method," Newton's method, or more easily, using the annuity buttons on a business calculator.

Solution by the "guess and check" method   By (3.4.2), the fact that i satisfies the above equation of value is equivalent with i being a root of the function

$$f(x) = 3,000 - 58,000x + (1 + x)^{-13}(91,813.54x - 3,000).$$

The investor contributes \$58,000 and has repayments totaling $13 \times \$7,000 = \$91,000$. The average time of repayment is seven years so looking for an initial guess of the yield rate, we might calculate $(91,000/58,000)^{\frac{1}{7}} - 1 \approx 6.6\%$.

But the investor's reinvestment rate is 9%, so a reasonable initial rate is somewhere in-between these rates, say their average, 7.8%. Now calculate that $f(.078) \approx 43.47790784$. Due to the equation of value, the fact that the evaluated value is positive leads us to consider a higher guess. Calculate that $f(.08) \approx -42.32192483$. So, the yield rate is in-between .078 and .08, and .079 is apt to be close. Check that $f(.079) \approx .866$ and $f(.0791) \approx -3.4$ so the yield rate is between 7.9% and 7.91%, probably closer to 7.9%. Continue to the needed degree of accuracy.

Solution by Newton's method   As in the previous solution, we begin with an initial guess $x_1 = .078$. Note that

$$f'(x) = -58{,}000 - 13(1 + x)^{-14}(91{,}813.54x - 3{,}000) + 91{,}813.54(1 + x)^{-13},$$

and $x_2 = x_1 - \dfrac{f(x_1)}{f'(x_1)} = .078 - \dfrac{43.47790784}{-42{,}319.78073} \approx .079027366$. This is already an excellent estimate, but we could repeat to find the yield rate with greater accuracy.

Solution using the TVM   In END mode with $C/Y = P/Y = 1$, push

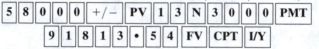

$$\boxed{5}\boxed{8}\boxed{0}\boxed{0}\boxed{0}\boxed{+/-}\boxed{\text{PV}}\boxed{1}\boxed{3}\boxed{\text{N}}\boxed{3}\boxed{0}\boxed{0}\boxed{0}\boxed{\text{PMT}}$$
$$\boxed{9}\boxed{1}\boxed{8}\boxed{1}\boxed{3}\boxed{\bullet}\boxed{5}\boxed{4}\boxed{\text{FV}}\boxed{\text{CPT}}\boxed{\text{I/Y}}$$

to obtain I/Y = 7.902018167. We note that for this last method to work, it was essential that Gian Carlo closed the savings account at time 13, the time of his last annuity payment.

Solution using the Cash Flow worksheet   Push

$$\boxed{\text{CF}}\boxed{\text{2ND}}\boxed{\text{CLR WORK}}$$

to open and clear the Cash Flow worksheet. Then press

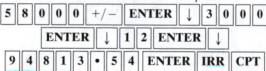

$$\boxed{5}\boxed{8}\boxed{0}\boxed{0}\boxed{0}\boxed{+/-}\boxed{\text{ENTER}}\boxed{\downarrow}\boxed{3}\boxed{0}\boxed{0}\boxed{0}$$
$$\boxed{\text{ENTER}}\boxed{\downarrow}\boxed{1}\boxed{2}\boxed{\text{ENTER}}\boxed{\downarrow}$$
$$\boxed{9}\boxed{4}\boxed{8}\boxed{1}\boxed{3}\boxed{\bullet}\boxed{5}\boxed{4}\boxed{\text{ENTER}}\boxed{\text{IRR}}\boxed{\text{CPT}}$$

to obtain IRR = 7.902018167. The justification for entering 12 for C01 is that the total inflows to the investor at times $1, 2, \ldots, 12$ are all \$3,000, but at time 13 the inflows total $\$3{,}000 + \$91{,}813.54 = \$94{,}813.54$. Thus the proper entry for C02 is 94,813.54. ∎

EXAMPLE 3.10.3

Problem: Serena invests a total of \$10,000. She uses part of the \$10,000 to purchase an annuity with payments of \$1,000 at the beginning of each year for ten years. The purchase price of the annuity is based on an effective interest rate of 8%. As annuity payments are received, they are reinvested at an annual effective interest rate of 7%. The balance of Serena's \$10,000 is invested in a ten-year certificate of deposit with a nominal annual interest rate of 9% convertible quarterly. Calculate the annual effective yield on the entire $10,000 investment over the ten-year period.

Solution   The value of the annuity is $1,000\ddot{a}_{\overline{10}|.08} \approx \$7,246.887911$. Since the purchase price must be an integral number of cents, it is $7,246.89. The amount invested in the certificate of deposit is therefore $10,000 − $7,246.89 = $2,753.11. At time 10, since Serena invests her annuity payments at 7%, she has accumulated $1,000\ddot{s}_{\overline{10}|.07} \approx \$14,783.60$ plus the value of the certificate of deposit which is $2,753.11(1 + \frac{.09}{4})^{40} \approx \$6,704.34$. The total accumulation is therefore $14,783.60 + $6,704.34 = $21,487.94, and her yield is i where $10,000(1 + i)^{10} = \$21,487.94$. So, $i = (21,487.94/10,000)^{\frac{1}{10}} - 1 \approx 7.949212452\%$. ∎

In the next example, it is the expenditures that form an annuity.

EXAMPLE 3.10.4

Problem: Hideo deposits $1,500 at the beginning of each year for sixteen years in a fund earning an annual effective rate of interest of 6%. The interest from this fund is paid out annually and can only be reinvested at an effective annual rate of 5.2%. At the end of twenty years, Hideo liquidates his assets. What is Hideo's yield rate for the twenty year-period?

Solution   The balance in the 6% account at time 20 is $16 \times \$1,500 = \$24,000$ since interest is paid out from this account. The balance in the 5.2% account is more interesting. At time k, $k \in \{1, 2, 3, ..., 20\}$, there is an interest payment from the 6% account deposited in the 5.2% account. If $k \in \{1, 2, 3, ..., 16\}$, the amount of this deposit is $(\$1,500k) \times .06 = \$90k$. This is because for $k \in \{1, 2, 3, ..., 16\}$, the amount on deposit in the 6% account in the k-th year is $\$1,500k$. On the other hand, if $k \in \{17, 18, 19, 20\}$, the amount on deposit in the 6% account in the k-th year stays constant at $16 \times \$1,500 = \$24,000$ and the amount deposited to the 5.2% account is $.06 \times \$24,000 = \$1,440$.

PAYMENT:	$1 \times \$90$	$2 \times \$90$	$\cdots$	$16 \times \$90$	$\$1,440$	$\cdots$	$\$1,440$
TIME:	0	1	2	$\cdots$	16	17	$\cdots$	20

Payments to the 5.2% account

To calculate the balance in the 5.2% account, we first note that the time 16 value of the first sixteen payments is $\$90(Is)_{\overline{16}|.052}$, whence the time 20 value of these payments is $(1.052)^4\$90(Is)_{\overline{16}|.052} = (1.052)^4\$90\left(\dfrac{s_{\overline{16}|.052}-16}{.052}\right) \approx$ $\$19,705.412$. The remaining four payments to the 5.2% account combine to form a level annuity, the time 20 accumulated value of which is $\$1,440 s_{\overline{4}|.052} \approx$ $\$6,225.058$. The time 20 balance in the 5.2% account is hence approximately $\$19,705.412 + \$6,225.058 = \$25,930.47$.

The liquidated value of Hideo's assets at the end of the twenty years is the sum of the balance in the 6% fund and the balance in the 5.2% account, namely $\$24,000 + \$25,930.47 = \$49,930.47$.

For the calculation of Hideo's yield rate, take our usual bottom line approach. Hideo pays $\$1,500$ at $t = 0, 1, 2, \dots, 15$, and gets $\$49,930.47$ at $t = 20$, and so has investments whose time 20 equation of value is

$$(1 + i)^5 1,500 s_{\overline{16}|i} = (1 + i)^4 1,500 \ddot{s}_{\overline{16}|i} = \$49,930.47.$$

The rate i may now be found using the "guess and check" method, Newton's method, or most easily the **Cash Flow worksheet** of the BA II Plus calculator. **Solution by the "guess and check" method** The time twenty equation of value may be rewritten as

$$\$1,500(1 + i)^{21} - \$1,500(1 + i)^5 - \$49,930.47i = 0.$$

The yield rate is therefore the root of the function

$$f(x) = 1,500(1 + x)^{21} - 1,500(1 + x)^5 - 49,930.47x.$$

Hideo's deposits total $16 \times \$1,500 = \$24,000$ and his liquidated value after twenty years is $\$49,930.47$. The average time of his deposit is at $t = 7.5$ so we might try i where $24,000(1 + i)^{20-7.5} \approx 49,930.47$, say $i \approx .06$. But $f(.06) \approx 96.2$, and consequently $i < .06$. Calculate that $f(.05) \approx -232$. We therefore know that $.05 < i < .06$, and it appears that i is closer to .06. Further calculations show $f(.057) \approx -20$ and $f(.058) \approx 16.6$, from which we conclude that $5.7\% < i < 5.8\%$. As usual, the method may be continued to get a smaller interval in which the yield rate is guaranteed to be located. **Solution by Newton's method** Let $f(x)$ be the same as in the "guess and check" method. Then

$$f'(x) = 31,500(1 + x)^{20} - 7,500(1 + x)^4 - 49,930.47.$$

As justified in the previous solution, we begin with an initial guess $x_1 = .06$. Then $x_2 = .06 - \dfrac{f(.06)}{f'(.06)} \approx .06 - \dfrac{96.17883442}{41,625.72017} \approx .057689437$. Continuing

$$x_3 \approx .057689437 - \frac{f(.057689437)}{f'(.057689437)} \approx .057689437 - \frac{4.923393435}{37,393.80106} \approx .057557774.$$

The rate is close to 5.76%.

Solution using the Cash Flow worksheet Push $\boxed{\text{CF}}$ $\boxed{\text{2ND}}$ $\boxed{\text{CLR WORK}}$ to open and clear the **Cash Flow worksheet**. Then press

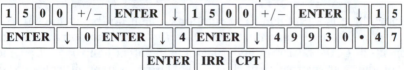

to obtain IRR = 5.755735603. Here, the reason we entered CF₀ = −1,500 is that at time 0, Hideo pays out $1,500. This is followed by fifteen successive payments of $1,500, so C01 = −1,500 and F01 = 15. But then there are no inflows or outflows at the four successive times 16, 17, 18, and 19, so we want C02 = 0 and F02 = 4. Finally, Hideo gets $49,930.47 at time 20, and therefore C03 = 49,930.47 and F03 = 1. ∎

3.11 ANNUITY SYMBOLS FOR NONINTEGRAL TERMS

We defined the annuity symbol $a_{\overline{n}|i}$ to be the value one period before the first payment of an annuity paying 1 at the end of each interest period for n periods. For this definition to make sense, n must be a nonnegative integer. On the other hand, we found

(3.2.4)
$$a_{\overline{n}|i} = \frac{1 - v^n}{i}.$$

Equation (3.2.4) suggests how we might define $a_{\overline{r}|i}$ for an arbitrary positive real number r.

(3.11.1)
$$\boxed{\,a_{\overline{r}|i} = \frac{1 - v^r}{i} \quad \text{for any positive real number } r.\,}$$

   To see how this symbol might arise in problem solving, consider the following situation. A lender loans a borrower an amount L. The two parties agree that the loan will be repaid by end-of-period payments of Q and that the effective interest for the payment period is to be i throughout the term of the loan. The borrower and the lender neglect the calculation of the number of payments r needed, then realize their omission and set out to calculate r. They realize that they should solve the equation

$$L = Qa_{\overline{r}|i}.$$

By (3.11.1), this equation is equivalent to the equation $L = Q(\frac{1-v^r}{i})$ and hence to the equation $v^r = 1 - \frac{iL}{Q}$. Taking natural logarithms of each side of this equation gives $r \ln v = \ln (1 - \frac{iL}{Q})$. Therefore

(3.11.2)
$$r = \frac{\ln(1 - \frac{iL}{Q})}{\ln v} = -\frac{\ln(1 - \frac{iL}{Q})}{\ln(1 + i)}.$$

But now there is a problem. The solution r might not be an integer. Write $r = n + f$ where n is an integer and $0 \leqslant f < 1$. If the borrower makes n payments of Q, the borrower underpays. On the other hand, $n + 1$ payments of Q produces an overpayment. To resolve this repayment difficulty, consider the following.

$$L = Qa_{\overline{r}|i} = Q\left(\frac{1 - v^r}{i}\right) = Q\left(\frac{1 - v^n + v^n - v^r}{i}\right)$$

$$= Q\left(\frac{1 - v^n}{i}\right) + Q\left(\frac{v^n - v^r}{i}\right)$$

$$= Qa_{\overline{n}|i} + Qv^r\left(\frac{(1 + i)^f - 1}{i}\right)$$

$$= Qa_{\overline{n}|i} + Q\left(\frac{(1 + i)^f - 1}{i}\right)v^r.$$

So, L is the sum of the present value of an annuity with n end-of-period payments of Q and the present value of a payment of amount $Q\left(\frac{(1+i)^f - 1}{i}\right)$ at time r periods. Therefore, the borrower can pay off the loan by making payments of Q at the ends of periods 1 through n and then one payment of $Q\left(\frac{(1+i)^f - 1}{i}\right)$ an fth of the way through the $(n + 1)$-st period.

Alternatively, the borrower can make n payments of Q at the ends of periods 1 through n and then, at the end of the $(n + 1)$-st period, make a payment of

(3.11.3)
$$Q\left(\frac{(1 + i)^f - 1}{i}\right)(1 + i)^{1-f}.$$

This $(n + 1)$-st payment is called a **drop payment**.

A second alternative is for the borrower to make $(n - 1)$ payments of Q and then a payment at the end of period n of

(3.11.4)
$$Q + Q\left(\frac{(1 + i)^f - 1}{i}\right)v^f.$$

This one-time larger payment is called a **balloon payment**.

EXAMPLE 3.11.5

Problem: Jamaar borrows $9,600 from Friendly Financing in order to purchase a car. The account representative who handles Jamaar's loan application determines that Jamaar can afford monthly payments of $345. Morover, the nominal interest rate convertible monthly on Jamaar's loan will be 5.4%.

(a) If the Friendly account representative arranges the loan so that there are N end-of-month payments of $345, followed by a drop payment, find N and the amount of the drop payment? (Note that the drop payment is the $(N + 1)$-st payment.)

(b) If the loan is structured so that it has $(N - 1)$ level monthly payments of $345, followed by a balloon payment, find the amount of the balloon payment. Here, N is as in part (a).

(c) If the loan has twenty-three level payments of $345, followed by a larger final payment at the end of the 24th month, find the amount of the final payment. This final payment is also called a balloon payment.

Solution (a) With $L = \$9,600$, $Q = \$345$, and an interest rate of $\frac{5.4\%}{12} = .45\%$, Equation (3.11.2) gives

$$r = -\ln\left(1 - \frac{(.0045)(\$9,600)}{\$345}\right)\Big/\ln(1.0045) \approx 29.79570004.$$

Therefore, there are $N = 29$ level payments of $345 followed by the drop payment. Using Equation (3.11.3), we calculate that Jamaar's 30th payment is for approximately

$$\$345\left(\frac{(1.0045)^{.79570004} - 1}{.0045}\right)(1.0045)^{1-.79570004} \approx \$274.6423618.$$

The amount of the drop payment is $274.64. Alternatively, you can note that if you made thirty payments of $345, you would have a credit of $345s_{\overline{30}|.0045} - (1.0045)^{30}\$9,600 \approx \$70.36$. So, the 30-th payment need only be for $345 - \$70.36 = \$274.64$.

(b) We now wish to calculate the amount of the 29-th payment, assuming that this payment follows twenty-eight level end-of-month payments of $345, and that it is exactly sufficient to pay off the loan. According to formula (3.11.4), the amount of this larger payment is

$$\$345 + \$345\left(\frac{(1.0045)^{.79570004} - 1}{.0045}\right)(1.0045)^{-.79570004} \approx \$618.4120078.$$

So, the amount of the balloon payment is $618.41.

(c) Using the retrospective method, the balance at time 24 months just prior to any final payment is $9,600(1.0045)^{24} - (1.0045)\$345s_{\overline{23}|.0045} \approx$ $10,692.26759 - \$8,377.97033 \approx \$2,314.297258$. The time 24 months balloon payment is $2,314.30.

Note that, as we did in part (c), the retrospective method could also have been used to figure the balances in parts (a) and (b). So, there is no need to memorize formulas (3.11.3) and (3.11.4).   ■

There are also times when it is useful to have definitions of our other annuity symbols for nonintegral terms. In each case we avoid defining the symbol as the value of an annuity with specified payments. Instead, we define the annuity symbol by an algebraic expression. For example, we define

(3.11.6)
$$s_{\overline{r}|i} = \frac{(1+i)^r - 1}{i} \quad \text{for any positive real number } r,$$

(3.11.7)
$$\ddot{a}_{\overline{r}|i} = \frac{1 - v^r}{d} \quad \text{for any positive real number } r,$$

and

(3.11.8)
$$\ddot{s}_{\overline{r}|i} = \frac{(1+i)^r - 1}{d} \quad \text{for any positive real number } r.$$

The annuity symbols of Section (3.9) may also be extended to nonintegral terms using the algebraic expressions given by (3.9.3) and (3.9.4).

Of course calculators that have built-in annuity buttons (such as the BA II Plus calculator) are programmed to make calculations with formulas such as (3.11.1), (3.11.6), (3.11.7), and (3.11.8), whether or not the number of periods is integral.

3.12   ANNUITIES GOVERNED BY GENERAL ACCUMULATION FUNCTIONS

We now consider our annuity symbols when there is a general accumulation function governing the growth of money. This assertion means that we have a single accumulation function $a(t)$ rather than having multiple ones dependent on the time of deposit. [So that you will appreciate the importance of having a single accumulation function, at the end of this section we include Example (3.12.10) where this assumption is dropped. We also see multiple accumulation functions, each of which is a compound interest accumulation function, in the investment year method studied in Section (3.13).] With our assumption that there is a single accumulation function $a(t)$ governing investments, there is a single discount function $v(t) = \frac{1}{a(t)}$. Then we have the present value equations

(3.12.1) $$a_{\overline{n}|} = v(1) + v(2) + \cdots + v(n),$$

and

(3.12.2) $$\ddot{a}_{\overline{n}|} = 1 + v(1) + v(2) + \cdots + v(n-1).$$

The annuity-immediate and annuity-due symbols are related by the equation

(3.12.3) $$\ddot{a}_{\overline{n}|} = 1 + a_{\overline{n-1}|}.$$

Those familiar with calculus should note that there are corresponding equations for the basic perpetuities, but convergence of the series appearing in them is by no means assured. In general there need not be a simple formula like (3.2.4) or (3.3.5) for the sums in (3.12.1) and (3.12.2), but in some special cases there are.

EXAMPLE 3.12.4

Problem: Given that $a(t) = \frac{1}{2}t^2 + \frac{3}{2}t + 1$, find formulas for $\ddot{a}_{\overline{n}|}$ and $a_{\overline{n}|}$.

Solution   There is a "trick" that works in this special case; namely using a "partial fractions decomposition." First note that $a(t) = \frac{1}{2}t^2 + \frac{3}{2}t + 1 = \frac{1}{2}(t+1)(t+2)$. Therefore,

$$v(t) = \frac{2}{(t+1)(t+2)} = \frac{2}{t+1} - \frac{2}{t+2}.$$

Equation (3.12.1) tells us that

$$a_{\overline{n}|} = \left(\frac{2}{2} - \frac{2}{3}\right) + \left(\frac{2}{3} - \frac{2}{4}\right) + \left(\frac{2}{4} - \frac{2}{5}\right) + \cdots + \left(\frac{2}{n+1} - \frac{2}{n+2}\right).$$

This is a "telescoping series," and therefore we may rearrange the above terms to obtain a simpler expression for $a_{\overline{n}|}$. We have

$$a_{\overline{n}|} = \frac{2}{2} + \left(-\frac{2}{3} + \frac{2}{3}\right) + \left(-\frac{2}{4} + \frac{2}{4}\right) + \left(-\frac{2}{5} + \frac{2}{5}\right)$$
$$+ \cdots + \left(-\frac{2}{n+1} + \frac{2}{n+1}\right) - \frac{2}{n+2}$$
$$= 1 - \frac{2}{n+2}.$$

Replacing n with $n - 1$, we find $a_{\overline{n-1}|} = 1 - \dfrac{2}{n+1}$. Consequently, Equation (3.12.3) gives $\ddot{a}_{\overline{n}|} = 2 - \dfrac{2}{n+1}$.

Note involving calculus: The sequences $\{a_{\overline{n}|}\}$ and $\{\ddot{a}_{\overline{n}|}\}$ converge to 1 and 2, respectively, as n approaches infinity. So, $a_{\overline{\infty}|} = 1$ and $\ddot{a}_{\overline{\infty}|} = 2$. ∎

The accumulated value annuity symbols $s_{\overline{n}|}$ and $\ddot{s}_{\overline{n}|}$ may be obtained from the present value annuity symbols. More specifically,

(3.12.5)
$$s_{\overline{n}|} = a(n)a_{\overline{n}|} \quad \text{and} \quad \ddot{s}_{\overline{n}|} = a(n)\ddot{a}_{\overline{n}|}.$$

Alternatively, these symbols are given by the series

(3.12.6)
$$s_{\overline{n}|} = \frac{a(n)}{a(1)} + \frac{a(n)}{a(2)} + \cdots + \frac{a(n)}{a(n)},$$

and

(3.12.7)
$$\ddot{s}_{\overline{n}|} = \frac{a(n)}{a(0)} + \frac{a(n)}{a(1)} + \cdots + \frac{a(n)}{a(n-1)}.$$

EXAMPLE 3.12.8

Problem: Suppose that deposits of \$500 are made at the end of each year for five years into a fund that is governed by the simple discount accumulation function $a(t) = \dfrac{1}{1-.04t}$. Find the accumulated value at the time of the last deposit.

Solution   The accumulation is given by

$$\$500 s_{\overline{5}|} = \$500 a(5) a_{\overline{5}|} = \$500(1.25) \sum_{k=1}^{5} v(k)$$

$$= \$625 \sum_{k=1}^{5} (1 - .04k) = \$625(.96 + .92 + .88 + .84 + .8)$$

$$= \$2{,}750.$$

∎

Accumulation functions governing investments that grow at compound interest with a fluctuating rate of compound interest occur commonly. We work with such an accumulation function in our next example.

EXAMPLE 3.12.9

Problem: During the years 1994 through 1998, Sylvia Bank paid compound interest on all funds on deposit. The following interest rates applied:

YEAR	INTEREST RATE
1994	6%
1995	7%
1996	6.5%
1997	6%
1998	5.25%

If Bob deposited $10,000 at the beginning of each of the years 1994 through 1998, what was his balance at the end of 1998?

Solution   We need to compute $\ddot{s}_{\overline{6}|}$.   Note that it is easiest to work this problem from basic principles, without any special formulas.   The balance in Bob's account at the beginning of 1995, just after his $10,000 deposit, was $10,000(1.06) + $10,000 = $20,600.   Immediately after his 1996 deposit, the balance was $20,600(1.07) + $10,000 = $32,042.   The balance just after his 1997 contribution was $32,042(1.065) + $10,000 = $44,124.73, and immediately following the 1998 deposit the account had $44,124.73(1.06) + $10,000 = $56,772.2138. Finally, the end-of-1998 balance was $56,772.2138(1.0525) \approx $59,752.76.   ∎

We finish this section with an example showing how different the investment picture may look if there is a new accumulation function for each accumulation time.

EXAMPLE 3.12.10

Problem: Suppose that each deposit earns simple interest at an annual rate r from the time of that deposit. Further suppose that 1 is deposited at the end of each year for n years. Find the accumulated value at the end of n years as a function of the rate r and the number of deposits n.

Solution A deposit of 1 made at time k, $k \in \{1, 2, ..., n\}$, grows to $1 + (n - k)r$ at time n. So, the total accumulation at time n is

$$\sum_{k=1}^{n}[1 + (n - k)r] = n + r((n - 1) + (n - 2) + \cdots + 1 + 0)$$

$$= n + r\left(\frac{n(n - 1)}{2}\right).$$

Note: This accumulation is higher than the accumulation you would have if there was a single accumulation function $a(t) = 1 + rt$ in effect from the time $t = 0$ of the first deposit. The reason for this is that simple interest has a decreasing force of interest function. For instance, if $r = 10\%$ and $n = 8$, the above formula produces an accumulation of $8 + .1\left(\frac{8(8-1)}{2}\right) = 10.8$, while the single simple interest accumulation function $a(t) = 1 + .1t$ produces an accumulation of

$$\sum_{k=1}^{n}\frac{1 + (.10)(8)}{1 + (.10)k} = 1.8\left(\frac{1}{1.1} + \frac{1}{1.2} + \frac{1}{1.3} + \frac{1}{1.4} + \frac{1}{1.5} + \frac{1}{1.6} + \frac{1}{1.7} + \frac{1}{1.8}\right)$$

$$\approx 10.19051684.$$

■

3.13   THE INVESTMENT YEAR METHOD

In Example (3.12.9), a sequence of annual interest rates was given. These rates were paid in the specified year without reference to the times funds were deposited. This has the advantage of simplicity. However, the managers of a fund might feel that it is more appropriate to index money by the time in which it was invested. After all, the holdings of the fund might include mortgages made at rates fluctuating over their origination time and bonds paying returns that looked advantageous at the times they were bought. If interest rates have gone up, the rates paid on these old investments will not be sufficient to attract new investors. Should rates have gone down, the investors of long standing should not have to share their rights to these high rates with newcomers. Usually funds that are sufficiently old no longer earn a **select rate** or **new money rate** based on the investment time. Instead, they earn a so-called **ultimate** or **portfolio rate**.

Suppose that an investment is made in year y and that select rates are in effect during the first N years that the money is part of the fund. If $k \in \{1, 2, ... N\}$, it is customary to denote the **select rate** earned by the investment during the k-th year of investment by i_k^y. Note that the superscript y in the notation i_k^y is not an exponent; rather it indicates the year the money was deposited. With this notation, if an amount X is contributed to the fund at the beginning of year y and left in the fund for an integral number of years $j \leqslant N$, it grows to

$$X(1 + i_1^y)(1 + i_2^y)(1 + i_3^y) \cdots (1 + i_j^y).$$

The **ultimate rate** used during the year z, for funds that have been on deposit during more than N years, is denoted i^z. Again, the superscript in not an exponent. This time the superscript displays the year during which the rate might be used. If X is deposited into the fund at the beginning of year y and left on deposit for j years where $j = N + r$, r a positive integer, at the end of the j years the initial investment of X will have grown to

$$\underbrace{X(1+i_1^y)(1+i_2^y) \cdots (1+i_N^y)}_{\text{select rates for } N \text{ years}} \underbrace{(1+i^{y+N})(1+i^{y+N+1}) \cdots (1+i^{y+N+r-1})}_{\text{ultimate rates for } r \text{ years}}.$$

For easy reference, we summarize this as Fact 3.13.1.

FACT 3.13.1: Suppose that the **investment year method** is in effect with select interest rates for the first N years the money is on deposit and ultimate interest rates thereafter. Let j denote a positive integer. Then X contributed at the beginning of year y and left on deposit for j years grows to

$$\begin{cases} X(1 + i_1^y)(1 + i_2^y)(1 + i_3^y) \cdots (1 + i_j^y) \\ \qquad\qquad\qquad \text{for } j \in \{1, 2, \dots, N\}, \\ \underbrace{X(1 + i_1^y) \cdots (1 + i_N^y)}_{\text{select rates for first } N \text{ years}} \underbrace{(1 + i^{y+N}) \cdots (1 + i^{y+N+r-1})}_{\text{ultimate rates for next } r \text{ years}} \\ \qquad\qquad\qquad \text{for } j = N + r > N. \end{cases}$$

It is quite common to present the select and ultimate rates in a table. We illustrate this for a fund using select rates for N years, for a period beginning with year y and extending through year $y + n$ where n is an a positive integer. (Do not confuse n with N, the number of years after investment that select rates apply.)

Year of Investment	Rate in Year 1	Rate in Year 2		Rate in Year N	Rate in Year N + 1	Year of first ultimate rate
y	i_1^y	i_2^y	$\cdots$	i_N^y	i^y	$y + N$
$y + 1$	i_1^{y+1}	i_2^{y+1}	$\cdots$	i_N^{y+1}	i^{y+1}	$y + 1 + N$
$y + 2$	i_1^{y+2}	i_2^{y+2}	$\cdots$	i_N^{y+2}	i^{y+2}	$y + 2 + N$
			$\vdots$			
$y + n$	i_1^{y+n}	i_2^{y+n}	$\cdots$	i_N^{y+n}	i^{y+n}	$y + n + N$

EXAMPLE 3.13.2

Problem: You are given a table of select rates and ultimate.

Year of Investment	Rate in Year 1	Rate in Year 2	Rate in Year 3	Rate in Year 4	First ultimate rate	Year of first ultimate rate
1965	.05	.055	.0475	.05	.055	1969
1966	.06	.525	.05	.06	.0575	1970
1967	.05	.05	.0625	.06	.0625	1971
1968	.0675	.07	.07	.0675	.0615	1972

Derartu deposits $3,000 in the fund at the beginning of 1966 and leaves it on deposit through 1972. What is his balance at the end of 1972?

Solution   Derartu's deposit was made at the beginning of 1966 and stayed in the fund through 1972. We therefore need to find seven interest rates, applicable to this deposit for the years 1966–1972 inclusive. Those for the years 1966–1970 are found by proceeding from left to right across the row for investment year 1966. They are 6%, 5.25%, 5%, 6%, and 5.75%, respectively. The first four of these are select rates, while the last is the portfolio rate for the year 1970. We also need the portfolio rates for 1971 and 1972, and these are found in the rows for the investment years 1967 and 1968, just below the ultimate rate for 1970. They are 6.25% and 6.15%. The entire sequence of rates is found by reading across the row for the investment year 1966 as far as possible and then reading down the column of ultimate rates. Using the seven rates we have made reference to, Derartu's balance at the end of 1972 is

$$\$3,000(1+.06)(1+.0525)(1+.05)(1+.06)(1+.0575)(1+.0625)(1+.0615)$$
$$\approx \$4,442.97.$$

■

EXAMPLE 3.13.3

Problem: Consider again a fund that originated in 1965 and was governed by the table of Example 3.13.2. What interest rates were in effect for the year 1968? How about for the year 1967?

Solution   The rate in effect in 1968 for money invested that year was 6.75%, as is seen from the entry in the bottom row, second column. To find the 1968 interest rate for 1967 investments, we look at the third column of the next to last row and find the interest rate 5%. Continuing on the upward diagonal between these two rates, we find the 1968 rate 5% for 1966 investments and again for 1965 investments. The rates in effect for 1967 investments lie on a line parallel to the 1968 rates. In 1967, for 1967 investments the rate is 5%, for 1966 investments the rate is 5.25%, and for 1965 investments the rate is 4.75%.

■

The investment year method is not always applied, even when a table of select rates (and ultimate rates) is available. Under the **portfolio yield method**, ultimate rates are used to figure accumulation from the start. That is to say, you neglect the select rates and use an ultimate rate for each year of investment.

EXAMPLE 3.13.4

Problem: Consider again Derartu's investment of Example (3.13.2). What would his balance have been at the end of 1972 if the portfolio yield method was used instead of the investment year method, and ultimate rates of 5% applied in each of the three years 1966, 1967, and 1968.

Solution   The sequence of ultimate years 5%, 5%, 5%, 5.5%, 5.75%, 6.25%, and 6.15% is applicable for the seven years of Derartu's investment. Since he made a single deposit of $3,000 at the beginning of the seven-year period, his final balance is

$$\$3,000(1.05)^3(1.055)(1.0575)(1.0625)(1.0615) \approx \$4,369.89.$$

■

3.14   PROBLEMS, CHAPTER 3

(3.0) Chapter 3 writing problems

(1) [following Section(3.1)] Find examples of real life annuities, and describe them using the vocabulary of Section (3.1). As a starting point, consider consumer loans, retirement plans, insurance payments, security holdings, and lottery winnings.

(2) [following Section (3.2)] Learn from a car dealer and from at least one bank or other financial institution what financing would be available on a specific model car (after you have priced the car, bargaining if possible). Describe your experiences, mentioning how credit worthiness must be established and what down payments are needed. Be sure to include the loan amount, payment periods (probably monthly), duration of the loan, and annual effective rate for each possibility you describe.

(3) [following Section (3.2)] Talk to one or more mortgage officers, and learn about what is needed to qualify for a mortgage on a home. Ask about income requirements and down payment requirements. Learn what closing costs there will be, and which of these can be rolled into your loan as opposed to being paid upfront. Your monthly payments will of course include payments based on the total of the loan amount but ask for all other payments that will be made through an escrow account. Ask under what conditions (say a higher down payment) an escrow account can be avoided. Now write a newspaper column for potential homeowners, educating them about what you have learned.

(4) [following Section (3.6)] Discuss usages of the words "prospective" and "retrospective" in the English language. Give examples of other words with the same roots.

(3.2) Annuities-immediate

(1) Find the present value (one period before the first payment) of an annuity-immediate that lasts five years and pays \$3,000 at the end of each month, using a nominal interest rate of 3% convertible monthly. Then repeat the problem using an annual effective discount rate of 3%. Which is higher? Why?

(2) Tracy receives payments of \$X at the end of each year for n years. The present value of her annuity is \$493. Gary receives payments of \$3X at the end of each year for $2n$ years. The present value of his annuity is \$2,748. Both present values are calculated at the same annual effective interest rate. Find v^n.

(3) The Browns wish to accumulate at least $150,000 at the time of their last deposit in a college fund for their daughter by contributing an amount A into the account at the end of each year for eighteen years. What is the smallest annual payment A that will suffice if the college fund earns a level annual effective interest rate of 5%? If at the end of ten years, it is announced that the annual effective interest rate will drop to 4.5%, how much must the Browns increase their payments in order to reach their accumulation goal? Assume that the Browns wish to continue to make level payments except for a slightly reduced final payment.

(4) Elwood wishes to purchase a home. She has saved up $13,200 for a down payment. Based on her earnings, she qualifies for a thirty-year mortgage with level monthly payments of $820 including escrow and a nominal interest rate convertible monthly of 5.85%. Her payments are due at the end of each month. From each payment, $240 will be put aside in an escrow account for the payment of taxes and homeowners insurance. What is the most expensive house Elwood can buy if her aunt has promised to give her the money needed for loan applications, inspections, and all other required buyer's closing costs?

——(5) Matt inherited as a trust a fifteen-year annuity-immediate with annual payments. He has been told that the annuity payments earn compound interest at a level rate and that at the end of fifteen years, their accumulated value will be $37,804.39. He has further been assured that figured at this same rate of interest, the value of his inheritance was $15,077.10. The trust executor will not reveal the amount of the annual payments. Determine this amount and also the annual effective interest rate earned by the annuity payments.

(6) Mrs. Williams finds that she has two options for investing $32,000.02 for fifteen years. The first option is to deposit the $32,000.02 into a fund earning a nominal rate of discount $d^{(4)}$ payable quarterly. The second option is to purchase an annuity-immediate with 15 level annual payments, the annuity payments computed using an annual effective rate of 7%, and then when she gets an annuity payment, to immediately invest it into a fund earning an annual effective rate of 5%. Mrs. Williams calculates that the second option produces an accumulated value that is $1,500 more than the accumulated value yielded by the first option. Calculate $d^{(4)}$.

(7) A buyer of a 2003 Protege S Hatchback has a choice of 0% financing for 60 months or a $3,600 rebate. He plans to make no down payment. The buyer is able to qualify for 7% annual effective financing through his credit union and thereby take advantage of the rebate. Let Y denote his negotiated price for the Protege S Hatchback. How large must Y be in order for the 0% dealer financing to be preferable?

(8) Sigmund and Karl each borrowed an identical amount from Ludwig at a nominal rate of discount of 5.4% convertible quarterly. Sigmund repays his loan by making payments of $2,000 at the end of each year for six years. Karl makes payments of $3,200 at four equally spaced times T, $2T, 3T$, and $4T$. Find T. [HINT: You will need to find the interest rate I for a period of length T.]

(3.3) Annuities-due

(1) April received an inheritance from her grandmother in the form of an annuity. The annuity pays $3,000 on January 1st from 1966 through 1984. Find the value of this annuity on January 1, 1966 using an annual effective interest rate of 5% and represent this value by an appropriate annuity symbol.

(2) Suppose $i = 3\%$. Find the value one month before the first payment of a level annuity-due paying $200 at the beginning of each month for five years.

(3) (a) Describe in words what the difference $a_{\overline{n+1}|i} - a_{\overline{n}|i}$ is measuring.

— (b) Given that $a_{\overline{n+1}|i} - a_{\overline{n}|i} = .177208656$ and $\ddot{a}_{\overline{n+1}|i} - \ddot{a}_{\overline{n}|i} = .185248436$, find the integer n.

(4) Steven Wong wishes to save for his retirement by depositing $1,200 at the beginning of each year for thirty years. Exactly one year after his last deposit, he wishes to begin making annual level withdrawals until he has made twenty withdrawals and exhausted the savings. Find the amount of each withdrawal if the effective interest rate is 5% during the first thirty years but only 4% after that.

(5) Starting on his 25th birthday and continuing through his 60th birthday, Fred deposits $7,500 each year on his birthday into a retirement fund earning an annual effective rate of 5%. Immediately after the last deposit, the accumulated value of the fund is transferred to a fund earning an annual effective rate of j. Five years later, a twenty-five year annuity-due paying $5,800 each month is purchased with the funds. The purchase price of the annuity was determined using an annual effective rate of interest of 4%. Find j.

— (6) Given $\ddot{a}_{\overline{n}|} = 12$ and $\ddot{a}_{\overline{2n}|} = 21$, find $a_{\overline{4n}|}$.

(7) Given $\ddot{a}_{\overline{n}|} = 31.61667882$ and $s_{\overline{n+1}|} = 64,024.90944$, find i and n.

(8) Given $a_{\overline{n-1}|} = 11.38229339$ and $s_{\overline{n+1}|} = 414.3137296$, find i and n.

(9) If you multiply Equation (3.2.18) by $\frac{1}{1+i}$, you may easily derive the equation $d + \frac{1}{s_{\overline{n}|i}} = \frac{1}{a_{\overline{n}|i}}$. Following our derivation of (3.2.18), we gave a second demonstration of why this equation must hold based on two

alternate scenarios of paying back a loan of 1 made for n interest periods. Think of another pair of ways of repaying a loan which give an alternative demonstration of $d + \dfrac{1}{s_{\overline{n}|i}} = \dfrac{1}{a_{\overline{n}|i}}$.

(10) (a) Show algebraically that $\ddot{s}_{\overline{n}|i} = s_{\overline{n}|i} - 1 + (1 + i)^n$.
   (b) Verbally explain why the equality in (a) must hold.
   (c) Given that $\ddot{s}_{\overline{n}|i} = 551.40$ and $a_{\overline{n}|i} = 315.85$, find n.

(3.4) Perpetuities

(1) Athlete Kalen wishes to retire at age forty-five and receive annual birth-day payments of $40,000 beginning on his forty-fifth birthday. After his death, the payments on the anniversary of his birth should go to his heirs. In order for Kalen to be able to carry out his plan, he makes contributions to a savings account with a guaranteed annual effective interest rate of 4%. How much money will Kalen need to have accumulated at age forty-five, just prior to his first $40,000 birthday payment?

(2) Graham receives $640,000 at his retirement.   He invests X in a twenty-year annuity-immediate with annual payments and the remaining $640,000 − X is used to purchase a perpetuity-immediate with annual payments. His total annual payments received during the first twenty years are twice as large as those received thereafter. The annual effective interest rate is 5%. Find X.

(3) Suppose $40,000 was invested on January 1, 1980 at an annual effec-tive interest rate of 7% in order to provide an annual (calendar-year) scholarship of $5,000 each year forever, the scholarships paid out each January 1.

   (a) In what year can the first $5,000 scholarship be made?
   (b) What smaller scholarship can be awarded the year prior to the first $5,000 scholarship?

(4) Svetlana won $1,000,000 in a contest, to be paid in twenty $50,000 payments at yearly intervals, the first payment paid at the time of the contest. (Of course, the present value of her winnings is less than $1,000,000.) Svetlana decided to keep X each year to spend and deposit the remaining $50,000 − X into an account earning an annual effective interest rate of 5%. She chose the value X to be as large as possible so that, at the moment of the 20th deposit, the account would have grown to such a size that it would provide Svetlana and her heirs at least X per year in interest forever. Find X.

(5) Given $54\ddot{a}_{\overline{\infty}|i} = 1,000$, find $s_{\overline{22}|i}$.

(6) Cheyenne wishes to endow a professorship at her alma mater. She learns that to do so, she must contribute $1,000,000. What annual salary would a $1,000,000 gift support forever if salary payments are made monthly, starting one month after the gift, and the money's growth is governed by an annual effective discount rate of 6%?

(3.5) Deferred annuities and annuity values on any date

(1) Sydney wins a prize. She has a choice of receiving a payment of $160,000 immediately or of receiving a deferred perpetuity with $10,000 annual payments, the first payment occurring in exactly four years. Which has a greater present value if the calculation is based on an annual effective interest rate is 5%? How about if the annual effective rate used is 6%? What real life considerations should enter into Sydney's choice besides maximizing her present value?

(2) A level perpetuity-immediate is to be shared by three charities providing medical research and a fourth charity providing assistance to children of veterans. For n years, the three research charities will receive the payments equally. Thereafter, all the payments will go to the charity aiding children of veterans. It is reported that the present value of each of the charities bequest is equal to a common amount when calculated with an annual effective interest rate of 12.25%. Find n. If the interest rate is a more modest 6%, what proportion of the total bequest is directed to the charity aiding children of veterans?

— (3) Alice owned an annuity which had level annual payments for twelve consecutive years, the first of these being in exactly twelve years. She sold it, and the selling price of $21,092.04 was based on a yield rate for the investor of 7.8%. What is the amount of the level payments?

(4) When computed using an effective interest rate of i, it is known that the present value of $2,000 at the end of each year for $2n$ years plus an additional $1,000 at the end of each of the first n years is $52,800. Using this same interest rate, the present value of an n year deferred annuity-immediate paying $4,000 per year for n years is $27,400. Find n. [HINT: First find $a_{\overline{n}|i}$ and $a_{\overline{2n}|i}$, then v^n and i.]

(5) Catfish Hunter's 1974 baseball contract with the Oakland Athletics called for half of his $100,000 salary to be paid to a life insurance company of his choice for the purchase of a deferred annuity. More precisely, there were to be semi-monthly contributions in Hunter's name to the Jefferson Insurance Company with the first payment on April 16 and the final payment on September 30.[1] We suppose that the first eleven of

---

[1] You may read about this in Marvin Miller's *A Whole Different Ballgame:The inside Story of Baseball's New Deal*, Simon and Schuster, New York, 1991.

these were to be for $4,0166.67, and the final payment was to be for four cents less. ($12 \times \$4,166.67 = \$50,000.04$.)

(a) Using an annual effective interest rate of 6% (a rate that figures in a six-year personal loan of $120,000 that Oakland's owner Charlie Finley had made to Hunter in 1969 and then promptly recalled), find the value of the specified payments to the insurance company at the scheduled time of the last payment. (Hunter wished to have such an annuity in lieu of immediate salary for tax reasons. Finley claimed that he was fulfilling the contract when he offered Hunter a $50,000 check at the end of the season. Finley's default on his contractual obligation led to Hunter's historic free agency. The New York Yankees signed Hunter to a five-year, $3,750,000 contract.)

(b) Suppose that the contracted payments had been made to the insurance company from April 16, 1974 through September 30, 1974, and that they accumulated at an annual effective interest rate of 6%. Further suppose that Hunter had drawn a level January 1st salary for twenty years, beginning on January 1, 1980, the first January after his retirement. Find the amount of the annual salary payments. (Hunter died on September 9, 1999, so he would not haven't have received a January 1, 2000 annuity payment.)

(3.6) Outstanding loan balances

(1) On May 12, Jemeel takes out a personal loan at an annual effective interest rate of 6%. The loan is to be repaid by payments on each of the next ten May 12s, the first six being for $822 and the final four being for $1,516.

(a) Find the loan balance immediately after the sixth payment.
(b) Find the outstanding loan balance immediately after the third payment.

(2) Olena loans her sister Irini $8,000. The loan is to be repaid at a nominal interest rate of 4.8% payable monthly. The monthly payments are to be for $100 except for a final smaller payment. How much does Irini owe Olena at the end of one year?

(3) Mr. Bell buys a home for $C. He pays a down payment of $20,000 and finances the remainder for 15 years with level end-of-month payments of $1,692. The annual effective interest rate for the first five years is 4%, and thereafter it is 6%. Mr. Bell sells the house just after making his 100th mortgage payment. The selling price is $258,000. How much money will Mr. Bell get at closing? (Remember, the loan holder is paid first, and then Mr. Bell receives the balance of the inflow from the resale.)

(4) A loan of $20,000 is being paid by installments of $1,000 at the end of each six-month period and a smaller final payment made half a year after the last regular payment. The interest rate for the loan is a nominal interest rate of 6% convertible semiannually. Find the outstanding loan balance when the borrower has made payments equal to 75% of the amount of the loan.

(5) Alice purchases a boat. The purchase price is $18,300 and she makes a down payment of $3,800. She finances the balance She finances for six years. There are level end-of-month payments (except for a slightly reduced final payment) and the loan is made at an annual effective rate of 5.2%.

   (a) Find the amount of the first seventy-one payments.
   (b) Find the outstanding loan balance just after the twenty-fourth payment.

(6) Mr. Johnson has a thirty-year mortgage with end-of-month payments. It has a nominal quarterly interest rate of 4.4% during the first three years and a nominal monthly interest rate of 5.7% for the remaining twenty-seven years. The repayment schedule shows level payments except for a final slightly reduced payment. The amount financed is $180,000. Find the amount of each of the first 359 payments. If Mr. Johnson sold the house just after his 36th payment for $200,000, how large a check did he receive?

(7) Janelle receives a home improvement loan of $14,508.97. The loan has a nominal interest rate convertible monthly of 4.8%. The term of the loan is two years and Janelle is expected to make level end-of-month payments, except that she is allowed to miss one payment so long as she then pays higher level payments for the remainder of the two years, so as to have repaid the loan at the end of the two-year period. Suppose Janelle misses the payment at the end of the nineth month. What must her new level payments be?

(3.7) Nonlevel annuities

(1) Stuart wishes to have $14,000 to buy a used car three years from now. He plans to accomplish this, through an account with a nominal interest rate of 3% compounded monthly, by depositing $300 at the end of each month during the first year, and Q at the end of each month during the second two years. What is the smallest Q that will suffice?

(2) On January 1, 1980, Suzanne received a twenty-year annuity-due that paid $100 each January 1 and $300 each July 1. What was the value of this annuity on January 1, 1980, calculated using an effective rate of interest of 6%?

(3) To accumulate $217,593.30 at the end of $5n$ months, deposits of $100 are made at the end of each of the first $2n$ months and $300 at the end of the next $3n$ months. Given that $(1 + i)^n = 92.372$, find n. (Be careful to note that i is an annual rate but payments are monthly.)

(4) Bill deposits $100 at the end of each year for thirteen years into fund A. Seth deposits $100 at the end of each year for thirteen years into fund B. Fund A earns an annual effective rate of 15% for the first five years and an annual effective rate of 6% thereafter. Fund B earns an annual effective rate of i throughout the thirteen years. The two funds have equal accumulated values at the end of the thirteen years. Find i.

(5) [BA II Plus calculator desirable]   Sean and Karl each borrowed an identical amount from their parents at an annual effective rate of interest of 5%. Sean repays his loan by making monthly payments at the end of each month for eight years. For the first four years these monthly payments are each $200 and for the next four years they are each $500. Karl makes payments of $7,000 at times T, $2T$, and $3T$, $3,000 at times $4T$ and $5T$, and a payment of $2,000 at time $7T$. Find T. [HINT: You will need to find the interest rate I for a period of length T.]

(6) A thirty-year annuity X has annual payments of $1,000 at the beginning of each year for twelve years, then annual payments of $2,000 at the beginning of each year for eighteen years. A perpetuity Y has payments of Q at the end of each year for twenty years, then payments of $3Q$ at the end of each year thereafter. The present values of X and Y are equal when calculated using an annual effective discount rate of 10%. Find Q.

(7) Lucy received a gift of a twenty-one year annuity on the day she was born. The annuity pays $500 on her odd birthdays and $700 on her even birthdays.

   (a) If the nominal rate of interest is 8% payable monthly, find the value of this annuity on the day she was born.

   (b) If the value of the annuity at the time of her birth was $6,000, find the annual effective discount rate as a percent, correct to the nearest one-hundredth of a percent.

(8) On January 1, Alex received an inheritance of a thirty-year annuity. Starting on the day of the inheritance, the annuity pays $1,000 each January 1, $2,000 each April 1, $1,500 each July 1, and $3,000 each October 1.

    (a) Using $i = 4\%$, calculate the value of this inheritance on the day Alex inherits it.

    (b) If the value on the day she inherits it is $130,000, find the annual effective rate i used to calculate this value to the nearest hundredth of a percent.

(3.8) Annuities with payments in geometric progressions

(1) Al and Sal are twins. Al is given a fifteen-year annuity with end-of-year payments. The first payment Al receives, precisely one year from the date he is given the annuity, is for $100, and then subsequent payments **decrease** by 4% annually. Sal is given an n-year level annuity that has the same present value as Al's when the present values are calculated using $i = 5\%$. Again calculated using $i = 5\%$, the accumulated value at the end of n years of Sal's annuity is $1,626.29. Find the common present value of the two annuities and then find n.

(2) On January 1, 1988, Wanda received a deferred perpetuity paying $3,000 on July 1 of even numbered years beginning on July 1, 1996 and $1,200 on July 1 of odd years beginning on July 1, 1997. The interest rate is 4% effective for even numbered years and 5% effective for odd numbered years. (That is, the interest rate is 4% effective for 1988, 1990, 1992, ... and 5% effective for 1989, 1991, 1993, ...) Find the value of this perpetuity on January 1, 1988.

(3) Consider an annuity-immediate with monthly payments for twenty years. The payments are level in the course of each year, then increase by 2% for the next year. Find the present value of this annuity if the initial payment is $1,000 and $i = 4\%$.

(4) Annual end-of-year deposits are made to a fund paying an annual effective rate of interest of 6%. The first deposit is $1,000 and then they go up by 3% annually. Interest from the 6% account is paid out annually and is reinvested at an annual effective discount rate of 4%. At the end of the thirty years, the funds are liquidated. Find the investor's accumulated value at the time of the liquidation.

(5) On January 1, 1988, Felix inherited a perpetuity with annual payments beginning in six months. The first payment was $3,000, and after that the payments increased by 3% each year. Find the value of this perpetuity on January 1, 1995 if the annual effective interest rate was 6% from January 1, 1988 through January 1, 1996 and 4% thereafter.

(3.9) Annuities with payments in arithmetic progressions

(1) Suppose that the effective interest rate per interest period is 3%.
Describe what the following annuity symbols mean and **calculate** them
to the nearest $\dfrac{1}{100}$.

(a) $(Ds)_{\overline{28}|}$ 　　　　 (b) $(I\ddot{a})_{\overline{\infty}|}$ 　　　　 (c) $(I_{100,10}a)_{\overline{15}|}$

(2) Payments of $5,000 are made into a fund at the beginning of each year
for ten years. The fund is invested at an annual effective rate of i. The
interest generated is reinvested at 10%. The total accumulated value at
the end of the ten years is $100,000. Find i.

(3) A perpetuity has annual payments. The first payment is for $320 and
then payments increase by $30 each year until they become level at $980.
Find the value of this annuity at the time of the first payment using an
annual effective interest of 4%.

(4) [calculus needed] Perpetuity X has level payments of $300 at the end of
each year. Perpetuity Y also has end-of-year payments but they begin at
$45 and increase by $45 each year. Find the rate of interest which will
make the difference in present values between these two perpetuities a
maximum.

(5) Bob deposits $11,000 at the beginning of each year for six years in a
fund earning an effective annual rate of 7.5%. The interest from this
fund can only be reinvested at an effective annual rate of 5%. Find the
accumulated value of Bob's investments at the end of thirteen years.

(6) Erik receives an eight-year annuity-immediate with monthly payments.
The first payment is $300 and payments increase by $6 each month. The
payments are deposited in an account earning interest at a nominal rate
of 6% convertible monthly. What is the balance in the account at the
end of the eight years?

(7) An n-year annuity-immediate has a payment of k^2 at time k, $k \in
\{1, 2, \dots, n\}$. Derive a formula for the present value of this annuity in
terms of the symbol $(I_{3,2}a)_{\overline{n-2}|i}$, assuming money grows according to the
accumulation function $a(t) = (1 + i)^t$. Evaluate numerically if $n = 30$
and $i = .04$. [HINT: $a^2 - b^2 = (a - b)(a + b)$.]

(8) Use the equation $(Ia)_{\overline{n}|i} + (Da)_{\overline{n}|i} = (n + 1)a_{\overline{n}|i}$ to derive formula
(3.9.7) from (3.9.6).

(3.10) Yield rate examples involving annuities

(1) An investor invests $58,000 and receives an annuity of $7,000 at the end of each year for twelve years and an additional payment of $15,000 at the end of the thirteenth year. Each time he gets a $7,000 payment, he immediately deposits $4,000 in a savings account that earns 9%. Find the annual yield received by the investor over the thirteen years.

(2) An investor purchased an eight-year financial instrument having the following features. The investor receives payments of $10,000 at the end of each year for eight years. These payments earn interest at an effective rate of 5% per year but the interest can only be reinvested at a rate of 3.5% per year. Calculate the purchase price to produce an annual yield of 6% over the eight years.

(3) John invests a total of $10,000. He purchases an annuity with payments of $1,000 at the beginning of each year for ten years at an effective interest rate of 8%. As annuity payments are received, they are reinvested at an effective annual rate of 7%. The balance of the $10,000 is invested in a ten-year certificate of deposit with a nominal annual interest rate of 9% compounded quarterly. Calculate the annual effective yield on the entire $10,000 investment over the ten-year period.

(4) An investor pays $250 at the beginning of years 1 - 6 and $750 at the beginnings of years 13 - 18. In return, he gets $900 at the beginnings of years 7 -12. Find all positive yield rates for this investment.

(5) Companies A, B, and C enter into a financial arrangement. Established Company A will today pay expanding firms B and C $300,000 and $500,000, respectively. One year from today, B will pay C $100,000, and four years from today B will pay A $250,000. Three years from today C will pay A $700,000. Calculate the yield rate, correct to the nearest thousandth of a percent, that each company experiences over the period of their involvement (4 years for A and B, 3 years for C).

(6) [BA II Plus calculator desirable] Fernando invests $85,000. In return, he receives $4,000 monthly starting with a payment 6 months from the date of investment and concluding with a payment 30 months from the date of investment. From each of these $4,000 payments, he invests $3,000 in an account with an annual effective interest rate of 5%. Fernando liquidates his 5% account exactly five years after he invested his $85,000. Find the annual yield rate earned by Fernando over this five-year period.

(7) Tom invests $80,000 and earns $12,000 on the first five anniversaries of his investment. Upon receipt of each $12,000 payment, he immediately deposits $5,000 into a savings account earning 5% payable annually. If Tom withdraws the accumulated money in the savings account a year after the last payment, find the yearly yield rate earned by Tom during

the six-year period. Note that this investment includes two parts, the investment of the original $80,000 and the reinvestment at 5%. Check that your yield rate is between the rates for these two parts.

(8) [BA II Plus calculator] An investor pays $20,000 at $t = 0$. In return, he gets $3,000 at $t = 1, 2, ..., 8$ and an additional payment of Q at time T.

   (a) Find the yield if $T = 8$ and $Q = \$2,000$ using the **TVM** and again using the **Cash Flow worksheet**.
   (b) Find the yield if $T = 9$ and $Q = \$2,000$ using the **TVM** and again using the **Cash Flow worksheet**.
   (c) If the yield is 5% and $Q = \$2,000$, find T.
   (d) If the yield is 5% and $T = 12$, find Q.

(9) [BA II Plus calculator desirable] An investor pays $10,000. In return, he gets end-of-quarter payments for three years of $1,000 and an additional payment of $300 at the end of 61 months. Find the annual yield correct to the nearest one-hundredth of a percent.

(3.11) Annuity-symbols for nonintegral terms

(1) Alex loans Nomar $200,000 at a rate of 4% nominal interest convertible quarterly. They agree that Nomar will repay the loan by making quarterly payments. These payments will each be $25,000 except for the last payment which will be a drop payment. Find the total length of the loan and the amount of the final payment.

(2) In the previous problem, suppose that the parties decide they would prefer to decrease the loan term by one period by opting for a balloon payment rather than a drop payment. Find the amount of the balloon payment.

(3) Brandt loans Lori $5,000 at an annual effective interest rate of 5%. They agree that Lori will make annual end-of-year payments of $600 to repay the loan, then realize that they must figure the term of the loan. If the term T is to be such that $\$5,000 = \$600a_{\overline{T}|5\%}$, find T and the amount of the last payment at time T.

(3.12) Annuities governed by general accumulation functions

(1) Suppose that the accumulation in an investment fund is governed by the accumulation function $a(t) = 1 + .04t$ where time is measured in years from January 1, 1990.

   (a) Find the accumulation $\$100s_{\overline{6}|}$ at the time of the last deposit if deposits of $100 are made at the end of each year for six years, the first deposit being on December 31, 1990.

(b) What effective rate of compound interest would produce the same accumulated value (on December 31, 1995) for the sequence of six $100 deposits?

(2) Suppose that accumulation is governed by compound interest and the annual effective rate of interest in the k-th year is $.02 + .01k$. Find $a_{\overline{5}|}$ and $s_{\overline{5}|}$. Find T so that $s_{\overline{5}|}/a_{\overline{5}|} = a(T)$.

(3) [calculus needed] Find the value at $t = 0$ of a perpetuity that pays 1,000 at the end of each year starting at $t = 3$ assuming that $a(t) = \frac{1}{3}(t + 1)(t + 3)$. [HINT: Consider Example (3.12.4).]

(3.13) The investment year Method

(1) Frank invests $1,000 at the beginning of 1989, 1990, and 1991 in Experience Investment Fund which credits money according to the investment year method using the following chart. There are no withdrawals or further deposits. What would his balance have been on January 1, 1994?

Year of Original Investment	i_1^y	i_2^y	i_3^y	i^{y+3}	Year of Portfolio Rate
1988	.05	.045	.04	.04	1991
1989	.06	.055	.05	.045	1992
1990	.065	.06	.055	.05	1993
1991	.06	.055	.05	.05	1994
1992	.065	.055	.05	.05	1995
1993	.06	.06	.06	.05	1996

(2) Suppose that money was equally likely to be invested in the Experience Investment Fund of Problem (3.13.1) in any of the years 1988–1993. What was the average rate of return paid by the fund during 1993?

Chapter 3 review problems

(1) Pierre receives a thirty-year annuity paying $200(1.03)^t + \$100(1.02)^{2t} + \$80t - \$60$ at the beginning of the t-th year. Find the value of this annuity at the end of the tenth year if the annual effective interest rate remains 4% during the thirty years.

(2) Mrs. Chen has saved $22,000 for a down payment and qualifies for a thirty-year mortgage at an annual effective interest rate of 6%. In addition, Mrs. Chen is opening a retirement account that has a 5.2% annual effective interest rate. Mrs. Chen wants to be able to retire in thirty-eight years, at which point she needs to be able to withdraw $4,100 at the beginning of each month for twenty-five years. Mrs. Chen has $2,600 at the end of each month to use for a combination of mortgage payments and retirement contributions. How expensive a house can she buy? [HINT: Let $M denote Mrs. Chen's monthly mortgage payments. Her contributions to the retirement fund are $2,600 − $M at the end of each month during the first thirty years and $2,600 per month each during the next eight years.]

(3) Cheryl receives a sixty-year annuity-due. The annual payments are $3,000 during the first twenty years and $6,000 during the next forty years. David receives a perpetuity-immediate with annual payments. The odd numbered payments are P and the even numbered payments are $2P$. The present values of Cheryl's annuity and David's perpetuity are equal if they are calculated using $i = 5\%$. Find P.

(4) A fifty-year annuity-due is level except for the mth payment. The m-th payment is $3,000 and each of the other payments is $5,000. When calculated using an annual effective interest rate of 4%, the value of the annuity five years after the first payment is $134,707.94. Find m.

(5) On January 1, 1980, Julie received an annuity-certain. The annuity promised to pay $100 on each September 30th through 1997 and $200 on each December 31st through 1997. If the value of the annuity on January 1, 1980 was calculated to be $3,150, what annual effective rate of interest was used?

(6) Derek receives a perpetuity. It has annual payments with the first payment in exactly eight years. The first payment is $10,000, the second is $15,000, and then the payments alternate between $10,000 and $15,000 until there have been a total of thirty payments. After that, the payments are all $10,000. Find the present value of this perpetuity if $i = 4\%$.

(7) Serena receives a fifty-year annuity-due that has payments that start at $2,000 and increase by 3% per year through the twenty-fourth payment, then stay level at $4,000. Find the accumulated value of this annuity at the end of fifty years if the annual effective rate of interest remains 4.2% throughout the time of the annuity.

C H A P T E R   4

Annuities with different payment and conversion periods

4.1 INTRODUCTION

In Chapter 3 we studied annuities with regularly spaced payments in which the payment period and interest period coincided. These included, but were not limited to, annuities with level payments, annuities with payments in arithmetic progression, and annuities with payments in geometric progression. In this chapter we again study annuities with regularly spaced payments and a variety of payment patterns, but now we drop the requirement that the interest payment and payment periods coincide. Our consideration of level annuities with unsynchronized interest and payment periods includes work with ratios of familiar annuity symbols when the payment period consists of several interest periods (Section 4.2) and the introduction of new annuity symbols when there are multiple payments per interest period (Section 4.3). This is the traditional treatment and is needed in some areas of actuarial science. However, with our inclusion of instruction using the BA II Plus calculator, we are also able to focus on efficient numerical computation. Sections (4.4) and (4.5) look at annuities with increasing payment patterns. In Section (4.6) we discuss how to pass from progressively shorter payment periods to so-called continuously paying annuities. This includes continuously paying level and continuously paying increasing annuities. In real life there are no continuously paying annuities, since for this one would need a continuously dripping money faucet. However, such annuities are used in actuarial science and are theoretically appealing. As usual, our primary concern is with growth by compound interest, but some consideration is given to other accumulation functions. For such accumulation functions, the approximations one can obtain may be particularly helpful. This chapter moves a bit more quickly than previous ones, since the material is less essential except for Section (4.3)'s use in actuarial science.

4.2   LEVEL ANNUITIES WITH PAYMENTS LESS FREQUENT THAN EACH INTEREST PERIOD

Imagine that a monthly interest rate is given, and an annuity with annual payments is to be studied. One possibility, and this is the approach of Chapter 3 (except when the BA II Plus calculator is used and an alternative method is presented), is to convert the monthly interest rate to an annual one. However, there is a second approach that is aesthetically appealing, and is also sometimes computationally more efficient. This new method is to work with a formula that has been derived by working **symbolically** rather than numerically. Not surprisingly, the derivation of the formula again uses the technique of finding an equivalent interest rate for the payment period. (Recall that the interval between payments of an annuity with regularly spaced payments is called a payment period.) We now introduce it.

Suppose that an interest rate is given for a specified period. We refer to this interval as "the interest period." Further suppose that we have a level annuity lasting n interest periods, that $n = rk$ where r and k are integers, and that the annuity has a payment of 1 at the end of each k interest periods.

PAYMENT:		1	1	1	$\cdots$	1
TIME:	0	k	$2k$	$3k$	$\cdots$	$rk = n$

FIGURE (4.2.1) Annuity-immediate

Let i denote the effective interest rate for the interest period. Then, since the payment period lasts k interest periods, the payment period has effective interest rate $I = (1 + i)^k - 1$, and if a numerical value for i is given we can calculate I. In addition, the present value of the annuity is then $a_{\overline{r}|I}$. We next give an example illustrating this approach numerically.

EXAMPLE 4.2.2   Interest rate for payment period found

Problem: Chinh is presented an annuity that is to pay $4,000 at the end of each six months for six years. Using compound interest at a nominal monthly interest rate of 3%, what is the value of Chinh's reward at the time of the announcement (six months before the first $4,000 payment)?

Solution   The monthly effective interest rate is $\frac{.03}{12} = .0025$. Therefore, the effective semiannual interest rate (the interest rate for a six-month period) is $I = (1.0025)^6 - 1 \approx 1.509406309\%$. The annuity lasts six years and has two payments each year so it has a total of $6 \times 2 = 12$ payments. The value of Chinh's annuity at the time of the reward is $\$4,000 a_{\overline{12}|I} \approx \$43,604.47.$ ∎

Returning to the general annuity described in this section and depicted in Figure (4.2.1) and proceeding symbolically, the annuity lasts n interest periods, and this is $\frac{n}{k} = r$ payment periods. Therefore, the present value of the annuity is $a_{\overline{r}|I}$. Using $I = (1 + i)^k - 1$, and (3.2.4), we find

$$(4.2.3) \quad a_{\overline{r}|I} = \frac{1 - (1 + I)^{-r}}{I} = \frac{1 - (1 + i)^{-rk}}{(1 + i)^k - 1} = \frac{\frac{1-(1+i)^{-n}}{i}}{\frac{(1+i)^k-1}{i}} = \frac{a_{\overline{n}|i}}{s_{\overline{k}|i}}.$$

Multiplying (4.2.3) by $(1 + I)^r = (1 + i)^n$ yields

$$(4.2.4) \qquad\qquad s_{\overline{r}|I} = \frac{s_{\overline{n}|i}}{s_{\overline{k}|i}}.$$

We thus have shown

FACT (4.2.5): Suppose k divides n. The present value of an annuity lasting n interest periods and paying 1 at the end of each k interest periods is $\frac{a_{\overline{n}|i}}{s_{\overline{k}|i}}$. The accumulated value of this annuity at the time of the last payment is $\frac{s_{\overline{n}|i}}{s_{\overline{k}|i}}$.

A good way to view this result is as follows. Imagine that you had an annuity with level end-of-interest period payments of $\frac{1}{s_{\overline{k}|i}}$. If you looked at the accumulated value at the end of a payment period of the k payments made during that period, it would be 1. So, the present value of our annuity (with payments of 1 at the end of each k interest periods) must be equal to that of the level annuity with payments of $\frac{1}{s_{\overline{k}|i}}$, and this present value is $\frac{a_{\overline{n}|i}}{s_{\overline{k}|i}}$.

EXAMPLE 4.2.6   Applying Fact (4.2.5)

Problem: Return to Chinh's award of Example (4.2.2). Use Fact (4.2.5) to check that the value of the annuity at the time of the presentation was about $43,604.47.

Solution Since there are 72 months in a six-year period, Chinh's annuity pays $4,000 at the end of each six months, and the monthly interest rate is .25%, Fact (4.2.5) tells us that the value at the time Chinh is awarded the annuity is

$$\$4,000\frac{a_{\overline{72}|.25\%}}{s_{\overline{6}|.25\%}} \approx \$4,000\left(\frac{65.81685774}{6.037625235}\right) \approx \$43,604.47. \qquad \blacksquare$$

We have been working with the basic annuity-immediate of Figure (4.2.1). Let us now turn our attention to a basic annuity-due with payments once every k interest periods.

PAYMENT:	1	1	1	$\cdots$	1	
TIME:	0	k	$2k$	$\cdots$	$(r-1)k$	$rk = n$

FIGURE (4.2.7) Annuity-due

The following result is similar to Fact (4.2.5), but we see present values in the denominators rather than accumulated values.

FACT (4.2.8): Suppose k divides n. The present value of an annuity lasting n interest periods and paying 1 at the beginning of each k interest periods is $\frac{a_{\overline{n}|i}}{a_{\overline{k}|i}}$. The accumulated value of this annuity at the time of the last payment is $\frac{s_{\overline{n}|i}}{a_{\overline{k}|i}}$.

To see that Fact (4.2.8) is true, note that on account of (4.2.3), the described annuity-due has present value

$$\ddot{a}_{\overline{r}|I} = (1+I)a_{\overline{r}|I} = (1+i)^k\left(\frac{a_{\overline{n}|i}}{s_{\overline{k}|i}}\right) = \frac{a_{\overline{n}|i}}{v^k s_{\overline{k}|i}} = \frac{a_{\overline{n}|i}}{a_{\overline{k}|i}},$$

and its accumulated value is obtained by multiplying this by $(1+I)^r = (1+i)^n$. Another approach is to imagine that you had an annuity with level end-of-interest period payments of $\frac{1}{a_{\overline{k}|i}}$. For this annuity, consider the value at the beginning of a payment period of the sequence of k payments occurring during that period. This value is 1. So, the present value of our annuity (with payments of 1 at the beginning of each k interest periods) must be equal to that of the level annuity with payments of $\dfrac{1}{a_{\overline{k}|i}}$, and this present value is $\dfrac{a_{\overline{n}|i}}{a_{\overline{k}|i}}$.

EXAMPLE 4.2.9   Two methods: Use interest rate for payment period or apply Fact (4.2.8)

Problem: Josh makes annual beginning-of-year contributions of $3,000 to a retirement account. The account has a nominal interest rate of 4.8% convertible monthly. Find Josh's balance just before his nineth contribution by finding the annual effective interest rate and then check your answer using (4.2.8).

Solution   The annual effective interest rate is

$$I = \left(1 + \frac{.048}{12}\right)^{12} - 1 \approx .049070208.$$

The accumulated value is $3,000\ddot{s}_{\overline{8}|I} \approx \$29,953.30$.

Alternatively, since there are $12 \times 8 = 96$ months in eight years and the monthly interest rate is $\dfrac{.048}{12} = .004$, (4.2.8) gives this accumulated value as equal to $\$3,000\dfrac{s_{\overline{96}|.004}}{a_{\overline{12}|.004}} \approx \$3,000\left(\dfrac{116.7553336}{11.69373775}\right) \approx \$29,953.30$.

We note that the BA II Plus calculator provides yet another way to solve the problem of Example (4.2.9). Put the calculator in BGN mode, and set P/Y = 1 and C/Y = 12 by keying

| 2ND | P/Y | 1 | ENTER | ↓ | 1 | 2 | ENTER | 2ND | QUIT |.

Follow this by pressing

| 8 | N | 4 | • | 8 | I/Y | 0 | PV | 3 | 0 | 0 | 0 | +/− | PMT | CPT | FV |.

The calculator display then shows "FV = 29,953.29709".   ∎

The present value of a perpetuity having payments at the end of each k interest periods is

(4.2.10) $$a_{\overline{\infty}|I} = \frac{1}{I} = \frac{1}{(1+i)^k - 1} = \frac{1}{is_{\overline{k}|i}},$$

and the present value of a perpetuity having payments at the beginning of each k interest periods is

(4.2.11) $$\ddot{a}_{\overline{\infty}|I} = \frac{1}{I/(1+I)} = \frac{(1+i)^k}{(1+i)^k - 1} \times \frac{v^k}{v^k} = \frac{1}{1 - v^k} = \frac{1}{ia_{\overline{k}|i}}.$$

Those who know calculus may prefer to derive (4.2.10) and (4.2.11) by taking the limit as r approaches infinity in (4.2.3) and (4.2.8), respectively, noting that the number k is a constant (the constant k gives the number of interest periods between payments), so n approaches infinity when r does.

EXAMPLE 4.2.12   Perpetuity

Problem: A perpetuity paying $1,650 at the end of each year is to be replaced by another paying X at the beginning of each two years. The present value of each perpetuity is $31,250. Find X.

Solution   The original perpetuity has present value $\frac{\$1,650}{i}$. So, $31,250 = \frac{\$1,650}{i}$ and $i = \frac{\$1,650}{\$31,250} = .0528$. According to (4.2.11), the replacement perpetuity has present value $\frac{X}{.0528 a_{\overline{2}|.0528}}$. Since this is equal to $31,250,

$$X = \$31,250(.0528)a_{\overline{2}|.0528} = \$1,650a_{\overline{2}|.0528} \approx \$3,055.90.$$

If you prefer to convert interest rates numerically as was our approach in Chapter 3, the two-year effective interest rate is $I = (1.0528)^2 - 1$, the equivalent effective discount rate is $D = \frac{I}{1+I} = \frac{(1.0528)^2-1}{(1.0528)^2}$, $31,250 = \frac{X}{D}$, and

$$X = \$31,250\left(\frac{(1.0528)^2-1}{(1.0528)^2}\right) \approx \$3,055.90. \qquad \blacksquare$$

4.3   LEVEL ANNUITIES WITH PAYMENTS MORE FREQUENT THAN EACH INTEREST PERIOD

Last section we considered annuities whose payment period consisted of several interest periods. In contrast, in this section we focus on annuities whose interest period is made up of multiple payment periods. Before looking at the general situation, we will consider an example.

EXAMPLE 4.3.1   Interest rate for payment period found

Problem: Mahpee was offered a scholarship with level monthly payments for four years. The first payment would be made on September 30, 2000, and each year the payments would total $5,904. Mahpee, who was contemplating actuarial studies, wished to compare this scholarship to a fellowship offered by another college. The latter would pay $5,800 on September 1st of years 2000, 2001, 2002, and 2003. Assume that the growth of money is governed by compound interest at an annual effective interest rate of 4%, and compare the values of the two grants on September 1, 2000.

Solution   The September 1, 2000 value of the grant with annual payments of $5,800 was $5,800\ddot{a}_{\overline{4}|4\%} \approx \$21,895.53$. To compute the September 1,

2000 value of the grant with monthly payments, we first determine the effective monthly interest rate. It is $J = (1.04)^{\frac{1}{12}} - 1 \approx .327373978\%$. The September 1, 2000 value of the scholarship with monthly payments was $(\frac{1}{12} \times \$5,904)a_{\overline{4\times12}|J} \approx \$21,821.02$. Mahpee explained to her family that the financial aid offer with annual payments totaling \$5,904 had a lower monetary value than the one with beginning-of-year payments of \$5,800, at least if one calculated using a 4% annual effective interest rate. ∎

In the solution to Example (4.3.1), when contemplating the annuity with monthly payments, we numerically calculated the interest rate J per payment period and considered the number of payments m per year (m equaled 12), and the duration n of the annuity in interest periods. (The interest period was a year and n equaled 4.) We then found that the present value was equal to $(\frac{1}{m} \times \$5,904)\$a_{\overline{nm}|J}$. Put another way, for each annual dollar of scholarship payment we had a present value (in dollars) equal to $\frac{1}{m}a_{\overline{nm}|J}$. More generally, we have

---

FACT (4.3.2):

The expression $\frac{1}{m}a_{\overline{nm}|J}$ gives the present value of an annuity that pays $\frac{1}{m}$ at the end of each m-th of an interest period, for a total of 1 per interest period. Here $J = (1 + i)^{\frac{1}{m}} - 1 = \frac{i^{(m)}}{m}$ is the interest rate for the payment period.

Of course, we could have valued the annuity at the end of the nm payment periods (n interest periods) rather than at the beginning, and this would have changed our results by a factor $(1 + J)^{nm}$. So, the annuity of (4.3.2) has an accumulated value of $\frac{1}{m}s_{\overline{nm}|J}$ at the time of the last payment.

Let us now take a somewhat different approach. We again consider a level annuity lasting n interest periods and paying $\frac{1}{m}$ at the end of each m-th of an interest period. As previously noted, this annuity has payments totaling 1 each interest period. Introduce the new annuity symbols $a_{\overline{n}|i}^{(m)}$ for the present value of this annuity and $s_{\overline{n}|i}^{(m)}$ for its accumulated value at the time of the last payment.

PAYMENT: $\dfrac{1}{m}$ $\dfrac{1}{m}$ ... $\dfrac{1}{m}$ $\dfrac{1}{m}$ $\dfrac{1}{m}$ ... $\dfrac{1}{m}$ ... $\dfrac{1}{m}$ $\dfrac{1}{m}$ ... $\dfrac{1}{m}$

TIME: 0 $\dfrac{1}{m}$ $\dfrac{2}{m}$ ... 1 $\dfrac{m+1}{m}$ $\dfrac{m+2}{m}$ ... 2 ... $\dfrac{(n-1)m}{m}$ $\dfrac{(n-1)m+1}{m}$ ... n

interest period 1   interest period 2   interest period n

VALUE: $a_{\overline{n}|i}^{(m)}$ $s_{\overline{n}|i}^{(m)}$

FIGURE (4.3.3) Annuity-immediate

If $m = 1$, then $a_{\overline{n}|i}^{(m)} = a_{\overline{n}|i}$ and $s_{\overline{n}|i}^{(m)} = s_{\overline{n}|i}$, so the symbols $a_{\overline{n}|i}^{(m)}$ and $s_{\overline{n}|i}^{(m)}$ generalize $a_{\overline{n}|i}$ and $s_{\overline{n}|i}$.

We already know that

(4.3.4)
$$a_{\overline{n}|i}^{(m)} = \frac{1}{m}a_{\overline{nm}|J} \text{ and } s_{\overline{n}|i}^{(m)} = \frac{1}{m}s_{\overline{nm}|J}$$
$$\text{where } J = (1 + i)^{\frac{1}{m}} - 1 = \frac{i^{(m)}}{m},$$

but we now wish to derive alternate formulas for $a_{\overline{n}|i}^{(m)}$ and $s_{\overline{n}|i}^{(m)}$ which generalize (3.2.4) and (3.2.10). Using the formula (3.2.2) for the sum of a geometric series, we find

(4.3.5)
$$a_{\overline{n}|i}^{(m)} = \frac{1}{m}v^{\frac{1}{m}} + \frac{1}{m}v^{\frac{2}{m}} + \dots + \frac{1}{m}v^{\frac{nm}{m}}$$
$$= \frac{\frac{1}{m}v^{\frac{1}{m}}\left(1 - (v^{\frac{1}{m}})^{nm}\right)}{1 - v^{\frac{1}{m}}} = \frac{v^{\frac{1}{m}}(1 - v^n)}{m(1 - v^{\frac{1}{m}})} = \frac{1 - v^n}{m[(1 + i)^{\frac{1}{m}} - 1]}.$$

[In Problem (4.3.7), you are asked to show that this last formula for $a_{\overline{n}|i}^{(m)}$ may also be derived starting with Fact (4.3.2).] Recalling (1.10.2) and (3.2.4), it follows from (4.3.5) that

(4.3.6)
$$a_{\overline{n}|i}^{(m)} = \frac{1 - v^n}{i^{(m)}} = a_{\overline{n}|i}\left(\frac{i}{i^{(m)}}\right).$$

Multiplying (4.3.6) by $(1 + i)^n$ yields

(4.3.7)
$$s_{\overline{n}|i}^{(m)} = \frac{(1 + i)^n - 1}{i^{(m)}} = s_{\overline{n}|i}\left(\frac{i}{i^{(m)}}\right).$$

The quotients in Equations (4.3.6) and (4.3.7) giving the annuity symbols $a_{\overline{n}|i}^{(m)}$ and $s_{\overline{n}|i}^{(m)}$ are just like those in (3.2.4) and (3.2.10) *except* that in the *denominator* the effective interest rate i has been replaced by the nominal rate $i^{(m)}$. The superscript "(m)" in $a_{\overline{n}|i}^{(m)}$ and $s_{\overline{n}|i}^{(m)}$ will hopefully help you to remember you need a superscript "(m)" on the interest rate in the denominator.

EXAMPLE 4.3.8   Three methods

Problem: Shirley invests $200 at the end of each quarter for eight years in an account having an annual effective interest rate of 4%. Find Shirley's accumulated value at the time of her last deposit.

Solution   One solution to the problem is obtained by noting that the quarterly interest rate is $J = \dfrac{i^{(4)}}{4} = (1.04)^{\frac{1}{4}} - 1$, and there are $8 \times 4 = 32$ quarters in an eight-year period. Therefore, the accumulated value is $200s_{\overline{32}|J} \approx \$7{,}481.05$.

Our most recent discussion suggests that we solve the problem as follows. Note that $i^{(4)} = 4((1.04)^{\frac{1}{4}} - 1)$. Shirley's quarterly deposits are each $\$800 \times \frac{1}{4}$. Therefore, her accumulated value is $\$800s_{\overline{8}|i}^{(4)} = \$800\dfrac{(1.04)^8 - 1}{4[(1.04)^{\frac{1}{4}} - 1]} \approx \$7{,}481.05$.

A third approach is to use the BA II Plus calculator's **TVM** with P/Y = 4 and C/Y =1. Key

$$\boxed{2}\boxed{0}\boxed{0}\boxed{+/-}\;\boxed{\text{PMT}}\;\boxed{3}\boxed{2}\;\boxed{\text{N}}\;\boxed{4}\;\boxed{\text{I/Y}}\;\boxed{0}\;\boxed{\text{PV}}\;\boxed{\text{CPT}}\;\boxed{\text{FV}}.$$

This results in FV = 7,481.048277 being displayed.   ∎

We primarily consider symbols $a_{\overline{n}|i}^{(m)}$ and $s_{\overline{n}|i}^{(m)}$ when m is a positive integer greater than 1, and this is the situation in Figure (4.3.3). However, we might also consider these symbols for m equaling the reciprocal $\frac{1}{k}$ of a positive integer k. Then we are no longer considering an annuity with multiple payments per interest period. In fact, if $m = \frac{1}{k}$, then Equation (4.3.5) yields k times the annuity whose present value is given in (4.2.3).

Annuities-due with m payments per interest period also warrant our attention. Focus on an annuity lasting n interest periods and paying $\frac{1}{m}$ at the beginning of each m-th of an interest period. We define $\ddot{a}_{\overline{n}|i}^{(m)}$ to be the present value of this annuity and $\ddot{s}_{\overline{n}|i}^{(m)}$ to be its accumulated value at the end of n interest periods.

PAYMENT: $\frac{1}{m}$ $\frac{1}{m}$ $\frac{1}{m}$ ... $\frac{1}{m}$ $\frac{1}{m}$ $\frac{1}{m}$ ... $\frac{1}{m}$ ... $\frac{1}{m}$ ... $\frac{1}{m}$

TIME: 0 $\frac{1}{m}$ $\frac{2}{m}$ ... 1 $\frac{m+1}{m}$ $\frac{m+2}{m}$ ... 2 ... $\frac{(n-1)m}{m}$ $\frac{(n-1)m+1}{m}$ ... $\frac{nm-1}{m}$ n

interest period 1   interest period 2   interest period n

VALUE: $\ddot{a}^{(m)}_{\overline{n}|i}$ $\ddot{s}^{(m)}_{\overline{n}|i}$

$$\textbf{FIGURE (4.3.9) Annuity-due}$$

The symbols $\ddot{a}^{(m)}_{\overline{n}|i}$, $\ddot{s}^{(m)}_{\overline{n}|i}$, $a^{(m)}_{\overline{n}|i}$, and $s^{(m)}_{\overline{n}|i}$ are widely used in actuarial science.

Formulas for $\ddot{a}^{(m)}_{\overline{n}|i}$ may be derived directly from the definition of the symbol using the formula (3.2.2) for the sum of a geometric series. Alternatively, since $\frac{i^{(m)}}{m}$ is the effective interest rate for an m-th of an interest period and Equation (1.10.13) gives $1 + \frac{i^{(m)}}{m} = \frac{i^{(m)}}{d^{(m)}}$, from (4.3.6) we obtain

$$(4.3.10) \qquad \ddot{a}^{(m)}_{\overline{n}|i} = \left(1 + \frac{i^{(m)}}{m}\right)a^{(m)}_{\overline{n}|i} = \frac{i^{(m)}}{d^{(m)}}a^{(m)}_{\overline{n}|i} = a_{\overline{n}|i}\left(\frac{i}{d^{(m)}}\right) = \frac{1 - v^n}{d^{(m)}}.$$

Starting with (4.3.10) and multiplying by $(1 + i)^n$, we find

$$(4.3.11) \qquad \begin{aligned} \ddot{s}^{(m)}_{\overline{n}|i} &= \left(1 + \frac{i^{(m)}}{m}\right)s^{(m)}_{\overline{n}|i} = \frac{i^{(m)}}{d^{(m)}}s^{(m)}_{\overline{n}|i} \\ &= s_{\overline{n}|i}\left(\frac{i}{d^{(m)}}\right) = \frac{(1+i)^n - 1}{d^{(m)}}. \end{aligned}$$

Note also that, just as Equation (4.3.10) was obtained from (4.3.6) by multiplying by $1 + \frac{i^{(m)}}{m} = 1 + J$, multiplying the annuity equations of (4.3.4) by $1 + J$ yields

$$(4.3.12) \qquad \begin{aligned} \ddot{a}^{(m)}_{\overline{n}|i} &= \frac{1}{m}\ddot{a}_{\overline{nm}|J} \text{ and } \ddot{s}^{(m)}_{\overline{n}|i} = \frac{1}{m}\ddot{s}_{\overline{nm}|J} \\ \text{where } J &= (1 + i)^{\frac{1}{m}} - 1 = \frac{i^{(m)}}{m}. \end{aligned}$$

With so many new equations giving the values of the annuity symbols $a^{(m)}_{\overline{n}|i}$, $s^{(m)}_{\overline{n}|i}$, and their annuity-due counterparts, it behooves us to now present a numerical example showing their use.

EXAMPLE 4.3.13   Computing $a_{\overline{n}|i}^{(m)}$ and $\ddot{a}_{\overline{n}|i}^{(m)}$

Problem: Compute the values of the annuity symbols $a_{\overline{48}|5\%}^{(12)}$ and $\ddot{a}_{\overline{48}|5\%}^{(12)}$. Describe what each of them measures.

Equation (4.3.4) solution   We calculate $a_{\overline{48}|5\%}^{(12)}$ by applying equation (4.3.4) with $n = 48$, $m = 12$, and $J = (1.05)^{\frac{1}{12}} - 1 \approx .407412378\%$. Calculation shows that $a_{\overline{48}|5\%}^{(12)} = \frac{1}{12}a_{\overline{48\times12}|J} = \frac{1}{12}a_{\overline{576}|J} \approx 18.48777674$. Then $\ddot{a}_{\overline{48}|5\%}^{(12)} = (1 + J)a_{\overline{48}|5\%}^{(12)} \approx (1.00407412378)(18.48777674) \approx 18.56309823$. Of course, we might alternatively calculate $\ddot{a}_{\overline{48}|5\%}^{(12)}$ by using Equation (4.3.12) with n, m, and J as specified; that is, using $\ddot{a}_{\overline{48}|5\%}^{(12)} = \frac{1}{12}\ddot{a}_{\overline{576}|J}$. With your BA II Plus calculator in BGN mode and P/Y = C/Y = 1, you quickly find $\ddot{a}_{\overline{48}|5\%}^{(12)} = \frac{1}{12}\ddot{a}_{\overline{576}|J} \approx 18.56309823$; in END mode with P/Y = C/Y = 1, use N = 576, I/Y = J as a percent, PMT = -1, and FV = 0.

Equation (4.3.6) solution   In order to use equation (4.3.6) with $n = 48$ and $m = 12$, we need to calcuate $i^{(12)}$. But $i^{(12)} = 12[(1.05)^{\frac{1}{12}} - 1] \approx .048889485$. Then $a_{\overline{48}|5\%}^{(12)} = \frac{1-(1.05)^{-48}}{i^{(12)}} \approx \frac{.90385789}{.048889485} \approx 18.48777674$. Then equation (4.3.10) can be used to calculate $\ddot{a}_{\overline{48}|5\%}^{(12)}$, and since $\frac{i^{(12)}}{12} = J$, this is the same calculation as we saw in our first solution.

TVM worksheet solution   Push

| 2ND | P/Y | 1 | 2 | ENTER | ↓ | 1 | ENTER | 2ND | QUIT |.

Assuming you are in END mode, keying

| 5 | 7 | 6 | N | 5 | I/Y | 1 | ÷ | 1 | 2 | = | +/− | PMT | 0 | FV | CPT | PV |

should result in the display reading "PV = 18.48777674". On the other hand, if you change to BGN mode by keying | 2ND | BGN | 2ND | SET | 2ND | QUIT | pushing | CPT | PV | results in the display showing "PV = 18.56309823". ∎

If you look at equations (3.2.4), (3.3.5), (4.3.6), and (4.3.10), you note that each of them gives an annuity symbol, representing a present value of an annuity lasting n periods, as a quotient of $1 - v^n$ by an interest measure. Similarly, equations (3.2.10), (3.3.6), (4.3.7), and (4.3.11) each present a formula for an annuity symbol giving the accumulated value at time n of an annuity lasting from time 0 to time n, and these formulas each show the symbol as a quotient of $(1 + i)^n - 1$ by an interest measure. It follows from these formulas and Important Fact (1.10.16), that

(4.3.14)
$$\ddot{a}_{\overline{n}|i} > \ddot{a}_{\overline{n}|i}^{(m)} > a_{\overline{n}|i}^{(m)} > a_{\overline{n}|i}.$$

Equation (4.3.14) may also be obtained by considering descriptions of the annuities underlying the annuity symbols. This approach is taken in Section (4.6) where the string of inequalities of (4.3.14) is extended to include one additional annuity symbol.

EXAMPLE 4.3.15   Comparing annuity symbols

Problem: Find the numerical values of the four annuity symbols in inequality (4.3.14) when $n = 48$, $m = 12$, and $i = 5\%$.

Solution In Example (4.3.13) we found that $a_{\overline{48}|5\%}^{(12)} \approx 18.48777674$ and $\ddot{a}_{\overline{48}|5\%}^{(12)} \approx 18.56309823$. One calculates that $a_{\overline{48}|5\%} \approx 18.07715782$ and $\ddot{a}_{\overline{48}|5\%} \approx 18.98101571$.

Substituting these numerical values into (4.3.14), the inequality reads

$$18.98101571 > 18.56309823 > 18.48777674 > 18.07715782.$$

■

Having studied annuities with payments once every m-th of an interest period, let us move on to consider perpetuities with this payment pattern. If the perpetuity is immediate, the present value of the perpetuity is denoted $a_{\overline{\infty}|i}^{(m)}$, and if it is due, the present value is denoted $\ddot{a}_{\overline{\infty}|i}^{(m)}$. Once again, let J denote the effective interest rate for an m-th of an interest period. Then $J = (1 + i)^{\frac{1}{m}} - 1$. The perpetuity under consideration has payments of $\frac{1}{m}$, whence (3.4.2) yields

(4.3.16)
$$a_{\overline{\infty}|i}^{(m)} = \frac{1}{m}\frac{1}{J} = \frac{1}{m((1 + i)^{\frac{1}{m}} - 1)} = \frac{1}{i^{(m)}}.$$

Then, again recalling (1.10.13),

(4.3.17)
$$\ddot{a}_{\overline{\infty}|i}^{(m)} = \left(1 + \frac{i^{(m)}}{m}\right) a_{\overline{\infty}|i}^{(m)} = \frac{i^{(m)}}{d^{(m)}}\frac{1}{i^{(m)}} = \frac{1}{d^{(m)}}.$$

If you are familiar with calculus, note that formulas (4.3.16) and (4.3.17) follow from (4.3.6) and (4.3.10) since $\lim_{n \to \infty} v^n = 0$.

EXAMPLE 4.3.18   Perpetuity

Problem: Pavol purchases a perpetuity-due. The perpetuity has monthly payments of $100 during the first five years, and then these increase to $300. Determine the purchase price if it is based on an annual effective discount rate of 3.6%.

Solution   The purchase price is equal to the sum of the present value of a perpetuity-due with payments of $100 monthly and the present value of a deferred perpetuity-due with payments of $200 monthly and a deferment period of five years. We note using (1.10.9) that $d^{(12)} = 12[1 - (1 - .036)^{\frac{1}{12}}] \approx .036608031$, and the perpetuities have annual payments totaling $1,200 and $2,400. Therefore, using (4.3.17), the purchase price is equal to $\frac{\$1,200}{d^{(12)}} + (1 - .036)^5 \frac{\$2,400}{d^{(12)}} \approx \$32,779.692 + \$54,578.305 \approx \$87,358.00.$ ■

4.4   ANNUITIES WITH PAYMENTS LESS FREQUENT THAN EACH INTEREST PERIOD AND PAYMENTS IN ARITHMETIC PROGRESSION

As in Section (4.2), we consider an annuity with payments once every k interest periods where k a positive integer. However, we suppose that the annuity has payments in arithmetic progression rather than constant payments. Payments in arithmetic progression were considered in (3.9), but then the interest period and payment period coincided. We assume that the first payment occurs at the end of k interest periods and is for an amount P and that subsequent payments each increase by an amount Q.

PAYMENT:		P	$P + Q$	$P + 2Q$	$\cdots$	$P + (r-1)Q$
TIME:	0	k	$2k$	$3k$	$\cdots$	$rk = n$

FIGURE (4.4.1)

We wish to discover a formula for the annuity's accumulated value S. We do so by noting that the interest rate for the payment period is $I = (1 + i)^k - 1 = is_{\overline{k}|i}$ and the annuity under consideration has $r = \frac{n}{k}$ payments. Therefore, by (3.9.3),

$$S = Ps_{\overline{r}|I} + \frac{Q}{I}(s_{\overline{r}|I} - r).$$

Using (4.2.4), we find

$$S = P\frac{s_{\overline{n}|i}}{s_{\overline{k}|i}} + \frac{Q}{is_{\overline{k}|i}}\left(\frac{s_{\overline{n}|i}}{s_{\overline{k}|i}} - \frac{n}{k}\right).$$

Should we want the present value A of this annuity, we multiply S by v^n. Therefore,

$$A = P\frac{a_{\overline{n}|i}}{s_{\overline{k}|i}} + \frac{Q}{is_{\overline{k}|i}}\left(\frac{a_{\overline{n}|i}}{s_{\overline{k}|i}} - \frac{n}{k}v^n\right).$$

We gather these findings below.

FACT 4.4.2: Let k divide n. The accumulated value at the time of the last payment of an annuity lasting n interest periods and paying $P + Q(j - 1)$ at the end of the j-th interest period is

$P\frac{s_{\overline{n}|i}}{s_{\overline{k}|i}} + \frac{Q}{is_{\overline{k}|i}}\left(\frac{s_{\overline{n}|i}}{s_{\overline{k}|i}} - \frac{n}{k}\right).$ The value of this annuity one payment

period before the first payment is $P\frac{a_{\overline{n}|i}}{s_{\overline{k}|i}} + \frac{Q}{is_{\overline{k}|i}}\left(\frac{a_{\overline{n}|i}}{s_{\overline{k}|i}} - \frac{n}{k}v^n\right).$

Fact (4.4.2) generalizes (3.9.3) and its companion formula (3.9.4).

As in Equation (3.9.5), it is interesting to consider the special case where $P = Q = 1$. Then, recalling (3.9.5) and (4.2.4)

$$S = \frac{\ddot{s}_{\overline{n}|I} - r}{I} = \frac{\ddot{s}_{\overline{n}|I} - r}{is_{\overline{k}|i}} = \frac{(1 + i)^k s_{\overline{n}|I} - r}{is_{\overline{k}|i}}$$

$$= \frac{(1 + i)^k \frac{s_{\overline{n}|i}}{s_{\overline{k}|i}} - r}{is_{\overline{k}|i}} = \frac{\frac{s_{\overline{n}|i}}{a_{\overline{k}|i}} - \frac{n}{k}}{is_{\overline{k}|i}}.$$

Therefore, we have

FACT 4.4.3: Let k divide n. The accumulated value at the time of the last payment of an annuity lasting n interest periods, having a payment at the end of each k interest periods, and having the j-th

payment be for an amount j is $\dfrac{\frac{s_{\overline{n}|i}}{a_{\overline{k}|i}} - \frac{n}{k}}{is_{\overline{k}|i}}$. The value of this annuity

one payment period before the first payment is $\dfrac{\frac{a_{\overline{n}|i}}{a_{\overline{k}|i}} - \frac{n}{k}v^n}{is_{\overline{k}|i}}$.

This generalizes (3.9.5) and (3.9.6).

EXAMPLE 4.4.4   Applying Fact (4.4.3)

Problem: On April 1, 1998, Mordekai opened a savings account with a nominal interest rate of 3.6% payable monthly by making a $100 deposit. At the same time he set up an investment fund with a nominal interest rate of 4.4% payable quarterly. Mordekai deposited $500 into the investment fund at the end of each quarter with his first payment being on July 1, 1998. The quarterly interest payments from the fund were directly deposited into the 3.6% savings account. Find the balance in the 3.6% savings account on July 1, 2003, just after the twentieth interest deposit is received from the 4.4% fund and monthly interest has been credited.

Solution   First note that the quarterly interest rate on the investment fund is $\frac{4.4\%}{4} = 1.1\% = .011$, and the monthly interest rate on the savings account is $\frac{3.6\%}{12} = .3\% = .003$.

The balance in Mordekai's investment fund for the quarter prior to the j-th interest payment is $500j$. Therefore, the amount of the j-th interest payment to the savings account is $500j(.011) = \$5.50j$.

We now focus on the 3.6% nominal account that receives these deposits. The first deposit is $P = \$5.50$, the increment between payments is $Q = \$5.50 = P$. These payments are received at the end of every $k = 3$ months (interest periods for this account) for a period of $n = 3 \times 20 = 60$ months. Using Fact (4.4.3) and recalling that the monthly interest rate is .3%, we find the accumulation from these contributions, at the time of the twentieth one, to be

$$\$5.50\frac{\frac{s_{\overline{60}|.003}}{a_{\overline{3}|.003}} - \frac{60}{3}}{.003s_{\overline{3}|.003}}.$$

Recollecting that the initial $100 has been on deposit for 63 months, we determine that the balance in the 3.6% savings account on July 1, 2003 is

$$\$100(1.003)^{63} + \$5.50\frac{\frac{s_{\overline{60}|.003}}{a_{\overline{3}|.003}} - \frac{60}{3}}{.003s_{\overline{3}|.003}} \approx \$1,344.57.$$

Of course this example could also have been solved by finding a quarterly interest rate for the 3.6% account, namely $I = (1.003)^3 - 1$, and then using Equation (3.9.5) to find that part of the accumulation resulting from the interest payments from the investment fund. ∎

4.5   ANNUITIES WITH PAYMENTS MORE FREQUENT THAN EACH INTEREST PERIOD AND PAYMENTS IN ARITHMETIC PROGRESSION

In Section (4.4) we considered annuities with the same schedule of payment times as in (4.2), namely once every k interest periods, but the annuities had payments in arithmetic progression rather than being level. The schedule of payments in (4.3) was more frequent than in (4.2) and (4.4), namely m times per interest period. Here we wish to work with this same, more rapid payment schedule, first when the payments increase in arithmetic progression so there is a constant increase each *payment period*, second when the payments stay level each interest period but there is a constant increase each *interest period*. In Section (4.4) we did not define any new annuity symbols, but here we will define a new set, used somewhat in actuarial science, for each of the two scenarios.

A constant increase each payment period

Consider the annuity that has payments at the end of each m-th of an interest payment, and the j-th payment is $\frac{j}{m^2}$, $j = 1, 2, \ldots, nm$. This annuity is an example of one that has a constant increase each payment period. Denote the present value of this annuity by $(I^{(m)}a)_{\overline{n}|i}^{(m)}$ and its accumulated value at the time of the last payment by $(I^{(m)}s)_{\overline{n}|i}^{(m)}$.

PAYMENT:		$\dfrac{1}{m^2}$	$\dfrac{2}{m^2}$	$\dfrac{3}{m^2}$	$\cdots$	$\dfrac{nm}{m^2}$
TIME:	0	$\dfrac{1}{m}$	$\dfrac{2}{m}$	$\dfrac{3}{m}$	$\cdots$	n
VALUE:	$(I^{(m)}a)_{\overline{n}|i}^{(m)}$					$(I^{(m)}s)_{\overline{n}|i}^{(m)}$

FIGURE (4.5.1)

Let J denote the effective interest rate for an m-th of an interest period. Then $J = (1 + i)^{\frac{1}{m}} - 1$ and $mJ = i^{(m)}$. Recalling (3.9.6), we have

$$(I^{(m)}a)_{\overline{n}|i}^{(m)} = \frac{1}{m^2}(Ia)_{\overline{nm}|J} = \frac{1}{m^2}\left(\frac{\ddot{a}_{\overline{nm}|J} - nm(\frac{1}{1+J})^{nm}}{J}\right) = \frac{\frac{1}{m}\ddot{a}_{\overline{nm}|J} - nv^n}{mJ}.$$

Now recall the definition of the annuity symbol $\ddot{a}_{\overline{n}|i}^{(m)}$, and conclude that

$$(4.5.2) \qquad \boxed{(I^{(m)}a)_{\overline{n}|i}^{(m)} = \frac{\ddot{a}_{\overline{n}|i}^{(m)} - nv^n}{i^{(m)}}.}$$

Multiplying by $(1 + i)^n$, gives the companion formula

$$(4.5.3) \qquad \boxed{(I^{(m)}s)_{\overline{n}|i}^{(m)} = \frac{\ddot{s}_{\overline{n}|i}^{(m)} - n}{i^{(m)}}.}$$

We note that the notation $(I^{(m)}a)_{\overline{n}|i}^{(m)}$ has m appearing twice. This may help you to remember that the right-hand side of formula (4.5.2) is identical to that of (3.9.6) except for two references to m. First, in the denominator the effective interest rate i is replaced by the nominal interest rate $i^{(m)}$. Second, in the numerator the annuity symbol $\ddot{a}_{\overline{n}|i}$ is replaced by the annuity symbol $\ddot{a}_{\overline{n}|i}^{(m)}$. Likewise, $(I^{(m)}s)_{\overline{n}|i}^{(m)}$ has two occurrences of m, and the right-hand side of formula (4.5.3) is the same as the last expression in (3.9.5) except for two m's.

The annuity symbol $(I^{(m)}a)_{\overline{\infty}|i}^{(m)}$ is used to denote the present value of the perpetuity paying $\frac{j}{m^2}$ at the end of the j-th m-th of an interest period. Recalling (4.3.17), just as (3.9.4) gives rise to (3.9.16), it follows from (4.5.2) that

$$(4.5.4) \qquad \boxed{(I^{(m)}a)_{\overline{\infty}|i}^{(m)} = \frac{\ddot{a}_{\overline{\infty}|i}^{(m)}}{i^{(m)}} = \frac{1}{i^{(m)}d^{(m)}}.}$$

Now suppose that an annuity pays at the end of each m-th of an interest payment for n interest conversion periods, and the j-th payment is $P + (j - 1)Q$. Then from (4.5.3) or from (3.9.3), one can derive that this annuity has accumulated value $mPs_{\overline{n}|i}^{(m)} + m^2\frac{Q}{i^{(m)}}(s_{\overline{n}|i}^{(m)} - n)$ [see Problem (4.5.8)].

EXAMPLE 4.5.5   Payments increase each payment period

Problem: Consider an eight-year annuity-immediate with monthly payments, the first payment being $25 and the monthly increase being $2. Calculate the present value using an annual effective interest rate of 3.3%.

Solution This annuity may be thought of as consisting of a level annuity-immediate with monthly payments of $23 for eight years together with an annuity-immediate with monthly payments that begin at $2 and increase by $2 monthly. Because the annuity symbol $(I^{(12)}a)^{(12)}_{\overline{8}|3.3\%}$ has an initial payment $\frac{1}{144}$ and increases by $\frac{1}{144}$ monthly, the present value of the given annuity is equal to

$$(23 \times 12)a^{(12)}_{\overline{8}|3.3\%} + (2 \times 144)(I^{(12)}a)^{(12)}_{\overline{8}|3.3\%}.$$

This is equal to

$$276a_{\overline{8}|3.3\%}\frac{.033}{i^{(12)}} + 288\left(\frac{\ddot{a}^{(12)}_{\overline{8}|3.3\%} - 8(1.033)^{-8}}{i^{(12)}}\right)$$

$$= 276a_{\overline{8}|3.3\%}\frac{.033}{i^{(12)}} + 288\left(\frac{a_{\overline{8}|3.3\%}\frac{.033}{d^{(12)}} - 8(1.033)^{-8}}{i^{(12)}}\right).$$

Note that $i^{(12)} \approx .0325111514$, $d^{(12)} = i^{(12)}/\left(1 + \frac{i^{(12)}}{12}\right) \approx .032423308$, and our present value is approximately $\$1,941.915 + \$7,839.369 \approx \$9,781.28$. ■

A constant increase each <u>interest</u> period

Let us now move on to consider an annuity with payments occurring at the end of each m-th of an *interest* period, level payments each *interest* period, and a constant increase each *interest* period. Consider the annuity that lasts n interest periods and in the j-th interest period has payments of $\frac{j}{m}$ at the end of each m-th of an interest period, $j = 1, 2, ..., n$. This annuity has payments totaling j in the j-th interest period. Denote the present value of this annuity by $(Ia)^{(m)}_{\overline{n}|i}$ and its accumulated value at the time of the last payment by $(Is)^{(m)}_{\overline{n}|i}$.

FIGURE (4.5.6)

A formula for $(Ia)_{\overline{n}|i}^{(m)}$, whose somewhat technical derivation is given later, is

(4.5.7)
$$\boxed{(Ia)_{\overline{n}|i}^{(m)} = \frac{\ddot{a}_{\overline{n}|i} - nv^n}{i^{(m)}}.}$$

Naturally, there is also a corresponding formula

(4.5.8)
$$\boxed{(Is)_{\overline{n}|i}^{(m)} = \frac{\ddot{s}_{\overline{n}|i} - n}{i^{(m)}}.}$$

In contrast to the notation $(I^{(m)}a)_{\overline{n}|i}^{(m)}$, $(Ia)_{\overline{n}|i}^{(m)}$ has only one reference to m. The right-hand side of formula (4.5.7) is identical to that of (3.9.6) except that in the denominator the effective interest rate i is replaced by the nominal interest rate $i^{(m)}$. Thus, the one reference to m corresponds to one change (involving m) in the formula.

The annuity symbol $(Ia)_{\overline{\infty}|i}^{(m)}$ is used to denote the present value of the perpetuity paying $\frac{j}{m}$ at the end of each m-th of an interest period during the j-th interest period, $j = 1, 2, \ldots, n, \ldots$. Just as (3.9.4) gives rise to (3.9.16), (4.5.7) gives

(4.5.9)
$$\boxed{(Ia)_{\overline{\infty}|i}^{(m)} = \frac{1}{di^{(m)}}.}$$

For those wishing to see the details, a derivation of (4.5.7) is as follows:

$$(Ia)_{\overline{n}|i}^{(m)} = \frac{1}{m}(v^{\frac{1}{m}} + v^{\frac{2}{m}} + \cdots + v^{\frac{m}{m}}) + \frac{2}{m}v(v^{\frac{1}{m}} + v^{\frac{2}{m}} + \cdots + v^{\frac{m}{m}})$$

$$+ \frac{3}{m}v^2(v^{\frac{1}{m}} + v^{\frac{2}{m}} + \cdots v^{\frac{m}{m}}) + \cdots + \frac{n}{m}v^{n-1}(v^{\frac{1}{m}} + v^{\frac{2}{m}} + \cdots + v^{\frac{m}{m}})$$

$$= \frac{1}{m}(v^{\frac{1}{m}} + v^{\frac{2}{m}} + \cdots + v^{\frac{m}{m}})(1 + 2v + 3v^2 + \cdots + nv^{n-1})$$

$$= a_{\overline{1}|i}^{(m)}(1 + 2v + 3v^2 + \cdots + nv^{n-1})$$

$$= a_{\overline{1}|i}\frac{i}{i^{(m)}}(1 + 2v + 3v^2 + \cdots + nv^{n-1})$$

$$= v\frac{i}{i^{(m)}}(1 + 2v + 3v^2 + \cdots + nv^{n-1}) = \frac{i}{i^{(m)}}(v + 2v^2 + 3v^3 + \cdots + nv^n)$$

$$= \frac{i}{i^{(m)}}(Ia)_{\overline{n}|i} = \frac{i}{i^{(m)}}\left(\frac{\ddot{a}_{\overline{n}|i} - nv^n}{i}\right) = \frac{\ddot{a}_{\overline{n}|i} - nv^n}{i^{(m)}}.$$

We note that an annuity lasting n interest periods with payments of $P + (j - 1)Q$ at the end of each m-th of the j-th interest period has accumulated value $m\left[Ps^{(m)}_{\overline{n}|i} + \frac{Q}{i^{(m)}}(s_{\overline{n}|i} - n)\right]$ at the time of the last payment. [see Problem (4.5.9)].

EXAMPLE 4.5.10   Payments level within each interest period and increasing from one interest period to the next

Problem: Dontrelle earns $2,000 at the end of each month during the first year of his employment. His monthly rate of compensation increases by $400 for each year of his employment. Therefore, it is $2,400 in the second year, $2,800 in the third year, and $2,000 + $400(20 − 1) = $9,600 in the twentieth year. If Dontrelle is employed for exactly twenty years, what level monthly rate of compensation has the same present value if the present values are computed at an annual effective interest rate of 5%?

Solution   Dontrelle receives salary payments at the end of each twelfth of a year. The first year these are each $2,000, and then each year they increase by $400. The monthly salary payments can be thought of as $1,600 plus 400j$ in the j-th year, Since the annuity symbol $a^{(12)}_{\overline{20}|.05}$ gives the present value of an annuity having level payments of $\frac{1}{12}$ each month, and the annuity underlying the symbol $(Ia)^{(12)}_{\overline{20}|.05}$ has payments of $\frac{j}{12}$ in the j-th year, the present value of Dontrelle's salary over the twenty-year period is

$$(\$1,600 \times 12)a^{(12)}_{\overline{20}|.05} + (\$400 \times 12)(Ia)^{(12)}_{\overline{20}|.05} .$$

Since $i^{(12)} = 12((1.05)^{\frac{1}{12}} - 1) \approx .048889485, (4.3.6)$ and $(4.5.7)$ allow us to calculate this to be approximately $244,709.5082 + $544,660.0542 \approx $789,369.56.

The monthly interest rate is $j = (1.05)^{\frac{1}{12}} - 1$ and a twenty-year period consists of 240 months. We therefore need to find X such that $Xa_{\overline{240}|j} = \$789,369.57$. We calculate that $X \approx \$5,161.19$.   ∎

4.6   CONTINUOUSLY PAYING ANNUITIES

(calculus needed here)

In Sections (4.3) and (4.5) we considered various annuities that have payments at the end of each m-th of an interest period. Here we discuss the limiting case where the payment frequency increases so that payments are continuous. You might think of this situation as one in which there is a magic faucet delivering a continuous trickle of payments. This description suggests that such annuities do not actually occur, and this is indeed the case, but they are of theoretical interest, are used extensively in actuarial science, and may be useful for approximating annuities payable with great frequency.

Consider the continuous payment of money at the uniform and constant rate of 1 per interest period for n interest periods, the effective interest rate per interest period being a constant rate i. The annuity symbol $\bar{a}_{\overline{n}|i}$ denotes the present value of the payments at the start of the n interest periods and the symbol $\bar{s}_{\overline{n}|i}$ denotes the accumulated value at the end. Recall that the annuity underlying the symbols $a_{\overline{n}|i}^{(m)}$ and $s_{\overline{n}|i}^{(m)}$ also paid 1 per interest period, and the payments came in quantities of $\frac{1}{m}$ once every m-th of an interest period. So, $\bar{a}_{\overline{n}|i}$ and $\bar{s}_{\overline{n}|i}$ may be obtained by taking limits of $a_{\overline{n}|i}^{(m)}$ and of $s_{\overline{n}|i}^{(m)}$, respectively, as m tends to infinity. Recall the constant force of interest $\delta = \lim_{m \to \infty} i^{(m)} = \ln(1 + i)$; this is formula (1.11.1), and δ is the nominal interest rate convertible continuously and equivalent to i. We then have

(4.6.1)
$$\bar{a}_{\overline{n}|i} = \lim_{m \to \infty} a_{\overline{n}|i}^{(m)} = \lim_{m \to \infty} \frac{1 - v^n}{i^{(m)}} = \frac{1 - v^n}{\delta} = \frac{1 - e^{-\delta n}}{\delta},$$

and

(4.6.2)
$$\bar{s}_{\overline{n}|i} = \lim_{m \to \infty} s_{\overline{n}|i}^{(m)} = \lim_{m \to \infty} \frac{(1 + i)^n - 1}{i^{(m)}} = \frac{(1 + i)^n - 1}{\delta} = \frac{e^{\delta n} - 1}{\delta}.$$

EXAMPLE 4.6.3   Annuity with level continuous withdrawals

Problem: Martin deposits $12,000 in a savings account with a discount rate of 10% convertible quarterly. He leaves his money in this account to accumulate for fifteen years, then moves it to a fund that is accumulating at 8% per annum convertible continuously. If, starting at time 15 when he invests in the new fund, money is withdrawn continuously at a rate of $6,000 per annum, how long will Martin's money last?

Solution   At the end of fifteen years (sixty quarters), Martin's $12,000 has grown to $12,000\left(1 - \frac{.1}{4}\right)^{-60} \approx \$54,815.62$. This is the amount that is transferred into an account with a constant force of interest $\delta = .08$. Note that when $\delta = .08$, $i = e^{.08} - 1$. Since money is withdrawn continuously at a rate of $6,000 per annum, if T denotes the length of time in years until the account with $\delta = .08$ is exhausted, we have $\$54,815.62 = \$6,000 \bar{a}_{\overline{T}|(e^{.08}-1)}$. Recalling (4.6.1), $\$54,815.62 = \$6,000\left(\frac{1-e^{-.08T}}{.08}\right) = \$75,000(1 - e^{-.08T})$. Equivalently, $\$20,184.38 = \$75,000 e^{-.08T}$. So, $T = \dfrac{\ln\left(\frac{20,184.38}{75,000}\right)}{-.08} \approx 16.40723844$. Since $16.41 + 15 = 31.41$, Martin's money lasts about 31.41 years from the time of the initial $12,000 deposit. ■

We note that the annuity symbols $a_{\overline{n}|i}$, $\ddot{a}_{\overline{n}|i}$, $a_{\overline{n}|i}^{(m)}$, $\ddot{a}_{\overline{n}|i}^{(m)}$, and $\overline{a}_{\overline{n}|i}$ all give the present value of an annuity lasting from time 0 to time n and paying 1 per interest period. The present value $\ddot{a}_{\overline{n}|i}$ is highest because you get 1 at the very beginning of each interest period. In contrast, the annuity underlying $\ddot{a}_{\overline{n}|i}^{(m)}$ pays $\frac{1}{m}$ at the beginning of each m-th of an interest period, and the annuity underlying $\overline{a}_{\overline{n}|i}$ has 1 trickling in levelly over an m-th of an interest period, so at time $\frac{1}{m}$ you have received a total of $\frac{1}{m}$. Therefore, $\ddot{a}_{\overline{n}|i} > \ddot{a}_{\overline{n}|i}^{(m)} > \overline{a}_{\overline{n}|i}$. We can continue this string of inequalities, if we note that of the five present values mentioned at the beginning of this paragraph, $a_{\overline{n}|i}$ is the lowest because you have to wait until the very end of an interest period to get any of the 1 due that period. The symbol $a_{\overline{n}|i}^{(m)}$ is the present value of an annuity paying $\frac{1}{m}$, but only at the very end of an m-th of an interest period. Therefore, it is less than $\overline{a}_{\overline{n}|i}$ since, as already mentioned, the annuity underlying $\overline{a}_{\overline{n}|i}$ has 1 trickling in levelly per interest period, hence $\frac{1}{m}$ coming in gradually over an m-th of an interest period. So $\overline{a}_{\overline{n}|i} > a_{\overline{n}|i}^{(m)} > a_{\overline{n}|i}$. Putting this all together, we have

(4.6.4)
$$\ddot{a}_{\overline{n}|i} > \ddot{a}_{\overline{n}|i}^{(m)} > \overline{a}_{\overline{n}|i} > a_{\overline{n}|i}^{(m)} > a_{\overline{n}|i}.$$

Formula (4.6.4) extends the string of inequalities of (4.3.14). Note that we established the inequalities of (4.3.14) algebraically using the fact that all of the symbols had numerators of $1 - v^n$. Equation (4.6.1) gives $\overline{a}_{\overline{n}|i}$ as a quotient of $1 - v^n$ divided by δ, so (4.6.4) has an algebraic verification as well. Just combine Equations (3.2.4), (3.3.5), (4.3.6), (4.3.10), and (4.6.1), each of which give one of our present values as a quotient of $1 - v^n$ divided by an interest measure, and Equation (1.11.5), which orders the denominators of these ratios.

Next suppose that we have a continuous payment of money at rate t per interest period at time t, this lasts for n interest periods, and that there is an annual effective rate of interest of i throughout the n interest periods. (The mythical faucet's drip increases at a steady rate.) The annuity symbol $(\overline{I}\overline{a})_{\overline{n}|i}$ denotes the present value of this continuous annuity and the symbol $(\overline{I}\overline{s})_{\overline{n}|i}$ gives the accumulated value at the end of the n interest periods. The annuity underlying the symbols $(I^{(m)}a)_{\overline{n}|i}^{(m)}$ and $(I^{(m)}s)_{\overline{n}|i}^{(m)}$ lasts n interest periods, and it pays at a rate of $\frac{j}{m}$ over the interval $[\frac{j-1}{m}, \frac{j}{m}]$ [see Problem (4.6.7)(a)].

(4.6.5)
$$(\overline{I}\overline{a})_{\overline{n}|} = \lim_{m \to \infty} (I^{(m)}a)_{\overline{n}|i}^{(m)} = \frac{\ddot{a}_{\overline{n}|i} - nv^n}{\delta}.$$

and

(4.6.6)
$$(\bar{I}\bar{s})_{\overline{n}|} = \lim_{m\to\infty} (I^{(m)}s)^{(m)}_{\overline{n}|i} = \frac{\bar{s}_{\overline{n}|i} - n}{\delta}.$$

In order to obtain (4.6.5) and (4.6.6), we have used the limit (1.11.1).

For the remainder of Section (4.6), let us drop the assumption that the growth of money is governed by compound interest. Suppose we have an arbitrary integrable discount function $v(t)$. Let the symbol $a^{(m)}_{\overline{n}|}$ denote the present value of an annuity lasting from time 0 to time n and paying $\frac{1}{m}$ at the end of each m-th of a unit of time. By $\bar{a}_{\overline{n}|}$ one indicates the present value of a continuously paying annuity lasting from time 0 to time n and paying at a level rate of 1 per unit of time. The present value $a^{(m)}_{\overline{n}|}$ is equal to the sum $\frac{1}{m}v(\frac{1}{m}) + \frac{1}{m}v(\frac{2}{m}) + \ldots + \frac{1}{m}v(\frac{nm}{m})$. Noting that the function $v(t)$ is a decreasing function, this is the lower sum for the partition $\{0, \frac{1}{m}, \frac{2}{m}, \ldots, \frac{nm}{m}\}$ of the interval $[0, n]$.

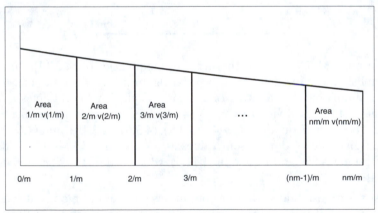

FIGURE (4.6.7) Area below graph of $v(t)$

As m increases to infinity, the width $\frac{1}{m}$ of the rectangles tends to zero. Therefore, the sum giving $a^{(m)}_{\overline{n}|i}$ approaches $\int_0^n v(t)\,dt$ and we have

(4.6.8)
$$\bar{a}_{\overline{n}|} = \lim_{m\to\infty} a^{(m)}_{\overline{n}|} = \int_0^n v(t)\,dt.$$

You may think of the integral $\int_0^n v(t)\,dt$ as the sum over all times t where payments are made $(0 \leqslant t \leqslant n)$ of the product of the amount paid at time t

(namely, dt) and the discount function $v(t)$. This is commonly done in later actuarial thinking.

In the special case where $v(t) = v^t$, computing the integral of (4.6.8) produces $\frac{1-v^n}{\delta}$ (since $\int_0^n v^t \, dt = \frac{1}{\ln v} v^t \big|_0^n = \frac{1}{\ln v}(v^n - 1) = \frac{1}{-\delta}(v^n - 1) = \frac{1}{\delta}(1 - v^n)$) and gives an alternate proof of (4.6.1).

To study more general, continuously paying annuities, let $f(t)$ denote a continuous function on the interval $[0, n]$. Focus on a continuously paying annuity lasting from time 0 to time n and paying at a rate of $f(t)$ per unit of time at time t. The value of this annuity at time 0 is $\int_0^n f(t)v(t) \, dt$, because the differential $f(t)v(t)dt$ is the present value of the payment $f(t) \, dt$ made at exact instant t.

FACT (4.6.9): Let $f(t)$ denote a continuous function on the interval $[0, n]$. The present value of the continuously paying annuity that pays from time 0 to time n at a rate of $f(t)$ at time t is $\int_0^n f(t)v(t) \, dt$.

If $f(t) = 1$, this is just (4.6.8) once again. If $f(t) = t$, then we are considering the present value of an annuity such as is in (4.6.5), although we are not restricted to compound interest.

EXAMPLE 4.6.10   Applying Fact (4.6.9) with accumulation other than by compound interest

Problem: Suppose that money grows according to simple discount at a rate of 3.6%. Find formulas for $\bar{a}_{\overline{n}|}$, $\bar{s}_{\overline{n}|}$, and $(\bar{I}\bar{a})_{\overline{n}|}$.

Solution The accumulation function governing the growth of money is $a(t) = \frac{1}{1-.036t}$, $0 \le t < \frac{1}{.036}$. Therefore, the discount function is $v(t) = 1 - .036t$, $0 \le t < \frac{1}{.036}$ and

$$\bar{a}_{\overline{n}|} = \int_0^n v(t) \, dt = \int_0^n (1 - .036t) \, dt = \left(t - \frac{.036t^2}{2}\right)\bigg|_0^n$$

$$= n - .018n^2 \text{ for } 0 \le n < \frac{1}{.036}.$$

The annuity underlying the symbol $\bar{s}_{\overline{n}|}$ is also the annuity underlying $\bar{a}_{\overline{n}|}$ so

$$\bar{s}_{\overline{n}|} = a(n)\bar{a}_{\overline{n}|} = \frac{n - .018n^2}{1 - .036n} \quad \text{for} \quad 0 \le n < \frac{1}{.036}.$$

We further observe that

$$(\bar{I}\bar{a})_{\overline{n}|} = \int_0^n tv(t)\, dt = \int_0^n t(1 - .036t)\, dt = \int_0^n (t - .036t^2)\, dt$$

$$= \left(\frac{t^2}{2} - .012t^3\right)\Big|_0^n = \frac{n^2}{2} - .012n^3 \quad \text{for} \quad 0 \le n < \frac{1}{.036}\,.$$

■

Let us now study an example in which $f(t)$ is neither constant nor equal to t.

EXAMPLE 4.6.11   Annuity with continuous payments and a varying rate of payment

Problem: An annuity is payable continuously for ten years. The rate of payment at time t is t^2 yearly. Assuming there is a constant force of interest $\delta = .04$, find the present value of the annuity.

Solution If $\delta = .04$, then $v(t) = e^{-.04t}$. Therefore, the present value is $\int_0^{10} t^2 e^{-.04t}\, dt$. An antiderivative of $t^2 e^{-.04t}$ is $-t^2(\frac{1}{.04})e^{-.04t} - 2t(\frac{1}{.04})^2 e^{-.04t} - 2(\frac{1}{.04})^3 e^{-.04t}$. Of course, you may differentiate this expression to check that it is indeed an antiderivative. As to the question of how you might discover this, you may use integration by parts (twice) or the related "tabloid method." Using this antiderivative, we find that the present value is $[-t^2(\frac{1}{.04})e^{-.04t} - 2t(\frac{1}{.04})^2 e^{-.04t} - 2(\frac{1}{.04})^3 e^{-.04t}]\Big|_0^{10} = [-100(\frac{1}{.04})e^{-.4} - 20(\frac{1}{.04})^2 e^{-.4} - 2(\frac{1}{.04})^3 e^{-.4}) - (0 + 0 - 2(\frac{1}{.04})^3)] \approx \$247.70.$ ■

4.7 A YIELD RATE EXAMPLE

EXAMPLE 4.7.1

Problem: Barbara deposits $1,800 at the beginning of each *quarter* for fifteen years in a fund earning a nominal rate of interest of 6% convertible monthly. The interest from this fund is paid out *monthly* and can only be reinvested at an effective annual rate of 4.2%. Find the accumulated value of Barbara's investments at the end of the twentieth year and her annual effective yield for the twenty-year period.

Solution

It is helpful to view a timeline showing Barbara's deposits into the fund.

no new deposits

PAYMENT:	$1,800	$1,800	$1,800	...	$1,800		

TIME:	0	1	2	...	59	60	...	80 quarters

Note that the unit of time is a quarter, and a quarter consists of three months. The fund pays out interest monthly, and has a monthly interest rate of $\frac{6\%}{12} = .005$. Furthermore, the amount on deposit in the fund increases by $1,800 at the end of each quarter for the first sixty quarters, and then it stays constant. Therefore, since $1,800 \times .005 = \$9$, a timeline showing the deposits to the account with a 4.2% annual effective interest rate is as follows.

monthly
deposits
of $540

PAYMENT:	$9	$9	$9	$18	$18	$18	$27	$27	$27...$540	$540	$540

TIME:	0	1	2	3	4	5	6	7	8	9 ... 178	179	180 ...240 months

TIME:	0	1	2	3	...	60 ... 80 quarters

For the 4.2% account, the effective quarterly rate of interest is $I = (1.042)^{\frac{1}{4}} - 1 \approx .010338563$, and the effective monthly interest rate is $J = (1.042)^{\frac{1}{12}} - 1 \approx .003434379$. Just as we use $i^{(3)}$ to denote the nominal interest rate convertible 3 times per year when i is the effective annual interest rate, we will use $I^{(3)}$ to signify the nominal interest rate convertible 3 times per quarter when I is the effective quarterly interest rate. Observe that $I^{(3)} = 3J \approx .010303138$. The accumulated value in the 4.2% account is

$$(\$9 \times 3)(Is)^{(3)}_{\overline{60}|I}(1.042)^5 + \$540s_{\overline{60}|J}$$

$$\approx (\$33.16670737)\frac{\ddot{s}_{\overline{60}|I} - 60}{I^{(3)}} + \$540s_{\overline{60}|J}$$

$$\approx \$75,385.1344 + \$35,911.6267 \approx \$111,296.76.$$

At the end of the twenty years, the 6% nominal fund has a balance of $60 \times \$1,800 = \$108,000$, so the accumulated value of Barbara's investments at the end of the twentieth year is $\$111,296.76 + \$108,000 = \$219,296.76$.

To calculate Barbara's annual effective yield i for the twenty-year period, observe that the only payments she made were $1,800 at the beginning of each quarter for the first fifteen years, and the only money she received was the accumulated value $219,296.76 at the end of the twenty years. Therefore, the the twenty equation of value is

(4.7.2) $(4 \times \$1,800)\ddot{s}_{\overline{15}|i}^{(4)}(1 + i)^5 = \$219,296.76.$

Using (4.3.11), (1.9.7), and (1.10.9), we find

$$\ddot{s}_{\overline{15}|i}^{(4)} = \frac{(1 + i)^{15} - 1}{d^{(4)}} = \frac{(1 + i)^{15} - 1}{4(1 - v^{\frac{1}{4}})} = \frac{(1 + i)^{15} - 1}{4(1 - (1 + i)^{-.25})} \quad .$$

Substituting the last expression into (4.7.2) gives

(4.7.3) $\$7,200 \left(\dfrac{(1 + i)^{15} - 1}{4\left[1 - (1 + i)^{-.25}\right]} \right) (1 + i)^5 = \$219,296.76,$

and we need to solve for i.

We now use the "guess and check" method. Since the interest rates involved in the two accounts are a nominal monthly rate of 6% (which corresponds to an annual rate of approximately 6.17% and an effective rate of 4.2%, we know that $4.2\% < i < 6.17\%$. We might then begin our "guess and check" by calculating the left-hand side of (4.7.3) when $i = .052$. Doing so, we obtain approximately $209,788.918. Therefore, the yield rate i is higher than .052. Successive modifications bring us to the rate $i = .055435235$ which gives a right-hand side of $219,296.76, the desired value. Of course you might content yourself to knowing i to fewer places of accuracy. Another approach is to approximate i using Newton's method [Problem(4.7.1)].

Cash Flow worksheet solution   Depress CF 2ND CLR WORK to open and clear the **Cash Flow worksheet**. Now key

1 8 0 0 +/− ENTER ↓ 1 8 0 0 +/− ENTER ↓ 5 9

ENTER ↓ 0 ENTER ↓ 2 0 ENTER ↓

2 1 9 2 9 6 • 7 6 ENTER IRR CPT .

Then IRR=1.357968437 is displayed. This gives the quarterly yield rate expressed as a percentage. We convert to the annual yield rate .055435235 by pushing

× • 0 1 = + 1 y^x 4 = − 1 = .

4.8   PROBLEMS, CHAPTER 4

(4.0) Chapter 4 writing problems

(1) The problems in this chapter involve new technical difficulties, but there are no new fundamental ideas. Carefully write two solutions to the following problem. These should be at a level appropriate for a fellow interest theory student who has read through Section (4.2) but who has not yet thought about the problem.

> Fred wishes to repay a loan of X by level payments each quarter of a year. The nominal interest rate is 6.2% convertible monthly. If the payments are not to exceed $X/20$, how many payments are needed?

One of your solutions should include finding the equivalent effective interest rate per quarter, and the other should involve a ratio of annuity symbols. Make sure that you write clearly. Correct grammar is important.

(2) The annuity symbol $(Ia)_{\overline{n}|i}^{(m)}$ gives the present value of an annuity that has payments each m-th of a year, the payments in the j-th year each being for an amount $\frac{j}{m}$. That is to say, the payments are level each year and total j in the j-th year. So, the annual payments increase in arithmetic progression. Introduce a new annuity symbol $(Ga)_{\overline{n}|i}^{(m)}(r)$ giving the present value of an annuity that has level payments each year, the payments in the j-th year totaling $(1 + r)^{j-1}$. Here the letter "G" stands for "geometric" and was chosen to reflect the fact that the underlying annuity has total annual payments that increase geometrically. Explain how you can calculate $(Ga)_{\overline{n}|i}^{(m)}(r)$.

(4.2) Level annuities with payments less frequent than each interest period

(1) An annuity pays $100 at the end of each quarter for ten years. The payments are made directly to a savings account with a nominal interest rate of 4.85% payable monthly, and they are left in the account.

 (a) Find the effective interest rate for a quarter, and use it to compute the balance in the savings account immediately after the last payment.

 (b) Use formula (4.2.4) to recalculate the balance in the savings account immediately after the last payment. Make sure that your answer agrees with your answer to part (a), and note which method you found easier.

 (c) [BA II Plus calculator required] Recalculate the balance in the savings account immediately after the last payment by setting P/Y = 4 and

C/Y=12. Once again, you should get the same answer.

(2) Anurag receives an annuity that pays $1,000 at the end of each month. He wishes to replace it with an annuity that has the same term and has only one payment each year, and that payment should be at the beginning of the year. How much should the payments be if the exchange is based on a nominal discount rate of 3% payable quarterly?

(3) Renee buys a perpetuity paying $1,000 every two years, starting immediately. She deposits the payments into a savings account earning interest at an effective annual interest rate of 6%. Ten years later, before receiving the sixth payment, Renee sells the perpetuity based on an effective annual interest rate of 6%. Using proceeds from the sale plus the money in the savings account, Renee purchases an annuity paying P at the end of every three years for thirty years at an annual effective interest rate of 6%. Find P. Do this problem twice, once using Chapter 3 methods and once using Section (4.2) techniques.

(4) It is reported that the present value of an annuity paying $1,000 at the end of each year for Y years is $12,692.58, and the present value of an annuity paying $300 each year for four years is $1,069.12. Find the present value of an annuity paying $500 at the end of each two years for Y years.

(5) The present value of an annuity lasting 72 interest periods and paying $1,000 at the beginning of each k interest periods is $4,769.30. Given that the effective interest rate per interest period is 3.6575%, find k.

(6) Show that the value at time 1 of 1 payable at times $7, 10, 13, 16, \cdots, 40$ is $\dfrac{a_{\overline{39}|i} - a_{\overline{3}|i}}{s_{\overline{3}|i}}$. Use this to establish that the time 2 value of these payments is $\dfrac{a_{\overline{39}|i} - a_{\overline{3}|i}}{s_{\overline{2}|i} + a_{\overline{1}|i}}$ and that the time 0 value is $\dfrac{a_{\overline{41}|i} - a_{\overline{5}|i}}{s_{\overline{2}|i} + a_{\overline{1}|i}}$.

(7) A perpetuity paying $1,000 at the beginning of each two years has the same present value as another perpetuity with level payments, this one having payments at the end of each three years. Express the level payment amount of the second annuity as a function of the annual effective interest rate i.

(4.3) Level annuities with payments more frequent than each interest period

(1) Calculate the annuity symbols $\ddot{s}^{(4)}_{\overline{23}|2.25\%}$ and $a^{(12)}_{\overline{\infty}|4\%}$. Carefully describe what each one measures.

(2) Sylvestre receives an annuity-immediate with monthly payments of $100. Susan receives an annuity-due with annual payments of $1,165 and the same term. The value of Sylvestre's annuity is 97.09% times the value of Susan's. Calculate $\frac{(1-v)}{i^{(12)}}$.

(3) Find an expression in terms of the nominal interest rate convertible quarterly for the accumulated value at the end of the twenty-one years of an annuity that pays $640 at the beginning of each four-month period for twenty-one years. Your expression should NOT be a sum of terms.

(4) Ibama received a ten-year annuity. It paid $100 at the end of each quarter for the first four years, and $35 each month for the remaining six years. Express the value at the time of the last payment in terms of annuity symbols introduced in this section if the annual effective interst rate is i for the first four years and j for the following six years. Evaluate if $i = 3\%$ and $j = 4.5\%$.

(5) Jason purchases a deferred perpetuity for $13,520. The perpetuity has quarterly payments of $750. Express the waiting time until the first payment as a function of the annual effective interest rate i.

(6) A perpetuity-due has monthly payments. The payments in the k-th year are each $(1 + 2 + \cdots + k)/12$. Show that the present value of this perpetuity due is $(\ddot{a}^{(12)}_{\overline{\infty}|i})(I\ddot{a})_{\overline{\infty}|i}$.

(7) Use Fact (4.3.2) to obtain an alternate proof of the formula $a^{(m)}_{\overline{n}|i} = \dfrac{1-v^n}{m[(1+i)^{\frac{1}{m}}-1]}$.

(4.4) Annuities with payments less frequent than each interest period and payments in arithmetic progression

(1) An annuity has end of quarter payments for fifteen years, and the payment at the end of the j-th quarter ($j = 1, 2, \ldots, 60$) is $100j$. The payments are made directly into a savings account with a nominal interest rate of 6% payable monthly, and they are left in the account.

   (a) Find the effective interest rate for a quarter, and use it to compute the balance in the savings account immediately after the last payment.

   (b) Use Fact (4.4.3) to recalculate the balance in the savings account immediately after the last payment. Make sure that your answer agrees with your answer to part (a), and note which method you found easier.

(2) It is reported that an annuity-immediate with $100 annual payments for s years has an accumulated value of $933.52 at the time of its last payment.

Furthermore, an annuity-immediate that has $40 annual payments and a term four times as long accumulates to $2,680.11 at the time of its last payment. Now consider an annuity that has the same term as the second of these annuities but only has a payment at the end of each four years. Suppose its first payment is $2,400 and each further payment is $300 more than its predecessor. Express the accumulated value of this third annuity at the time of its last payment as a function of the annual effective interest rate i. Make sure any annuity symbols appearing in the function are to be evaluated at the rate i used in valuing the first two annuities.

(3) Jasper is bequeathed a thirty-year deferred annuity that has a payment at the end of each third year. The first payment is for $15,000 and is made five years after she receives the inheritance. There is always an increase of $4,000 from one payment to the next. Find the value of her legacy at the time of the bequest, assuming that the annual effective interest rate remains level at 5%. Do this problem twice, once using Chapter 3 methods and once using Section (4.4) techniques.

(4) A perpetuity-due has a payment once every k years. The j-th payment is j. Show that this perpetuity has present value equal to $(\frac{1}{i a_{\overline{k}|i}})^2$.

(4.5) Annuities with payments more frequent than each interest period and payments in arithmetic progression

(1) Suppose that the effective interest rate per interest period is 3.2%. Describe what the following annuity $(I^{(4)}\ddot{a})^{(4)}_{\overline{\infty}|}$ means and calculate it to the nearest $\frac{1}{1,000}$.

(2) A thirty-year annuity has end-of-month payments. The first year the payments are each $120. In subsequent years each payment increases by $5 over what it was the previous year. Find the present value of the annuity if $i = 3\%$.

(3) A twenty-year annuity pays $2,400 + 300k$ on the first day of the k-th month of the year. Thus it pays $2,700 each January 1, $3,000 each February 1,..., $6,000 each December 1. The first payment is on a January 1. Find an expression for the value of this annuity just before the first payment and evaluate it if the annual effective interest rate is 5%.

(4) Cassandra receives an annuity-due with a payment each month. The annuity has its first payment on her 12th birthday and the last payment is on her 16th birthday. The amount of the payment is $100 times her age with no credit being given for fractions of a year. All the deposits are made to an account earning compound interest at an annual effective interest rate of 5.82%. Find the accumulated value on her 18th birthday.

(5) Bob deposits $1,500 at the beginning of each quarter for sixteen years in a fund earning a nominal rate of interest of 6% convertible monthly. The interest from this fund is paid out monthly and can only be reinvested at an effective annual rate of 5.2%. Find the accumulated value of Bob's investments at the end of twentieth year.

(6) Melanie receives an annuity paying $1,000 at the end of each month for eight years. This is directly deposited to a fund paying interest at an annual effective rate of 7.28%. Interest is paid out at the end of each year to a fund with an annual yield of 4%. Both funds are liquidated at the end of ten years. Find the total amount of Melanie's liquidation from the two funds.

(7) During a ten-year period, Darlene deposits $1,000 at the beginning of each quarter into a fund earning interest at a nominal interest rate of $i^{(12)} = 6\%$. The interest from this fund is paid out monthly and is reinvested at a quarterly interest rate of j. Darlene liquidates her money at the end of twelve years, $2\frac{1}{4}$ years after her last deposit of $1,000. Find an expression for the accumulated value of Darlene's investments in which any annuity symbols used are at interest rate j. [HINT: Begin by making a timeline showing the deposits into the account having the quarterly interest rate of j.]

(8) An annuity lasting n interest periods has a payment at the end of each m-th of an interest period. The first payment is for an amount P and further payments are each an amount Q more than their predecessor. Show that this annuity has accumulated value $m\,Ps_{\overline{n}|i}^{(m)} + m^2\frac{Q}{i^{(m)}}(s_{\overline{n}|i}^{(m)} - n)$ at the end of the n-th interest period.

(9) An annuity lasting n interest periods has a payment at the end of each m-th of an interest period. The first payment is for an amount P. Payments are level within each interest period, and the the individual payments increases by an amount Q from one interest period to the next. Show that the accumulated value at the time of the last payment is $m\left(Ps_{\overline{n}|i}^{(m)} + \frac{Q}{i^{(m)}}(s_{\overline{n}|i} - n)\right)$.

(4.6) Continuously paying annuities

(1) Suppose that the effective interest rate per interest period is 3.2%. Describe what the annuity symbol $\overline{a}_{\overline{18}|}$ means and calculate it to the nearest $\frac{1}{1,000}$.

(2) A continuously paying level annuity pays $50 each year for ten years. The force of interest is $(1 + t)^{-1}$. Find the present value of this annuity.

(3) An annuity is continuously varying and payable for ten years. The rate

of payment at time t is $(2 + t)^2$ and the force of interest is $(1 + t)^{-1}$. Find the present value of this annuity.

(4) Stacey and her husband David have a joint savings account that earns 3.5% interest payable continuously and has a current balance of $58,458. Each year, David wishes to withdraw $4,000 payable continuously at a level rate. Stacey wishes to deposit X at the beginning of each year (for thirty years) so that the account will last for thirty years. What is the least X that will work?

(5) A continuously paying level annuity pays $72 each year for twenty years. The force of interest at time t is $\frac{5}{3+2t}$. Find the present value of this annuity.

(6) Quang deposits $20,000 in a savings account with a discount rate of 4.8% convertible quarterly. He leaves his money in this account to accumulate for twelve years, then moves it to a fund which is accumulating at 5.4% per annum convertible continuously. If, starting at time 12 when he invests in the new fund, money is withdrawn levelly and continuously at a rate of $8,000 per annum, how long will Quang's money last?

(7) This problem concerns the annuity underlying the annuity symbols $(I^{(m)}a)_{\overline{n}|i}^{(m)}$.

(a) The rate of payment over an interval is the amount paid normalized by dividing by the length of the interval. Show that this annuity pays at a rate of $\frac{j}{m}$ over the interval $[\frac{j-1}{m}, \frac{j}{m}]$.

(b) Show that the payments of this annuity total $\frac{n^2}{2} + \frac{n}{2m}$.

(c) Show that the limit $\lim_{m\to\infty} \frac{n^2}{2} + \frac{n}{2m}$ of the total payment amounts found in (b) is equal to $\int_0^n t\, dt$, the total payments of the continuously paying annuity underlying the symbol $(\overline{I}\overline{a})_{\overline{n}|}$.

(4.7) A yield rate example

(1) Introduce a function $f(x)$ whose root is the yield rate desired in Example (4.7.1). Apply Newton's method with initial approximation $x_1 = .052$ to find x_2 and x_3. Discuss your results.

(2) Bob deposits $1,500 at the beginning of each quarter for sixteen years in a fund earning a nominal rate of interest of 6% convertible monthly. The interest from this fund is paid out monthly and can only be reinvested at an effective annual rate of 5.2%. This is just as in Problem (4.5.5). Find Bob's yield rate for the twenty-year period.

(3) Daphne deposits $100 at the beginning of each month for five years in a fund earning an annual effective rate of interest of 5.4%. The interest

from this fund is paid out annually, and can only be reinvested at a nominal interest rate of 4% convertible quarterly. Find the accumulated value of Daphne's investments at the end of eight years and the yield rate for the eight-year period.

Chapter 4 review problems

(1) Suppose that the effective interest rate per interest period is 3%. Describe what the following annuity symbols mean and calculate them to the nearest $\frac{1}{1000}$.

(a) $\ddot{s}_{\overline{30}|}$; (b) $(I^{(12)}a)_{\overline{\infty}|}^{(12)}$; (c) $(Ia)_{\overline{\infty}|}^{(4)}$.

(2) A level continuously paying annuity pays $1,500 each month for eight years. The force of interest is $\delta_t = \frac{2t}{t^2+5}$ where time is measured in years. Find the present value of this annuity.

(3) A fifteen-year annuity pays $1,400 + 300k$ on the first day of the k-th month of the year. Thus it pays $1,700 each January 1, $2,000 each February 1,..., $5,000 each December 1. The first payment is on a January 1. Find an expression for the value of this annuity just before the first payment and evaluate it if the annual effective interest rate is 3%.

(4) This problem concerns the accumulated value at the end of twenty-four years of an annuity which pays $500 at the beginning of each four-month period for twenty years. Express this accumulated value as a function of $i^{(4)}$ which does not involve any annuity symbols or series.

(5) Show that $\frac{1}{a_{\overline{n}|i}^{(m)}} = \frac{1}{s_{\overline{n}|i}^{(m)}} + i^{(m)}$.

(6) Shelly and her husband Bubba have a joint savings account that earns 2.5% interest payable continuously and has a current balance of $32,458. Each year, Bubba wishes to withdraw $3,000 payable continuously at a level rate. Starting now and continuing for twenty years, Shelly wishes to deposit $X at the beginning of each year so that the account will last for thirty years. What is the least $X that will work?

(7) Show that the accumulated value at the time of the last payment of a twenty-six-year annuity-immediate, which has payments of $100 at the end of years 3, 6, ... , 24, and $200 at the ends of the other years, is

$$200s_{\overline{26}|i} - 100\frac{s_{\overline{24}|i}}{s_{\overline{1}|i}+a_{\overline{2}|i}}.$$

(8) Amanda receives an N-year annuity with a payment at the beginning of each month. Here N is a positive integer. The payments the first year are each P and afterward they decrease by q% each year. The monthly interest rate is j%. Express the present value as a constant times a product of annuity symbols. Evaluate if $N = 15$, $P = $1,200, $q = 3$, and $j = .4$.

CHAPTER 5

Loan repayment

5.1 INTRODUCTION

Throughout this book, we have seen examples of loans. Some have been repaid by a single repayment or a few repayments, whose amounts and times were agreed upon at the time of the loan. In Chapter 3 we looked at loans with level payments (except perhaps for a slightly reduced final payment) at the end of each interest period. Favorite examples included standard mortgages and car loans. In the loans of Chapter 3, and in their generalizations appearing in Chapter 4, each payment consisted of interest due since the last payment and money to reduce the outstanding loan balance. In Section (5.2) we study such **amortized loans** and the question as to how each payment is divided into payment of interest and payment of principal. **Amortization Tables** are introduced as is the BA II Plus calculator **Amortization worksheet**. In Section (5.3) we introduce the concept of a **sinking fund account**. This is a savings account in which one accumulates the loan amount. While the sinking fund grows at one interest rate, the borrower keeps current on the interest on the loan, which may well be at a higher rate. In Section (5.4), we look at loans that are repaid by payments in arithmetic or geometric progressions. Finally, in Section (5.5) we look at yield rate problems where a sinking fund account is involved. The important concept of **replacement of capital** is presented.

5.2 AMORTIZED LOANS AND AMORTIZATION SCHEDULES

When a loan is an **amortized loan**, each time a payment is made, interest due on the outstanding loan balance is first deducted from the payment amount. This reduced amount, the payment amount less the interest due, is called the **principal** (payment). The principal payment is applied to cut the outstanding loan balance.

EXAMPLE 5.2.1   Amortization

Problem: LaTroy took out a loan for $2,000. The loan was governed by compound interest at a rate of 5%. He made a payment of $800 one year later and a payment of $1,000 three years after he took out the loan. Find the amount of principal in each payment and LaTroy's outstanding loan balance immediately after the second payment.

Solution   One year after the loan was made, the interest due was $2,000(.05) = $100. Therefore, the amount of principal in the $800 payment was $800 − $100 = $700, and the outstanding balance after this payment was $2,000 − $700 = $1,300. The next payment was made two years after the outstanding loan balance was $1,300. The interest due at the time of this payment was therefore $1,300((1.05)^2 − 1) = $133.25. So, the amount of principal in the $1,000 payment was $1,000 − $133.25 = $866.75, and the outstanding loan balance immediately after this payment was $1,300 − $866.75 = $433.25.   ∎

The mortgages and car loans we considered in Chapter 3 are amortized loans with payments at regular intervals, these being level except for a slightly reduced final payment.

Focus on an amortized loan. The payments each have two components, an interest component and a principal component. Since interest payments may be treated differently from principal repayments when determining taxes, it is important to know how much of each payment is interest. An **amortization schedule** is a chart that shows the time and amount of each payment, the division of the payments into interest and principal, and the outstanding loan balance immediately after each payment.

EXAMPLE 5.2.2   Amortization schedule for loan; principal payment beyond the required

Problem: Consider LaTroy's loan of Example (5.2.1). Suppose that he makes an additional payment four years after the time of the loan so that the loan is completely repaid. Make an amortization schedule for this loan.

Solution   From the solution to Example (5.2.1), we know that the balance after the $1,000 payment is $433.25. We are now given that there is one more payment, and it occurs one year after the $1,000 payment. The interest in this payment is therefore $433.25(.05) ≈ $21.66. Since this additional payment results in the loan being completely paid off, it must contain $433.25 as the principal payment. Therefore, the amount of the additional payment is $21.66 + $433.25 = $454.91, the sum of the interest due and the principal payment. We consequently have the following amortization table.

TIME (years since loan)	PAYMENT	INTEREST IN PAYMENT	PRINCIPAL IN PAYMENT	BALANCE AFTER PAYMENT
0	—	—	—	$2,000.00
1	$800.00	$100.00	$700.00	$1,300.00
3	$1,000.00	$133.25	$866.75	$433.25
4	$454.91	$21.66	$433.25	$0.00

■

EXAMPLE 5.2.3   Amortization schedule for loan; level payments with slightly reduced final payments

Problem: Make an amortization table for a loan at effective interest rate i per interest period that lasts n interest periods and is repaid by level end-of-interest period payments of Q except for one slightly reduced final payment of R.

Note: The payments for such a loan are found in Algorithm (3.2.19), but in that algorithm the level payments are denoted by Q_1 or Q_2.

Solution   The balance at the end of an interest period, immediately following the payment (if any), may be found by the prospective method. Recall that this method was introduced in Section (3.6), and it is based on the fact that the outstanding loan balance must equal the value of the required loan payments that have not yet been made. Using Equation (3.6.3), we see that if $k \in \{1, 2, ..., n\}$, at time $k - 1$ the outstanding loan balance is $Q a_{\overline{n-k}|i} + R v^{n-k+1}$. Since $a_{\overline{n-k}|i} = \frac{1-v^{n-k}}{i}$, the interest due at time k is $i \left(\frac{Q(1-v^{n-k})}{i} + R v^{n-k+1} \right) = Q(1 - v^{n-k}) + R i v^{n-k+1}$. The principal in a payment is found by subtracting the interest in the payment from the payment amount. So, we have the following amortization schedule:

TIME (yrs since loan)	PMT	INTEREST IN PAYMENT	PRINCIPAL IN PAYMENT	BALANCE AFTER PAYMENT
0	—	—	—	$Qa_{\overline{n-1}\rvert i} + Rv^n$
1	Q	$Q(1-v^{n-1}) + Riv^n$	$Qv^{n-1} - Riv^n$	$Qa_{\overline{n-2}\rvert i} + Rv^{n-1}$
2	Q	$Q(1-v^{n-2}) + Riv^{n-1}$	$Qv^{n-2} - Riv^{n-1}$	$Qa_{\overline{n-3}\rvert i} + Rv^{n-2}$
$\vdots$				
k	Q	$Q(1-v^{n-k}) + Riv^{n-k+1}$	$Qv^{n-k} - Riv^{n-k+1}$	$Qa_{\overline{n-k-1}\rvert i} + Rv^{n-k}$
$\vdots$				
$n-1$	Q	$Q(1-v) + Riv^2$	$Qv - Riv^2$	Rv
n	R	Riv	Rv	$0

■

Note that in the table of Example (5.2.3), starting with the line for time $t = 2$ and continuing through the line for time $t = n - 1$, the amount of principal is equal to $1 + i$ times the principal payment recorded in the previous line. This observation may be used when creating the amortization table for a specific example, for instance the following one.

EXAMPLE 5.2.4   Amortization schedule

Problem: Jared borrows $20,000 at an annual effective discount rate of 4.8%. The loan is to be repaid by annual end-of-year payments for four years, the first three of which are for the amount $5,645.75 and the time 4 payment is for $5,645.72. Make an amortization schedule for this loan.

Solution   The interest rate is $i = \frac{d}{1-d} \approx 5.042016807\%$, and the interest due at the end of the first year is $20,000 \times i \approx \$1008.403361$. Therefore, the principal in the first payment is approximately $5,645.75 - \$1,008.403361 \approx \$4,637.346639$.   So, the principal in the second year is approximately $4,637.346639 \times (1 + i) \approx \$4,871.162436$ and the principal in the third year is $4,871.162436 \times (1 + i) \approx \$5,116.767264$. The balance at the end of the third year is approximately $20,000 - \$4,637.346639 - \$4,871.162436 - \$5,116.767264 \approx \$5,374.723662$. Consequently, the interest in the final payment is about $5,374.723662(1 + i) \approx \$270.9944703$.

The remainder of the $5,645.72 payment, $5,645.72 − $270.9944703 ≈ $5,374.72553 is principal. Note that this exceeds the outstanding balance at the end of the third year by about .00186813, which is less than half a penny. This slight discrepency is due to the fact the payments must each be for an integral number of cents.

The values we have found have been rounded to the nearest cent and then recorded in the following amortization table, along with (rounded) outstanding loan balances that are obtained by subtracting the newly paid (rounded) principal from the previous outstanding loan balance.

TIME (years since loan)	PAYMENT	INTEREST IN PAYMENT	PRINCIPAL IN PAYMENT	BALANCE AFTER PAYMENT
0	—	—	—	$20,000.00
1	$5,645.75	$1,008.40	$4,637.35	$15,362.65
2	$5,645.75	$774.59	$4,871.16	$10,491.49
3	$5,645.75	$528.98	$5,116.77	$5,374.72
4	$5,645.72	$271.00	$5,374.72	$0.00

∎

The BA II Plus calculator has an **Amortization worksheet** (factory programmed using the formulas displayed in the table of Example (5.2.3) that may be used to find any of the interest, principal, or balance entries in Example (5.2.4), or to find amortization entries in other special cases of Example (5.2.3). We now illustrate this in detail for the loan of Example 5.2.4. Begin by checking that the calculator is in END mode with P/Y = C/Y = 1, and then push $\boxed{4}$ $\boxed{N}$ to enter the total number of equally spaced payments. Next depress $\boxed{2}$ $\boxed{0}$ $\boxed{0}$ $\boxed{0}$ $\boxed{0}$ $\boxed{+/-}$ $\boxed{PV}$ to place the negative of the loan amount in the PV register. Follow this by keying $\boxed{5}$ $\boxed{6}$ $\boxed{4}$ $\boxed{5}$ $\boxed{\cdot}$ $\boxed{7}$ $\boxed{5}$ $\boxed{PMT}$ so that the level payment amount has been entered. The final payment is 3 cents ($.03) less than the other payments, so push $\boxed{\cdot}$ $\boxed{0}$ $\boxed{3}$ $\boxed{+/-}$ $\boxed{FV}$ to enter −.03 in the FV register. Next push $\boxed{CPT}$ $\boxed{I/Y}$ so that we have a set of five consistent values filling the **TVM worksheet**. At this point "I/Y = 5.042020112" is displayed. (Note that this is close to the value of i computed above but not exactly the same. This results from the fact that the amount repaid is only $20,000 when considered to the nearest cent. If you prefer, rather than computing the interest rate from the other four entries in the **TVM**, you may enter the value we calculated from the given discount rate and then push $\boxed{\text{CPT}}$ $\boxed{\text{PV}}$. You will display "PV = −20,000.00153". Entering the interest rate may take you longer, but the numerical results given by the **Amortization worksheet** then round to exactly those we recorded in our amortization table, a perfect agreement that would not otherwise hold (e.g., the balance entry in the time 3 row). Now open the **Amortization worksheet** by pushing $\boxed{\text{2ND}}$ $\boxed{\text{AMORT}}$. The display will now show "P1 = ". If you desire to determine the interest or principal in the m-th payment or the outstanding loan balance immediately after the m-th payment, begin by keying $\boxed{m}$ $\boxed{\text{ENTER}}$ $\boxed{\downarrow}$ $\boxed{m}$ $\boxed{\text{ENTER}}$. If you push $\boxed{\downarrow}$ once, the outstanding loan balance immediately after the m-th payment will be displayed, and if you depress $\boxed{\downarrow}$ again, the amount of principal in the m-th payment will appear. A third keying of $\boxed{\downarrow}$ results in the amount of interest in the m-th payment being exhibited. For instance, if you wish to verify the entries in the time 2 row of the above amortization table, begin by keying $\boxed{2}$ $\boxed{\text{ENTER}}$ $\boxed{\downarrow}$ $\boxed{2}$ $\boxed{\text{ENTER}}$ $\boxed{\downarrow}$. The display now reads "BAL = −10,491.49213". Another $\boxed{\downarrow}$ brings to view "PRN = 4,871.161895". One more $\boxed{\downarrow}$ exhibits "INT = 774.5881055". Note that if you check the balance at time 4 by this method, you should view "BAL = .029999995". This would be the balance just after the fourth payment if Jared made a level payment of $5,645.75 and had not yet been refunded his three-cent overpayment.

The BA II Plus calculator **Amortization worksheet** is also ideal for computing the amount of principal or the amount of interest in a sequence of consecutive payments. This is demonstrated in the second solution to our next example.

EXAMPLE 5.2.5   Annual interest paid on a mortgage

Problem: On May 1, 1988, the Ramakrishnas purchased a home for $308,000. Their down payment was $46,000, and the remaining $262,000 was financed with a standard fifteen-year amortized loan at a nominal interest rate of 5.55% convertible monthly. How much interest did they pay in 1993, assuming that there payments have all been on the first of the month, the first payment was on June 1, 1988, and that there have been no additional payments beyond the scheduled ones?

Solution 1 This solution will be based upon the following observation: *The amount of interest paid during a given period is always equal to the total amount paid during that period less the decrease in the outstanding balance over the period.*

We first determine the total amount paid during 1993. Note that a standard fifteen-year amortized loan has $15 \times 12 = 180$ monthly payments,

and the Ramakrishna's loan has a monthly interest rate of $\frac{5.55\%}{12} = .4625\%$.
But $\frac{262{,}000}{a_{\overline{180}|.4625\%}} \approx 2{,}147.716532$, so the first 179 monthly mortgage payments
were each for \$2,147.72. Therefore, since the Ramakrishnas made 12 monthly
mortgage payments in 1993, they paid a total of $12 \times \$2{,}147.72 = \$25{,}772.64$
during 1993.

Next we determine the decrease in the outstanding balance during 1993.
Observe that the the Ramakrishnas made seven monthly mortgage payments
in 1988 (June 1, July 1, August 1, ... , December 1) and then twelve in
each of the years 1989, 1990, 1991, 1992, and 1993. Therefore, they have
made $7 + (4 \times 12) = 55$ payments prior to the beginning of 1993 and $55 +
12 = 67$ payments by the end of 1993. So, by the retrospective method, the
outstanding loan balance at the beginning of 1993 was $\$262{,}000(1.004625)^{55} -$
$\$2{,}147.72s_{\overline{55}|.4625\%} \approx \$203{,}534.9107$, and the outstanding loan balance at the
end of 1993 was $\$262{,}000(1.004625)^{67} - \$2{,}147.72s_{\overline{67}|.4625\%} \approx \$188{,}684.4769$.
The outstanding loan balance declined by $\$203{,}534.9107 - \$188{,}684.4769 \approx$
$\$14{,}850.4338$ in 1993.

Finally, using the observation underlined above, the amount of interest
paid by the Ramakrishnas in 1993 was $\$25{,}772.64 - \$14{,}850.43 = \$10{,}922.21$.

Solution 2 As determined in Solution 1, the payments made during 1993 were
the 56th through 67th payments. To calculate the amount of interest in these
payments, we note that according to the amortization table of Example (5.2.3),
this is equal to the sum

$$\sum_{k=56}^{67} \left(Q(1 - v^{n-k}) + Riv^{n-k+1} \right)$$

where $n = 180$, i is the effective interest rate per payment period (in this
case $\frac{5.55\%}{12} = .4625\%$), $v = \frac{1}{1+i}$, and Q and R are the level and final payment
respectively. Since $\frac{262{,}000}{a_{\overline{180}|.4625\%}} \approx 2{,}147.716532$, $Q = \$2{,}147.72$ (as already noted
in Solution 1), and Algorithm (3.2.14) yields $R = \$2{,}146.75$. Consequently,
the Ramakrishna's 2003 interest is

$$\sum_{k=56}^{67} \left(\$2{,}147.72(1 - v^{180-k}) + \$2{,}146.75iv^{181-k} \right)$$

$= (12 \times \$2{,}147.72) - \$2{,}147.72(v^{124} + v^{125} + \cdots + v^{113})$
$\quad + \$2{,}146.75i(v^{125} + v^{124} + \cdots + v^{114})$
$= \$25{,}772.64 - \$2{,}147.72(a_{\overline{124}|i} - a_{\overline{112}|i}) + \$2{,}146.75i(a_{\overline{125}|i} - a_{\overline{113}|i})$
$\approx \$25{,}772.64 - \$14{,}919.08591 + \$68.65209276.$
$\approx \$10{,}922.21.$

Amortization worksheet solution 1   With the BA II Plus calculator in END mode and P/Y=C/Y=1, push

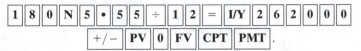

At this point the display shows "PMT = 2,147.716532" so the first 179 payments should be $2,147.72. Therefore, push $\boxed{2}\boxed{1}\boxed{4}\boxed{7}\boxed{\bullet}\boxed{7}\boxed{2}\boxed{\textbf{PMT}}$. Then depress $\boxed{\textbf{CPT}}\boxed{\textbf{FV}}$. The calculator will display "FV=−0.970771204". Now open the **Amortization worksheet** by keying $\boxed{\textbf{2ND}}\boxed{\textbf{AMORT}}$. Then push

$$\boxed{5}\boxed{6}\boxed{\textbf{ENTER}}\boxed{\downarrow}\boxed{6}\boxed{7}\boxed{\textbf{ENTER}}\boxed{\downarrow}\boxed{\downarrow}\boxed{\downarrow}.$$

Your display should now show "INT = 10,922.20616". Reporting the interest to the nearest penny (for instance to the Internal Revenue Service), the Ramakrishna's 2003 interest is $10,922.21.

Amortization worksheet solution 2

With the BA II Plus calculator in END mode and P/Y=C/Y=12, push

$$\boxed{1}\boxed{8}\boxed{0}\boxed{\textbf{N}}\boxed{5}\boxed{\bullet}\boxed{5}\boxed{5}\boxed{\textbf{I/Y}}\boxed{2}\boxed{6}\boxed{2}\boxed{0}\boxed{0}\boxed{0}$$
$$\boxed{+/-}\boxed{\textbf{PV}}\boxed{0}\boxed{\textbf{FV}}\boxed{\textbf{CPT}}\boxed{\textbf{PMT}}.$$

Your display should now show "PMT = 2,147.716532" just as it did in **Amortization worksheet** solution 1, and the remainder of this solution is a repeat of that solution. ∎

If $Q = R$ in Example (5.2.3), then all the payments are equal. Then if $k \in \{1, 2, ..., n\}$, at time $k - 1$ the outstanding loan balance is $Qa_{\overline{n-k+1}|i}$, and the interest at time k is $Q(1 - v^{n-k+1})$. The principal is therefore Qv^{n-k+1}. We thus have the following simpler amortization schedule.

AMORTIZATION SCHEDULE 5.2.6 (Schedule when $Q = R$)

TIME (years since loan)	PAYMENT	INTEREST IN PAYMENT	PRINCIPAL IN PAYMENT	BALANCE AFTER PAYMENT
0	—	—	—	$Qa_{\overline{n}|i}$
1	Q	$Q(1 - v^n)$	Qv^n	$Qa_{\overline{n-1}|i}$
2	Q	$Q(1 - v^{n-1})$	Qv^{n-1}	$Qa_{\overline{n-2}|i}$
$\vdots$				
k	Q	$Q(1 - v^{n-k+1})$	Qv^{n-k+1}	$Qa_{\overline{n-k}|i}$
$\vdots$				
n	Q	$Q(1 - v)$	Qv	$\$0$

Note that in Schedule (5.2.6), as in Example (5.2.3) excluding the final payment, each principal payment is $1 + i$ times the preceding payment.

EXAMPLE 5.2.7   Early payments of an amortized loan contain significantly more interest

Problem: An amortized loan made at an effective interest rate of 6.6% is to be repaid over a period of fifteen years by annual end-of-year payments of $1,800. What proportion of the loan's total interest is paid in the first five years?

Solution   The total amount paid is $15 \times \$1,800 = \$27,000$. The loan amount is $16,816.61 since $\$1,800a_{\overline{15}|6.6\%} \approx \$16,816.60871$. Therefore, the total interest for the fifteen years is approximately $\$27,000 - \$16,816.61 = \$10,183.39$. On the other hand, looking at the interest column of Amortization Schedule (5.2.6), we calculate that the total interest for the first five years is

$$\$1,800(1 - v^{15}) + \$1,800(1 - v^{14}) + \$1,800(1 - v^{13})$$
$$+ \$1,800(1 - v^{12}) + \$1,800(1 - v^{11})$$
$$= (5 \times \$1,800) - \$1,800(v^{15} + v^{14} + v^{13} + v^{12} + v^{11})$$
$$= \$9,000 - \$1,800(a_{\overline{15}|6.6\%} - a_{\overline{10}|6.6\%})$$
$$\approx \$5,062.946321,$$

and the proportion of the interest paid in the first five years is about

$$\frac{\$5,062.946321}{\$10,183.39} \approx .497176905.$$

Amortization worksheet solution   Make sure that the BA II Plus calculator is in END mode and P/Y=C/Y=1. Then, enter the given information for the loan by pressing

Now push | **CPT** | **PV** |. It is now time to open the Amortization worksheet by pushing | **2ND** | **AMORT** |. Then, with m representing the memory register of your choice, key

| 1 | **ENTER** | ↓ | 1 | 5 | **ENTER** | ↓ | ↓ | ↓ | **STO** | m |

| ↓ | ↓ | 5 | **ENTER** | ↓ | ↓ | ↓ | ÷ | **RCL** | m | = |.

The number 0.497176842 should then be seen on your calculator display. Once again, the proportion of the interest paid in the first five years is about .497177.

∎

5.3   THE SINKING FUND METHOD

A possible way to structure a loan's repayment is for the borrower to pay the interest on the loan periodically but to make no partial repayments of the loan amount. That is, the payments made prior to the end of the loan term contain no principal. Since the borrower keeps current on the interest due, a single, lump-sum payment of the loan amount will pay off the loan. Perhaps the borrower is required to accumulate the loan amount at the end of the loan term by making periodic deposits to a savings account, and then use the accumulated funds to erase the debt. We shall refer to a loan so structured as a loan repaid by the **sinking fund method**, and we call the savings account used to accumulate the loan amount a **sinking fund account**. In the sinking fund method, the loan is governed by an interest rate i, and the sinking fund account earns an interest rate j that is typically less than i. Or, more generally, the loan is governed by a sequence of interest rates $\{i_t\}$ and the sinking fund is governed by a sequence of interest rates $\{j_t\}$.

EXAMPLE 5.3.1   Sinking fund method and needed deposits

Problem: Teresa took out a loan of $10,000, to be paid back by the sinking fund method. The term of the loan is ten years, and interest is to be repaid annually at an annual effective interest rate of 6%. The sinking fund account earns 3% annual effective interest for the first four years and 4% interest for the next six years. Teresa deposits $800 in the sinking fund account at the end of each of the first nine years. What is the present value of the interest payments she must make, calculated using the rates earned by her sinking fund account? What deposit must she make at the end of the tenth year, so that she will have the requisite $10,000?

Solution   Teresa's interest payment at the end of each year is $.06 \times \$10,000 = \$600$. Thus, the present value of her sequence of interest payments, calculated using the rates earned by her sinking fund, is $\$600a_{\overline{4}|3\%} + \$600a_{\overline{6}|4\%}(1.03)^{-4} \approx \$5,024.80$.

Just prior to Teresa making a deposit at the end of the ten-year period, her balance is $\$800s_{\overline{4}|3\%}(1.04)^6 + \$800s_{\overline{5}|4\%}(1.04) \approx \$8,741.28$. Therefore, Theresa must deposit $\$10,000 - \$8,741.28 = \$1,258.72$ at the end of the tenth year. ∎

Just as an amortization schedule tracks an amortized loan, a **sinking fund schedule** shows the progress of a loan repaid by the sinking fund loan.

EXAMPLE 5.3.2   Sinking fund schedule

Problem: Suppose that LaTroy's $2,000 loan of Example (5.2.1) was not amortized but was made according to the sinking fund method. Further suppose that the annual effective interest rate on the loan is 5% but that LaTroy's sinking fund account has an annual effective interest rate of 2.5%. As in Example (5.2.1), assume that LaTroy made total payments of $800 one year after the loan origination and $1,000 three years after he took out the loan, and that he totally fulfilled the loan obligation at the end of the fourth year. Make a sinking fund schedule for LaTroy's loan.

Solution   The annual effective interest rate on the loan is 5% and the loan amount is $2,000. Therefore the interest for a one-year period is $.05 \times \$2,000 = \$100$ and for a two-year period is $\$2,000((1.05)^2 - 1) = \$205$. So, LaTroy deposits $\$800 - \$100 = \$700$ to the sinking fund at the end of the first year, and he deposits $\$1,000 - \$205 = \$795$ at the end of the third year. *The balance in the sinking fund is always found by adding the previous sinking fund balance, the new deposit to the sinking fund, and interest credited to the sinking fund since the previous payment.* So, at the end of one year LaTroy's balance in his sinking fund account is $\$0 + \$700 + \$0 = \$700$, and at the end of three years it is $\$700 + \$795 + \$700((1.025)^2 - 1) = \$700 + \$795 + \$35.4375 = \$1,530.4375$.

Additionally, at the end of four years LaTroy's balance in his sinking fund account is $\$1,530.4375 + X + \$1,530.4375(.025) = \$1,530.4375 + X + \$38.2609375 = \$1,568.6984375 + X$ where X denotes the sinking fund deposit at the end of the fourth year. Since the loan is totally repaid at the end of the four years, this balance must be $2,000. It follows that $X \approx \$2,000 - \$1,568.6984375 = \$431.3015625$. Since X is an integral number of cents, $X = \$431.30$, and LaTroy's total payment at the end of the fourth year is $431.30 + $100 = $531.30. He has a higher total payment than in Example (5.2.2), in which the payment at the end of the fourth year was $433.25, because he sacrifices interest by receiving a lower rate on his sinking fund balance than he pays on the $2,000 loan balance. We thus have the following sinking fund schedule where the **net balance on the loan** is the loan amount less the sinking fund balance. (In the table, we have rounded all entries to the nearest penny.)

TIME IN YEARS	INTEREST ON LOAN	SINKING FUND DEPOSIT	INTEREST ON S.F.	S.F. BAL. AFTER DEPOSIT	NET BALANCE ON LOAN
0	$0.00	$0.00	$0.00	$0.00	$2,000.00
1	$100.00	$700.00	$0.00	$700.00	$1,300.00
3	$205.00	$795.00	$35.44	$1,530.44	$469.56
4	$100.00	$431.30	$38.26	$2,000.00	$0.00

■

Suppose that you have a loan amount L that is financed by the sinking fund method, the effective interest rate per interest period on the loan amount is i, the effective interest rate per interest period on the sinking fund is j, and the loan is to be repaid by the sinking fund method with level payments for n interest periods. Then, at the end of each interest period, you have an interest payment of iL and a sinking fund deposit of $\dfrac{L}{s_{\overline{n}|j}}$. You thus have payments totaling $iL + \dfrac{L}{s_{\overline{n}|j}}$ at the end of each interest period. By contrast, for an amortized loan of amount L made at interest rate i' and repaid by level payments at the end of n interest periods, the payments are $\dfrac{L}{a_{\overline{n}|i'}}$. So, our sinking fund loan at interest rates i and j is equivalent to an amortized loan at rate i' where $i + \dfrac{1}{s_{\overline{n}|j}} = \dfrac{1}{a_{\overline{n}|i'}}$. If $i = j$, then, since $i + \dfrac{1}{s_{\overline{n}|i}} = \dfrac{1}{a_{\overline{n}|i}}$ by Equation (3.2.18), i' must equal $i = j$, so an amortized loan at interest rate i is equivalent to a sinking fund loan with $j = i$. If $i > j$, then $i' > i$ because, as in Example (5.3.2), the borrower sacrifices interest by paying interest on the loan amount at a higher rate than he receives on his sinking fund deposits. An estimate of i' may be obtained by considering the average balance in the sinking fund. If we take this average to be $\dfrac{L}{2}$, then the amount of interest sacrificed per unit borrowed is $\dfrac{1}{2}(i - j)$ and we estimate

(5.3.3) $$i' \approx i + \frac{1}{2}(i - j).$$

If $i < j$, then $i' < i$ and Equation (5.3.3) is again a reasonable estimate.

As already remarked, in *each interest period* the total payment on a loan of L by the sinking fund method at rates i and j is $iL + \dfrac{L}{s_{\overline{n}|j}}$, the sum of the interest payment and the deposit to the sinking fund account. So that this total payment resembles the payment formula $\dfrac{L}{a_{\overline{n}|j}}$ of an amortized loan, one defines $a_{\overline{n}|i \& j}$ by

$$(5.3.4) \qquad a_{\overline{n}|i \& j} = \frac{1}{i + \dfrac{1}{s_{\overline{n}|j}}}.$$

Then the total payment made each interest period by the sinking fund method is $\dfrac{L}{a_{\overline{n}|i \& j}}$. Furthermore, since by Equation (3.2.18) $j + \dfrac{1}{s_{\overline{n}|j}} = \dfrac{1}{a_{\overline{n}|j}}$,

$$(5.3.5) \quad a_{\overline{n}|i \& j} = \frac{1}{i + \left(\dfrac{1}{a_{\overline{n}|j}} - j\right)} = \frac{1}{(i - j) + \dfrac{1}{a_{\overline{n}|j}}} = \frac{a_{\overline{n}|j}}{(i - j)a_{\overline{n}|j} + 1}.$$

EXAMPLE 5.3.6   Sinking fund schedule with formulas

Problem: A loan of $\$10a_{\overline{3}|i \& j}$ is made for three years by the sinking fund method. The annual effective interest rate on the loan is i, and the annual effective interest rate earned by the sinking fund is j. Payments of interest are made annually, along with level sinking fund deposits. Make a sinking fund schedule for general i and j and include numerical values if $i = 5.2\%$ and $j = 3\%$. For the indicated interest rates, compute $\$10a_{\overline{3}|i \& j}$.

Solution We begin by noting that for the indicated interest rates, the loan amount is $\$10a_{\overline{3}|i \& j} \approx \$26.62900521$. We remark that in the following sinking fund schedule, in which dollar amounts are recorded to the nearest cent, were we to compute the final numerical entry of our chart from the previously recorded rounded entries, the final entry would be $\$26.64$ instead of $\$26.63$. Note also that, by Definition (5.3.4) and the fact that the balance in the sinking fund is always found by adding (1) the previous sinking fund balance, (2) the new deposit to the sinking fund, and (3) interest credited to the sinking fund since the previous payment, the final expression (last row, second-to-last column) in the chart results from the previous ones. The relevant algebra includes

$$\frac{\dfrac{\$10}{s_{\overline{3}|j}}}{i + \dfrac{1}{s_{\overline{3}|j}}}\left[1 + (2 + j) + (2 + j)j\right]$$

$$= \frac{\dfrac{\$10}{s_{\overline{3}|j}}}{i + \dfrac{1}{s_{\overline{3}|j}}}(3 + 3j + j^2) = \frac{\dfrac{\$10}{s_{\overline{3}|j}}}{i + \dfrac{1}{s_{\overline{3}|j}}}\left[1 + (1 + j) + (1 + j)^2\right]$$

$$= \frac{\dfrac{\$10}{s_{\overline{3}|j}}s_{\overline{3}|j}}{i + \dfrac{1}{s_{\overline{3}|j}}} = \frac{\$10}{i + \dfrac{1}{s_{\overline{3}|j}}} = \$10\,a_{\overline{n}|i \& j}.$$

TIME IN YEARS	TOTAL PAYMENT	INTEREST ON LOAN	SINKING FUND DEPOSIT	INTEREST ON S.F.	S.F. BAL. AFTER DEPOSIT
0	—	—	—	—	$0
1	$10	$\dfrac{\$10i}{i+\frac{1}{s_{\overline{3}|j}}}$	$\dfrac{\frac{\$10}{s_{\overline{3}|j}}}{i+\frac{1}{s_{\overline{3}|j}}}$	$0	$\dfrac{\frac{\$10}{s_{\overline{3}|j}}}{i+\frac{1}{s_{\overline{3}|j}}}$
	($10)	($1.38)	($8.62)	($0)	($8.62)
2	$10	$\dfrac{\$10i}{i+\frac{1}{s_{\overline{3}|j}}}$	$\dfrac{\frac{\$10}{s_{\overline{3}|j}}}{i+\frac{1}{s_{\overline{3}|j}}}$	$\dfrac{\frac{\$10}{s_{\overline{3}|j}}(j)}{i+\frac{1}{s_{\overline{3}|j}}}$	$\dfrac{\frac{\$10}{s_{\overline{3}|j}}(2+j)}{i+\frac{1}{s_{\overline{3}|j}}}$
	($10)	($1.38)	($8.62)	($.26)	($17.49)
3	$10	$\dfrac{\$10i}{i+\frac{1}{s_{\overline{3}|j}}}$	$\dfrac{\frac{\$10}{s_{\overline{3}|j}}}{i+\frac{1}{s_{\overline{3}|j}}}$	$\dfrac{\frac{\$10}{s_{\overline{3}|j}}(2+j)j}{i+\frac{1}{s_{\overline{3}|j}}}$	$\$10\,a_{\overline{n}|i \& j}$
	($10)	($1.38)	($8.62)	($.52)	($26.63)

The numerical entries of this sinking fund schedule may seem of little importance if they are as reported. However, things seem rather different if the units are changed from dollars to millions of dollars.   ∎

The next example gives a closer look at how loans made by the sinking fund method contrast with amortized loans.

EXAMPLE 5.3.7   Sinking fund method and amortization method

Problem: A $2,000,000 loan lasts five years and is repaid by the sinking fund method. Interest is paid annually at an effective interest rate of 8%, and the sinking fund savings account, to which there are level deposits, earns an effective interest rate of 5%. First, make a sinking fund table for this loan. Then show that the given loan is equivalent to an amortized loan at an annual effective interest rate of approximately 9.580261711%, and make an amortization table for a $2,000,000 five-year loan with level payments and effective interest rate of 9.580261711%.

Solution   The annual interest payment on the sinking fund loan is .08 × $2,000,000 = $160,000. Since $\frac{\$2,000,000}{s_{\overline{5}|5\%}} \approx \$361,949.5963$, the level deposits are each $361,949.60. (These actually accumulate to two cents more than the loan amount.) Recall that the sinking fund interest each year is 5% of the balance for the previous year, and the sinking fund balance is found by adding the previous sinking fund balance, the new deposit to the sinking fund, and interest credited to the sinking fund since the previous payment. Further note that the net balance on the loan is the loan amount minus the sinking fund balance. The following is therefore a sinking fund schedule for the given loan.

TIME IN YEARS	INTEREST ON LOAN	SINKING FUND DEPOSIT	INTEREST ON S.F.	S.F. BAL. AFTER DEPOSIT	NET BALANCE ON LOAN
0	$0.00	$0.00	$0.00	$0.00	$2,000,000.00
1	$160,000.00	$361,949.60	$0.00	$361,949.60	$1,638,050.40
2	$160,000.00	$361,949.60	$18,097.48	$741,996.68	$1,258,003.32
3	$160,000.00	$361,949.60	$37,099.83	$1,141,046.11	$858,953.89
4	$160,000.00	$361,949.60	$57,052.31	$1,560,048.02	$439,951.98
5	$160,000.00	$361,949.60	$78,002.40	$2,000,000.02	−$.02

The total payment each year is $160,000 + $361,949.60 = $521,949.60. These repay a loan of $2,000,000 and give the borrower 2 cents at the end of the five years, so the loan is equivalent to an amortized loan at rate i' where

$$\$2,000,000 = \$521,949.60 a_{\overline{5}|i'} - \$.02(1 + i')^{-5}.$$

If $i' = 9.580261711\%$, the right-hand side is approximately $2,000,000.000013[1], so the given effective interest rate is an excellent approximation! (This interest

[1]When you calculate the right-hand side, your display will show $2,000,000. However, if you then subtract $2,000,000, you will obtain .000013 rather than 0.

rate was found using the BA II Plus calculator. Of course it can also be found using the "guess and check" or Newton's methods, but these calculations are fairly time consuming. Those in possession of a BA II Plus calculator are encouraged to repeat the calculation of the unknown amortization rate i'.)

Next make an amortization table in the standard manner. (That is, first multiply the loan balance by .09580261711 to obtain the interest at time 1. Subtract that interest amount from the payment amount $521,949.60 to obtain the principal. Obtain the new balance by subtracting the just calculated principal from the previous balance. The time 1 line is now completed. Repeat this process to complete the chart, one line at a time.) So doing, you should obtain

TIME (years since loan)	PAYMENT	INTEREST IN PAYMENT	PRINCIPAL IN PAYMENT	BALANCE AFTER PAYMENT
0	—	—	—	$2,000,000.00
1	$521,949.60	$191,605.23	$330,344.37	$1,669,655.63
2	$521,949.60	$159,957.38	$361,992.22	$1,307,663.41
3	$521,949.60	$125,277.58	$396,672.02	$910,991.39
4	$521,949.60	$87,275.36	$434,674.24	$476,317.15
5	$521,949.60	$45,632.43	$476,317.17	$-$.02

Note: It is worth remarking that while we proclaimed the amortized loan at rate $i' = 9.580261711\%$ equivalent to the original sinking fund loan, meaning that the total cashflows at all times were equal, there is a difference between the original loan and the amortized loan that may be significant should taxes be considered. Specifically, the sequence of annual interest payments for the amortized loan is equal to neither the sequence of annual interest payments nor the sequence of annual net interest payments for the original sinking fund method loan. The **net interest payment** for a loan repaid by the sinking fund method is defined to be the amount of interest paid on the loan minus the amount of interest paid to the borrower on all sinking fund accounts established for the accumulation of the loan amount. It is, however, true that for the amortized loan as well as for the original loan by the sinking fund method, the *total* net interest paid is $609,747.98. ■

We end this section with another example that involves an amortization loan as well as a loan made by the sinking fund method.

EXAMPLE 5.3.8 **A hybrid loan, half by the sinking fund method and half by the amortization method**

Problem: A borrower is repaying a loan with eighteen annual payments of $1,200, the first payment occurring one year after the loan is made. Half of the loan (an amount $\frac{L}{2}$) is an amortized loan at an annual effective interest rate of 8%. The other half (another $\frac{L}{2}$) is to be repaid by the sinking fund method with the lender receiving an annual effective interest rate of 7.5%, and the sinking fund account earning 3% annual effective interest. Find the amount L of the loan.

Solution The payments on the amortized loan are $\dfrac{\frac{L}{2}}{a_{\overline{18}|8\%}} = \dfrac{L}{2a_{\overline{18}|8\%}}$.

The payments on the sinking fund loan are $.075(\frac{L}{2}) + \dfrac{\frac{L}{2}}{s_{\overline{18}|3\%}} = \dfrac{.075L}{2} + \dfrac{L}{2s_{\overline{18}|3\%}}$.

Therefore, $1,200 = \dfrac{.075L}{2} + \dfrac{L}{2s_{\overline{18}|3\%}} + \dfrac{L}{2a_{\overline{18}|8\%}}$, and

$$L = \frac{\$2,400}{.075 + \dfrac{1}{s_{\overline{18}|3\%}} + \dfrac{1}{a_{\overline{18}|8\%}}} \approx \$10,694.67.$$

■

5.4 LOANS WITH OTHER REPAYMENT PATTERNS

A loan can be repaid by any schedule that is agreeable to the lender and the borrower. We next study several loans with unorthodox repayment schedules, and in our discussion we will find the following notation convenient. L denotes the loan amount, and B_t gives the balance at time t, immediately following any payment at time t. P_t signifies the payment amount at time t (possibly zero). If i_t denotes the interest rate for the interval $(t - 1, t]$, then

(5.4.1) $$B_t = (B_{t-1})(1 + i_t) - P_t.$$

If n denotes the duration of the loan in interest periods, then $B_n = 0$.

EXAMPLE 5.4.2 **An amortized loan with level payments of principal**

Problem: Adam takes out a $20,000 loan for ten years at a level annual effective interest rate of 5%. At the end of each year, he pays $2,000 in principal, which is $\frac{1}{10}$th of the loan amount, along with the interest due. Find a formula for P_t.

Solution If $k \in \{0, 1, 2, \ldots, 10\}$, $B_k = \$20,000 - \$2,000k$. Therefore, by (5.4.1),

$$
\begin{aligned}
P_t &= (B_{t-1})(1.05) - B_t \\
&= (\$20,000 - \$2,000(t-1))(1.05) - (\$20,000 - \$2,000t) \\
&= (\$21,000 - \$2,000t(1.05) + \$2,100) - \$20,000 + \$2,000t \\
&= \$3,100 - \$100t.
\end{aligned}
$$

∎

The payments in Example (5.4.2) form an arithmetic progression. This is also true in the next example.

EXAMPLE 5.4.3   An amortized loan with payments in arithmetic progression

Problem: Beatrice takes out a $20,000 loan for ten years at a level annual effective interest rate of 5%. Her original payment is P and than her payments increase by $200 each year. Find P.

Solution   Here it is simpler to use Equation (3.9.4) for the present value of an arithmetic progression rather than relying on Equation (5.4.1). Observe that

$$
\begin{aligned}
\$20,000 = L &= (I_{P,\$200}a)_{\overline{10}|5\%} \\
&= Pa_{\overline{10}|5\%} + \frac{\$200}{.05}\left(a_{\overline{10}|5\%} - 10(1.05)^{-10}\right) \\
&= (P + \$4,000)a_{\overline{10}|5\%} - \$40,000(1.05)^{-10}.
\end{aligned}
$$

Since $a_{\overline{10}|5\%} = \frac{1-(1.05)^{-10}}{.05}$,

$$
P = \frac{.05(\$20,000 + \$40,000(1.05)^{-10})}{1 - (1.05)^{-10}} - \$4,000 \approx \$1,770.274498.
$$

So, $P = \$1,770.28$. In fact, $(I_{\$1770.28,\$200}a)_{\overline{10}|5\%} \approx \$20,000.04$. ∎

In our next example, the payments grow geometrically.

EXAMPLE 5.4.4   An amortized loan and payments in geometric progression

Problem: Cedric also takes out a $20,000 loan for ten years at a level annual effective interest rate of 5%. His payments grow by 20% each year. Find the amount of his first payment.

Solution Again (this time because we are skilled at summing geometric series), it is advantageous to avoid use of Equation (5.4.1). Cedric's payments are P_1, $P_2 = P_1(1.2)$, $P_3 = P_1(1.2)^2 \ldots$, $P_{10} = P_1(1.2)^9$. Using the 5% interest rate, these have present values $\frac{P_1}{1.05}$, $\frac{P_1}{1.05}(\frac{1.2}{1.05})$, $\frac{P_1}{1.05}(\frac{1.2}{1.05})^2$, $\ldots$, $\frac{P_1}{1.05}(\frac{1.2}{1.05})^9$.

Therefore, $\$20{,}000 = \frac{P_1}{1.05} + \frac{P_1}{1.05}(\frac{1.2}{1.05}) + \frac{P_1}{1.05}(\frac{1.2}{1.05})^2 + \cdots + \frac{P_1}{1.05}(\frac{1.2}{1.05})^9$.

Recalling formula (3.2.2) for the sum of a geometric series, one finds

$$\$20{,}000 = \frac{P_1}{1.05}\left[\frac{1 - (\frac{1.2}{1.05})^{10}}{1 - \frac{1.2}{1.05}}\right].$$

Therefore,

$$P_1 = \$21{,}000\left(\frac{1 - \frac{1.2}{1.05}}{1 - (\frac{1.2}{1.05})^{10}}\right) \approx \$1{,}070.97.$$

∎

Except for the final payment, the payments again grow geometrically in the next example.

EXAMPLE 5.4.5   Repaying a loan by paying four times the interest due at each payment time

Problem: Daphne also takes out a $20,000 loan at a level annual effective interest rate of 5%. She makes annual payments that are four times the interest due until she can make a final end-of-year payment of no more than $2,500 to pay off the loan. When does she make this final payment, and what is its amount?

Solution (This time, Formula (5.4.1) takes a starring role in our solution.) Daphne's payment at time t is $P_t = 4(.05B_{t-1}) = .2B_{t-1}$. Therefore, Equation (5.4.1) yields $B_t = B_{t-1}(1.05) - P_t = B_{t-1}(1.05) - .2B_{t-1} = .85B_{t-1}$. Since $B_0 = \$20{,}000$, this forces $B_1 = (.85)(\$20{,}000)$, $B_2 = (.85)^2(\$20{,}000)$, and, in general, $B_{t-1} = (.85)^{t-1}(\$20{,}000)$. The balance just prior to the payment at time t is $B_{t-1}(1.05) = (.85)^{t-1}(\$20{,}000)(1.05) = (.85)^{t-1}(\$21{,}000)$. Since the final payment is to be at most $2,500, we want the smallest positive integer t so that $(.85)^{t-1}(\$21{,}000) \leq \$2{,}500$. The natural logarithm is an increasing function so this inequality is equivalent to $(t - 1)\ln(.85) \leq \ln(\frac{2{,}500}{21{,}000}) = \ln(\frac{5}{42})$. Since $\ln(.85) < 0$, this latest inequality has the same solutions as $(t - 1) \geq \ln(\frac{5}{42})/\ln(.85)$. But $1 + [\ln(\frac{5}{42})/\ln(.85)] \approx 14.095$. The smallest integer t making our inequalities hold is therefore $t = 15$. The payment at $t = 15$ is $(.85)^{15-1}(\$21,000) \approx \$2,158.16$.   ∎

5.5 YIELD RATE EXAMPLES AND REPLACEMENT OF CAPITAL

We have already seen examples of yield rate problems with amortized loans in Section (3.10). We now focus on problems involving yield rates where a sinking fund is involved. A variant of the yield rate with reinvestment method [illustrated in Examples (3.10.2)–(3.10.4)] occurs if one reinvests only that amount needed so as to accumulate the initial investment amount. This is called reinvestment for **replacement of capital** and is demonstrated in the examples of this section. The last two of these [Examples (5.5.7) and (5.5.8)] involve amortized loans where the amount loaned out is recovered by the lender in a sinking fund account.

EXAMPLE 5.5.1   Annual yield for an investor who replaces capital

Problem: An investor makes a loan of $120,000 to be repaid by thirty end-of-year payments of $9,132. The investor replaces his capital (the $120,000) by making deposits, at the time of each payment, into a sinking fund account earning 5% annual effective interest. What is the investor's annual yield rate, assuming that this replacement of capital is accomplished using level sinking fund deposits?

Solution 1   Since $\frac{\$120,000}{s_{\overline{30}|5\%}} \approx \$1,806.17221$, the deposits to the sinking fund are each taken to be for $1,806.17. Observe that $\$1,806.17s_{\overline{30}|5\%} \approx \$199,999.85$ so that the investor is 15 cents shy of recovering the $120,000 he loaned out. (Had he deposited $1,806.18, it would accumulate $\$1,806.18s_{\overline{30}|5\%} \approx \$120,000.51$.) So, the investor has the following contributions of money.

The investor pays $120,000 at time $t = 0$.

The investor keeps $9,132 - \$1,806.17 = \$7,325.83$ of each payment at $t = 1, 2, \ldots, 30$.

The investor gets an additional $199.999.85 at $t = 30$ when he closes the sinking fund account.

So, if i is the investor's yield rate, a time 0 equation of value describing the investor's experience on this transaction is

$$\$120,000 = \$7,325.83a_{\overline{30}|i} + \$119,999.85(1 + i)^{-30}.$$

Using a financial calculator, one finds $i \approx 6.104856781\%$. (Had the sinking fund deposits been $1806.18 instead of $1,806.17, the equation of value $\$120,000 = \$7,325.82a_{\overline{30}|i} + \$120,000.51(1 + i)^{-30}$ would lead to $i \approx 6.104855277\%$.)

Solution 2 Let i denote the investor's yield rate, and suppose that the investor makes level deposits to the sinking fund account of amount X to accumulate exactly \$120,000, the amount of capital expended. Then $X = \dfrac{\$120,000}{s_{\overline{30}|5\%}}$.

The investor pays \$120,000 at time $t = 0$.

The investor keeps $\$9,132 - X = \$9,132 - \dfrac{\$120,000}{s_{\overline{30}|5\%}}$ of each payment

at $t = 1, 2, \ldots, 30$.

The investor gets \$120,000 (the sinking fund accumulation) at $t = 30$.

A time $t = 0$ equation corresponding to this experience is

$$\$120,000 = \left(\$9,132 - \frac{\$120,000}{s_{\overline{30}|5\%}} \right) a_{\overline{30}|i} + \$120,000(1 + i)^{-30}.$$

This may be rewritten as

$$\$120,000 - \$120,000(1 + i)^{-30} = \left(\$9,132 - \frac{\$120,000}{s_{\overline{30}|5\%}} \right) \left(\frac{1 - (1 + i)^{-30}}{i} \right),$$

and therefore as

$$\$120,000i = \$9,132 - \frac{\$120,000}{s_{\overline{30}|5\%}}.$$

Intuitively, this last equation is right. The investor's net cash inflow each year, $\$9,132 - \dfrac{\$120,000}{s_{\overline{30}|5\%}}$, should be the product of his yield rate and his capital expenditure \$120,000. Then

$$X = \frac{\$120,000}{s_{\overline{30}|5\%}} = \$9,132 - \$120,000i.$$

But we know (see Solution 1) that the sinking fund deposits are for an amount \$1,806.17. Therefore, $\$9,132 - \$120,000i = \$1,806.17$. It follows that $i = \frac{\$9,132 - \$1,806.17}{\$120,000} \approx 6.1048583\%$. The discrepancy with the answer in Solution 1 results from the fact that the investor did not replace 15 cents of capital.   ∎

In the second solution given in Example (5.5.1), we introduced an important intuitive method for finding the yield rate when there is replacement of capital and the returns are level throughout the investment period.

INTUITIVE METHOD (5.5.2):

Suppose that an investor expends an amount of capital X at time 0. Further suppose that the investor makes no further outlays of capital, that a level amount R is paid to the investor at the end of each of the following n years, and that from each of these payments the investor immediately deposits $\dfrac{X}{s_{\overline{n}|j}}$ into a savings account with an annual effective interest rate of j, thereby accumulating exactly X at time n. Then the investor's (annual) yield rate i satisfies the equation $Xi = R - \dfrac{X}{s_{\overline{n}|j}}$.

Intuitive Method (5.5.2) will be utilized in Examples (5.5.5), (5.5.7), and (5.5.8). However, we first consider an example that is similar to Example (5.5.1) and has a solution resembling Solution 1 of (5.5.1), but for which Intuitive Method (5.5.2) is inapplicable.

EXAMPLE 5.5.3   Average annual yield for an investor who replaces capital

Problem: An investor makes a loan of $120,000 to be repaid by end-of-year payments for thirty years. The payments at the end of each of the first ten years are $7,000, while the remaining payments are each $12,000. The investor replaces his capital (the $120,000) by making deposits into a sinking fund account earning 5% annual effective interest. What is the investor's average annual yield rate over the thirty-year period assuming that this replacement of capital is accomplished using level sinking fund deposits?

Solution   As in Solution 1 of Example (5.5.1), the annual deposits to the sinking fund are $1,806.17, and these will accumulate to $119,999.85. The investor therefore has the following contributions:

The investor pays $120,000 at time $t = 0$.
The investor keeps $7,000 - \$1,806.17 = \$5,193.83$ at $t = 1, 2, \ldots, 10$.
The investor keeps $12,000 - \$1,806.17 = \$10,193.83$ at $t = 11, 12, \ldots, 30$.
The investor gets an additional $199.999.85 at $t = 30$.

So, if i is the investor's yield rate, a time 0 equation of value describing the investor's experience on this transaction is

$$(5.5.4) \qquad \$120,000 = \$5,193.83 a_{\overline{10}|i} + (1 + i)^{-10}(\$10,193.83 a_{\overline{20}|i})$$
$$+ \$119,999.85(1 + i)^{-30}.$$

"Guess and check" solution It is not obvious whether the investor of this example has a higher or lower yield than the investor of Example (5.5.1). So, it is reasonable to begin our "guess and check" with an approximation to the yield found in Example (5.5.1), say 6.1%. If $i = .061$, then the right-hand side of Equation (5.5.4) is approximately $122,511.35. This is greater than the left-hand side $120,000, so we need a higher estimate of i. Check that if $i = .062$, the right-hand side of Equation (5.5.4) is about $120,654.17, while a guess of .0625 produces approximately $119,740.47. Thus, the yield rate is between 6.2% and 6.25%, probably closer to 6.25%. Trying .06235 for i produces close to $120,013.55 and .06236 gives approximately $119,995.31. The yield is therefore sandwiched in between 6.235% and 6.236%.

Solution by Newton's Method As in the "guess and check method," we need to begin with an initial guess of the yield. Set $i_1 = .061$. Define

$$f(x) = 120{,}000 - 5{,}193.83\left(\frac{1 - (1 + x)^{-10}}{x}\right)$$

$$- 10{,}193.83\left(\frac{1 - (1 + x)^{-20}}{x}\right)(1 + x)^{-10}$$

$$- 119{,}999.85(1 + x)^{-30}.$$

Then Equation (5.5.4) is equivalent to $f(i) = 0$. Next note that

$$f'(x) = -5{,}193.83\left(\frac{x\left(10(1 + x)^{-11}\right) - \left(1 - (1 + x)^{-10}\right)}{x^2}\right)$$

$$- 10{,}193.83\left(\frac{x\left(20(1 + x)^{-21}\right) - \left(1 - (1 + x)^{-20}\right)}{x^2}\right)(1 + x)^{-10}$$

$$- 10{,}193.83\left(\frac{1 - (1 + x)^{-20}}{x}\right)\left(-10(1 + x)^{-11}\right)$$

$$+ 3{,}599{,}995.5(1 + x)^{-31}.$$

Careful calculation shows that $i_2 = i_1 - \frac{f(i_1)}{f'(i_1)} \approx .062337747$. Further iterations would produce even better estimates of the yield rate i.

Solution by BA II Plus calculator Cash Flow worksheet Open and clear the
Cash Flow worksheet by pushing CF   2ND   CLR WORK . Then key

1 2 0 0 0 0 +/− ENTER ↓ 5 1 9 3 • 8 3 ENTER ↓

1 0 ENTER ↓ 1 0 1 9 3 • 8 3 ENTER ↓ 1 9 ENTER

↓ 1 0 1 9 3 • 8 3 + 1 1 9 9 9 9 • 8 5 =

ENTER IRR CPT .

This should result in the display reading "IRR = 6.235743029". Note that
the resulting estimate of the yield rate $i \approx .06235743029$ is consistent with our
previous estimates and was obtained with much less work!

The reason that Example (5.5.3) does not have a solution by the Intuitive
Method (5.5.2) is that it would not give a level yield over the thirty-year
period. In fact, there is one yield y_1 for the first ten years (namely, $y_1 = \frac{\$5,193.83}{\$120,000} \approx .043281917$) and a second yield y_2 for the next twenty (namely,
$y_2 = \frac{\$10,193.83}{\$120,000} \approx .084948583$). ∎

Example (5.5.1) is a replacement of capital example where the unknown
is the yield rate, and Example (5.5.3) concerns replacement of capital where
the unknown is an average yield rate. In the next two examples, the yield rate
is given, but there are other unknowns to find.

EXAMPLE 5.5.5   Replacement of capital and unknown investment term

Problem: Mr. Guillen pays $4,909.10 to purchase an n-year annuity with
end-of-year payments of $456.73. His price will allow the replacement of the
original investment in a sinking fund earning 3.5% annual effective interest
and will produce an annual yield of 6%. Find n.

Solution Let X denote the amount of the sinking fund deposits. Then the
annual return on the $4,909.10 is $456.73 − X$. Mr. Guillen's annual yield rate
was given to be 6%, so the Intuitive Method (5.5.3) gives rise to the equation

$$(.06)\$4,909.10 = \$456.73 - X.$$

[Another approach is to derive this equation by looking at Mr. Guillen's
investment experience, thereby obtaining the time zero equation of value
$\$4,909.10 = (\$456.73 - X)a_{\overline{n}|6\%} + \$4,909.10(1.06)^{-n}$. Then use the fact
that $a_{\overline{n}|6\%} = \frac{1-(1.06)^{-n}}{.06}$ to conclude that $(.06)\$4,909.10 = \$456.73 - X$.]
Solving for X, we find $X \approx \$162.184$. Since X must be an integral number

of cents, $X = 162.18$. Because Mr. Guillen accumulates $\$4,909.10$ in his sinking fund account, $\$4,909.10 = Xs_{\overline{n}|3.5\%} = \$162.18s_{\overline{n}|3.5\%}$. This forces $n \approx 20.99999161$, so the integer n is 21. ∎

Next we have an example where there are varying rates used to calculate the present value and also governing the accumulation in the sinking fund.

EXAMPLE 5.5.6   Replacement of capital and varying rates

Problem: A thirty-year annuity-immediate (with level end-of-year payments P) has a present value of $\$120,000$ when calculated using an annual effective interest rate of 5% for the first ten years and an annual effective interest rate of 6% for the following twenty years. An investor buys this annuity at a price Q that over the entire thirty-year period yields 8% annually on the purchase price and further allows for the replacement of capital by making level deposits R to a sinking fund account that has an annual effective interest rate of 3% for the first ten years, followed by an annual effective interest rate of 4% for the next twenty years. Find P, Q, and R.

Solution   Note that $\$120,000 = Pa_{\overline{10}|5\%} + (1.05)^{-10}Pa_{\overline{20}|6\%}$. Therefore,

$$P = \frac{\$120,000}{a_{\overline{10}|5\%} + (1.05)^{-10}a_{\overline{20}|6\%}} \approx \$8,128.279652.$$

Since P is an integral number of cents, $P = \$8,128.28$. (This actually results in a present value of about $\$120,000.0051$.)

The investor pays Q at time $t = 0$, keeps $\$8,128.28 - R$ at times $t = 1, 2, \ldots, 30$, and gets an additional $\$120,000$ at $t = 30$. Since the investor's yield is given to be 8%, we have the time zero equation of value

$$Q = (\$8,128.28 - R)a_{\overline{30}|8\%} + Q(1.08)^{-30}.$$

Since $a_{\overline{30}|8\%} = \frac{1-(1.08)^{-30}}{.08}$, this tells us that

$$.08Q = \$8,128.28 - R \quad \text{and} \quad R = \$8,128.28 - .08Q.$$

Recall that the investor buys the annuity at a price Q and replaces the capital by making level deposits R to a sinking fund account that has an annual effective interest rate of 3% for the first ten years, followed by an annual effective interest rate of 4% for the next twenty years. Therefore,

$$
\begin{aligned}
Q &= Rs_{\overline{20}|4\%} + Rs_{\overline{10}|3\%}(1.04)^{20}\\
&= R(s_{\overline{20}|4\%} + s_{\overline{10}|3\%}(1.04)^{20})\\
&= (\$8,128.28 - .08Q)(s_{\overline{20}|4\%} + s_{\overline{10}|3\%}(1.04)^{20}).
\end{aligned}
$$

As a result,

$$Q = \frac{\$8{,}128.28\left[s_{\overline{20}|4\%} + s_{\overline{10}|3\%}(1.04)^{20}\right]}{1 + .08\left[s_{\overline{20}|4\%} + s_{\overline{10}|3\%}(1.04)^{20}\right]} \approx \$82{,}759.24017.$$

Since Q is an integral number of cents, $Q = \$82{,}759.24$ and $R = \$8{,}128.28 - .08Q \approx \$1{,}507.5408$. Again, we need an integral number of cents, so $R = \$1{,}507.54$. (With this rounding, the sinking fund balance is approximately $\$82{,}759.20$ which is 4 cents less than the purchase price Q.) ∎

We now turn to the promised examples in which the lender of an amortized loan recovers his capital in a sinking fund account.

EXAMPLE 5.5.7   Unknown yield; lender of an amortized loan replacing capital using a sinking fund account

Problem: Mr. Ng takes out a thirty-year mortgage from Sunset Mortgage Corporation. The amount financed is $\$82{,}311.66$, and the loan has a nominal interest rate of 6% convertible monthly. Sunset calculates its yield rate by assuming that it will replace the $\$82{,}311.66$ in a sinking fund earning a nominal interest rate of 4.8% convertible monthly. What effective annual yield does Sunset calculate?

Solution   Mr. Ng's monthly payment is $\$493.50$ since

$$\frac{\$82{,}311.66}{a_{\overline{360}|.5\%}} \approx \$493.4999898.$$

Let y denote Sunset's monthly yield rate. Then, the amount Sunset receives each month for unrestricted use is $(\$82{,}311.66)y$, the product of the investment amount and the monthly yield rate. The balance of Mr. Ng's $\$493.50$ payment, $\$493.50 - (\$82{,}311.66)y$, is needed for investment into the sinking fund so that at the end of the thirty years, Sunset Mortgage Corporation will recoup exactly its $\$82{,}311.66$ investment. Noting that the sinking fund has a monthly interest rate of $\frac{4.8\%}{4} = .4\%$, the equation

$$y = \frac{493.50}{82{,}311.66} - \frac{1}{s_{\overline{360}|.4\%}} \approx .004748852$$

must be satisfied. So, the annual yield is $(1 + y)^{12} - 1 \approx .058498442$. ∎

EXAMPLE 5.5.8   Unknown yield; lender of an amortized loan replacing capital using a sinking fund account

Problem: Sylvia took out a $30,000 amortized loan from Medical Corporation. The annual effective interest rate on the loan is stated as 5%, and the duration of the loan is eight years. There were payments of $3,500 at the end of each of the first three years and higher level payments at the end of each of the next five years. Medical Corporation replaced its $30,000 capital by means of a sinking fund earning an annual effective interest rate of 4%. Each time Medical Corporation received a payment from Sylvia, it deposited the portion representing principal into the sinking fund. What was Medical Corporation's annual yield on this eight-year transaction?

Solution Let E denote the excess amount over $3,500 in each of the last five payments. Then,

$$\$30,000 = \$3,500a_{\overline{8}|5\%} + Ea_{\overline{5}|5\%}(1.05)^{-3}.$$

Solving for E, we find

$$E = \frac{\$30,000 - \$3,500a_{\overline{8}|5\%}}{a_{\overline{5}|5\%}(1.05)^{-3}} \approx \$1,972.94784.$$

Since E is an integral number of cents, $E = \$1,972.95$. (With the last five payments being $3,500 + \$1,972.95$, Sylvia pays off $30,000.01, and the annual effective interest rate on the loan was actually slightly above 5% (approximately 5.000006073%).

Let i denote Medical Corporation's annual effective yield. Then, using Intuitive Method (5.5.3), Medical Corporation contributed all but $30,000i$ of each payment it received to the sinking fund. But the first three payments were for $3,500 and the next five were for $3,500 + \$1,972.95$. Thus, we can think of there having been level end-of-year sinking fund deposits of $3,500 - \$30,000i$ for eight years and additional end-of-year sinking fund deposits of $1,972.95 during the last five years. The sinking fund had an annual effective interest rate of 4%, so the sinking fund balance at the end of the eight years, immediately following the last of the deposits, was $(\$3,500 - \$30,000i)s_{\overline{8}|4\%} + \$1,972.95s_{\overline{5}|4\%}$. It was already noted that Medical Corporation was making sinking fund deposits so that this balance would be $30,000. Therefore,

$$\$30,000 = (\$3,500 - \$30,000i)s_{\overline{8}|4\%} + \$1,972.95s_{\overline{5}|4\%}.$$

It follows that

$$\frac{\$30,000 - \$1,972.95s_{\overline{5}|4\%}}{s_{\overline{8}|4\%}} = \$3,500 - \$30,000i,$$

and

$$i = \frac{1}{30,000}\left(3,500 - \left(\frac{30,000 - 1,972.95s_{\overline{5}|4\%}}{s_{\overline{8}|4\%}}\right)\right) \approx .046796932.$$

■

5.6   PROBLEMS, CHAPTER 5

(5.0) Chapter 5 writing problems

(1) For calendar year 2004, the Johnsons are eligible to claim a standard deduction of $9,700 on their income taxes. Alternatively, they may itemize deductions. Their eligible deductions are $600 of charitable contributions, $6,643 of property taxes, and interest on their mortgage. Their mortgage is a standard fifteen-year amortized loan with a nominal monthly interest rate of 5.85%, and the amount financed was $216,000. The loan was initiated on April 1 of calendar-year N. Explain to the Johnsons how small N must be so that they would pay less income tax by taking the standard deduction. That is, explain to the Johnsons when the $9,700 standard deduction is lower than their total itemized deductions? Note that the Johnsons are highly educated people who would appreciate a carefully written explanation of this tax issue.

(2) Call a local lending institution and learn what types of consumer loans they offer. Inquire whether they have any for which the borrower makes regular periodic payments of interest but does not have to repay any of the capital until the due date. For any such loans, find out how the lending institution protects itself against potential default on the part of the borrower. For instance, does the lender require a sinking fund account or securities held as collateral? Describe what you have learned.

(3) [following Section 5.5] In previous chapters we have considered yield rates where reinvestment is stipulated. Now we have introduced yield rates when there is replacement of capital. How are these topics related?

(5.2) Amortized loans and amortization schedules

(1) Copy and complete the following amortization table. Be careful to note that there is no payment at $t = 2$.

TIME (years since loan)	PAYMENT	INTEREST IN PAYMENT	PRINCIPAL IN PAYMENT	BALANCE AFTER PAYMENT
0	—		—	—
1	$8,000			$22,342
3		$1,916.14		$9,908
4				$0

(2) Ellen has a thirty-year mortgage with level monthly payments. The amount of principal in her 82nd payment is $259.34, and the amount of principal in her 56th payment is $230.19. Find the amount of interest in her 133rd payment.

(3) An amortized loan is repaid with annual payments which start at $400 at the end of the first year and increase by $45 each year until a payment of $1,480 is made, after which they cease. If interest is 4% effective, find the amount of principal in the fourteenth payment.

(4) A fifteen-year adjustable-rate mortgage of $117,134.80 is being repaid with monthly payments of $988.45 based upon a nominal interest rate of 6% convertible monthly. Immediately after the 60th payment, the interest rate is increased to a nominal interest rate of 7.5% convertible monthly. The monthly payments remain at $988.45, and there will be an additional balloon payment at the end of the fifteen years to pay the outstanding loan balance.

  (a) Calculate the loan balance immediately after the 84th payment.

  (b) Calculate the amount of interest in the 84th payment.

  (c) Calculate the amount of the balloon payment.

(5) Arlen buys a home for $328,000 and makes a down payment of $33,000. The balance he finances with a fifteen-year mortgage with monthly payments and an annual effective interest rate of 5.8%. There will be level payments followed by a final slightly reduced payment. Calculate the amount of interest that Arlen pays in the first five years of the loan.

(6) Let r and k denote positive integers, and set $n = rk$. An amortized loan lasting n interest periods has a payment of P at the end of each k interest periods. The effective interest rate per interest period is i.

(a) Explain why the outstanding loan balance at time jk, just after the payment of P is equal to $P\frac{a_{\overline{n-jk}|i}}{s_{\overline{k}|i}}$, $j \in \{0, 1, 2, \dots, r\}$ [HINT: Look at Section (4.2).]

(b) Use the result of (a) to establish that the interest in the payment at time $(j+1)k$ is $P(1 - v^{n-jk})$.

(7) An amortized loan lasting n interest periods has a level payment of P at the end of each m-th of an interest period. The effective interest rate per interest period is i.

(a) Let $k \in \{1, 2, \dots, nm\}$. Define $a^{(m)}_{\overline{n-k/m}|i} = \frac{1-v^{(n-k/m)}}{i^{(m)}}$. Explain why the outstanding loan balance at time $\frac{k}{m}$, just after the payment P, is $m\,Pa^{(m)}_{\overline{n-k/m}|i}$. [HINT: You may find it helpful to look at Section (3.11).]

(b) Use the result of (a) to show that the interest at time $\frac{k+1}{m}$ is

$$P(1 - v^{n-\frac{k}{m}}), k \in \{0, 1, 2, \dots, nm - 1\}.$$

(5.3) The sinking fund method

(1) A $14,000 loan is to be repaid by the sinking fund method, with irregular payments into the sinking fund. The table below is a partially completed sinking fund table for this situation. Find the missing entries, noting that there was no payment at the end of the 3rd year.

TIME IN YEARS	INTEREST ON LOAN	SINKING FUND DEPOSIT	INTEREST ON S.F.	S.F. BAL. AFTER DEPOSIT	NET BALANCE ON LOAN
0	$0	$0	$0.00	$0	$14,000
1	$889		$0.00	$5,200	
2		$3,000	$218.40		
4					$0

(2) The borrower in a $238,000 loan makes interest payments at the end of each six months for eight years. These are computed using an annual effective discount rate of 6.5%. Each time he makes an interest payment, the borrower also makes a deposit into a sinking fund earning a nominal interest rate of 4.2% convertible monthly. The amount of each sinking

fund deposit is D in the first three years and $2D$ in the remaining five years, and the sinking fund balance at the end of the eight years is equal to the loan amount. Find D.

(3) Alan borrows $18,000 for eight years and agrees to make quarterly payments of $770. Each of these payments consists of interest for the just completed quarter and a deposit to a sinking fund that has a nominal interest rate of 6% convertible quarterly. For the first six years, each year the lender receives 8% nominal interest convertible quarterly. For the remaining two years, the lender receives 12% nominal interest convertible quarterly. Find the amount by which the sinking fund is short of repaying the loan at the end of the eight years.

(4) Cindy borrows $13,500 for twelve years at an annual effective interest rate of i. She accumulates the amount necessary to repay the loan by a sinking fund. Cindy makes twelve payments of P at the end of each year, which includes payment on the loan at an annual effective interest rate of i and payment into a sinking fund on which the annual effective interest rate is 4%. If the annual effective rate on the loan had been $2i$ instead of i, Cindy's total annual payment would have been $1.2P$. Find the amount P.

(5) Bob and Barbara are friends. Bob takes out a $10,000 loan and agrees to repay it over twelve years by making annual level payments at an effective rate of 5.62499%. At the same time, Barbara takes out a $10,000 loan and agrees to repay it by making annual interest payments at an annual effective interest rate of i. She also agrees to make annual level deposits into a sinking fund that earns 4% annual effective interest so as to accumulate $10,000 at the end of the twelve years. Bob and Barbara discover they have the same total annual expenditures resulting from their loans. Find the rate i.

(5.4) Loans with other repayment patterns

(1) A loan of $39,999.85 is to be repaid by payments at the end of each quarter for eight years. Each payment is 2% higher than its predecessor. The loan is made at a nominal rate of discount of 4% payable quarterly. Find the balance just after the 20th payment, the amount of interest in the twentieth payment, and the amount of principal in the twentieth payment.

(2) A loan of $12,500 is made at an effective interest rate of 8.5%. Payments are made at the end of each interest period. Each payment equals twice the interest due until the borrower pays off the outstanding debt with a final payment of, at most, $1,800. Find the number of payments n and the amount of the final payment.

(3) A loan is repaid with monthly payments that start at $320 at the end of the first month and increase by $5 each month until a payment of $950 is made, after which they cease. If interest is 4% effective, find the amount of principal in the sixtieth payment.

(4) Eliza takes out a $36,000 loan at an annual effective interest rate of 6%. It is agreed that at the end of each of the first six years she will pay $1,800 in principal, along with the interest due, and that at the end of each of the next eight years she will make level payments of $2,500. Eliza will make one final payment at the end of fifteen years to exactly complete her loan obligation. Calculate the amount of Eliza's fifth payment, the amount of her tenth payment, and the amount of her fifteenth payment.

(5) Mr. Beltram takes out a $100,000 loan for twelve years. The applicable annual effective interest rate is a promotional rate of 2% for the first two years and 6% for the remainder of the loan term. Mr. Beltram's payments increase by 10% each year. Find the balance on the loan immediately following his fifth payment.

(6) Marilyn Ho borrowed $24,000 from Stewart Financial Enterprises. She was required to make sixteen equal payments of principal. These were to be made annually with the first payment due exactly one year after she received the $24,000. Ms. Ho also had to make annual payments of interest to the loan holder at an annual effective rate of 8.5%. Immediately after the loan is made, Brady Investment Corporation purchases the right to receive all of Ms. Ho's payments from Stewart Financial Enterprises. The price paid by Brady Investment Corporation resulted in Brady having a 7% annual yield on their sixteen-year investment. Find the price that Brady Investment Corporation paid Stewart Financial Enterprises.

(5.5) Yield rate examples and replacement of capital

(1) Admiral Trust Company makes an amortized loan of $47,000, to be repaid by annual end-of-year payments of $4,675 for eighteen years. In order to replenish its capital, the company will make level annual payments into a sinking fund account earning 5% interest.

    (a) If the sinking fund account is held by a third party, what is Admiral Trust Company's yield rate assuming complete replacement of capital?

    (b) If the sinking fund account is held by Admiral Trust Company, what is its yield rate assuming complete replacement of capital?

(2) Gerry pays W to buy a ten-year annuity with end-of-year payments of $1,400. This purchase price allows her to replace her capital by means

of a savings account that has an annual effective interest rate of 3% and also to earn an overall annual yield of 6% for the ten years of the annuity. Find W.

(3) A bank makes a package of three loans to a small business.

(a) $120,000 amortized monthly for ten years at a nominal discount rate of 6.8% convertible monthly.

(b) $100,000 to be repaid by monthly sinking fund payments for ten years where interest is assessed at a rate of 5.4% nominal convertible monthly and the sinking fund earns 4% nominal interest convertible monthly. The bank receives the sinking fund deposits.

(c) $200,000 to be repaid with interest at the end of ten years with an effective rate of discount of 8.2% throughout the ten years.

Find the bank's annual effective yield on each of these loans individually and on the package of loans over the ten-year period.

(4) A twenty-year loan of $25,000 is negotiated with the borrower agreeing to repay principal and interest at 5%. A level payment of $1,500 will apply during the first ten years, and a higher level payment will apply over the remaining ten years. Each time the lender receives a payment from the borrower, he will deposit the portion representing principal into a sinking fund with an annual effective interest rate of 4%. (This is the amount for replacement of capital.) What is the lender's yield rate on this entire investment?

(5) Sheryl Tran pays $245 at the end of each month for ten years in order to repay a loan of $22,000. The lender makes level deposits to a sinking fund account that is held by a third party and that pays an annual effective interest rate of 3% during the first four years and a nominal monthly discount rate of 4% thereafter. Find the difference between the annual effective interest rate charged to Ms. Tran and the lender's annual yield.

(6) [BA II Plus calculator] Signet Sign Company will repay a $240,000 loan from Small Business Bank by the sinking fund method with monthly payments. The first payment will be one year after the loan is made and the last will be eight years after the loan date. So there will be a total of eighty-five payments, each consisting of interest and a sinking fund deposit. Interest will be repaid at a nominal interest rate of 4% convertible **quarterly** during the first five years of the loan and at a nominal interest rate of 5% convertible **monthly** during the remaining three years of the loan. Deposits to the sinking fund account (again at Small Business Bank) will earn a nominal interest rate of 3% convertible **monthly** throughout the eight years. Deposits to the sinking fund will be level except for a slightly reduced final deposit as necessary. Find the bank's yield rate for this loan over the eight-year period.

Chapter 5 review problems

(1) Dustin borrows $100,000. Half of the loan is repaid by the amortization method at an annual effective interest rate of 6% with payments due at the end of each year for twelve years. The other half is to be repaid by the sinking fund method with interest on the loan balance assessed at an annual effective interest rate of 5% and the sinking fund earning an annual effective interest rate of 3%. The deposits to the sinking fund are to be made at the end of each year for twelve years and to increase by 4% each year. Find the total amount Dustin pays exactly five years after he takes out the $100,000 loan.

(2) A loan is being repaid by the amortization method with an installment at the end of each of 80 quarters at 6% annual effective interest, the first payment one quarter after the loan is made. In which payment are the principal and the interest most nearly equal to each other?

(3) A loan of $10,000 is negotiated, with the borrower agreeing to repay the principal over ten years as well as to make annual end-of-year payments of interest at 4% effective per annum. A $1,100 total payment will be due at the end of each year during the first five years, and a higher level end-of-year payment will be required during the second five years. The lender will replace his capital by means of a sinking fund earning 5% per annum. Each time he receives a payment from the borrower, he will deposit that portion representing principal into the sinking fund. What will be the lender's yield on the whole transaction, assuming all payments are made as scheduled?

(4) A loan is to be repaid by monthly installments of $800 for thirty-six months, each paid at the end of the month. The interest contained in the twelfth payment is $2.81 less than the interest contained in the eleventh payment. Find the effective monthly rate of interest.

(5) A loan is to be repaid by monthly payments of $674 for five years, the first of the sixty payments occurring exactly one year after the loan is made.   Find the total amount of interest at the times of the odd numbered payments if the annual effective interest rate for the loan is 4.8%. [HINT: Be very careful when considering the interest at the time of the first payment, and note that this interest exceeds the amount of the payment. An amortization table would therefore show a negative amount of principal paid at the first payment time.]

(6) A loan of $32,000 is to be repaid by the sinking fund method with annual payments. Interest on the loan is paid at a 6% annual effective rate of interest, and the sinking fund earns 4% annual effective interest. The total annual payments will be level at $3,300 until a final smaller annual payment suffices to pay off the loan. Find the amount of the final sinking fund deposit.

(7) An amortized loan of $75,000 has annual payments for eighteen years, the first occurring exactly one year after the loan is made. The first four payments will be for only half as much as the remaining fourteen. The annual effective interest rate for the loan is 6%. Calculate the amount of principal repaid in the seventh payment.

(8) On February 1, 1973 Consolidated Physicians Alliance took out a loan for $1,375,000 from Captain Financial. The loan was amortized using an annual effective interest rate of 6% and had end-of-month payments for five years, the payment amounts being level except for a slightly reduced final payment. At the time of the loan, Captain Financial calculated its yield assuming that it will replace its capital via level deposits in a sinking fund (at a bank) earning a nominal rate of interest of 5% convertible monthly. What annual yield rate did Captain Financial anticipate?

(9) A loan is repaid with monthly payments for six years, the payments beginning exactly one year after the loan is made. The payments are each $100 during the first year, and each year there is a $10 increase in the monthly payments. If the interest rate on the loan is a nominal rate of 4.5% convertible monthly, find the amount of principal in the fortieth payment.

(10) Tirunesh takes out a thirty-year mortgage for $126,523 with the nominal interest rate convertible monthly set at 6.75%. The scheduled monthly payment amounts are level except for a slightly reduced final payment. After making payments on schedule for five years, Tirunesh decided to increase his monthly payments by $300 in order to pay off the mortgage more quickly. Assuming these increased payments are made, calculate the amount of interest Tirunesh pays during the seventh year.

C H A P T E R 6

Bonds

6.1 INTRODUCTION

Investors seek a variety of low-risk growth opportunities. Among those often considered by an astute trader are **bonds**. A bond is a security, issued by a government entity or corporation, that promises certain payments at future dates. The last of these payments occurs at the **maturity date** or **redemption date** for the bond, and the **issue date** is the time that the investor loans the bond issuer money by "purchasing the bond." The **term** of the bond refers to either the interval from the issue date until the maturity date or the length of this interval. If the maturity date is fixed, the bond is said to be **noncallable**. The bonds we consider prior to Section (6.9) will all be noncallable. Callable bonds usually involve more risk considerations.

If a bond promises only a single payment at a fixed maturity date, it corresponds to a simple loan that we were already prepared to deal with in Chapter 1. Such bonds are sometimes called **zero-coupon bonds** or **(pure) discount bonds**. Examples of these include United States Treasury bills. [1]

---

[1]Technically, United States securities are classified by their terms. **Bills** have terms of one year or less, **notes** have terms of one to ten years, and **bonds** have terms of more than ten years. However, we will call all of these bonds.The available terms vary over time. For instance, the

The bonds that we wish to focus on are so-called **coupon bonds**. These promise a sequence of payments prior to and at the redemption date, as well as an additional redemption payment. The redemption payment is usually the largest payment. It is common for the preredemption payments to be level and regularly spaced, the length of the interval between coupon payments being referred to as the **coupon period**, and unless otherwise stated, we restrict ourselves to these bonds. Quite commonly the coupon period is half-a-year, so there are semiannual coupons.

The **indenture** is the legal document that specifies the terms and conditions of a particular bond.

Language and notation commonly used when discussing bonds are introduced in Section (6.2). We may then use basic principles to find an equation, the so-called basic price formula, for the price of a bond to yield the investor a specified yield. Further pricing formulas are discussed in Sections (6.3) and (6.4). The premium-discount formula is particularly useful if you want to create bond amortization schedules; this is a special case of loan amortization (which was introduced in Chapter 5) and is the subject of Section (6.5).

In bond amortization, the relevant interest rate is the yield rate that the bond purchaser would achieve if he held the bond until maturity, and valuing a bond at this interest rate is discussed in Section (6.6); more subtle is the question of what value you should assign between cashflow dates. Bonds are liquid investments and need not be held until their redemption date; any new buyer is likely to earn a different interest rate. In Section (6.7), we discuss the pricing or valuing of bonds at arbitrary interest rates. Again, this is more complicated at dates when there is no cashflow, and several different types of prices are discussed.

Bonds present new and interesting situations in which you may wish to calculate yields. Examples of these are discussed in Section (6.8).

Bonds with novel features are introduced in Sections (6.9) and (6.10). In particular, the bonds of Section (6.9) provide either the buyer or the seller with a choice as to when the bond is redeemed, while the bonds of Section (6.10) provide cashflows that are tied to prevailing interest rates at the time of the cashflow.

The chapter's final section [Section (6.11)] is just for users of the BA II Plus calculator. Here is a chance for you to become familiar with the calculator's **Bond worksheet**.

6.2   BOND ALPHABET SOUP AND THE BASIC PRICE FORMULA

The title of this section is a bit in jest. To study bonds with coupons, it is customary to introduce a large number of letters, each referring to a different quantity connected to a given coupon bond. With a little practice the

---

Treasury did not issue any thirty year bonds between August 9, 2001 and February 9, 2006.

terminology is easy, but at first the experience may be reminiscent of looking at a bowl of alphabet soup with randomly floating letters.

Printed on a bond is its **face** (or **par) value** F. In general, this need not be either the price the investor pays for the bond nor the redemption value of the bond at maturity. Rather, it is a number used to calculate the size of the coupon payments. The calculation of these coupon amounts also depends on a nominal rate α convertible m times per year where m is the number of coupons each year. Set $r = \frac{\alpha}{m}$. Then r is an effective rate for the coupon period. Both the nominal rate α and the effective rate r are sometimes called **coupon rates**.

The amount of each coupon payment is $Fr = \frac{F\alpha}{m}$, F the face amount and α the nominal coupon rate.

Let n denote the number of coupon periods in the term of the bond. If the bond is an N-year bond, $n = Nm$. The **redemption amount** is the amount that the bond holder receives at the bond's maturity date, excluding the regular coupon amount. The redemption amount at the end of n coupon periods is denoted by C. A synonym for face value is **par**, and if $F = C$, the bond is said to be a **par-value bond** or to be **redeemable at par**. *If no redemption is specified, either directly or indirectly, then you should assume a bond is a par-value bond.*

The redemption amount C is of importance to the bond participants, so the coupon amount might better be expressed in terms of it. Solely with this goal in mind, the **modified coupon rate** g is defined by

$$(6.2.1) \qquad\qquad g = \frac{Fr}{C}.$$

The coupon amount is then $Fr = Cg$.

The rates r and g are only of interest to the investor as a tool to calculate the coupon payment. In contrast, the investor may be deeply interested in the yield rate. We customarily reserve the letter i for an annual effective yield rate. So, *we will use the letter j for the investor's effective yield rate per coupon period.* Of course $i = (1 + j)^m - 1$. Usually, the yield rate is specified as a nominal rate I convertible m times per year. That is,

$$j = \frac{I}{m} \qquad \text{and} \qquad i = \left(1 + \frac{I}{m}\right)^m - 1.$$

The letter G is introduced to allow us to express the coupon amount in terms of the yield rate per coupon period j. Define $G = \frac{Fr}{j}$ and call G the **base amount**. Then, the amount of each coupon payment is $Fr = Cg = Gj$. The base amount is the amount you would need to invest now, at interest rate j, to create a perpetuity paying an amount of interest at the end of each coupon period that is equal to the coupon amount.

Finally, in our introduction of letters related to the bond, we let P denote the price paid for the bond to yield the investor j per interest period, $v_j = (1 + j)^{-1}$, and we set $K = Cv_j^n$, the value of the redemption at the issue date figured using compound interest and the yield rate received by the investor. Then, measuring time in coupon periods beginning at the issue date, the purchaser pays P at $t = 0$ and gets Fr at times $t = 1, 2, \ldots, n$ as well as the redemption value C at time n. So the yield j satisfies the time 0 equation of value

$$(6.2.2) \qquad P = (Fr)a_{\overline{n}|j} + Cv_j^n = (Fr)a_{\overline{n}|j} + K.$$

Equation (6.2.2) is called the **basic price formula** for a bond.

We next consider three basic examples that should help you become familiar with our terminology, along with a table of the variables we have introduced in this section.

EXAMPLE 6.2.3   Finding a bond's base amount and the price

Problem: An eight-year $3,000 10% bond with semiannual coupons and redemption value $2,800 is bought at a price to give the investor an annual effective yield rate of 12%. Find the base amount and the price of the bond.

Solution For the given bond, you have face value $F = \$3,000$, nominal coupon rate $\alpha = 10\%$, coupons per year $m = 2$, total number of coupons $n = 2 \times 8 = 16$, coupon rate per coupon period $r = \dfrac{10\%}{2} = 5\%$, redemption amount $C = \$2,800$, and annual effective yield rate $i = 12\%$. Therefore, the effective yield rate per coupon period is $j = (1 + i)^{\frac{1}{2}} - 1 \approx .058300524$. It follows that the coupon amount is $Fr = \$3,000(.05) = \$150$, and the base amount is

$$G = \frac{\$150}{j} \approx \$2,572.88.$$

The basic price formula gives

$$P = (\$150)a_{\overline{16}|j} + \$2,800v_j^{16} \approx \$2,664.61.$$

■

TYPE	LETTER	MEANING
extent	N	number of years in the bond term
frequency	m	number of coupons per year
extent	n	number of coupons, $n = Nm$
Coupon rate	α	nominal coupon rate, annual coupons total $F\alpha$
Coupon rate	r	coupon rate per coupon period, amount of each coupon is Fr
Coupon rate	g	modified coupon rate, coupon amount is Cg
Yield rate	I	nominal yield rate convertible m times per year
Yield rate	j	effective yield rate for the coupon period
Yield rate	i	annual effective yield rate
Dollar amount	F	face (or par) value, coupon amount is Fr
Dollar amount	C	redemption amount
Dollar amount	G	base amount, coupon amount is Gj
Dollar amount	P	price at issue
Dollar amount	K	present value of the redemption amount

TABLE (6.2.4) Bond variables

EXAMPLE 6.2.5   Finding a bond's yield rate

Problem: A twelve-year $2,000 8% par-value bond with quarterly coupons is bought for $2,200. Find the effective yield rate per coupon period j, the annual effective yield i, and the nominal yield rate I convertible quarterly.

Solution We are given that $N = 12$ and $m = 4$. Therefore, there are $n = 4 \times 12 = 48$ coupons. These are paid at the end of each quarter. The bond has a face value of $2,000 and $r = \frac{\alpha}{m} = \frac{8\%}{4} = 2\%$, so each coupon payment is for $2,000 × .02 = $40. Since the bond is purchased for $2,200 and redeemed for the face value $2,000, the effective quarterly yield rate j satisfies

$$\$2,200 = \$40a_{\overline{48}|j} + \$2,000(1 + j)^{-48}.$$

It follows (use the **TVM worksheet** if you have a BA II Plus calculator,) that $j \approx 1.693949325\%$. Then $i = (1 + j)^4 - 1 \approx 6.949917684\%$. The nominal rate I convertible $m = 4$ times per year is $4j \approx 6.7757973\%$. (Since $m > 1$, $I < i$.)   ■

EXAMPLE 6.2.6   Finding a bond's redemption amount

Problem: A nine-year $5,000 7% bond with semiannual coupons is purchased for $4,986. This price allows the purchaser an annual effective yield of 6%. Calculate the redemption amount.

Solution   The coupon amount is $\$5,000 \times \frac{.07}{2} = \$175$, and there are $9 \times 2 = 18$ coupons. These are paid at the end of each half-year, and the effective yield per half-year is $j = (1.06)^{\frac{1}{2}} - 1$. The basic price formula therefore gives

$$\$4,986 = \$175a_{\overline{18}|(1.06)^{\frac{1}{2}} -1} + C(1.06)^{-9}.$$

It follows that the redemption amount C satisfies

$$C = (\$4,986 - \$175a_{\overline{18}|(1.06)^{\frac{1}{2}} -1})(1.06)^9 \approx \$4,342.330857.$$

The redemption amount must be an integral number of cents, so it is $4,342.33. We note that this rounding causes the actual annual yield rate to be about 5.9999984%.

TVM solution   Make sure your calculator is in END mode with P/Y = C/Y = 1. You may find the redemption amount C by keying

| 1 | 8 | N | 1 | • | 0 | 6 | √x | − | 1 | = | × | 1 | 0 | 0 | = | I/Y |

| 4 | 9 | 8 | 6 | +/− | PV | 1 | 7 | 5 | PMT | CPT | FV |.

■

We have seen that determining the price of a bond at a given yield rate amounts to finding the net present value of the payments the bond entitles the holder to receive. This may be done for bonds, or packages of bonds, and these may have nonlevel payments or level payments.

We end the section with two more examples. In each of them, there is a new complication to address.

EXAMPLE 6.2.7

Problem: A twenty-year $1,000 bond with semiannual coupons is redeemable at par and has a nominal coupon rate of 8% for the first five years, 9% for the next five years, and 10% for the final ten years. Amy Astacio purchases the bond so as to receive a nominal yield of 9.2%. How much does she pay?

Solution The bond has $20 \times 2 = 40$ coupons. The first $5 \times 2 = 10$ of these are for $1,000\left(\frac{.08}{2}\right) = \$40$, the next $5 \times 2 = 10$ of these are for $1,000\left(\frac{.09}{2}\right) = \$45$, and the remaining $10 \times 2 = 20$ of these are for $1,000\left(\frac{.10}{2}\right) = \$50$. Since Amy puchased the bond to receive a yield of $\frac{9.2\%}{2} = 4.6\%$ per coupon period, and the bond is redeemable for $1,000, we have

$$P = \$50a_{\overline{40}|4.6\%} - \$5a_{\overline{20}|4.6\%} - \$5a_{\overline{10}|4.6\%} + \$1,000(1.046)^{-40}$$
$$\approx \$968.7178505 \approx \$968.72.$$

∎

EXAMPLE 6.2.8   Inflation

Problem: Sydney and Sons Tool Corporation has structured its bond offering to appeal to the inflation conscious investor. This is done projecting an annual rate of inflation of 2.5%. An eight-year $1,000 face value bond with annual coupons has a redemption amount equal to $1,218.40 \approx \$1,000(1.025)^8$, an initial coupon of $80, and coupons that increase by 2.5% annually. How much should an investor who wishes to earn a noninflation-adjusted rate of 8% be willing to pay for the bond? If the actual rate of inflation is 2%, what is the investor's inflation-adjusted yield?

Solution The price to yield the investor 8% is equal to the sum of the geometric series

$$80(1.08)^{-1} + \$80(1.025)(1.08)^{-2} + \$80(1.025)^2(1.08)^{-3}$$
$$+ \cdots + \$80(1.025)^7(1.08)^{-8}$$

$$= \$80(1.08)^{-1}\left(\frac{1 - \left(\frac{1.025}{1.08}\right)^8}{1 - \frac{1.025}{1.08}}\right) = \$80\left(\frac{1 - \left(\frac{1.025}{1.08}\right)^8}{1.08 - 1.025}\right) \approx \$497.0688374,$$

and the present value $1,218.40(1.08)^{-8} \approx \$658.2636089$ of the redemption payment. So, the price should be about $1,155.33. In fact, each of the coupons must be an integral number of cents and consequently the geometric series $\{\$80(1.025)^k\}_{k=0}^{k=7}$ is replaced by

$$\{.01\lfloor \$8{,}000(1.025)^k \rfloor\}_{k=0}^{k=7}$$
$$= \{\$80, \$82, \$84.05, \$86.15, \$88.31, \$90.51, \$92.78, \$95.09\}.$$

Direct computation, perhaps using the **Cash Flow worksheet** and $\boxed{\text{NPV}}$, shows that the net present value of the just given sequence (the one defined using the floor function [see Section (1.5)]) is approximately \$497.0696412. Adding in the present value of the redemption amount, $\$1{,}218.40(1.08)^{-8}$, you obtain that the price is about \$1,155.33325. Thus, the price is \$1,155.33. It is worth noting that Sydney and Sons Tool Corporation is figuring that a nominal annual interest rate of 8% results in an inflation-adjusted interest rate of $\frac{.08-.025}{1.025}$ [see Equation (1.14.3)]. At this yield rate, a \$1,000 eight-year bond redeemable at par and having level coupons of $\$78.05 \approx \frac{\$80}{1.025}$ will have a price of \$1,155.34. (If you do not round $\frac{\$80}{1.025}$ to the nearest cent, you get \$1,155.33.)

If the actual rate of inflation is 2%, then the real values of the coupons increase by a factor of $\frac{1.025}{1.02}$ each year. Therefore, if we denote the investor's inflation-adjusted yield rate by y, then the price \$1,155.34 is equal to

$$\left[\left(\frac{\$80}{1.02}\right)(1+y)^{-1} + \left(\frac{\$80}{1.02}\right)\left(\frac{1.025}{1.02}\right)(1+y)^{-2} + \cdots + \left(\frac{\$80}{1.02}\right)\left(\frac{1.025}{1.02}\right)^{7}(1+y)^{-8}\right] +$$

$$\left(\frac{\$1{,}218.40}{(1.02)^8}\right)(1+y)^{-8} = \frac{\$80}{(1.02)(1+y)}\left(\frac{1-(\frac{1.025}{(1.02)(1+y)})^8}{1-\frac{1.025}{(1.02)(1+y)}}\right) + \left(\frac{\$1{,}218.40}{(1.02)(1+y)^8}\right).$$

But we calculated the price \$1,155.33 by summing two terms that were identical to these except for the fact that the product $(1.02)(1+y)$ was replaced by 1.08. Therefore, $(1.02)(1+y) \approx 1.08$ and $y \approx \frac{1.08}{1.02} - 1 = \frac{.06}{1.02} \approx 5.8823529\%$. By Equation (1.14.3), the yield rate y is just the inflation-adjusted interest rate corresponding to a stated interest rate of 8% and an inflation rate of 2%. More generally, you may calculate the inflation-adjusted yield for any bond using Equation (1.14.3). ∎

6.3 THE PREMIUM-DISCOUNT FORMULA

The basic price formula (6.2.2) contains the expression $a_{\overline{n}|j}$ and also the expression $v_j{}^n$. But, as in Equation (3.2.4),

$$a_{\overline{n}|j} = \frac{1 - v_j{}^n}{j}.$$

Equivalently

(6.3.1) $$v_j^n = 1 - ja_{\overline{n}|j}.$$

Using (6.3.1) along with the basic price formula (6.2.2) and the fact that $Fr = Cg$, we have

$$P = (Fr)a_{\overline{n}|j} + Cv_j^n = (Cg)a_{\overline{n}|j} + C(1 - ja_{\overline{n}|j}).$$

Therefore,

(6.3.2) $$\boxed{P = C(g - j)a_{\overline{n}|j} + C.}$$

Equation (6.3.2) is known as the **premium-discount** pricing **formula** for a bond. It will be very useful in Section (6.5) when we consider bond amortization.

A bond is said to **sell at a premium** if the price P is greater than the redemption value C. The difference $P - C$ is called the **premium** or **amount of premium** for the bond. From Equation (6.3.2), we see that a bond sells at a premium if and only if $g > j$, and if that occurs, the premium is $C(g - j)a_{\overline{n}|j}$.

(6.3.3) $$\boxed{\text{premium} = P - C = C(g - j)a_{\overline{n}|j}.}$$

When the yield rate is less than the modified coupon rate ($j < g$), the coupons Cg overpay interest (figured at rate j) on the redemption amount C. To make-up for this, you buy the bond at a premium.

When a bond's price P is less than the redemption amount C, we say the bond **sells at a discount**. The difference $C - P$ is called the **discount** or **amount of discount** on the bond. By Equation (6.3.2), the bond sells at a discount when $g < j$, and in this case, the amount of the discount is $C(j - g)a_{\overline{n}|j}$.

(6.3.4) $$\boxed{\text{discount} = C - P = C(j - g)a_{\overline{n}|j}.}$$

If the yield rate is higher than the modified coupon rate ($j > g$), the coupons Cg underpay interest (figured at rate j) on the redemption amount C. To compensate for this, you buy this bond at a discount.

EXAMPLE 6.3.5   Determining the amount of a bond's premium or discount

Problem: A bond with a face value of $6,000 and an annual coupon rate of 12% convertible semiannually will mature in ten years for its face value. If the bond is priced using a nominal yield rate of 6% convertible semiannually, what is the amount of premium or discount in this bond?

Solution   Note that $m = 2, n = 10 \times 2 = 20, C = F = \$6,000$ and $g = r = \frac{12\%}{2} = .06$. Moreover, $j = \frac{6\%}{2} = .03$. Since $j < g$, the bond sells at a premium and by (6.3.3), the premium is $C(g - j)a_{\overline{n}|j} = \$6{,}000(.06 - .03)a_{\overline{20}|3\%} = \$2{,}677.95$. We note that you could alternatively find the solution by calculating the price P by the basic price formula (6.2.2) **(or TVM worksheet)** and then subtracting the redemption amount of \$6,000.   ∎

EXAMPLE 6.3.6   Finding a bond's price when the discount is known

Problem: A $7\frac{1}{2}$-year 14% bond with a face value of \$2,500 has semiannual coupons and is sold to yield 7.2% convertible semiannually. The discount on the bond is \$283.12. Find the price of the bond.

Solution   Observe that $F = \$2{,}500$, $\alpha = 14\%$, and $m = 2$. Each end-of-half-year coupon thus pays $\$2{,}500 \times \frac{14}{2} = \$175$, and the bond's term is for $7\frac{1}{2} \times 2 = 15$ coupon periods. Since $I = 7.2\%$, the yield rate per coupon period is $j = \frac{7.2\%}{2} = 3.6\%$. The discount is \$283.12, so [by (6.3.4)] $P = C - \$283.12$. The **basic** price formula now gives $C - \$283.12 = P = \$175a_{\overline{15}|3.6\%} + C(1.036)^{-15}$. It follows that $C\left[1 - (1.036)^{-15}\right] = \$175a_{\overline{15}|3.6\%} + \$283.12$, and

$$C = \frac{\$175a_{\overline{15}|3.6\%} + \$283.12}{1 - (1.036)^{-15}} \approx \$5{,}548.807226.$$

But C is an actual payment amount. Therefore, $C = \$5{,}548.81$ and $P = C - \$283.12 = \$5{,}265.69$. Note that this problem concerned a bond sold at a discount, but we utilized the basic price formula rather than the premium-discount formula.   ∎

6.4   OTHER PRICING FORMULAS FOR BONDS

The formulas of this section follow easily from the basic price formula (6.2.2) and the relationship

$$(6.4.1) \qquad\qquad a_{\overline{n}|j} = \frac{1 - v_j{}^n}{j}.$$

Equality (6.4.1) is just Equation (3.2.4) except that here our effective interest rate per payment period is called j rather than i. Combining the basic price formula with Equation (6.4.1), we have

$$(6.4.2) \qquad\qquad P = Fr\left(\frac{1 - v_j{}^n}{j}\right) + Cv_j{}^n.$$

If we substitute the alternate expression Gj for the coupon amount Fr, simplification occurs since we then have j in the numerator as well as in the denominator. In fact, we obtain the **base amount formula**

$$(6.4.3) \qquad P = G - Gv_j^n + Cv_j^n = (C - G)v_j^n + G.$$

Note that G is the investment needed at interest rate j to produce payments of Cg forever, and Gv_j^n is the present value of a deferred perpetuity paying Cg forever, the payments beginning at time $n - 1$. So, the difference $G - Gv_j^n$ is the present value of the n coupon payments.

If we substitute the product Cg for Fr in Equation (6.4.2), we find

$$P = Cg\left(\frac{1 - v_j^n}{j}\right) + Cv_j^n = \frac{g}{j}C(1 - v_j^n) + Cv_j^n.$$

Recalling the notation $K = Cv_j^n$ and applying the distributive law, we have **Makeham's formula**

$$(6.4.4) \qquad\qquad P = \frac{g}{j}(C - K) + K.$$

Formula (6.4.4) is named for the nineteenth century British actuary William Matthew Makeham, and is useful when the present value K of the redemption amount is known but the number of coupons n has not been given.

EXAMPLE 6.4.5   Using Makeham's formula

Problem: A $2,000 bond with coupon rate of 10% payable quarterly is redeemable after an unspecified number of periods for $2,250. The bond is bought to yield 8% convertible quarterly. If the present value of the redemption is $869.71, find the purchase price.

Solution We are given $F = \$2{,}000$, $m = 4$, $\alpha = 10\%$, $C = \$2{,}250$, $j = \frac{8\%}{4} = .02$, and $K = \$869.71$. Then $r = \frac{\alpha}{m} = \frac{10\%}{4} = .025$, the coupon amount is $Fr = \$2{,}000 \times .025 = \$50$, and $g = \frac{\$50}{\$2{,}250} = \frac{1}{45}$. So, Makeham's formula gives us

$$P = \frac{1/45}{.02}(\$2{,}250 - \$869.71) + \$869.71 \approx \$2{,}403.37.$$

We note that we could have also found the price P by first calculating n and then using the basic price formula (6.6.2). Since $K = Cv_j^n$, $\frac{C}{K} = (1 + j)^n$ and $n = \frac{\ln\left(\frac{C}{K}\right)}{\ln(1+j)} \approx 47.99997789$. Since n is the number of coupons paid, it must be the integer 48. Then $P = \$50a_{\overline{48}|2\%} + \$869.71 \approx \$2{,}403.37$. With our readily available calculators with logarithms, the advantage of using Makeham's formula has diminished. ∎

EXAMPLE 6.4.6   Using the base amount formula

Problem: An eight-year bond has annual coupons and a redemption value of $2,338. It is purchased to yield 9% and each coupon is for $63. Use the base amount formula to calculate the price of the bond.

Solution   Since there are annual coupons, $n = 8$ and $j = i = 9\%$. Therefore, the base amount G equals $\frac{\$63}{.09} = \$700$, and the base amount formula gives $P = (\$2,338 - \$700)(1.09)^{-8} + \$700 \approx \$1{,}522.056966$. The price is $1,522.06.
∎

6.5   BOND AMORTIZATION SCHEDULES

A bond is a certificate of indebtedness that promises the holder a certain sequence of payments, namely the coupon payments and the redemption payment. These payments consist of interest and principal, so we have an opportunity to make an amortization table as in Section (5.2). The time elapsed between successive payments is one coupon period, and the applicable interest rate for each interest period is j, the investor's effective yield rate per coupon period.

Prior to making any amortization tables, we introduce some standard notation and develop some formulas for the quantities introduced. First, set

$B_t =$ **the balance of the debt at time** t, immediately after any

time t coupon payment (but before any redemption payment),

where

> the balance of debt B_t is calculated using the investor's yield rate j.

Measuring in coupon periods,

(6.5.1) $B_0 = P$ and $B_n = C.$

The balance of debt B_t is commonly called the **book value** of the bond at time t. The book values $B_0, B_1, B_2, \dots, B_n$ are the values kept on the books, and *they give a smooth transition from the purchase price $P = B_0$ to the redemption amount $C = B_n$.* **The book values do not take into account market forces**, and B_t is therefore usually *not* equal to the price for which you could sell the bond at time t.

We denote the **interest due** at the time of the t-th coupon by I_t. That is

(6.5.2) $I_t = jB_{t-1}, \quad t = 1, 2, \dots, n.$

The amount I_t may be important for tax calculations.

The **amount for adjustment of principal** P_t in the t-th coupon is defined by

(6.5.3)
$$\boxed{P_t = B_{t-1} - B_t, \quad t = 1, 2, \ldots, n.}$$

Note that should I_t exceed the amount of the coupon payment at time t, then $B_t > B_{t-1}$ (since there is additional debt caused by being behind in your interest payments), and P_t is negative. We have already observed that this occurs whenever the bond is bought at a discount, that is when $B_0 < B_n$. (The book values smoothly transition between $P = B_0$ and $C = B_n$. So, if $P < C$, they form an increasing sequence and the P_t are negative.)

If $t \in \{0, 1, 2, \ldots, n\}$, the basic price formula or the premium-discount formula may be used to find B_t. The reason for this is that at time t, right after the t-th coupon has been paid, what remains of the bond is a bond with level coupons Cg at the end of each coupon period for $(n - t)$ coupon periods and the redemption amount C paid $(n - t)$ periods in the future. So, if the investor's yield on the bond each interest period is j, we have the **basic price formula**

(6.5.4)
$$B_t = Fra_{\overline{n-t}|j} + Cv_j^{n-t}, \quad t = 0, 1, 2, \ldots, n.$$

and the **premium-discount Formula**

(6.5.5)
$$B_t = C(g - j)a_{\overline{n-t}|j} + C, \quad t = 0, 1, 2, \ldots, n.$$

Replacing t by $t - 1$ in Equation (6.5.5), we find

$$B_{t-1} = C(g - j)a_{\overline{n-t+1}|j} + C, \quad t = 1, 2, \ldots, n + 1.$$

Therefore, recalling (6.5.5) as well as $a_{\overline{k}|j} = \dfrac{1 - v_j^k}{j}$, we have

(6.5.6)
$$
\begin{aligned}
I_t = jB_{t-1} &= j\left(C(g - j)a_{\overline{n-t+1}|j} + C\right) \\
&= j\left(C(g - j)\left(\frac{1 - v_j^{n-t+1}}{j}\right) + C\right) \\
&= C(g - j)(1 - v_j^{n-t+1}) + Cj \\
&= Cg - C(g - j)v_j^{n-t+1}.
\end{aligned}
$$

Equation (6.5.5) also gives rise to an important expression for the adjustment to principal P_t.

$$P_t = B_{t-1} - B_t$$

(6.5.7)
$$= \left(C(g - j)a_{\overline{n-t+1}|j} + C\right) - \left(C(g - j)a_{\overline{n-t}|j} + C\right)$$

$$= C(g - j)\left(a_{\overline{n-t+1}|j} - a_{\overline{n-t}|j}\right)$$

$$= C(g - j)v_j^{n-t+1}.$$

(To obtain the last equality of (6.5.7), we have used the series expression for the annuity symbols $a_{\overline{n-t+1}|j}$ and $a_{\overline{n-t}|j}$ [see Equation (3.2.4)].

It is immediate from equations (6.5.6) and (6.5.7) that

(6.5.8) $$\boxed{I_t + P_t = Cg.}$$

Since the coupon amount is Cg, this states that the coupon is split into payment of interest due and repayment of principal.

Since the redemption amount C is positive as is the discount factor v_j, it follows from Equation (6.5.7) that P_t has the same sign as the difference $g - j$. But, as we noted in Section (6.3), $g - j$ is positive when the bond sells at a premium, and $g - j$ is negative if the bond is sold at a discount. Therefore,

---

- If a bond sells at a premium, the amount P_t for adjustment of principal in the t-th coupon is positive. In this case, a part of each coupon should compensate the investor for the premium paid, and the remainder should be viewed as interest.
- If a bond sells at a discount, the amount P_t for adjustment of principal in the t-th coupon is negative. In this case, the coupon is not sufficient to pay all the interest due and the balance must be borrowed; the outstanding loan balance thus increases.
- If the price of a bond equals the redemption value, $P_t = 0$.

When the bond sells for a premium, the amount for adjustment of principal is usually referred to as as the **amount for amortization of premium** in the t-th coupon. The balances of debt $\{B_t\}$ form a decreasing sequence and the process of finding the intermediate balances of debt (book values) is called **writing down** the bond. Since $0 < v_j < 1$, the amounts for amortization of principal $\{P_t\}$ form an increasing sequence writing-down the book values, and we have the following picture.

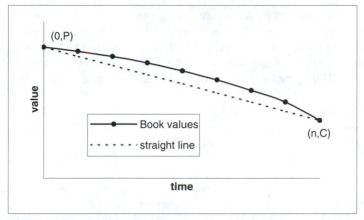

FIGURE (6.5.9) Balances of debt for a bond sold at a premium

If we connect the book values in a natural manner, the resulting graph is concave down. A straight-line approach would underestimate the book values.

The picture looks rather different for a bond sold at a discount. In this case the amount for adjustment of principal is negative. One refers to its *negative* $- P_t = C(j - g)v_j^{n-t+1}$ as the **amount for accumulation of discount**.

> When $g > j$, $-P_t =$ the amount for accumulation of discount in the t-th coupon.

The sequence of amounts for accumulation of discount form an increasing sequence $\{-P_t\}$. The balances of debt $\{B_t\}$ form an increasing sequence, and the process of finding the intermediate balances of debt is called **writing up the bond**. The graph of balances of debt over time (book values) appears as follows.

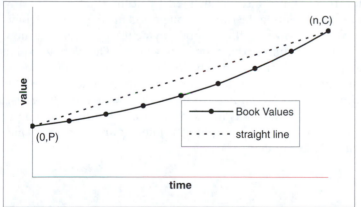

FIGURE (6.5.10) Balances of debt for a bond sold at a discount

This time a graph obtained by connecting our points in a natural manner is concave up. A straight-line approach would overestimate the book values.

Before moving on to several examples illustrating how the just-derived formulas might be used, we note that Equation (6.5.1) and Equations (6.5.5)–(6.5.7) may be brought together in the following amortization table.

TIME (in coupon periods)	COUPON PMT	I_t (by (6.5.6))	P_t (by (6.5.7))	B_t (by (6.5.5))
0	—	—	—	$P = C(g-j)a_{\overline{n}\rceil j} + C$
1	Cg	$Cg - C(g-j)v_j^n$	$C(g-j)v_j^n$	$C(g-j)a_{\overline{n-1}\rceil j} + C$
2	Cg	$Cg - C(g-j)v_j^{n-1}$	$C(g-j)v_j^{n-1}$	$C(g-j)a_{\overline{n-2}\rceil j} + C$
$\vdots$				
k	Cg	$Cg - C(g-j)v_j^{n-k+1}$	$C(g-j)v_j^{n-k+1}$	$C(g-j)a_{\overline{n-k}\rceil j} + C$
$\vdots$				
n	Cg	$Cg - C(g-j)v_j$	$C(g-j)v_j$	C

FIGURE (6.5.11) Amortization table for a bond with level coupons

Built into this table, and easily derived from Equations (6.5.2), (6.5.3), and (6.5.8) is the useful recursion formula

$(6.5.12)$
$$\boxed{B_t = (1 + j)B_{t-1} - Cg.}$$

Equation (6.5.12) may also be justified by noting that without any coupon payment at time t, during the t-th coupon period the balance of debt would rise to $(1 + j)B_{t-1}$. However, the coupon payment Cg reduces this new balance of debt by the amount of the coupon payment. This verbal explanation of Equation (6.5.12) might be said to be by the retrospective method.

Note that any entry in a bond amortization table can be obtained by using formulas. *You do not need to make the whole table if you need only a few entries.* We illustrate this in Examples (6.5.13)–(6.5.16). The first three of these examples concern the split into interest and principal while the last focuses on balances of debt.

EXAMPLE 6.5.13   Interest – Principal allocation

Problem: A six-year $1,800 8.5% bond with semiannual coupons is bought for $1,918 and is redeemable for $1,860. Find the amount for amortization of premium at the end of two-and-a-half years and the interest due at that time.

Solution 1   The fifth coupon is payable at the end of two-and-a-half years, so the desired amortization of premium is P_5. Note that $F = \$1,800$, $\alpha = 8.5\%$, $m = 2$, $Fr = \$1,800 \times \frac{.085}{2} = \$76.50$, and $n = 6 \times 2 = 12$. Therefore, the basic price formula gives $\$1,918 = \$76.50a_{\overline{12}|j} + \$1,860v_j^{-12}$, and this is satisfied by $j \approx .03784779646$. Moreover, $C = \$1,860$ and $g = \frac{\$76.50}{\$1860} \approx$ .041129032. Consequently, $P_5 = C(g - j)v_j^{n-5+1} \approx \$1,860(.041129032 -$ .03784779646$)(1.03784779646)^{-8} \approx \$4.533995075 \approx \$4.53$. The interest due at the end of two-and-a-half years is $\$76.50 - \$4.53 = \$71.97$.

Solution 2   Again note the "alphabet soup" as in solution 1, but there is no need to calculate g. The basic price formula gives $B_4 = \$76.50a_{\overline{12-4}|j} + \$1,860v_j^{12-4} \approx \$1,901.46$. Then $I_5 = jB_4 \approx 71.96600492 \approx 71.97$, and $P_5 \approx \$76.50 - \$71.97 = \$4.53$.

BA II Plus Solution   The nominal yield rate convertible semiannually of the bond may be found using the **TVM worksheet** in END Mode with $P/Y = C/Y = 1$. Key

| 1 | 2 | N | 1 | 9 | 1 | 8 | +/- | PV | 7 | 6 | • | 5 | PMT |

| 1 | 8 | 6 | 0 | FV | CPT | I/Y |.

This should result in "I/Y = 3.784779646" being displayed. Now, you have two good choices. One is to key [8] [N] [CPT] [PV] to obtain B_4 on the display and as the new entry in the PV register. Then you may procede similarly to the way you did in solution 2, remembering that the yield rate is stored for you in the I/Y register as a percent. A more novel choice (with the entry of the N-register still 12) is to use the **amortization worksheet**. Push

| 2ND | AMORT | 5 | ENTER | ↓ | 5 | ENTER | ↓ | ↓ |.

The amount of premium in the fifth payment is now shown. It is about $4.53. Depressing [↓] yet again will result in the amount of interest, about $71.96, being displayed. ∎

EXAMPLE 6.5.14   Amounts for amortization of premium

Problem: A twelve-year bond with semiannual level coupons is bought at a premium to yield 7.5% convertible semiannually. If the amount for amortization of premium in the fourth to the last payment is $8.02, find the premium and the total amount for amortization of premium in the first three years.

Solution   Note that $m = 2$, $n = 12 \times 2 = 24$, and $j = \frac{7.5\%}{2} = .0375$. We are given that $P_{21} = \$8.02$. On the other hand, by (6.5.3), $P_{21} = C(g - j)v_j^{24-21+1} = C(g - j)v_j^4$. Therefore, $\$8.02 = C(g - j)v_j^4$. The premium for the bond is then

$$C(g - j)a_{\overline{24}|j} = (C(g - j)v_j^4)(1.0375)^4 a_{\overline{24}|3.75\%}$$

$$= \$8.02(1.0375)^4 a_{\overline{24}|3.75\%} \approx \$145.38.$$

By the premium-discount formula, the total amount for amortization of premium in the first three years (first six coupon periods) is

$$B_0 - B_6 = ((C(g - j)a_{\overline{24}|j} + C) - ((C(g - j)a_{\overline{24-6}|j} + C)$$

$$= (C(g - j)v_j^4)(1.0375)^4(a_{\overline{24}|3.75\%} - a_{\overline{18}|3.75\%})$$

$$= \$8.02(1.0375)^4(a_{\overline{24}|3.75\%} - a_{\overline{18}|3.75\%})$$

$$\approx \$25.32.$$

∎

EXAMPLE 6.5.15   Amount for accumulation of discount

Problem: A \$10,000 par-value twenty-year 14% bond with annual coupons is bought for \$9,562. Find the amount for accumulation of discount in the third coupon and the amount of interest for the third year.

Solution 1   The amount of each annual coupon is $\$10,000 \times .14 = \$1,400$ and the redemption value is the face value \$10,000. Since the bond was purchased for \$9,562 and the bond has twenty annual coupons, $\$9,562 = \$1,400a_{\overline{20}|j} + \$10,000v_j^{20}$ and $j \approx 14.68768719\%$. The amount of accumulation of discount in the third coupon is

$$-P_3 = C(j - g)v_j^{20-3+1} = \$10,000(j - .14)v_j^{18} \approx \$5.835642 \approx \$5.84.$$

It follows that the amount of interest for the third year is $\$1,400 - P_3 = \$1,400 + \$5.84 = \$1,405.84$.

Solution 2   The balance of debt at the end of two years is $\$1,400a_{\overline{20-2}|j} + \$10,000v_j^{20-2} \approx \$9,571.524938$, and the interest in the third coupon is $jB_2 \approx \$1,405.84$. The amount for accumulation of discount in the third coupon is $-P_3 = -(\$1,400 - I_3) \approx -(\$1,400 - \$1,405.84) = \$5.84$.

BA II Plus Solution   The **TVM worksheet** and **Amortization worksheet** are ideal for this problem. Set your calculator in END Mode with P/Y=C/Y=1. Key

$$\boxed{2}\,\boxed{0}\,\boxed{\text{N}}\,\boxed{9}\,\boxed{5}\,\boxed{6}\,\boxed{2}\,\boxed{+/-}\,\boxed{\text{PV}}\,\boxed{1}\,\boxed{0}\,\boxed{0}\,\boxed{0}\,\boxed{0}\,\boxed{\text{FV}}$$

$$\boxed{1}\,\boxed{4}\,\boxed{0}\,\boxed{0}\,\boxed{\text{PMT}}\,\boxed{\text{CPT}}\,\boxed{\text{I/Y}}\,.$$

You may find the amount for accumulation of discount in the third coupon by pressing

$$\boxed{\text{2ND}}\,\boxed{\text{AMORT}}\,\boxed{3}\,\boxed{\text{ENTER}}\,\boxed{\downarrow}\,\boxed{3}\,\boxed{\text{ENTER}}\,\boxed{\downarrow}\,\boxed{\downarrow}\,\boxed{+/-}\,.$$

It is about $5.83. Pushing $\boxed{\downarrow}$ causes the amount of interest in the third coupon to be displayed, and it is about $1,405.84.   ■

EXAMPLE 6.5.16   Book values and the price of a bond

Problem: An n-year $1,000 par-value bond with 8% annual coupons has an annual effective yield of i, $1 + i > 0$. The book value of the bond at the end of the third year is $990.92 and the book value of the bond at the end of the fifth year is $995.26. Find the price of the bond.

Solution   The bond has $F = \$1,000$, $m = 1$, and $r = \alpha = 8\%$, so the amount of each annual coupon is $\$1,000(.08) = \$80$. In particular, the coupon amounts at $t = 4$ and $t = 5$ are each $80, so the book value at time 3 and the book value at time 5 are related by the time 5 equation of value

$$B_5 = B_3(1 + i)^2 - \$80(1 + i) - \$80.$$

(This time 5 equation of value is most easily justified by the retrospective method, but using Equation (6.5.12) for $t = 5$ and again for $t = 4$ will also work.)   The problem includes the information that $B_5 = \$995.26$ and $B_3 = \$990.92$. Accordingly,

$$990.92(1 + i)^2 - 80(1 + i) - 1,075.26 = 0.$$

and

$$1,075.26 - 990.92(1 + i)^2 + 80(1 + i) = 0.$$

This equation is quadratic in $(1 + i)$ which is given to be positive, whence the quadratic formula yields

$$1 + i = \frac{80 + \sqrt{(80)^2 + 4(990.92)(1,075.26)}}{2(990.92)} \approx 1.082835847.$$

Therefore $i \approx .082835847$ and, since the coupons at $t = 1$, $t = 2$, and $t = 3$ are each $80,

$$P = \$80a_{\overline{3}|i} + B_3v^3 = \$80a_{\overline{3}|i} + \$990.92v^3 \approx \$985.58.$$

■

We end this section by making two amortization tables. In Example (6.5.17) we see the writing down of a bond bought at a premium, and Example (6.5.18) illustrates the writing up a bond that was bought at a discount.

EXAMPLE 6.5.17   Writing down

Problem: An 8% par-value bond with face amount $100,000 and semiannual coupons matures after four years. Construct the amortization schedule over its term if the investor's nominal yield rate is 6% convertible semiannually.

Solution   First note that $F = C = \$100,000$, $m = 2$, $\alpha = 8\%$, and $g = r = \frac{8\%}{2} = 4\%$, so the coupon amount is $\$100,000 \times .04 = \$4,000$. Also observe that since $n = 4 \times 2 = 8$ and $j = \frac{6\%}{2} = 3\%$, the basic price formula gives $B_0 = P = \$4,000a_{\overline{8}|3\%} + \$100,000(1.03)^{-8}$. From this formula we calculate that $B_0 \approx \$107,019.6922$, so the purchase price $B_0 = \$107,019.69$. We now form the amortization schedule, on each line calculating the interest due by multiplying the previous balance of debt by the semiannual interest rate .03, the principal by subtracting the interest due from the coupon amount $4,000, and the new balance of debt by subtracting the principal from the previous balance of debt. If we do not round in our calculations (so start with $B_0 \approx \$107,019.6922$) but report all monetary amounts to the nearest cent, we find:

TIME (half-years)	COUPON PAYMENT	I_t	P_t	B_t
0	—	—	—	$107,019.69
1	$4,000	$3,210.59	$789.41	$106,230.28
2	$4,000	$3,186.91	$813.09	$105,417.19
3	$4,000	$3,162.52	$837.48	$104,579.70
4	$4,000	$3,137.39	$862.61	$103,717.10
5	$4,000	$3,111.51	$888.49	$102,828.61
6	$4,000	$3,084.86	$915.14	$101,913.47
7	$4,000	$3,057.40	$942.60	$100,970.87
8	$4,000	$3,029.13	$970.87	$100,000.00

The yield rate of Example (6.5.17) is less than the coupon rate so the bond was purchased at a premium. We next repeat the problem of Example (6.5.17) with the one change being a yield rate of 10%, which is higher than the coupon rate, resulting in the bond being bought at a discount.

EXAMPLE 6.5.18   Writing Up

Problem: An 8% par-value bond with face amount $100,000 and semiannual coupons matures after four years. Construct the amortization schedule over its term if the investor's nominal yield rate is 10% convertible semiannually.

Solution   The bond "alphabet soup" is the same as in the previous solution except that $j = \frac{10\%}{2} = 5\%$, and hence $B_0 = P = \$4,000a_{\overline{8}|5\%} + \$100,000(1.05)^{-8} \approx \$93,536.78724$. So, the purchase price is $B_0 \approx \$93,536.79$. The amortization schedule is constructed as in the last example except that the interest rate 5% replaces 3%. If we do not round in our calculations (so start with $B_0 \approx \$93,536.78724$) but report all monetary amounts to the nearest cent, the following schedule results.

TIME (half-years)	COUPON PAYMENT	I_t	P_t	B_t
0	—	—	—	$93,536.79
1	$4,000	$4,676.84	−$676.84	$94,213.63
2	$4,000	$4,710.68	−$710.68	$94,924.31
3	$4,000	$4,746.22	−$746.22	$95,670.52
4	$4,000	$4,783.53	−$783.53	$96,454.05
5	$4,000	$4,822.70	−$822.70	$97,276.75
6	$4,000	$4,863.84	−$863.84	$98,140.59
7	$4,000	$4,907.03	−$907.03	$99,047.62
8	$4,000	$4,952.38	−$952.38	$100,000.00

6.6   VALUING A BOND AFTER ITS DATE OF ISSUE

When an investor purchases a bond, the price paid for the bond might reasonably be referred to as its **value**. Of course, it would be more precise to call this the **value** *with respect to interest rate* j, the investor's effective yield rate per coupon period. You should be aware that this yield is not necessarily an intrinsic part of the bond. In some cases, certain buyers obtain higher yield rates than others because they are offered the bond at a lower price.[2]

In the last section we introduced the book values B_t. These were introduced to smoothly transition between the original price $P = B_0$ and the redemption amount $C = B_n$, and they were again defined with respect to the investor's yield j. So, we have "on the books" a sequence of values at the beginning of each coupon period (as well as one just preceding the bond's redemption).

Note that each B_t was defined as the present value, calculated with respect to the rate j, of the remaining payments promised by the bond. So, if we wish to value the bond in a **consistent** manner at an arbitrary time T given in coupon periods, $0 \leqslant T \leqslant n$, we might *consider* again using the present value of the remaining payments of the bond with respect to interest rate j. If $T = \lfloor T \rfloor + f$ where $0 \leqslant f < 1$, there are no coupon or other bond payments on the interval $(\lfloor T \rfloor, T]$ so we define

$$(6.6.1) \qquad \mathcal{D}_T = (1 + j)^f B_{\lfloor T \rfloor}.$$

We will refer to $\mathcal{D}_T$ as the **dirty value** or **theoretical dirty value** of the bond at time T. (Elsewhere you may see this still called the book value.) We observe that

> The **theoretical dirty value** $\mathcal{D}_T$ gives the present value of the bond at time T, just after the coupon payment if one occurs at that time. $\mathcal{D}_T$ is computed using the investor's yield rate j. $B_T = \mathcal{D}_T$ for $T \in \{0, 1, 2, \dots, n\}$.

Why do we give $\mathcal{D}_T$ this disparaging name? Because, as we now explain, the dirty values fail to transition nicely between the price P and the redemption amount C. In fact, if we define the dirty value function $F(t) = \mathcal{D}_t$, the function $F(t)$ has a discontinuity at each coupon date with a jump down equal to the coupon amount Cg. The reason for this is that an instant before a coupon is paid, the dirty value includes the value of that coupon, but as soon as the coupon has been paid, the dirty value excludes its value. For those knowing a

---

[2]Very often all purchasers at issue receive the same price, but variation is more common on the secondary market. A simple example where not all bond holders receive the same yield at issue are United States Treasury securities. Some potential lenders make noncompetitive bids, and these lenders receive the average rate paid on competitive bids.

little calculus, the jumps are as stated since if $k \in \{0, 1, 2, \ldots, n-1\}$,

$$\lim_{f \to 1^-} F(k+f) - F(k+1) = \lim_{f \to 1^-} B_k(1+j)^f - B_{k+1}$$
$$= B_k(1+j) - B_{k+1} = (B_k - B_{k+1}) + jB_k$$
$$= P_{k+1} + I_{k+1} = Cg.$$

EXAMPLE 6.6.2   Theoretical dirty values

Problem: Alvin Young purchased a $1,000 8% five-year bond with annual coupons from Hungary County at a price to yield 10%. The bond is redeemable at its face value. Graph the theoretical dirty value function $F(t) = \mathcal{D}_t$.

Solution   We have $C = F = \$1,000$, $g = r = .08$, $Cg = \$80$, and $n = 5$. The price of the bond is $C(g-j)a_{\overline{5}|10\%} + C = -\$20a_{\overline{5}|10\%} + \$1,000 \approx \$924.18$. We therefore have $F(0) = B_0 = 924.18$. We further calculate that $F(1) = B_1 = -\$20a_{\overline{4}|10\%} + \$1,000 \approx \$936.60$, $F(2) = B_2 = -\$20a_{\overline{3}|10\%} + \$1,000 \approx \$950.26$, $F(3) = B_3 = -\$20a_{\overline{2}|10\%} + \$1,000 \approx \$965.29$, and $F(4) = B_4 = -\$20a_{\overline{1}|10\%} + \$1,000 \approx \$981.82$. Also, $F(5) = B_5 = \$1,000$, the redemption value. Next note that if $k \in \{0, 1, 2, 3, 4\}$ and $0 \le f < 1$, $F(k+f) = B_k(1+j)^f = B_k(1.1)^f$. The function $(1.1)^x$ is concave up and increases over the interval $[0, 1)$. The graph therefore is as shown.

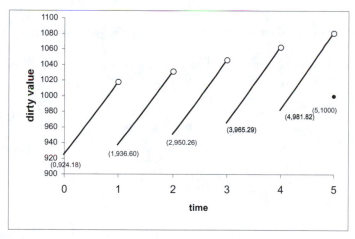

Graph of $F(t) = \mathcal{D}_t$

We have now observed that the dirty values do not provide a smooth transition between the purchase price $P = B_0$ and the redemption amount

$C = B_n$. Furthermore, we know the reason for that. The dirty values increase as one approaches the end of the coupon period, in anticipation of the next coupon payment as well as future ones, and then the part of the increase due to the next coupon payment is suddenly taken out at the moment the coupon is paid. So, in our quest for smoothly transitioning values, we should define **clean values** so that the value of the next coupon is gradually removed. Moreover, the clean values should agree with the book values when the latter are defined.

An easy way to specify clean values would be to define a continuous function $G(T)$ by specifying that $G(T) = B_T$ for $T \in \{0, 1, 2, \ldots, n\}$ and also stipulating that $G(T)$ is linear on each interval $[T - 1, T], T \in \{, 1, 2, \ldots, n\}$. The function $G(T)$ then defines **practical clean values** C_T^{prac}. In fact, if $T = \lfloor T \rfloor + f, 0 \leq f < 1$, then

(6.6.3)
$$\begin{aligned} C_T^{\text{prac}} &= B_{\lfloor T \rfloor} + f(B_{\lfloor T \rfloor + 1} - B_{\lfloor T \rfloor}) \\ &= B_{\lfloor T \rfloor} - f P_{\lfloor T \rfloor + 1} \quad \text{where} \quad P_{\lfloor T \rfloor + 1} \overset{\text{def}}{=} B_{\lfloor T \rfloor + 1} - B_{\lfloor T \rfloor}. \end{aligned}$$

In this linear mindset, to go along with these practical clean values, we might wish to use simple interest to define a dirty value. This is in contrast to using compound interest as in Equation (6.6.1). Specifically, we define the **practical dirty value** $\mathcal{D}_T^{\text{prac}}$ by

(6.6.4)
$$\mathcal{D}_T^{\text{prac}} = (1 + fj)B_{\lfloor T \rfloor}.$$

The practical dirty values transition between successive book values in a linear manner.

We next observe that Equations (6.5.2), (6.5.3), and (6.5.8), along with the definitions (6.6.3) and (6.6.4), may be combined as follows

(6.6.5)
$$\begin{aligned} \mathcal{D}_T^{\text{prac}} - C_T^{\text{prac}} &= (1 + jf)B_{\lfloor T \rfloor} - (B_{\lfloor T \rfloor} - f P_{\lfloor T \rfloor + 1}) \\ &= jf B_{\lfloor T \rfloor} + f P_{\lfloor T \rfloor + 1} \\ &= f(j B_{\lfloor T \rfloor} + P_{\lfloor T \rfloor + 1}) \\ &= f(I_{\lfloor T \rfloor + 1} + P_{\lfloor T \rfloor + 1}) \\ &= fCg \end{aligned}$$

For a typical bond purchased at a premium, we therefore have the following graph of practical clean and practical dirty values.

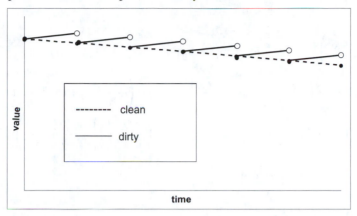

FIGURE (6.6.6) Practical clean and dirty values

In Problem (6.6.1) you are asked to make a similar graph for a typical bond purchased at a discount.

While the practical clean and dirty values have the virtue of simplicity, we prefer values having a strong theoretical basis reflecting *accumulation by compound interest.* Imagine that our actual bond is replaced by one that has the same value but which pays continuously and levelly, except for the one time redemption payment of C at time n. Ignoring the redemption amount C, this fictitious replacement bond will pay $Cg\delta_j/j = Cg\ln(1+j)/j$ each period since the original bond paid Cg at the end of each coupon period and

$$\left(\frac{Cg\delta_j}{j}\right)\overline{a}_{\overline{n}|j} = \left(\frac{Cg\delta_j}{j}\right)\left(a_{\overline{n}|j}\frac{j}{\delta_j}\right) = Cga_{\overline{n}|j}.$$

We define the **theoretical clean value** C_T at time T *of our original bond* to be the time T value of the remaining payments of our replacement continuously-paying bond.

But now we want a simple mathematical formula for C_T. Write $T = \lfloor T \rfloor + f, 0 \leqslant f < 1$. Then

$$C_T = \left(\frac{Cg\delta}{j}\right)\bar{a}_{\overline{n-T}|j} + Cv_j{}^{n-T}$$

$$= \left(\frac{Cg\delta}{j}\right)(1+j)^f(\bar{a}_{\overline{n-\lfloor T \rfloor}|j} - \bar{a}_{\overline{f}|j}) + Cv_j{}^{n-\lfloor T \rfloor}(1+j)^f$$

$$= (1+j)^f\left(\frac{Cg\delta}{j}\right)\bar{a}_{\overline{n-\lfloor T \rfloor}|j} - \left(\frac{Cg\delta}{j}\right)(1+j)^f\left(\frac{1-v_j{}^f}{\delta}\right) + Cv_j{}^{n-\lfloor T \rfloor}(1+j)^f$$

$$= (1+j)^f\left((Cg)a_{\overline{n-\lfloor T \rfloor}|j} + Cv_j{}^{n-\lfloor T \rfloor}\right) - Cg\left(\frac{(1+j)^f - 1}{j}\right)$$

$$= (1+j)^f B_{\lfloor T \rfloor} - Cg\left(\frac{(1+j)^f - 1}{j}\right)$$

$$= \mathcal{D}_T - Cg\left(\frac{(1+j)^f - 1}{j}\right).$$

The derivation was a bit complicated but the final result is quite simple.

FACT (6.6.7): If $T = \lfloor T \rfloor + f, 0 \le f < 1$, then the **theoretical clean value** C_T is equal to

$$\mathcal{D}_T - Cg\left(\frac{(1+j)^f - 1}{j}\right),$$

the difference of the dirty value $(1+j)^f B_{\lfloor T \rfloor}$ and $Cgs_{\overline{f}|j}$. In particular, *if T is an integer, the clean value, dirty value, and book value are all equal.* This common value is also equal to the practical clean and practical dirty values.

EXAMPLE 6.6.8   Theoretical clean values

Problem: Alvin Young purchased a $1,000 8% five-year bond with annual coupons from Hungary County at a price to yield 10%. The bond is redeemable at its face value. Graph the theoretical clean value function $H(t) = C_t$.

Solution   As noted in Example (6.6.2), $C = F = \$1,000$, $g = r = .08$, $Cg = \$80$, $n = 5$, $B_0 \approx 924.18$, $B_1 \approx \$936.60$, $F(2)$, $B_2 \approx \$950.26$, $B_3 \approx \$965.29$, $B_4 \approx \$981.82$, and $B_5 = \$1,000$.

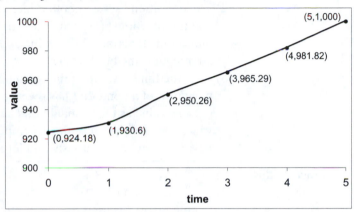

The values introduced in this section are all computed using the purchaser's original yield rate j per coupon period. The next example gives us an opportunity to review how these dirty and clean values are calculated, but first, we summarize basic facts about them in Table (6.6.9).

SYMBOL	NAME	NOTES
$\mathcal{D}_T$	Dirty value	(1) $\mathcal{D}_T = (1+j)^f B_{\lfloor T \rfloor}$. (2) $\mathcal{D}_T$ gives the time T value of the remaining bond payments.
$\mathcal{C}_T$	Clean value	(1) $\mathcal{C}_T = \mathcal{D}_T - Cg\left(\frac{(1+j)^f - 1}{j}\right) = \mathcal{D}_T - Cgs_{\overline{f}\mid j}$. (2) The definition uses compound interest and the function $F(T) = \mathcal{C}_T$ is a strictly monotonic function whose value at an integer is the book value. (3) $\mathcal{C}_T$ is the time T value of the remaining payments of a bond that has the same present value as the original bond, and which, when stripped of the redemption payment, pays continuously and levelly.
$\mathcal{D}_T^{\text{prac}}$	Practical dirty value	(1) $\mathcal{D}_T^{\text{prac}} = (1 + fj)B_{\lfloor T \rfloor}$. (2) $\mathcal{D}_T^{\text{prac}}$ approximates $\mathcal{D}_T$. Simple interest is used for the fractional period instead of compound interest.
C_T^{prac}	Practical clean value	(1) $C_T^{\text{prac}} = \mathcal{D}_T^{\text{prac}} - Cgf$. (2) (T, C_T^{prac}) lies on the line between $(\lfloor T \rfloor, B_{\lfloor T \rfloor})$ and $(\lfloor T \rfloor + 1, B_{\lfloor T \rfloor + 1})$. (3) C_T^{prac} approximates $\mathcal{C}_T$.

TABLE (6.6.9) Valuing at the original yield rate j

EXAMPLE 6.6.10   Computing dirty and clean values

Problem: A $1,000 three-year 6% bond with semiannual coupons is redeemable for $1,125. It was purchased at issue for its face value. Find its clean and dirty values seven months after issue by the practical method and also by the theoretical method. Use the "30/360" method to determine f.

Solution   Note that $P = F = \$1,000$, $C = \$1,125$, $m = 2$, $\alpha = 6\%$, $r = \frac{\alpha}{m} = 3\%$, and the coupon amount is $Fr = \$30$. So, by the basic price formula, $\$1,000 = \$30a_{\overline{6}\mid j} + \$1,125v_j^6$. It follows that $j \approx 4.844903819\%$.

According to the "30/360" method, each month is taken to have 30 days, and a half-year (the coupon period) is 180 days. Therefore, seven months is

$1\frac{1}{6}$ coupon periods, and we are interested in time $T = 1 + f$ where $f = \frac{1}{6}$.

We calculate that $B_1 = \$30a_{\overline{5}|j} + \$1,125v_j^5 \approx \$1,018.449038$. It follows that

$$D^{prac}_{1\frac{1}{6}} = \left(1 + \frac{1}{6}j\right)B_1 \approx \left(1 + \frac{1}{6}(.04844903819)\right)(\$1,018.449038)$$

$$\approx \$1,026.672851 \approx \$1,026.67,$$

and

$$D_{1\frac{1}{6}} = (1 + j)^{\frac{1}{6}}B_1 \approx (1.04844903819)^{\frac{1}{6}}(\$1,018.449038)$$

$$\approx \$1,026.511589 \approx \$1,026.51.$$

The clean values are $C^{prac}_{1\frac{1}{6}} = D^{prac}_{1\frac{1}{6}} - fCg \approx \$1,026.672851 - \frac{1}{6}(\$30) \approx$ $\$1,021.67$ and

$$C_{1\frac{1}{6}} = D_{1\frac{1}{6}} - \left(\frac{(1 + j)^f - 1}{j}\right)Cg$$

$$\approx \$1,026.511589 - \left(\frac{(1.04844903819)^{\frac{1}{6}} - 1}{.04844903819}\right)\$30$$

$$\approx \$1,026.511589 - \$4,901954085 \approx \$1021.61.$$

■

6.7   SELLING A BOND AFTER ITS DATE OF ISSUE

Bonds are negotiable securities and may be sold at any date. A bond's original yield j is determined based on market conditions at the time of its issue, while the resale price of a bond is determined by the market conditions when the bond is traded. A buyer of a bond purchased at time T to yield $\tilde{j}$ per coupon period, pays the time T net value of the bond's future payments to which he gains rights, calculated at rate $\tilde{j}$ per coupon period. Usually, at least in the United States, the bond is sold **cum dividend**. This means that the buyer receives the right to the coupon for the coupon period in progress, along with all other remaining coupons and the redemption amount.[3] We denote the price of the bond purchased cum dividend at time T, to afford the buyer a yield rate $\tilde{j}$ per coupon period, by $D^{\tilde{j}}_T$. Note that if $\tilde{j} = j$, then $D^{\tilde{j}}_T = D_T$,

---

[3] If the seller retains the rights to the coupon payment at the end of a coupon period in progress, the bond is said to be sold **ex dividend**. We will not discuss ex dividend sales further.

the theoretical dirty value at time T introduced in Section (6.6). The price $\mathcal{D}_T^{\tilde{j}}$ is called the **theoretical dirty price**, **dirty price**, **invoice price**, **flat price**, or **full price** at time T to yield $\tilde{j}$ to the bond purchaser.

A bond sale typically involves transaction costs, most notably a broker's mark-up. These costs, which commonly range from 1% to 5% of the bond's value, clearly must be taken into account when making investment decisions. However, we will disregard transaction costs and therefore *assume* that the seller of the bond receives the full amount paid by the buyer.

EXAMPLE 6.7.1   Reselling a bond at a new interest rate

Problem: Morton purchases a $2,000 6% eight-year par-value bond having annual coupons for $1,212. Three years later, just after receipt of the third coupon, he sells the bond (that is to say the right to all future coupons and the redemption amount) to Maheen for $1,600. Find Morton's balance of debt B_3, Maheen's yield $\tilde{j}$ on her five-year investment, and for Morton's original eight-year bond calculate $\mathcal{D}_3^{\tilde{j}}$. Also calculate Morton's actual yield y for his three-year investment.

Solution   Morton purchases the bond at a discount for $1,212. If he had held the bond for its eight-year term, his yield j would have satisfied the basic price equation $\$1,212 = (.06)(\$2,000)a_{\overline{8}|j} + \$2,000(1 + j)^{-8}$. So, his yield rate would have been approximately 14.69136626%. Morton's balance of debt B_3 at time 3 (calculated as usual using the yield the investor would realize if the bond were held to maturity) is $(.06)(\$2,000)a_{\overline{5}|j} + \$2,000(1 + j)^{-5} = \$1,413.02$. On the other hand, Maheen's yield $\tilde{j}$ on her five-year investment satisfies $\$1,600 = (.06)(\$2,000)a_{\overline{5}|\tilde{j}} + \$2,000(1 + \tilde{j})^{-5}$. Therefore, $\tilde{j} \approx 11.47640878\%$. We have $\mathcal{D}_3^{\tilde{j}} = \$1,600$ and $\mathcal{D}_3^{\tilde{j}} > B_3$. (This inequality is an immediate consequence of the inequality $\tilde{j} < j$.) Finally, Morton's actual yield y satisfies $\$1,212 = (.06)(\$2000)a_{\overline{3}|y} + \$1,600(1 + y)^{-3}$, and $y \approx 18.79553891\%$. Note that $y > j$ since $\mathcal{D}_3^{\tilde{j}} > B_3$.   ∎

The purchaser of a new bond cum dividend receives the coupon payment at the end of the coupon period during which the sale was made. That is, the buyer receives the *whole* coupon, even though he or she may only have held the bond for a portion $1 - f$ of the coupon period. Now recall that in the bond amortization, the coupon contains interest. For tax purposes, it may be of some importance to know what part of the coupon payment is **accrued interest** for which any tax liability is the responsibility of the previous owner.

The issue of accrued interest is fairly simple when the bond is sold to yield the new buyer the same yield j that the buyer received. In this case, if the bond trades at time T, the purchase price $\mathcal{D}_T^j = \mathcal{D}_T$. On the other hand, the difference $\mathcal{D}_T - \mathcal{C}_T$ between the theoretical dirty and clean values equals the amount of the coupon payment that the seller "deserves" at time T. This is the **accrued interest by the theoretical method at interest rate** j (or the accrued interest by compound interest at rate j), and we denote it by $\mathcal{A}_T^j$. Recalling Fact (6.6.7), we have

$$(6.7.2) \qquad \mathcal{A}_T^j = Cg\left(\frac{(1+j)^f - 1}{j}\right) = Cgs_{\overline{f}|j}, \quad T = \lfloor T \rfloor + f, \quad 0 \leqslant f < 1.$$

Now drop the assumption that the bond was sold to yield the new buyer the same yield rate that the seller would have earned had he held the bond to maturity. Then, there is a somewhat philosophical issue to resolve. You might argue that the fact that the sale is made at a different rate should not alter the amount of interest that has accrued *prior* to that date. On the other hand, if $\tilde{j} \neq j$, the seller of the bond did not end up realizing a yield rate j, so a case can be made for using the seller's actual yield to compute the accrued interest. Yet again, if the buyer's taxes are impacted by the amount of the coupon payment that is the responsibility of the seller, perhaps the rate $\tilde{j}$ is the appropriate one to use in computing the accrued interest. Heeding this last remark, we define

$$(6.7.3) \qquad \mathcal{A}_T^{\tilde{j}} = Cg\left(\frac{(1 + \tilde{j})^f - 1}{\tilde{j}}\right) = Cgs_{\overline{f}|\tilde{j}}, \quad T = \lfloor T \rfloor + f, \quad 0 \leqslant f < 1,$$

and refer to $\mathcal{A}_T^{\tilde{j}}$ as the **accrued interest by the theoretical method at interest rate** $\tilde{j}$.

Very frequently, accrued interest is computed using an alternative method that does not involve the choice of a rate. In this alternative method, the so-called **practical method**, the accrued interest is computed by designating that the seller's accrued interest is proportional to the fraction of the coupon period for which he or she owned the bond. More specifically, the **accrued interest by the practical method** is given by

$$(6.7.4) \qquad A_T^{\text{prac}} = Cgf, \quad T = \lfloor T \rfloor + f, 0 \leqslant f < 1.$$

Recalling Equation (6.6.5), we see that

$$(6.7.5) \qquad A_T^{\text{prac}} = \mathcal{D}_T^{\text{prac}} - \mathcal{C}_T^{\text{prac}} \quad \text{and} \quad \mathcal{C}_T^{\text{prac}} = \mathcal{D}_T^{\text{prac}} - Cgf.$$

Computing interest by the practical method is extremely simple, and in the United States it is the method used for bonds having more than six months until maturity.[4]

If the dirty price is computed using the theoretical method and we use Equation (6.7.3) to determine the accrued interest, so that compound interest at rate $\tilde{j}$ is used for both, then we define the **theoretical clean price** or the **clean price by the theoretical method at rate** $\tilde{j}$, to be their difference. More precisely, set

$$(6.7.6) \qquad \boxed{\mathcal{C}_T^{\tilde{j}} = \mathcal{D}_T^{\tilde{j}} - \mathcal{A}_T^{\tilde{j}} = \mathcal{D}_T^{\tilde{j}} - Cg\left(\frac{(1 + \tilde{j})^f - 1}{\tilde{j}}\right) = \mathcal{D}_T^{\tilde{j}} - Cgs_{\overline{f}|\tilde{j}}\,.}$$

The term **market price by compound interest** is also used for $\mathcal{C}_T^{\tilde{j}}$.

Should you calculate the dirty price by the theoretical method and the accrued interest by the practical method, then you are said to be using the **semipractical method**. We define a **semipractical clean price theoretical clean price** as the difference between the theoretical dirty price $\mathcal{D}_T^{\tilde{j}}$ and the practical accrued interest;

$$(6.7.7) \qquad \boxed{\mathcal{C}_T^{\tilde{j},\,\text{semiprac}} = \mathcal{D}_T^{\tilde{j}} - \mathcal{A}_T^{\text{prac}} = \mathcal{D}_T^{\tilde{j}} - Cgf.}$$

The semipractical clean price has the rather unpleasant feature that it is not constant over time when all the book values are constant. Otherwise put, if the bond were purchased at its redemption price, the clean price by the semipractical method would not be constant even though all the book values are equal to the redemption value. A semipractical clean price is part of the BA II Plus calculator **Bond worksheet** considered in Section (6.11). Clean prices are the prices reported by the newspapers in the bond tables. These are likely to be computed by the semipractical method.

> Whichever method is used to calculate the dirty price, accrued interest, and clean price, it is standard to calculate time on a "30/360" basis for corporate and municipal bonds and on an "actual/actual" basis for government bonds.

[The "30/360" and "actual/actual" bases for counting time were introduced in Section (1.4).]

We record the quantities introduced so far in this section in Table (6.7.8). We stress that this is a table of prices and adjustments to price, and it uses a yield rate $\tilde{j}$ which may be different from the yield to the original buyer j. In contrast, the entries of Table (6.6.9) each are calculated using the rate j.

---

[4]See the Standard Securities Calculation Methods, published by the Securities Industry Association.

SYMBOL	NAME	NOTES
$\mathcal{D}_T^{\tilde{\jmath}}$	Dirty price	(1) This is the price paid at time T to yield the new investor $\tilde{\jmath}$. (2) $\mathcal{D}_T^{\tilde{\jmath}} = \mathcal{D}_{\lfloor T \rfloor}^{\tilde{\jmath}}(1 + \tilde{\jmath})^f$. Note that this equation uses the dirty price $\mathcal{D}_{\lfloor T \rfloor}^{\tilde{\jmath}}$, which is not equal to the book value $B_{\lfloor T \rfloor}^{\tilde{\jmath}}$ unless $\tilde{\jmath} = j$. (3) This is also called the invoice price, flat price, full price, and theoretical dirty price.
$\mathcal{C}_T^{\tilde{\jmath}}$	Theoretical clean price	(1) $\mathcal{C}_T^{\tilde{\jmath}} = \mathcal{D}_T^{\tilde{\jmath}} - Cgs_{\overline{f}|\tilde{\jmath}}$. (2) Compound interest is specified, and the accrued interest $\mathcal{A}_T^{\tilde{\jmath}} = Cgs_{\overline{f}|\tilde{\jmath}}$ is computed using the yield rate $\tilde{\jmath}$ which the new investor will receive if he holds the bond until maturity. (3) This may be called the market price.
$\mathcal{C}_T^{\tilde{\jmath},\,\text{semiprac}}$	semipractical clean price	(1) $\mathcal{C}_T^{\tilde{\jmath},\,\text{semiprac}} = \mathcal{D}_T^{\tilde{\jmath}} - Cgf$ (2) The amount Cgf that is subtracted is the accrued coupon (by the practical method) $\mathcal{A}_T^{\text{prac}}$. (3) This may also be called the market price, but compound interest is not now used.
$\mathcal{A}_T^{\tilde{\jmath}}$	Accrued interest by the theoretical method at interest rate $\tilde{\jmath}$	$\mathcal{A}_T^{\tilde{\jmath}} = Cg\left(\dfrac{(1+\tilde{\jmath})^f - 1}{\tilde{\jmath}}\right) = Cgs_{\overline{f}|\tilde{\jmath}}$
$\mathcal{A}_T^{\text{prac}}$	Accrued interest by the practical method	$\mathcal{A}_T^{\text{prac}} = Cgf$

TABLE (6.7.8) Prices and Adjustments to price, at rate $\tilde{\jmath}$

EXAMPLE 6.7.9   Computing accrued interest and theoretical dirty price

Problem: A $1,000 three-year 6% bond with semiannual coupons is redeemable for $1,125. It was originally purchased at issue for its face value.

Sirkka purchases the bond seven months later at a price to yield her a nominal rate of 11% convertible semiannually. Find the invoice price, accrued interest by the theoretical method at Sirkka's yield rate, and the accrued interest by the practical method. [This bond is the same as the bond in Example (6.6.10).]

Solution   Example (6.6.10) involved the same bond and we use values computed in the solution to that example when convenient. Once again, we are interested in time seven months ($1\frac{1}{6}$ coupon periods) and $f = \frac{1}{6}$. Sirkka buys the bond to yield $\tilde{\jmath} = \frac{11\%}{2} = 5.5\%$ per coupon period so, recalling the value of B_1 from Example (6.6.10), the invoice price is $D_T^{\tilde{\jmath}} = B_1(1 + \tilde{\jmath})^f \approx$ $\$1{,}018.449038(1.055)^{\frac{1}{6}} \approx \$1{,}027.577798 \approx \$1{,}027.58$. The accrued interest by the theoretical method is $A_T^{\tilde{\jmath}} = Cg\left(\frac{(1+\tilde{\jmath})^f - 1}{\tilde{\jmath}}\right) \approx \$30\left(\frac{(1.055)^{\frac{1}{6}} - 1}{.055}\right) \approx \$4.89$.

[We remark that this is just one cent less than A_T^j which may be computed using (6.7.2) or by $A_T^j = D_T - C_T$, D_T and C_T as reported in Example (6.6.10).] The accrued interest by the practical method is $A_T^{\text{prac}} = \$30\left(\frac{1}{6}\right) = \$5$. ∎

A bond is purchased at a **premium** if the yield $\tilde{\jmath}$ is less than the coupon rate g, and the amount of the premium is $C_T^{\tilde{\jmath}} - C$. If the new buyer's yield rate $\tilde{\jmath}$ is greater than the coupon rate g then, to compensate, the redemption amount C must be more than the theoretical clean price $C_T^{\tilde{\jmath}}$. In this case the bond has a **discount** $C - C_T^{\tilde{\jmath}}$.

EXAMPLE 6.7.10   Premiums and discounts for bonds sold between coupon dates

Problem: Gilah Rosenberg purchased a fifteen-year $10,000 8% par-value bond with annual coupons at a price to yield her 8%. She held it for sixty-five months, then sold it to Uzi Schwartz at a price to yield Uzi 9% should he hold the bond to maturity. Three years later, Uzi sold the bond to Osnat Treisman at a price to yield Osnat 7%. Find the three prices paid for the bond and the amount of discount or premium (if any) in each of them.

Solution   The bond has $C = F = \$10{,}000$, $m = 1$, and $g = r = \alpha = 8\%$. So, the payments guaranteed by the original bond are annual coupon payments of $800 and a redemption payment of $10,000 at the end of the fifteen years. Since Gilah buys the bond to yield the coupon rate, there is no premium or discount in her price. That is, she pays $10,000.

Uzi purchases the bond to yield a higher rate, so his price will include a discount. Specificallly, Uzi purchases the bond five months after the fifth annual coupon has been paid, and we refer to this as time $5\frac{5}{12}$. So that Uzi has a yield rate to maturity of 9%, his price will be the sum of the present value of the remaining coupons and the present value of the redemption amount, computed using compound interest at 9% annual effective interest. Therefore, Uzi should pay $(1.09)^{\frac{5}{12}} \left(\$800a_{\overline{10}|9\%} + \$10,000(1.09)^{-10} \right) \approx \$9,700.369674$. So, Uzi's price is \$9,700.37. The accrued coupon figured using a 9% rate is $\$800 \left(\frac{(1.09)^{\frac{5}{12}} - 1}{.09} \right) \approx \$324.9762584 \approx \$324.98$. So, the theoretical clean price at 9% is \$9,700.37 $-$ \$324.98 $=$ \$9,375.39, and the discount is \$10,000 $-$ \$9,375.39 $=$ \$624.61.

Osnat's purchase of the bond is made at a price to give him a yield rate of 7% provided he holds the bond until maturity. This price is

$$(1.07)^{\frac{5}{12}} \left(\$800a_{\overline{7}|7\%} + \$10,000(1.07)^{-7} \right) \approx \$10,840.26.$$

The accrued coupon at 7% is $\$800 \left(\frac{(1.07)^{\frac{5}{12}} - 1}{.07} \right) \approx \$326.77$. Consequently, Osnat's clean price is \$10,513.49, and he purchases the bond with a premium of \$513.49. ∎

We have previously [see Section (6.5)] considered the allocation of each coupon between interest and adjustment to principal. How should this be extended to any accrued interest, which would have been better called "accrued coupon"? Suppose $T = \lfloor T \rfloor + f, 0 < f < 1$. (We disallow $f = 0$ since in that case there is no accrued interest.) If we have computed accrued interest by the practical method, it is natural to calculate the interest in the accrued interest A_T^{prac} as $f I_{\lfloor T \rfloor + 1}$ and the adjustment to principal as $f P_{\lfloor T \rfloor + 1}$. [The notations I_T and P_T were introduced in Section (6.5) for integral T.]

Not surprisingly, this division into interest and adjustment of principal is a bit more complicated when we use the theoretical method. Let's see how it works if the bond is sold again to yield j. The theoretical accrued interest at time $T = \lfloor T \rfloor + f$, computed using the original yield rate j per coupon period, is equal to $Cg s_{\overline{f}|j} = Cg \left(\frac{(1+j)^f - 1}{j} \right)$. On the other hand, the seller of the bond at time T is responsible for the interest earned during $[\lfloor T \rfloor, T]$, as well as during $[0, \lfloor T \rfloor]$. But the value of the bond at time $\lfloor T \rfloor$ was the book value $B_{\lfloor T \rfloor} = C(g - j)a_{\overline{n - \lfloor T \rfloor}|j} + C$. Thus, again using the interest rate j, we find that the amount of interest $\mathcal{I}_T$ that should be credited to the seller (for tax or other accounting purposes) is

$$
\begin{aligned}
\mathcal{I}_T &= [(1 + j)^f - 1]B_{\lfloor T \rfloor} \\
&= [(1 + j)^f - 1](C(g - j)a_{\overline{n - \lfloor T \rfloor}|j} + C) \\
&= [(1 + j)^f - 1]\left(\frac{C(g-j)(1-v_j^{n-\lfloor T \rfloor})}{j} + \frac{Cj}{j}\right) \\
&= \frac{(1+j)^f - 1}{j}(C(g - j)(1 - v_j^{n-\lfloor T \rfloor}) + Cj) \\
&= s_{\overline{f}|j}(Cg - Cj - C(g - j)v_j^{n-\lfloor T \rfloor} + Cj) \\
&= s_{\overline{f}|j}(Cg - C(g - j)v_j^{n-\lfloor T \rfloor}).
\end{aligned}
$$

(6.7.11)

Therefore, the time T accrued interest by the theoretical method, which might better be called the time T "accrued coupon," is equal to the sum of $\mathcal{I}_T$ and $s_{\overline{f}|j}C(g - j)v_j^{n-\lfloor T \rfloor}$. We might reasonably set

(6.7.12) $$\mathcal{P}_T = s_{\overline{f}|j}C(g - j)v_j^{n-\lfloor T \rfloor},$$

and call it the **accrued adjustment to principal** for the interval $\left[\lfloor T \rfloor, T\right]$. Then

(6.7.13) $$A_T^j = \mathcal{I}_T + \mathcal{P}_T.$$

We thus have the split of the accrued interest into interest and a **capital gain** (or **capital loss**).[5] Note that if T is an integer or if the bond was bought for its redemption value, then $\mathcal{P}_T = 0$, and there is no capital gain or loss. On the other hand, when $f > 0$, $\mathcal{P}_T$ is positive if the bond was purchased at a premium, and it is negative if the bond was purchased at a discount. So, there is a capital loss when the bond is purchased at a discount and a capital gain when it is purchased at a premium. Those knowing calculus should note that

$$
\lim_{f \to 1} \mathcal{P}_T = C(g - j)v_j^{n-\lfloor T \rfloor} = P_{\lfloor T \rfloor + 1} = B_{\lfloor T \rfloor} - B_{\lfloor T \rfloor + 1}.
$$

EXAMPLE 6.7.14   Interest and adjustment to principal in the accrued interest

Problem: Talbot Richardson purchases a ten-year 7% $1,000 par-value bond with annual coupons thirty-nine months after its issue. The original purchaser paid a price to yield 7.5% if the bond was held until maturity, and Mr. Richardson buys it at a price to yield 6.2%, this again being a yield calculation based on the bond being held until its maturity. Compute the accrued interest by the practical method and then (three times) by the theoretical method, specifically at 7.5%, 6.2%, and at the yield rate earned by the original bond holder over

---

[5]You should not assume this is consistent with the usage by the Internal Revenue Service.

the thirty-nine month period prior to the resale. Find the split of the accrued interest by the theoretical method at 7.5% into interest and principal.

Solution   We have $F = C = \$1,000$, $m = 1$, $\alpha = r = g = 7\%$, and coupon amount $Cg = Fr = \$70$. Thirty-nine months is $T = 3\frac{1}{4}$ years. So, $f = \frac{1}{4}$. By the basic price formula, the original price was $\$70a_{\overline{10}|7.5\%} + \$1,000(1.075)^{-10}$ and we thus calculate that the original price was $\$965.68$. Mr. Richardson's yield price is $(1.062)^{\frac{1}{4}}(\$70a_{\overline{6}|6.2\%} + \$1,000(1.062)^{-7}) \approx \$1,060.17$. The actual yield y received by the original holder of the bond satisfies $\$965.68 = \$70a_{\overline{3}|y} +$ $\$1,060.17(1 + y)^{-3\frac{1}{4}}$, and therefore $y \approx 9.474076776\%$. (We found this using the BA II Plus calculator **Cash Flow worksheet**. Use CFo $= -965.68$, C01 $=$ C03 $=$ C05 $= 0$, F01 $=$ F03 $=$ F05 $= 3$, C02 $=$ C04 $=$ C06 $= 70$, F02 $=$ F04 $=$ F06 $= 1$, C07 $= 1,060.17$, and F07 $= 1$. Then key $\boxed{\text{IRR}}$ $\boxed{\text{CPT}}$, and remember to change this rate, which is a quarterly rate given as a percent, to the equivalant annual rate.)

The accrued interest by the practical method A_T^{prac} is $\frac{1}{4}(\$70) = \$17.50$. Calculated using the theoretical method at the original yield rate 7.5%, the accrued interest A_T is $\$70\left(\dfrac{(1.075)^{\frac{1}{4}}-1}{.075}\right) \approx \$17.02829436 \approx \$17.03$. At Mr. Richardson's lower interest rate of 6.2%, it is $\$70\left(\dfrac{(1.062)^{\frac{1}{4}}-1}{.062}\right) \approx \$17.10724088$ $\approx \$17.11$. At the rate y calculated above, the accrued interest is $\$70\left(\dfrac{(1+y)^{\frac{1}{4}}-1}{y}\right)$ $\approx \$16.91053416 \approx \$16.91$.

Now consider what part of this accrued interest is accrued adjustment to principal. In the entire fourth coupon, according to Equation (6.5.7), the amount for adjustment to principal is $P_4 = C(g - j)(v_j)^{10-4+1} =$ $-\$5(1.075)^{-7} \approx -3.013774504$. So, by the practical method, in a quarter of the coupon period we have an adjustment of principal of $\frac{1}{4}(-3.013774504) \approx$ $-\$.75$. Therefore, according to the practical method, the interest in the accrued interest (better called the accrued coupon!) is $\$17.50 - (-\$.75) = \$18.25$. On the other hand, if we use the theoretical method at interest rate 7.5%, according to Equation (6.7.12) the amount for adjustment of principal in the accrued interest is $s_{\overline{1/4}|7.5\%}\$1,000(.07 - .075)(1.075)^{-7} \approx -\$.73$. Similarly, at interest rate 6.2% it is $s_{\overline{1/4}|6.2\%}\$1,000(.07 - .062)(1.062)^{-7} \approx \$.1.28$, and at interest rate y it is $-\$3.17$. The corresponding amounts of interest in the accrued interest, by the theoretical method at rates 7.5%, 6.2%, and y, are then $\$17.03 - (-\$.73) = \$17.76$, $\$17.11 - \$1.28 = \$15.83$, and $\$16.91 - (-\$3.17) = \$20.08$, respectively.   ■

As we have discussed, if you purchase a bond and hold it to maturity, you receive the yield rate that you contracted when you first purchased the bond. However, if you choose to sell the bond prior to maturity, it may happen that interest rates have gone up and consequently the bond sells for less than its book value; as a result your yield for the investment would be less than you anticipated.

We now look briefly at another type of investment for which the seller rather than the buyer takes on the risk resulting from interest rate shifts. A **guaranteed investment contract** (GIC) is similar to a bond, but if you sell the contract, the issuer will pay you the amount that the book value exceeds the market value. So, GICs are very safe investments provided that the issuing coorporation, an insurance company, is not at risk of defaulting. GICs play an important role in many retirement plans. They tend to be structured to fit the needs of the participant, and hence there is considerable variation from contract to contract including the schedule, flexibility of cashflows, and whether interest may be reinvested at a fixed or floating rate. The fact that GICs are customized tends to limit their liquidity. GICs are considered to be insurance contracts, and this may have regulatory or tax consequences.

6.8 YIELD RATE EXAMPLES

Bonds provide a series of cashflows. While we have concentrated on bonds with level coupons, a portfolio of bonds may give a nonlevel cashflow pattern due to staggered purchase or redemption times [see Examples (6.8.2) and (6.8.3)]. Bonds may be resold prior to maturity one or more times [See Example (6.8.4)], and coupons may be partially or fully reinvested [see Example (6.8.5)]. An investor may purchase a bond at a premium and desire to recoup the amount of the premium in a sinking-fund account, as in Example (6.8.1). These practical situations provide a wealth of new problems involving yield rates.

EXAMPLE 6.8.1   Recouping a premium in a sinking fund account

Problem: Alicia bought a newly issued $1,000 20% ten-year bond, redeemable at $1,100 and having yearly coupons. It was bought at a premium at a price of $1,400.02. Alicia immediately took a constant amount D from each coupon and deposited it in a savings account earning 8% effective annual interest, so as to accumulate the full amount of the premium a moment after the final deposit. At the end of the ten years, Alicia closed out her savings account. Find the yearly effective yield rate earned by Alicia for her combined ten-year investment.

Solution   In order to accumulate the premium $1,400.02 - $1,100 = $300.02 in her savings account, each year Alicia needs to deposit $\frac{\$300.02}{s_{\overline{10}|8\%}} \approx \$20.7102272$. So, if we let $D = \$20.71$, at the end of the ten years Alicia will receive $300.02 when she closes out her savings account. Since the annual coupon amount is $.2 \times \$1,000 = \$200$, she keeps $200 - $20.71 = $179.29 at the ends of years $1, 2, \ldots, 10$. But at time 10 she also gets the redemption $1,100 and the savings account liquidation $300.02, and these total $1,400.02. Thus she earns $179.29 each year while keeping her principal of $1,400.02 intact and her yield is $\frac{\$179.29}{1,400.02} \approx 12.8062456\%$. Of course, since $a_{\overline{10}|i} = \frac{1-(1+i)^{-10}}{i}$, this yield could also be obtained by solving for i in the time 0 equation of value $\$1,400.02 = \$179.29 a_{\overline{10}|i} + \$1,400.02(1 + i)^{-10}$. ∎

EXAMPLE 6.8.2   Finding the yield of a laddered portfolio of bonds

Problem: On January 1, 1978, Lenny Ladderman purchased a portfolio of twelve $1,000 par-value bonds for $10,568.61. The portfolio consisted of 6% bonds, each with annual coupons, The twelve maturity dates for the bonds were January 1 of the years from 1983 through 1994. Find Lenny's yield rate.

Solution   The annual coupon payment resulting from each bond not yet redeemed is $(.06)(\$1,000) = \$60$. Through January 1, 1983, when the first redemption takes place, Lenny had twelve $60 coupons for a total of $720, and after that the contribution from coupons decreases by $60 annually. Each year from 1983 through 1994 there is a $1,000 redemption in addition to the coupon contribution. So, we have cashflows as indicated in the following figure.

PAYMENT:	$720	$720 $\cdots$	$1,720	$1,660	$1,600 $\cdots$	$1,060
TIME: (Jan. 1) 78	79	80	83	84	85	94

The yield rate j satisfies the time 0 equation of value

$$\$10,568.61 = \$720 a_{\overline{4}|j} + (I_{1,720,-60}) a_{\overline{12}|j}(1 + j)^{-4}$$

$$= \$720 a_{\overline{4}|j} + \left(\$1,720 a_{\overline{12}|j} + \frac{-60}{j}\left(a_{\overline{12}|j} - 12(1 + j)^{-12}\right)\right)(1 + j)^{-4}.$$

You may now use the "guess and check" method to find the yield rate j, beginning with a rate somewhat greater than the coupon rate, say 7%, since the package of bonds was bought at a discount (due to the fact that the purchase price was less than the sum of the redemption values). In fact, $j \approx .07750904074$.

The **Cash Flow worksheet** of the BA II Plus calculator may be used to obtain this yield rate relatively quickly. The cashflows shown in the above timeline may be entered by keying

| CF | 2ND | CLR WORK | 1 | 0 | 5 | 6 | 8 | • | 6 | 1 | +/− | ENTER |

| ↓ | 7 | 2 | 0 | ENTER | ↓ | 4 | ENTER | ↓ | 1 | 7 | 2 | 0 | ENTER |

| ↓ | ↓ | 1 | 6 | 6 | 0 | ENTER | ↓ | ↓ | 1 | 6 | 0 | 0 | ENTER |

| ↓ | ↓ | 1 | 5 | 4 | 0 | ENTER | ↓ | ↓ | 1 | 4 | 8 | 0 | ENTER |

| ↓ | ↓ | 1 | 4 | 2 | 0 | ENTER | ↓ | ↓ | 1 | 3 | 6 | 0 | ENTER |

| ↓ | ↓ | 1 | 3 | 0 | 0 | ENTER | ↓ | ↓ | 1 | 2 | 4 | 0 | ENTER |

| ↓ | ↓ | 1 | 1 | 8 | 0 | ENTER | ↓ | ↓ | 1 | 1 | 2 | 0 | ENTER |

| ↓ | ↓ | 1 | 0 | 6 | 0 | ENTER | IRR | CPT |

This sequence of keystrokes sets CF0= −10,568.61, C01=720, F01= 4, C02=1,720, F02=1, C03=1,660, F03=1, C04=1,600, F04=1, C05=1,540, F05=1, C06=1,480, F06=1, C07=1,420, F07=1, C08=1,360, F08=1, C09=1,300, F09=1, C10=1,240, F10=1, C11=1,180, F11=1, C12=1,120, F12=1, C13=1,060, and F13=1, and then the calculator computes the yield rate to be 7.750904074%.   ∎

EXAMPLE 6.8.3   Pricing a laddered portfolio of bonds

Problem: Nicolae Miloslav prefers a laddered portfolio of bonds. He purchases eight $20,000 7% par-value bonds, each having semiannual coupons. The terms of these bonds are 3, 4, 5, 6, 7, 8 , 9, and 10 years. Calculate the total price of these bonds to yield Mr. Miloslav a nominal rate of 8% convertible semiannually.

Solution   Each coupon of a $20,000 7%, bond with semiannual coupons is for $20,000$\left(\frac{.07}{2}\right)$ ≈ $700. Therefore, for the first three years, at the end of each half-year Mr. Miloslav receives coupons totaling $8 \times \$700 = \$5,600$. During the subsequent seven years, each year he holds one fewer bond and hence the total of the semiannual coupons he receives goes down by $700. In addition to coupons, Mr. Miloslav receives redemption amounts of $20,000 at the end of years 3 through 10. So Mr. Miloslav's incoming cashflows consist of

(1) semiannual coupons totaling $5,600 at each of the times $\frac{1}{2}, 1, 1\frac{1}{2}$, and 2. (Calculated using a nominal interest rate of 8% convertible semiannually, these have a time 0 value of $5,600a_{\overline{4}|4\%}$.)

(2) Mid-year coupon payments totaling $5,600, $4,900, $4,200, $3,500, $2,800, $2,100, $1,400, and $700 at times $t = 2\frac{1}{2}, 3\frac{1}{2}, 4\frac{1}{2}, 5\frac{1}{2}, 6\frac{1}{2}, 7\frac{1}{2}, 8\frac{1}{2}$, and $9\frac{1}{2}$, respectively. (Calculated using a nominal interest rate of 8% convertible

semiannually, these have a time 0 value given by the expression of
$(1.04)^{-3}(I_{\$5,600,-\$700}a)_{\overline{8}|(1.04)^2-1}\cdot)$

(3) End-of-year coupon payments totaling $5,600, $4,900, $4,200, $3,500, $2,800, $2,100, $1,400, and $700 at times 3, 4, 5, 6, 7, 8, 9, 10 respectively. (Calculated using a nominal interest rate of 8% convertible semiannually, these have a time 0 value of $(1.04)^{-4}(I_{\$5,600,-\$700}a)_{\overline{8}|(1.04)^2-1}\cdot)$

(4) Redemption payments of $20,000 at times 3, 4, 5, 6, 7, 8, 9, 10. (Calculated using a nominal interest rate of 8% convertible semiannually, these have a time 0 value of $(1.04)^{-4}\$20,000a_{\overline{8}|(1.04)^2-1}\cdot)$

Therefore, the total price of Mr. Miloslav's portfolio of bonds to provide him with a nominal yield of 8% convertible semiannually is

$$\$5,600a_{\overline{4}|4\%} + (1.04 + 1)(1.04)^{-4}(I_{\$5,600,-\$700}a)_{\overline{8}|8.16\%}$$
$$+ (1.04)^{-4}\$20,000a_{\overline{8}|8.16\%}$$
$$\approx \$20,327.41326 + (2.04)(1.04)^{-4}(\$19,628.22637) + \$97,651.2855$$
$$\approx \$152,206.41.$$

■

EXAMPLE 6.8.4   Multiple resellings of a bond

Problem: Audrie purchases a $10,000 fifteen-year 6% bond with semiannual coupons and a redemption value of $9,800. He pays a price P_A. After six-and-a-half years, just after the thirteenth coupon, Audrie sells the bond to Barry at a price P_B. Barry holds the bond for five years (ten coupon payments), then sells it to Carl for a price P_C, and Carl holds the bond through its redemption. Denote the semiannual yield rates earned by Audrie, Barry, and Carl by j_A, j_B, and j_C respectively. It is known that $j_A = 2.9\%$, $j_B = 4.2\%$, and $j_C = 3.5\%$. Find the discount or premium on the bond purchased by Audrie.

Solution   The bonds purchased by all three investors have semiannual coupons of $(\$10,000)(\frac{.06}{2}) = \$300$. Since Audrie gets thirteen coupons and Barry receives ten, Carl's bond has $30 - 13 - 10 = 7$ coupons. So, according to the basic price formula, Carl's price P_C satisfies $P_C = \$300a_{\overline{7}|3.5\%} + \$9,800(1.035)^{-7}$. Calculating and rounding, one finds $P_C = \$9,537.07$. Since Carl pays this amount to Barry at the time Barry receives his tenth coupon payment, Barry's time 0 equation is $P_B = \$300a_{\overline{10}|4.2\%} + \$9,537.07(1.042)^{-10}$ and thus $P_B = \$8,729.52$. Finally, Audrie receives $8,729.52 at the time of his thirteenth coupon, and $P_A = \$300a_{\overline{13}|2.9\%} + \$8,729.52(1.029)^{-13}$. So, $P_A = \$9,230.90$ and the discount is $9,800 - \$9,230.90 = \$569.10$. ■

EXAMPLE 6.8.5   Partial reinvestment of coupons

Problem: Ivanon paid $2,180 to purchase a $2,000 13% eight-year bond with semiannual coupons and redemption amount $1,820. Each time he received a coupon, he made a deposit of part of the coupon into a savings account with a 4% annual effective interest rate. His deposits were each $50 during the first five years and then they increased to $100. At the end of the eight years Ivanon received the bond's redemption payment and liquidated his 4% savings account. What was Ivanon's annual yield for this eight-year period?

Solution   Ivanon's 4% account had a six-month interest rate of $I = (1.04)^{\frac{1}{2}} - 1 \approx 1.980390272\%$, and his accumulated value at the end of the eight years was therefore $\$50s_{\overline{10}|I}(1.04)^3 + \$100s_{\overline{6}|I} \approx \$1,245.7975$. So, Ivanon received $1,245.80 when he closed his savings account. The coupon amount is $\$2,000(\frac{.13}{2}) = \$130$. Thus, with a basic time unit of six months, Ivanon's cashflow experience was as follows:

$2,180 is paid out at $t = 0$.
$130 − $50 = $80 is received at times $t = 1, 2, ... , 10$.
$130 − $100 = $30 is received at times $t = 11, 12, ... , 15$.
$30 + $1,820 + $1,245.80 = $3,095.80 is received at time $t = 16$.

This allows us to calculate Ivanon's half-yearly yield rate $j \approx 4.75336856\%$, a calculation that is done most easily using a **Cash Flow worksheet**, but also results from the equation of value corresponding to Ivanon's cashflow. Ivanon's annual yield rate is $i = (1 + j)^2 - 1 \approx 9.732682246\%$. ∎

6.9 CALLABLE BONDS

Some bonds, referred to as **callable bonds**, are issued with an agreement or **call provision** that allows the *issuer* to hasten the repayment of the debt. The issuer has the option of redeeming the bond, or perhaps a portion of the bond, at any of a set of designated **call dates**, and for each date a redemption value is specified. Often there is a period of years, called the **lockout period**, before the first call date. Since mortgages usually allow early repayment, mortgage-backed bonds are among the bonds that are apt to be callable as are longer-term bonds.

A callable bond gives the issuer more flexibility. If the corporation or government entity is able to reduce its debt, it may call in bonds to cover that decrease. Also, if interest rates decline, the issuer may issue new bonds at a lower rate and call in the old ones, thereby reducing its borrowing costs.

From the bond purchaser's perspective, a call provision makes the bond less desirable. The call provision creates difficulty in investment planning and increases the likelihood that there will be a redemption when interest rates are relatively low. Call provisions also tend to reduce the appreciation in the price of the bond when there is a decrease in interest rates.

Since a call provision makes a bond more attractive to the issuer and less so to a potential purchaser, a compensating feature must be added. Callable bonds tend to offer higher yields then noncallable bonds. In addition, the redemption values specified at call dates are often higher than the redemption price at maturity. At a given call date, the bonus in redemption value over the redemption value at maturity is called a **call premium**.

A bond is said to have a **European option** if it has a single call date prior to maturity. A bond with a **Bermuda option** is callable at multiple specified dates, and these usually coincide with coupon dates. We say that a bond has an **American option** if it is callable at *any* date following the lockout period. Call premiums at dates other than coupon dates should include accrued interest. It should be noted that these names do not necessarily agree with the geographic origin of the bond, nor with the market in which the bond is traded. For instance, a bond with a European option may be issued by an American firm and be traded on U.S. financial markets.

When viewing a callable bond, a potential buyer should consider yields if the bond is redeemed at any of the potential call dates as well as the maturity date. Since the issuer is the one who decides when to redeem the bond, it may be sensible for a potential buyer to focus on the smallest of these yields, since that is the worst yield that might be earned.

EXAMPLE 6.9.1   Bond bought at a discount with zero call premiums

Problem: Callable Corporation is offering a \$1,000 bond with semiannual coupons and a nominal coupon rate of 8% convertible semiannually. It is a five-year bond with call dates at the end of years 2 and 3. If the redemption value at each of these dates and also at maturity is \$1,060 and the price is \$1,022, find the possible yields to the investor and note which is least.

Solution   We have $F = \$1{,}000$, $m = 2$, $\alpha = 8\%$, $r = 4\%$, and the coupons are each \$40. Moreover, the bond has ten coupons, $P = \$1{,}022$, and the redemption amount at any call date is \$1,060. Therefore, if the bond is called at the end of two years (four coupons), the yield rate y_4 (per coupon period) satisfies $\$1{,}022 = \$40a_{\overline{4}|y_4} + \$1{,}060(1 + y_4)^{-4}$. Consequently, $y_4 \approx$ 4.779395028% and the annual yield $(1 + y_4)^2 - 1$ is about 9.787%. If the bond is called at the end of three years (six coupons), the holder's yield y_6 (per coupon period) satisfies $\$1{,}022 = \$40a_{\overline{6}|y_6} + \$1{,}060(1 + y_6)^{-6}$. So, $y_6 \approx$ 4.467900953. The corresponding annual yield $(1 + y_6)^2 - 1$ is about 9.135%. If the bond is redeemed at maturity, the yield j satisfies $\$1{,}022 = \$40a_{\overline{10}|j} + \$1{,}060(1 + j)^{-10}$. Therefore, $j \approx$ 4.220434792%, and this corresponds to an annual yield $(1 + j)^2 - 1$ of about 8.619%. We have $y_4 > y_6 > j$. The yield rate decreases as the time to redemption increases. The yield to the bond holder is smallest if the bond is held to maturity. ∎

In Example (6.9.1), the bond was bought at a discount and the lowest yield rate corresponds to the longest wait until redemption. This seems reasonable because if the bond is redeemed prior to the maturity date, the bonus or discount is obtained more quickly, apparently to the advantage of the investor. The issuer has to make a large redemption payment (large in comparison with the funds received at issue) and delays this payment as long as possible. We now show that for a bond bought at discount with a level redemption value C, redeeming at a later coupon date *always* results in a lower yield, not taking into account reinvestment rates.

CLAIM 6.9.2: Suppose a bond is bought at a discount and is callable on any coupon date with redemption amount C. Let y_k denote the yield per coupon period if the bond is redeemed just after the k-th coupon payment. Then $\{y_k\}$ is a decreasing sequence. Equivalently, if $k \in \{2,3,...,n\}$ then $y_{k-1} > y_k$. So, the smallest yield occurs when the bond is not called prior to maturity.

Proof: Fix $k \in \{2,3,...,n\}$. According to the basic price equation (6.2.2),
$$P = (Cg)a_{\overline{k}|y_k} + C(1 + y_k)^{-k}.$$

Since $a_{\overline{k}|y_k} = a_{\overline{k-1}|y_k} + (1 + y_k)^{-k}$, we have
$$P = (Cg)a_{\overline{k-1}|y_k} + (Cg)(1 + y_k)^{-k} + C(1 + y_k)^{-k}$$
$$= (Cg)a_{\overline{k-1}|y_k} + C(g + 1)(1 + y_k)^{-k}$$
$$= (Cg)a_{\overline{k-1}|y_k} + C\left(\frac{1 + g}{1 + y_k}\right)(1 + y_k)^{-(k-1)}.$$

We were given that $P < C$. Therefore, the premium-discount formula (6.3.2) gives $P = C(g - y_k)a_{\overline{k}|y_k} + C$, and we may conclude that $g < y_k$ and $\frac{g+1}{y_k+1} < 1$. So, we have the inequality
$$P < (Cg)a_{\overline{k-1}|y_k} + C(1 + y_k)^{-(k-1)}.$$

On the other hand, by the definition of y_{k-1} as the yield rate earned by the investor if the bond is redeemed immediately after the $(k - 1)$-st coupon payment,
$$P = (Cg)a_{\overline{k-1}|y_{k-1}} + C(1 + y_{k-1})^{-(k-1)}.$$

Thus,

$$(Cg)a_{\overline{k-1}|y_{k-1}} + C(1 + y_{k-1})^{-(k-1)} < (Cg)a_{\overline{k-1}|y_k} + C(1 + y_k)^{-(k-1)}.$$

The expressions on the two sides of this inequality are identical except for the fact that the left-hand side uses the interest rate y_{k-1} while the right-hand side uses y_k. Moreover, each represents the present value of a sequence of positive cashflows, so a higher present value corresponds to a lower interest rate. Therefore, $y_{k-1} > y_k$.

Let us now consider bonds purchased at a premium. If there are no call premiums, the borrower cancels the loan at any call date with a constant payment that is smaller than the original loan amount. Therefore, the issuer is inclined to call in the loan as soon as possible. Lacking call-premiums for early redemption, the bondholder will have his or her highest yield if the bond is not called. If there are call premiums, the yield rates at the call dates and at maturity will need to be determined to find the minimum and maximum possible yields.

CLAIM 6.9.3: Suppose that a bond is bought at a premium and is callable on any coupon date with redemption amount C. Let y_k denote the yield per coupon period if the bond is redeemed just after the k-th coupon payment. Then $\{y_k\}$ is an increasing sequence, so the smallest yield occurs if the bond is called at the earliest call date.

The proof of (6.9.3) is left as an exercise [Problem (6.9.4)].

EXAMPLE 6.9.4   Bond bought at a premium with positive call premiums

Problem: A $2,000 six-year 7% bond with annual coupons is purchased for $2,200. If held to maturity, it is redeemable at par. The issuer may redeem the bond at the end of any year once it has been held for at least three years. The call premiums are $150 at $t = 3$ years, $140 at $t = 4$ years, and $130 at $t = 5$ years. Determine the possible yields to the investor and note which is minimal and which is maximal.

Solution   There are annual coupons of $140. The possible redemption values are $2,000 + $150 = $2,150 at $t = 3$, $2,000 + $140 = $2,140 at $t = 4$, $2,000 + $130 = $2,130 at $t = 5$, and $2,000 at $t = 6$. The purchase price in all cases is $2,200, and the respective yield rates satisfy equations given by the basic price formula. For instance, if redemption takes place at the end of three years, the yield rate y_3 satisfies $2,200 = 140a_{\overline{3}|y_3} + 2,150(1 + y_3)^{-3}$. Therefore, $y_3 \approx 5.647276938\%$. If $t = 4$, the yield rate y_4 makes the equation $2,200 = 140a_{\overline{4}|y_4} + 2,140(1 + y_4)^{-4}$ hold, so $y_4 \approx 5.737774053\%$. With redemption at $t = 5$ one has a yield y_5 satisfying $\$2,200 = \$140a_{\overline{5}|y_5} + \$2,130(1 + y_5)^{-5}$, whence $y_5 \approx 5.796899653\%$. Finally, if the bond is held to maturity, the basic price formula gives a yield of about 5.028057188%. So, the minimum yield is achieved if none of the call options are exercised, and the maximum yield occurs if the bond is called at the end of five years. ∎

If the call premiums stay level for several call dates, Claim (6.9.3) can be modified to minimize the number of individual yield rates that need to be computed. This is illustrated in the next example.

EXAMPLE 6.9.5   Bond bought at a premium with level call premiums for several periods

Problem: A $\$1,000$ 10% ten-year bond with annual coupons is purchased for $\$1,100$. Its redemption amount at maturity is $\$980$ while it may be called in at the end of any year, the call premiums being $\$115$ at the ends of years 1–3, $\$80$ at the ends of years 4–6, and $\$40$ at the ends of years 7–9. Find the minimum yield rate to the purchaser.

Solution   Using the principle of Claim (6.9.3), the minimum yield will occur at either $t = 1$, $t = 4$, $t = 7$, or $t = 10$, because each of these times is the earliest of a group with level call premiums. The annual coupons are level at $\$100$. If $t = 1$, the redemption is $\$980 + \$115 = \$1,095$, and the yield rate can be computed to be approximately 8.636%. If $t = 4$, the redemption is $\$980 + \$80 = \$1,060$, and the yield rate can be computed to be approximately 8.287%. If $t = 7$, the redemption is $\$980 + \$40 = \$1,020$, and the yield rate can be computed to be approximately 8.283%. Finally, if the bond is held to maturity, in which case it has a redemption value of $\$980$, the yield can be computed to be approximately 8.350%. So, the minimum yield is about 8.283%, and this occurs if the bond is called at the end of seven years. ∎

A callable bond is one in which the *issuer* has discretion, subject to the call provisions spelled out in the indenture, to choose the redemption date. There also exists a type of bond, called a **putable bond**, where the *purchaser* has the option of determining the redemption date. At redemption, the purchaser is said to "put" the bond back to the issuer. In exchange for flexibility, the purchaser of a putable bond is apt to have a lower yield than is available on comparable bonds lacking a **put option**. If you are contemplating purchasing a callable bond, since the choice is the issuer's, you should view the yield rate as being the smallest (worst) of the possible yields. On the other hand, when considering a putable bond, consider the yield as being the greatest (best) of the possibilities. Holders of putable bonds will decide how long to hold the bonds based on their need for capital, interest rates, and projected interest rates.

6.10   FLOATING-RATE BONDS

Most bonds have a fixed rate of interest for the life of the bond. However, you should also be aware that there are bonds that have variable interest rates. The indenture spells out how these rates are to be determined, and commonly they are tied to interest rates on short-term Treasury bills or money market accounts. Recent history suggests that when interest rates are expected to go up, the prevalence of floating-rate bonds tends to increase. This is because investors shy away from the fixed rate issues, except if they have a short term or a put option.

Basic principles may be used to find the yield rate received on a floating-rate bond. Of course this may be done only after the various coupon rates have been declared.

EXAMPLE 6.10.1   Yield of a floating-rate bond

Problem: A six-year $1,000 par-value floating-rate bond with annual coupons had annual coupon rates of 5%, 5.4%, 5.6%, 6.2%, 5.8%, and 6.6% successively. Find the annual yield rate on the bond if it was bought at its face value.

Solution   The floating-rate bond cost $1,000 at $t = 0$ and then returned coupon payments of $50, $54, $56, $62, $58, and finally $66 at $t = 1, 2, \dots, 6$. At time $t = 6$ there was also a redemption of $1,000, so the total received at $t = 6$ is $1,066. Let $v = \frac{1}{1+i}$ where i is the annual yield rate for the floating-rate bond. Then

$$\$1,000 = \$50v + \$54v^2 + \$56v^3 + \$62v^4 + \$58v^5 + \$1,066v^6.$$

You may solve this equation to the desired degree of accuracy using the "guess and check" method or Newton's method, perhaps beginning with your initial estimate equal to the average of the annual yield rates, namely $\frac{5\% + 5.4\% + 5.6\% + 6.2\% + 5.8\% + 6.6\%}{6} \approx 5.77\%$. If available, the BA II Plus calculator **Cash Flow worksheet** along with $\boxed{\text{IRR}}$ $\boxed{\text{CPT}}$ quickly yields the solution. The yield rate is $i \approx 5.721240964$. ∎

We next consider a comparison between a fixed rate bond and a floating-rate bond.

EXAMPLE 6.10.2

Problem: Jacques Giraud has a choice of purchasing either of two ten-year $5,000 bonds. The first is a fixed rate 5% bond with annual coupons. The second is a floating-rate bond with semiannual coupons. Mr. Giraud's financial advisor predicts that the coupon rate in each of the coupon periods in the

k-th year will be $.04 + ck = (4 + 100ck)\%$. Both bonds sell at a discount for \$4,800. Find c so that if the bond is held until maturity and the financial advisor's estimates are correct, Mr. Giraud will receive the same yield no matter which bond is purchased.

Solution   Note that $(.05)(\$5,000) = \$250$, so the fixed rate bond has annual yield rate i_f where $\$4,800 = \$250a_{\overline{10}|i_f} + \$5,000(1 + i_f)^{-10}$. We calculate that $i_f \approx 5.531472472\%$. The rate i_f is equaivalent to a semiannual rate $(1 + i_f)^{\frac{1}{2}} - 1 \approx (1.05531472472)^{\frac{1}{2}} - 1 \approx 2.728512338\%$.

During year k, the rate $.04 + ck$ is the applicable nominal coupon rate convertible two times per year for the floating-rate bond. So, each of the coupons in year k are for the amount $\$5,000\frac{(.04+ck)}{2} = \$100 + \$2,500ck$. If we let j denote the yield rate on the floating-rate bond per coupon period (a semiannual rate), we have the time 0 equation of value

$$\$4,800 = \$100a_{\overline{20}|j} + (\$2,500c)(Ia)_{\overline{10}|(1+j)^2-1}(1 + (1 + j)) + \$5,000(1 + j)^{-20}.$$

This is equivalent to

$$c = \frac{\$4,800 - \$5,000(1 + j)^{-20} - \$100a_{\overline{20}|j}}{(\$2,500)(Ia)_{\overline{10}|(1+j)^2-1}(1 + (1 + j))}.$$

If $j = (1 + i_f)^{\frac{1}{2}} - 1 \approx 2.728512338\%$, then

$$c = \frac{\$4,800 - \$5,000(1 + i_f)^{-10} - \$100a_{\overline{20}|(1+i_f)^{\frac{1}{2}}-1}}{(\$2,500)(Ia)_{\overline{10}|i_f}(1 + (1 + i_f)^{\frac{1}{2}}))}$$

$$\approx \frac{355.7775059}{192,935.0787} \approx .001844027.$$

We note that with this value of c, Mr. Giraud's financial advisor is forecasting a nominal coupon rate of about 5.844% for the floating-rate bond in its final year. ∎

6.11   THE BA II PLUS CALCULATOR BOND WORKSHEET

Bond calculations are fairly simple. As we have seen already, the **TVM worksheet** and **Cash Flow worksheet** may be put to good use in many calculations. However, built into your BA II Plus calculator you have a special **Bond worksheet**. This worksheet is designed to handle standard bond calculations for bonds with annual or semiannual level coupons. It is of particular value when actual calendar dates are given for the settlement and redemption dates of the bond. Calculations may be performed on an "actual/actual" or "30/360" basis. In addition, the standard method for figuring prices for bonds having less than one coupon period until maturity is somewhat different from the way prices are determined for bonds with at least one coupon period until maturity, and formulas for the approved method are built into the **Bond worksheet.**[6]

The **Bond worksheet** is accessed by keying $\boxed{\text{2ND}}$ $\boxed{\text{BOND}}$. We begin by familiarizing you with the registers of this worksheet. Table (6.11.1) includes a complete list of the registers in this worksheet along with brief comments. More extensive remarks then follow to the extent we feel they are needed or helpful. Our discussion is predicated upon the bond being resold at time $T = \lfloor T \rfloor + f, 0 \le f < 1$ at a price to yield the new purchaser a yield rate of $\tilde{j}$ per coupon period. If in fact, you are interested in the original sale of the bond, then j should replace $\tilde{j}$, and $T = f = 0$. If the bond is purchased after issue, the starting date (in the **Bond worksheet**) is the trade date. In this and in all entries to the worksheet, for a bond traded after issue, we take the perspective of the new buyer.

The SDT, CPN, RDT, and RV registers must each have a value entered. The value entered in the RV register should be the sum of the redemption amount and the call premium (if any) if there is an early redemption. You must also choose a method for counting time periods and specify a number of coupons per year. Once the first six registers are properly set, you may either enter a price or a yield, then compute the other. You do this by scrolling down to the register you wish to specify (using $\boxed{\downarrow}$ repeatedly), keying the desired numerical value followed by $\boxed{\text{ENTER}}$, then scrolling to the unfilled register (using $\boxed{\downarrow}$ or $\boxed{\uparrow}$) and pushing $\boxed{\text{CPT}}$. Provided the other registers have been filled compatibly, the accrued interest will automatically be computed, as will the modified duration (see Chapter 9) if you have a BAII Plus Professional caculator. (The BA II Plus calculator does not have a modified duration register.)

You enter the starting date and redemption date for a bond in the same manner as you make entries into the **Date worksheet**. Specifically, you key the month followed by a decimal point, then the day of the month as a two-digit number, immediately followed by the last two digits of the year. (For example, enter July 9, 2006 by keying $\boxed{7}\boxed{\bullet}\boxed{0}\boxed{9}\boxed{0}\boxed{6}\boxed{\text{ENTER}}$.)

The entry in the PRI register is always a semipractical clean price as a percentage of par. So, the **Bond worksheet** is designed so that you can quickly read the price of bonds with $100 face value. The price of a bond with face value $F \ne 100$ is computed by multiplying the displayed price by $\frac{F}{100}$.

---

[6]See the Standard Securities Calculation Methods, published by the Securities Industry Association.

REGISTER	MEANING	NOTES
SDT	starting date	Use any date in 1950 - 2049.
CPN	nominal coupon rate α as a percentage	This is $100\alpha = 100rm$.
RDT	redemption or call date	Use any date in 1950 - 2049.
RV	redemption fraction $100\frac{C}{F}$	This is the redemption amount as a percentage of the face value.
360 or ACT	select the method of counting days	Change the basis if you need to by keying 2ND SET.
2/Y or 1/Y	select the number of coupons per year	Change the number if you need to by keying 2ND SET.
YLD	nominal yield I convertible m times as a percentage	This entry is $100m\,\tilde{\jmath}$.
PRI	$\frac{100}{F}(\mathcal{D}_T^{\tilde{\jmath}} - \mathcal{A}^{\mathrm{prac}})$, the semipractical clean price as a percentage of par	The method for computing the dirty price is special if less than one coupon period remains.
AI	accrued interest $\dfrac{\mathcal{A}_T^{\mathrm{prac}}}{100}$	This is given per \$100 of par-value. The practical method is used.
DUR (*Professional only*)	modified duration	This is defined in Chapter 9 and is also called volatility.

FIGURE (6.11.1) BAII Plus and BAII Plus Professional Bond worksheet

When a bond has more than one coupon period until redemption, the dirty price used in computing the price is computed using compound interest at interest rate $\tilde{\jmath} = \dfrac{\tilde{I}}{100m}$ per coupon period, where $\tilde{I}$ is the number put in the YLD register. That is, it is the present value of the remaining coupons and interest, figured using compound interest at interest rate $\tilde{\jmath}$. When the bond has less than one coupon period until redemption, rather than using compound interest to compute the present value of the remaining coupon plus the redemption payment, it is customary to use <u>simple interest at rate $\tilde{\jmath}$ from the time of the resale</u>. That is, we have:

FACT (6.11.2) When the bond has less than one coupon period until redemption, the clean price used in the Bond worksheet is

$$\frac{C + Cg}{1 + (1 - f)\tilde{j}} - fCg,$$

rather than the semipractical clean price $\frac{C+Cg}{(1+\tilde{j})^{1-f}} - fCg$ as would result using compound interest .

With settlement less than one coupon before redemption, if you choose to enter a numerical entry $Z = \frac{100X}{F}$ in the PRI register, then the calculator "interprets" it as $100\left(\frac{D_T^{\tilde{j}}-A_T^{prac}}{F}\right)$, then calculates $\tilde{I} = 100m\,\tilde{j}$ based on the equation

$$(6.11.3) \qquad \tilde{j} = \frac{(C + Cg) - (X + fCg)}{X + fCg} ? \frac{1}{1 - f} .$$

EXAMPLE 6.11.4   Bond purchased with <u>less</u> than one coupon payment remaining

Problem: Suppose a $1,000 5% bond with semiannual coupons and a redemption payment of $1,320 was purchased on March 11, 2005 at a price to yield the new buyer a nominal rate of 6% convertible semiannually. Further suppose that the redemption payment was due on June 13, 2005, and that the bond was a municipal bond so that day counts should be accomplished by the "30/360" method. Explain how the BA II Plus calculator **Bond worksheet** may be used to calculate the price paid at the March 11th settlement and the clean price for the sale. Check your answer by computing the price without using the Bond worksheet.

Solution Key 2ND BOND to open the Bond worksheet and display the last used settlement date. Substitute March 11, 2005 by keying 3 • 1 1 0 5 ENTER . Then key ↓ 5 ENTER to enter a coupon rate of 5 (percent). Next enter the redemption date June 13, 2005 by pushing ↓ 6 • 1 3 0 5 ENTER . The bond is redeemed at $132 per each $100 of face value, so you should fill the RV register with 132 by keying ↓ 1 3 2 ENTER . Pushing ↓ advances you to the basis for counting days and pushing ↓ again will move you to display the number of coupons. These should be set at "360" and "2/Y", and if either of these needs changing, depress 2ND SET before you scroll past. Push ↓ to arrive at the YLD register and then key 6 ENTER ↓ CPT . This should result in PRI $= 131.2465893$ being displayed. The price paid is the sum of this and the practical accrued interest AI $= 1.2222222222$. The latter is found by pushing $\boxed{\downarrow}$ yet again. To sum these two values, you should store one of them, then scroll to the other and perform the addition. The sum 132.4688116 represents the price per $100 of face value. So, the price of the $1,000 bond is $1324.69.

It is not difficult to check this without using the **Bond worksheet**. First note that on a "30/360" basis, March 11, 2005 – June 13, 2005 is two days over three months, and is therefore counted as 92 days. Furthermore, the bond was given to have semiannual coupons so, again on a "30/360" basis, the coupon period consists of 180 days. So, $180 - 92 = 88$ days of the coupon period have elapsed prior to settlement, and $f = \frac{88}{180}$. Note that $F = \$1,000$, $r = \frac{5}{2} = .025$, and the coupon payments are each for $25. Since the redemption amount is $1,320 and the bond is sold to provide the buyer a nominal yield of 6% convertible semiannually, Fact (6.11.2) tells us that the applicable clean price is

$$\frac{C + Cg}{1 + (1 - f)\tilde{\jmath}} - fCg = \frac{\$1,320 + \$25}{1 + \left(\frac{92}{180}\right)\left(\frac{.06}{2}\right)} - \left(\frac{88}{180}\right)(\$25) \approx \$1,312.47.$$

The invoice (dirty) price is $\frac{\$1,320+\$25}{1+\left(\frac{92}{180}\right)\left(\frac{.06}{2}\right)} \approx \$1,324.69.$ ∎

The bond of Example 6.11.4 was a municipal bond, so we used the "30/360" method when figuring the fraction of the coupon period that has elapsed. It is worth noting that if everything in the problem were unchanged except the bond were a government bond or other bond for which the "actual/actual" basis were applicable, we would change the day-count-method by keying $\boxed{\text{2ND}}$ $\boxed{\text{SET}}$. If we then scrolled to the AI entry, it would display "AI $= 1.208791209$". But this should be equal to $25f$, so the calculator has used $f \approx .048351648$ and $\frac{88}{182} \approx .048351648$. The actual number of days from December 13, 2004 to June 13, 2005 is 182, and there are 88 days from December 13, 2004 to March 11, 2005.

EXAMPLE 6.11.5   Bond purchased with <u>more</u> than one coupon payment remaining

Problem: Suppose a $2,500 8% bond with annual coupons and a redemption payment of $2,800 was purchased on October 3, 1998 for $2,912. Further suppose that the redemption payment was due on May 30, 2004, and that the bond was a government bond so that day counts should be accomplished by the "actual/actual" method. Explain how the BA II Plus calculator **Bond worksheet** may be used to calculate the yield to the October 3rd buyer. Also discuss how this calculation may be accomplished without the **Bond worksheet**.

Solution Key $\boxed{\text{2ND}}$ $\boxed{\text{BOND}}$ to open the Bond worksheet and display the last used settlement date. Enter October 3, 1998 by pushing $\boxed{1}\boxed{0}\boxed{\bullet}\boxed{0}\boxed{3}\boxed{9}\boxed{8}$ $\boxed{\text{ENTER}}$. Then key $\boxed{\downarrow}\boxed{8}\boxed{\text{ENTER}}$ to enter a coupon rate of 8 (percent), followed by $\boxed{\downarrow}\boxed{5}\boxed{\bullet}\boxed{3}\boxed{0}\boxed{0}\boxed{4}\boxed{\text{ENTER}}$ to enter the redemption date. The redemption is $\frac{28}{25}$ times the par-value and the Bond worksheet is designed to price bonds with $100 face value, so we wish to enter $\left(\frac{28}{25}\right)(\$100)$ in the RV register. This is accomplished by the sequence $\boxed{\downarrow}\boxed{2}\boxed{8}\boxed{\div}\boxed{2}\boxed{5}\boxed{\times}\boxed{1}\boxed{0}\boxed{0}\boxed{=}$ $\boxed{\text{ENTER}}$. Now set the day-count basis to "ACT" and the number of coupons per year to "1/Y". Scroll down to the AI entry. You should find "AI = 2.761643836". Store 2.761643836 in an available numbered register, say register m, by keying $\boxed{\text{STO}}\boxed{m}$. We have now stored the accrued interest for a $100 face bond in register m. Since we were given that the dirty price of a $2,500 bond is $2,912, the dirty price of a $100 bond is $\frac{\$2,912}{25} = \$116.48$, and its clean price $\$116.48 - \$2.761643836$. We enter this in the PRI register by scrolling to PRI, then keying $\boxed{1}\boxed{1}\boxed{6}\boxed{\bullet}\boxed{4}\boxed{8}\boxed{-}\boxed{\text{RCL}}\boxed{m}\boxed{=}\boxed{\text{ENTER}}$. Now, push $\boxed{\uparrow}\boxed{\text{CPT}}$ to compute the yield. The buyer's annual yield rate is about 6.795493346%.

This same rate may be obtained without relying on the **Bond worksheet**. We note that $F = \$2,500$ and $r = 8\%$, so the annual coupon payments are each $200. The price $2,912 is the sum of the October 3, 1998 values of the remaining six May 30th coupons and of the $2,800 redemption on May 30, 2004, figured by compound interest at the buyer's yield $\tilde{\jmath}$. This is equal to $(1 + \tilde{\jmath})^{f-1}\left(\$200\ddot{a}_{\overline{6}|\,\tilde{\jmath}} + \$2,800(1 + \tilde{\jmath})^{-5}\right)$ where f is the fraction of the coupon period May 30, 1998 May 30, 1999 that has elapsed at the October 3, 1998 settlement date, calculated on an "actual/actual" basis. There were 126 days from May 30, 1998 to October 3, 1998 and the year beginning on May 30, 1998 had 365 days, so $f = \frac{126}{365}$. We might then use the "guess and check" method to determine the yield rate $\tilde{\jmath}$. This may be avoided by using the **Cash Flow worksheet** to find a daily yield rate, which you then can convert to an equivalent annual yield. Noting that there is an extra day in the coupon periods that include a February 29th, you should enter

CFo $= -2,912$, C01$=0$, F01$= 238$ ($= 365 - 126 - 1$),
C02 $= 200$, F02 $= 1$, C03$=0$, F03$= 365$, C04 $= 200$, F04$=1$,
C05$=0$, F05$= 364$, C06 $= 200$, F06$=1$, C07$=0$, F07$= 364$,
C08 $= 200$, F08 $= 1$, C09$=0$, F09$= 364$, C10 $= 200$, F10 $= 1$,
C11$=0$, F11$= 365$, C12 $= 200 + 2,800 = 3,000$, and F12 $= 1$.

Pushing $\boxed{\text{IRR}}\boxed{\text{CPT}}$ results in "IRR = 0.017997019". This is a daily yield rate.

To obtain the equivalent annual yield, we need the average number of days in a year for the interval October 3, 1998 - May 30, 2004, but we will not be far off if we use 365.25. Since $(1.00017997019)^{365.25} - 1 \approx 6.793641073\%$, we again find that the yield is close to 6.79%.[7]   ∎

6.12   PROBLEMS, CHAPTER 6

(6.0) Chapter 6 writing problems

(1) The bond pricing formula (6.4.4) is named for the British actuary William Matthew Makeham. Write a short biography of him. Mention of Makeham's scholarly contributions is appropriate.

(2) Learn how to read a bond chart, and write a short explanation of how this is done. (A Google search can help you find this information.)

(3) Learn about income and capital gains taxes that apply to bond earnings and report your findings.

(6.2) Bond alphabet soup and the basic price formula

(1) A $1,000 10% ten-year bond has semiannual coupons. It is purchased new at $880 and is redeemable at $1,020. Find the coupon amount and the effective yield rate per coupon period.

(2) A $2,500 6.5% eight-year bond has annual coupons. If it is purchased for $2,590, the investor will anticipate a 5.4% annual yield for the eight-year investment. Find the redemption amount on this bond.

(3) A 6% $1,000 par-value bond maturing in eight years and having semi-annual coupons is to be replaced by a 5.5% $1,000 par bond, also with semiannual coupons. Both bonds are bought to yield 5% nominal interest convertible semiannually. In how many years should the new bond mature? (Both bonds have the same price as well as the same yield.) Answer to the nearest half-year.

(4) An investor will pay $2,318.63 for an n-year $2,000 par bond with a coupon rate of 10% compounded semiannually or he will pay $2,531.05 for an n-year $2,000 par bond with a coupon rate of 11% compounded semiannually. Assuming that the investor gets the same yield on the two bonds, find this yield rate expressed as a nominal rate convertible two times per year. Also find n.

(5) An investor owns a $3,000 par-value 12% bond with semiannual coupons. The bond will mature at par at the end of fourteen years. The investor

---

[7]The yearly rate will be equal to the rate produced by the Bond worksheet if we use 365.3463816 instead of 365.25. The interval October 3, 1998 - May 30, 2004 includes two February 29th's so the estimate 365.25 is slightly lower than the average number of days in a year during the period October 3, 1998 - May 30, 2004.

decides that a ten-year bond would be preferable. Current yield rates are 6% convertible semiannually. The investor uses the proceeds from the sale of the 12% bond to purchase an 8% bond with semiannual coupons, maturing at par at the end of ten years. Find the face value of the 8% bond.

(6) Christie DeLeon purchased a ten-year $1,000 bond with semiannual coupons for $982. The bond had a $1,100 redemption payment at maturity, a nominal coupon rate of 7% for the first five years, and a nominal coupon rate of $q\%$ for the final five years. Christie calculated that her annual effective yield for the ten-year period was 7.35%. Find q.

(7) Martin Maradiaga was considering two bond offerings for purchase on March 1, 1995. Each had a purchase price of $10,000. Bond A was an "inflation-adjusted" 4% ten-year $10,000 bond with annual coupons; the coupon payments were to be based on March 1, 1995 dollars so that the inflation-adjusted coupon rate was 4% and the bond would be redeemable at an amount worth $10,000 in March 1, 1995 dollars. Bond B was a 7% ten-year $10,000 par-value bond with annual coupons offered by Delta Diagnostics. Which should Mr. Maradiaga purchase if he forecasts that inflation will be at a level rate of 2.75%? Why? If inflation is actually at 2.2%, find the inflation-adjusted yield on each bond.

(6.3) The premium-discount formula

(1) A $3,000 9% twelve-year bond with annual coupons is purchased with a discount of $57 and yields 9.1% if held to maturity. Find the price.

(2) A $2,000 11% ten-year bond has semiannual coupons and is sold to yield 5.2% convertible semiannually. The discount on the bond is $83.28. Find the redemption amount.

(3) Alicia bought a newly issued $1,000 20% ten-year bond, redeemable at $1,100 and having yearly coupons. It was bought at a premium with a price of $1,400. Alicia immediately took a constant amount D from each coupon and deposited it in a savings account earning 8% effective annual interest, so as to accumulate the full amount of the premium by a moment after the final deposit. How much did Alicia deposit each year in the 8% account?

(6.4) Other pricing formulas for bonds

(1) A $1,000 bond with a coupon rate of 8% has quarterly coupons and is redeemable after an unspecified number of years at $957. The bond is bought to yield 12% convertible semiannually. If the present value of the redemption amount is $355.40, find the purchase price using the

Makeham formula. Then check your answer using another price formula.

(2) A $20,000 bond has annual coupons and is redeeemable at the end of fourteen years for $22,600. It has a base amount equal to $18,450 when purchased to yield 6%. Find its base amount if it were purchased to yield 7%.

(3) Lucia Gabrielli purchased an n-year par-value $2,000 bond that had a coupon rate of 9% convertible quarterly; here $4n$ is an integer. Her sister Elana purchased a par-value bond with an identical coupon rate but having a term of $2n$ years. The coupons that Lucia and Elana received for the n years they both held their bonds were identical, and the yield rate used to determine the prices of each of the bonds was a nominal rate of 6% convertible quarterly. Elana paid $233.02 more than Lucia. Calculate the number of coupons in Lucia's bond; use the premium-discount equation.

(6.5) Bond amortization schedules

(1) A $2,000 bond with semiannual coupons is redeemable for $2,100 in fifteen years. It has a coupon rate of 6.5%. The bond is purchased to yield 8% per annum compounded semiannually. Find the price of the bond, the amount for accumulation of discount in the tenth coupon, and the amount of interest in the tenth coupon payment.

(2) A bond with a face value of $6,000 and an annual coupon rate of 12% convertible semiannually will mature in ten years for its face value. If the bond is priced using a nominal yield rate of 6% convertible semiannually, what is the amount of premium in this bond? What is the amount for amortization of premium in the 7th coupon?

(3) A fifteen-year bond, which was purchased at a premium, has semiannual coupons. The amount for amortization of the premium in the second coupon is $977.19 and the amount for amortization of premium in the fourth coupon is $1,046.79. Find the amount of the premium.

(4) An n-year $2,000 par-value bond with 9% annual coupons has an annual effective yield of i, $i > 0$. The book value of the bond at the end of the fifth year is $1,902.63 and the book value of the bond at the end of the seventh year is $1,924.18. Find the price of the bond.

(5) A three-year $1,000 6% bond with semiannual coupons has redemption amount $1,040. Make amortization tables for this bond if it is bought to yield a nominal rate of 5% convertible semiannually respectively. Repeat for a nominal rate of 6% and then for a nominal rate of 7%, each convertible semiannually.

(6.6) Valuing a bond after its date of issue

(1) A $2,500 14% six-year bond with annual coupons is bought to yield 6% annually. The price is $3,432.26. Find its clean and dirty values at the end of each quarter of the fourth year after issue, by the practical method and also by the theoretical method.

(2) On April 30, 1990, April purchased a $1,000 10% par-value seven-year bond having semiannual coupons; these were payable at the end of each October as well as on the anniversaries of the purchase. April paid $1,120. On July 18, 1993, she wished to know the dirty and clean values of this bond, figured using the theoretical method and again by the practical method. Calculate them all for her, using the "actual/actual" method for figuring day counts.

(3) As in Problem (6.5.5), we are concerned with a three-year $1,000 6% bond with semiannual coupons and a redemption amount $1,040. Suppose that the bond was purchased on January 1, 2000. Make a chart showing the theoretical and practical dirty and clean values of the bond at the end of each quarter if the bond was purchased at a discount to yield a nominal rate of 7% convertible semiannually. Use the "30/360" basis for counting days.

(6.7) Selling a bond after its date of issue

(1) Miguel purchases a $22,000 9% fifteen-year par-value bond having annual coupons for a price to provide a 7% annual yield if the bond is held to maturity. Five years later, just after the receipt of the fifth coupon, he sells it at a price to provide the new purchaser a yield to maturity of 8%. Find the difference between Miguel's book value B_5 and the invoice price. What was Miguel's actual yield for the five-year period?

(2) A $1,000 seven-year 6% bond with semiannual coupons is redeemable for $1,065. It was originally purchased at issue for $970. It is sold after 45 months for $995. Find the accrued interest by the practical method and again by the theoretical method using the new yield to maturity.

(3) On May 27, 1994 Jen Mago purchased a new $18,000 fifteen-year 10% bond with annual coupons and a redemption payment of $19,000. Jen sold the bond to Edna Wilder on December 31, 2000. The semipractical clean price for the sale was $18,375; this was based on her market price and the "30/360" method for counting days. Still working on a "30/360" basis, find her annual yield rate $\tilde{j}$ and theoretical clean price $C_T^{\tilde{j}}$.

(4) Jiayin purchases a ten-year 8% $62,000 par-value bond with annual coupons eighty months after its issue. The original purchaser paid a price to yield 8.5% if the bond was held until maturity, as does Jiayin.

Compute the accrued interest by the theoretical method at 8.5% and also by the practical method. Find the split of the accrued interest by the theoretical method at 8.5% into interest and principal.

(5) (calculus needed) Consider the difference between the accrued interest by the theoretical method (at a fixed positive yield rate $\tilde{\jmath}$) and by the practical method. Find an expression, in terms of the interest rate $\tilde{\jmath}$, for the time when the difference is largest. That is to say, for what fraction f of an interest period is it largest? Calculate for $\tilde{\jmath} = 1\%$, $\tilde{\jmath} = 7\%$, and $\tilde{\jmath} = 21\%$.

(6.8) Yield rate examples

(1) A twenty-year par-value bond has 12% annual coupons and a par-value of $2,000. Coupons can be reinvested at a nominal interest rate of 6% convertible semiannually. P is the highest price that an investor, who reinvests each of the coupons, can pay for the bond and obtain an effective yield rate of at least 8%. Find P.

(2) An investor wishes to have an annual (effective) yield of 7%. With this goal in mind, he purchases a twelve-year $1,000 par-value bond with 10% coupons payable quarterly. The price he pays for this bond is based on the investor earning an annual yield of exactly 7% and the assumption that the reinvestment annual effective rate of interest on coupons will be 7%. In fact, the investor only earns 4% nominal interest convertible quarterly on the coupons, which are each reinvested at the moment they are paid. What is the investor's actual annual yield rate?

(3) Mitch bought a newly issued $1,000 par-value 13% eight-year bond with semiannual coupons. The bond was priced to yield a nominal 9%, convertible semiannually, so Mitch paid a premium. Mitch immediately took a constant amount D from each coupon and deposited it in a savings account with a 6% annual effective interest rate. This caused his actual yield to be less than 9%. The amount D was as small as possible so that the balance in the account immediately after the final deposit was at least equal to the premium. Find D and Mitch's annual yield for the eight-year period.

(4) A $3,000 par-value bond with 6% annual coupons is purchased at a premium ten years prior to its maturity date. The proceeds of the coupons are invested in a savings account with a 4% annual effective rate of interest. The annual effective yield on the ten-year investment (including the bond and the savings account) is 5%. What is the amount for amortization of premium in the third coupon the investor receives?

(5) On March 1, 1990 Juanita paid $6,317 to acquire a portfolio of six $1,000 par-value bonds. All the bonds had annual coupons. The portfolio

consisted of three 12% bonds with redemption dates of March 1, 1992, 1994, and 1996 and three 10% bonds with redemption dates of March 1, 1993, 1995, and 1997. Find Juanita's yield rate.

(6) Maureen purchased a laddered portfolio of par-value bonds with semi-annual coupons for $20,918. The bonds had common face amount X and coupon rate 6.8%. There was one bond redeemable at the end of each year for ten years. Maureen's yield on the portfolio was an annual effective rate of 8%. Find X.

(7) Eric purchases a $8,000 fifteen-year 7% bond with semiannual coupons and a redemption value of $9,000. After twenty-three months, he sells the bond to Pierre who holds the bond for seven years and then sells it to Irene for the same price at which he bought it. She holds the bond until maturity. The nominal yield rates convertible semiannually earned by Eric and Irene are 3.6% and 3.8% respectively. Find the price for the bond at issue and the price Pierre paid for it.

(6.9) Callable bonds

(1) Dominique LeBlanc is the owner of a new ten-year $50,000 8% par-value bond with a Bermuda option and annual coupons. Allowable call dates are at the ends of years 6 through 10, and the call premium at the end of year n is $300(10 - n)$. Dominique purchased the bond for $51,248.

   (a) Find the lowest yield that Dominique may receive during the period she holds the bond as well as the highest.

   (b) Upon receipt, Dominique deposits each coupon and the redemption amount in an account earning 6%. Find the lowest yield that Dominique may receive during the ten-year period and also the highest.

(2) Drew Jefferson purchases a new $20,000 9% twelve-year bond with semiannual coupons. If held to maturity, the redemption payment is $18,500, and the bond would yield Mr. Jefferson 8% convertible semiannually. The bond has an American option and is callable beginning at three years from issue. If the bond is called at time T, where T is measured in coupon periods, the call premium is $p(T)$. Find an expression for the amount $p(T)$ so that Mr. Jefferson's yield is 8% no matter when the bond is called.

(3) Sofia purchases a $6,000 7% eight-year par-value bond with annual coupons. If held to maturity, her yield is 6.6%. The bond is callable at the end of two years for $6,300 and at the end of years five, six, and seven for $6,200.

   (a) Find Sofia's minimal yield for the period she holds the bond.

(b) If the bond is called prior to maturity, Sofia has made arrangements to have the redemption amount accumulate in her $i = 5.5\%$ savings account until the end of the eight years. What is her minimal yield for the period she holds the bond? (The coupons are not reinvested.)

(4) Prove Claim (6.9.3). It may help you to look at the proof of (6.9.2).

(6.10) Floating-rate bonds

(1) A five-year $2,000 floating-rate bond with annual coupons had a $2,125 redemption amount and was sold at its face value. The coupon rates were 7%, 6.4%, 5.8%, 6.2%, and 6.6%, successively. What level coupon rate would result in the same yield rate to the investor for the five-year period?

(2) Jorge has carefully studied the prospectus for a ten-year $10,000 floating-rate par-value bond with annual coupons. He anticipates that the coupon rates will be level at 5.5% for the first five years, then go up by a factor of 1.04 each year. How much should he be willing to pay for this bond if he wishes a yield rate of at least 7% for the ten-year investment period?

(6.11) The BAII Plus calculator Bond worksheet

(1) Suppose a $1,500 5% bond with annual coupons and a redemption payment of $1,650 was purchased on February 23, 1995 at a price to yield the new buyer 8.2% annually. Further suppose that the redemption payment was due on August 1, 1995 and that the bond was a corporate bond so that day counts are made using the "30/360" method. Use the BA II Plus calculator **Bond worksheet** to calculate the price paid at the February 23rd settlement and the accrued interest included in the price. Check your answers without using the Bond worksheet.

(2) A $6,000 10% municipal bond with semiannual coupons and a redemption payment of $5,500 was purchased on November 13, 1988 at a price to yield the new buyer 11.5% nominal convertible semiannually. The redemption payment is due on December 2, 1992. Use the BA II Plus calculator **Bond worksheet** to calculate the price paid at the November 13th settlement and the accrued interest included in the price. Check your answers without using the **Bond worksheet**.

(3) A $3,000 6% government bond with semiannual coupons was traded on August 11, 2002. It is redeemable on December 15, 2002. Determine the amount of accrued interest in the invoice price. If the bond is purchased to yield 4.5% convertible semiannually, find the invoice price.

Chapter 6 review problems

(1) A $1,000 6% n-year par-value bond has annual coupons. Tabitha bought the bond to yield 5%. The amount of interest in the first coupon is $52.89. Calculate the amount of premium Tabitha paid for the bond.

(2) On March 2, 2000, Spencer purchased a $1,000 6% four-year bond having annual coupons and a $950 redemption. He paid $925. Find the July 5, 2002 dirty and clean values of this bond by the theoretical method, using an "actual/actual" count of days.

(3) An n-year $1,000 par-value bond with 8% annual coupons has an annual effective yield of i, $1 + i > 0$. The book value of the bond at the end of the third year is $990.92 and the book value of the bond at the end of the fifth year is $995.10. Find the price of the bond.

(4) An n-year $1,000 par-value bond with 8% annual coupons has an annual effective yield of i, $1 + i > 0$. The book value of the bond at the end of the third year is $990.92 and the book value of the bond at the end of the fifth year is $995.10. Find the price of the bond.

(5) On October 30, 1978, Mr. Cole purchased a laddered portfolio of eight $2,000 par-value bonds. Each bond in the package had annual coupons and a coupon rate of 7.3%. The redemption dates were October 30 of years 1980–1987. The package was priced to give Mr. Cole an 8% annual yield rate. Find the price of the portfolio.

(6) On July 6, 2004 Gayle purchased a $25,000 8% par-value bond with semiannual (March 3 and September 3) coupons and a September 3, 2007 maturity. How much did she pay if she purchased the bond to yield a nominal rate of 5.8%? What part of this price consisted of accrued interest? Answer this four times, using the theoretical and practical methods, along with the "actual/actual" and the "30/360" methods.

(7) An eight-year $5,000 6.75%, par-value bond with semiannual coupons has a one-time put option at the end of five years. The bond is sold for $5,250. The purchaser has no pressing needs for money and all cashflows received (coupon and redemption) are reinvested until the end of the eight-year period. Any money reinvested earns interest at an annual effective interest rate of 5.5% during each of the first three years and at an annual effective rate I for each of the remaining five years. For what values of I should the bondholder exercise the put option?

CHAPTER 7

Stocks and financial markets

7.1 COMMON AND PREFERRED STOCK

Corporations issue bonds to raise money. As we discussed in Chapter 6, bonds are certificates of indebtedness. Another way for a corporation to raise capital is for it to sell off assets. Notable among its assets are corporate ownership, and that is precisely what you get when you purchase shares of **common stock**. Each share of common stock represents a claim to a tiny bit of the company. If you own shares of common stock, you have **equity** — that is ownership — in the issuing corporation.

How might you benefit from possession if you own common stock? When a company makes money, as a holder of common stock in the company, you may receive a part of the profit in the form of a **dividend**. However, it is also possible that the corporation will hold on to whatever profits it can, with an eye to making further profits. If the value of a corporation in which you own common stock increases, you will experience capital appreciation, in the sense that should you decide to sell your shares, the price should be higher. As a common stock shareholder, you will have voting rights and therefore a say in company policy, but you will not have the opportunity to vote on whether a dividend should be declared or its amount (although you may vote for members of the the Board of Directors, the body that makes decisions on dividends). In summary, you might purchase common stock in a company because you believe that the corporation will increase in value or that it will pay attractive dividends. If the company loses money, your share will decline in value.

A second type of equity investment that a corporation may issue is **preferred stock**. If you purchase shares of preferred stock, then you are buying the right to certain future earnings, again called **dividends**, as spelled

316

out in the stock offering. In the United States, stock dividends are usually paid quarterly. Preferred stock dividends are most often fixed, although **participating preferred stock** pays extra dividends if the company has sufficient profits, and **adjustable preferred stock** pays dividends that change each period based on some set formula: Most likely the size of the dividend depends on the yield earned by holders of Treasury bills or on some other market rate.

The payment of dividends on preferred stock is guaranteed so long as the company has first paid its creditors. So, preferred stockholders may receive dividends when common stockholders receive none; thus, the name "preferred." In addition, if the preferred stock is **cumulative**, should dividends fail to be paid for some period, they accrue until the company can pay them. In case the issuing company declares bankruptcy, once again bondholders have priority over stockholders, but preferred stockholders' claims are met prior to common stockholders'. To compensate for the fact that preferred stockholders have less potential for capital appreciation than the holders of common stock, the dividends paid on preferred stocks tend to be higher than those received on common stock. In general, the performance of the issuing company has much less effect on the price of preferred stock than it does on common stock prices.

Some issues of preferred stock are **convertible**. This means that there is a time period during which they may be exchanged for common stock, the trade accomplished by a formula spelled out when the preferred stock shares are issued. Purchasers of preferred stocks should be careful to note what options the company has for buying back the shares of stock. The preferred stock may be **callable**, commonly after five years, at its issue price. As you can see, while preferred stocks are equity assets (representing ownership) rather than debt obligations, they have much in common with bonds. However, since a stock is a certificate of ownership, it does not have a maturity date.

EXAMPLE 7.1.1   Preferred stock

Problem: At the beginning of a quarter, Shermann DePew purchases 100 shares of preferred stock at a price to provide him with a yield rate equal to 5.5%, assuming all dividends are paid. The stock is nonadjustable and nonparticipating so the dividends are fixed at the level stipulated at issuance, namely $.36 per share each quarter. How much does Mr. DePew pay for the stock?

Solution   Since Mr. DePew purchases 100 shares, he will receive $36 at the end of each quarter. Thus, he should pay an amount $X such that $36 = $((1.055)^{\frac{1}{4}} - 1)X$. That is, $X = \dfrac{\$36}{(1.055)^{\frac{1}{4}} - 1} \approx \$2{,}671.58$. Note that this is the present value of a perpetuity-immediate paying $36 each quarter.   ■

Pricing by looking at the present value of the dividends, as in Example (7.1.1), is sometimes referred to as the **dividend discount model**.

Stock ownership gives you equity in a company. Of course it is possible that the company will go bankrupt. Then, you may lose your total investment or receive only a small fraction of what you paid for it. However, the ownership of common or preferred stock is a limited liability. The most you can lose is the amount you invested.

Stocks and bonds are examples of **securities**, legal claims representing financial value whose trade takes place in financial markets. These markets may be physical locations, called **exchanges**, or trades may be accomplished over the telephone or computer. These latter transactions are said to be **over the counter** (OTC). In the United States, the major national exchange where stocks are traded is the **New York Stock Exchange** (**NYSE**). The **American Exchange (AMEX)** and regional exchanges are of considerably lesser importance. Many stock transactions now take place over the counter using the network of computers known as **NASDAQ** (**National Association of Security Dealers Automated Quotation system**). Like the exchanges, NASDAQ has strict rules as to how trades may be performed. These are set out by the **National Association of Security Dealers** (**NASD**), the largest self-regulatory group of security dealers in the United States. Additionally, the **Securities and Exchange Commision (SEC)**, which was created by the Security and Exchange Act of 1934, regulates American security trading. The SEC is in charge of ensuring that potential investors are given access to information on all publicly traded securities.,

There are many newspapers and Internet sites that list stock prices. Readily available information also includes several indices that try to represent the health of the market, or a segment of the market, with a single number. In particular, you may have heard of the **Standard and Poor's 500** (**S&P500**), the **Dow Jones Industrial Average** (**DJIA**), and the **Nasdaq index**.[1]

The DJIA, introduced in 1896, is the oldest of these and its calculation is rather simple. Currently based on the stock of thirty very large companies that are subjectively selected by the *Wall Street Journal*'s managing editor and index editor, the DJIA is computed by adding the thirty share prices and dividing by the **Dow divisor**. The purpose of the divisor is to account for stock splits in the thirty component stocks, thereby allowing historical continuity.[2] Since the DJIA is proportional to the sum of the thirty share prices, a 1%

---

[1] The index, unlike the acronym NASDAQ, has lowercase letters after the first "N". Thus, Nasdaq refers to an index, NASDAQ to a network of computers.

[2] To understand the use of the divisor, consider a simplified example with three stocks trading at \$10, \$17, and \$27. Suppose that the \$27 stock undergoes a three-for-one stock split so that each share of the \$27 stock is replaced by three \$9 shares. Then the average of the share prices changes from $\frac{\$10+\$17+\$27}{3} = \$18$ to $\frac{\$10+\$17+\$9}{3} = \$12$ without any market forces at work. This is avoided if we change the second denominator (or divisor) from 3 to 2.

change in a share with a high price has a bigger impact than a 1% impact in the price of a stock with a lower share price.

In contrast, the S&P500 and Nasdaq are each **market-weighted** indices. This means that each stock's weight in these indices is proportional to the product of the share price by the number of shares outstanding. So, Microsoft may account for 3% of the index while a small company might account for less than .1% of the index. The S&P500 is based on five hundred stocks chosen for "market size, liquidity, and group representation." The index only uses American companies and excludes real-estate stocks and companies that mainly hold stock in other companies. The Nasdaq is based on the more than four thousand stocks that trade on the NASDAQ. While it is hard for an individual to own stocks in all the companies included in the indices, for the interested investor there are financial **index funds** that hold the entire index portfolio.

The legal claim granted by stock ownership is represented by a stock certificate. In this computer age, these fancy pieces of paper are usually not passed on to you when you buy stock. Rather, the brokerage firm (perhaps an online entity) that handles the trade usually holds the certificate and your claim is reflected on a brokerage statement. This eliminates the need for the seller to deliver the stock certificate so that a sale may be completed. The stocks are likely to be registered "in street name" rather than in your name although you may ask that they be registered in your name. This facilitates the brokerage firm borrowing its clients' stocks for the use of another client. [See Section (7.4) on short sales.] If your shares are held in street name, you may have to request your voting rights if you wish to exercise them.

The fact that you likely do not hold your stock certificates makes insurance against brokerage house failure especially important. Just as many banks have deposits covered by Federal Deposit Insurance Corporation (**FDIC**) insurance, clients of covered brokerage firms receive protection from the Securities Investor Protection Corporation (**SIPC**). Customers are insured for losses up to $500,000 with cash loss limited to $100,000. Additional private insurance is offered by some brokerage firms.

Now that you know what stocks and bonds are, we can tell you briefly what a mutual fund is. **Mutual funds** are investment companies that allow investors to invest their pooled money in a variety of securities. The investors are shareholders and each owns a part of the company. Generally, mutual funds sell their shares on a continuous basis, and mutual fund shares may be redeemed upon demand by the investor.[3] A mutual fund's portfolio is likely to contain a mixture of stocks, bonds, money market instruments, and

---

[3] **Closed-end funds** also are actively managed companies that pool money for security investments, but they trade on exchanges. A third type of investment company, a **unit investment trust**, has a fixed composition of redeemable securities, and is not actively managed.

other securities including derivative securities that are studied in Chapter 8. The particular objective of a mutual fund will help dictate the company's investment choices, and these goals will impact the level of risk. Examples of mutual fund objectives include steady income production and targeting a particular industry or type of company, say rapidly growing companies. Mutual funds offer investors the advantages of diversification and professional management. In the United States, regulation of mutual funds is generally provided by the SEC under the Investment Company Act of 1940.

7.2 BROKERAGE ACCOUNTS

If you are going to invest in stocks, you will almost certainly need to deal with an intermediary. The reason for this is that it would be a rare situtation in which you have direct access to the stocks you wish to purchase. You may not enter the trading floor on an exchange and even trading on NASDAQ requires a representative.

There are fees or commissions that are charged by the intermediary but we will largely ignore these in our discussion. For simplicity, we will refer to the intermediaries as "brokers" or "brokerage firms." Actually, you may deal with either a **broker** who just acts as your agent and collects a commision for services rendered, or with a **dealer** who holds an inventory of stocks. When you use a dealer, you may buy from the dealer's inventory or have the shares you sell added to this inventory. A dealer makes money by, at any time, buying shares at a lower price than he or she sells those shares for. This price difference is called a **bid-ask spread**. Many brokerage firms act as a broker on some trades and a dealer on others. With the advent of discount brokerages and the internet, many of the costs associated with trading have decreased considerably in recent years. In general, fees (per share traded) are higher for small investments than for large ones.

In order to be able to invest in stocks, you will need to open a **brokerage account**. If you select a **cash brokerage**, you pay for the securities you purchase with money from the account, and when you sell securities, the proceeds (less any commission) are credited to your account. You may withdraw money from your cash account, but the account does not serve as an instrument for borrowing money. In contrast, a **margin account** allows you to borrow from your broker, using assets in the account (money or securities) as collateral. The loan may be for the purpose of purchasing securities, which you then place in the account, but borrowing may not be restricted to this use.

When you use a margin account for a stock purchase, you are borrowing part of the money used to purchase the shares. The brokerage firm will hold onto the purchased shares for you. The Federal Reserve Board (FRB) imposes the **initial margin requirement** that, at the time of purchase, you must deposit a certain proportion m of the value of the securities purchased on margin into your margin account. This is your downpayment. [You may satisfy your downpayment with securities instead of with cash, but only a fraction of the value of the securities (perhaps again m) is counted: So the market value of securities needed to meet the initial margin requirement is greater than the cash amount needed.] The rate m has been constant at 50% from 1974 to 2006, and it has ranged from a low of 40% to a high of 100% since the Securities Exchange Act of 1934 required federal regulation of the buying of securities on margin. Furthermore, you must have on deposit at least $2,000[4] in your margin account. The amount that you deposit is your **initial equity**.

The loan balance is called the **debit balance** or **net debit amount**, and it is usually updated daily. Changes in the debit balance result from interest assessed by the broker, cash deposits or withdrawals (if allowed), and from trading securities. Of course the market value of the securities you hold is subject to frequent change, but this value does not affect your debit balance. In general, the difference between the market value and the net debit balance is referred to as **equity**. Equivalently,

$$\text{Market Value} = \text{Net Debit Balance} + \text{Equity}.$$

In addition to the FRB's initial margin requirement, the NYSE and NASD further impose a **maintenance margin** requirement: They insist that the equity in the account is at least 25% of the market price of the securities bought on margin. In fact, many brokerage houses impose higher maintenance margins (such as, 30%), and margin requirements may vary based on the size of the account, the type of securities purchased, and whether your account is diversified.

EXAMPLE 7.2.1   Margin requirements

Problem: Elvira Hamilton purchases 360 shares of a stock that is currently selling at $57.82 per share using her margin account. Her purchase is subject to the FRB's initial margin requirement and a 35% maintenance margin requirement. What is the dollar amount of the FRB's initial margin requirement? What is the highest per share price at which she would fall below the 35% maintenance margin requirement, assuming that the initial margin requirement is satisfied by a cash deposit and therefore has a constant dollar value? Assume that any interest on the margin account is immediately paid in cash so that the debit balance does not increase. Also assume that if the borrowed stock pays a dividend, which results in a reduction in the balance of the margin account, Elvira immediately makes a deposit to return the balance to the predividend amount.

---

[4]$25,000 for "daytraders"

Solution   The FRB's initial margin requirement on Elvira's stock purchase is $.5 \times 360 \times \$57.82 = \$10,407.60$. So, since Elvira makes the minimal allowable deposit to fullfill the FRB's requirement, she deposits $\$10,407.60$. The remaining $\$10,407.60$ of the purchase price is borrowed. So, the inital debit balance is $\$10,407.60$, and by assumption, this is also equal to the later debit balance. Therefore, if the price of the stock is p, the market value of the 360 shares of stock is $360p$ and Elvira's equity is $360p - \$10,407.60$. So, in order that there not be a margin call, the equity $360p - \$10,407.60$ must be at least $.35(360p)$. The inequality

$$360p - \$10,407.60 \geqslant .35(360p)$$

is equivalent to $p \geqslant \frac{\$10,407.60}{.65(360)} \approx \$44.4769$. This inequality is satisfied if $p = \$44.48$ but not if $p = \$44.47$. So, since the price is a whole number of cents, the largest per share price that will result in the 35% maintenance margin requirement being violated is $\$44.47$.   ∎

Failure to satisfy the margin requirements may result in a **margin call** or **maintenance call**, a demand that you immediately take action so that your account will again meet the requirements. Your response to a margin call may be to deposit money or marginable securities to your account so that your equity again meets the maintenance margin requirement. Another option is for you to instruct your broker to sell certain margined securities held in the account, thereby simultaneously increasing the cash in the account and reducing the amount of securities on margin. Margin buyers should be aware that margin calls are made at the discretion of the brokerage house. If your account dips below the margin requirement, the broker may liquidate securities *without consulting you* to bring your account into compliance. This may result in big losses! Therefore, it is important that a margin account holder keep an eye on the value of the assets in the account and the debit balance and take action *before* the broker. *Do not count on the broker issuing a margin call!* Margin buyers should further be aware that the brokerage firm may increase its maintenance margins at any time.

To analyze margin considerations more generally, we introduce some notation. Let B_t denote your debit balance at time t and P_t denote the market value of your margined holdings, also at time t. Then $E_t = P_t - B_t$ is your account equity at time t. The **margin** m_t **at time** t is

$$(7.2.2) \qquad\qquad m_t = \frac{P_t - B_t}{P_t} = \frac{E_t}{P_t},$$

the ratio of the net holdings, or equity, to the current price the margined stock would fetch. Suppose the maintenance requirement is m'. You may calculate the minimum price Q_t that your entire package of assets must command at time t in order for a **margin call** *not* to be made. More specifically, by Equation (7.2.2) we have $m' = \frac{Q_t - B_t}{Q_t}$. This is equivalent to the equation $m' Q_t = Q_t - B_t$ as well as to

$$(7.2.3) \qquad\qquad Q_t = \frac{B_t}{1 - m'}.$$

If the margin m_t is less than m', the account is said to be **undermargined** and is therefore subject to a margin call. If m_t is between the maintenance margin and the initial margin, the account is said to be **restricted**. This means that you may not withdraw money without selling stocks and getting back to the *initial* margin requirement. Furthermore, purchases are not allowed until the initial margin requirement is again met, and part of the proceeds from any sale must go to reducing the deficiency.

When m_t exceeds the initial margin requirement, you are free to withdraw assets from the margin account, as long as you leave sufficient funds to meet or exceed the *initial* margin requirement.

EXAMPLE 7.2.4   Avoiding a margin call, portfolio of two stocks bought on margin

Problem: On February 4, 2001, Seth Stubbs made a margin purchase of 200 shares of an oil stock for $78 per share, borrowing to the full extent allowable with a 50% margin requirement. Two months later, he again made a margin purchase using the same account. This time he bought 100 shares of a tech stock at $121 per share. In addition to the 50% initial margin requirement, the oil stock had an $m_1' = 30\%$ maintenance margin requirement, and the more volatile tech stock had an $m_2' = 40\%$ maintenance margin requirement. Suppose that the price of the oil stock at the time of the tech stock's purchase was $81, and that Seth took advantage of the $3 per share increase when he purchased the tech stock, using any extra unrestricted equity to reduce the amount of new money he needed to add to the margin account.

(1) How much new money did Seth have to deposit to accomplish the purchase of the tech stock?

(2) If the share price of the tech stock two months after Seth purchased the stock was $101, how high must the price of the oil stock have been so that Seth's account is not subject to a margin call? Assume that there was no account activity other than what has been reported and that (unrealistically) there was no growth in the debit balance.

Solution (1) Originally, Seth purchased $200 \times \$78 = \$15,600$ of stock using $7,800 of his own funds and $7,800 of borrowed money. Two months later, the oil stock was worth $200 \times \$81 = \$16,200$. But, according to Equation (7.2.2), this resulted in a margin of $\frac{\$16,200 - \$7,800}{\$16,200} \approx 51.85\%$. To achieve a margin of

50%, as is required for funds to be unrestricted, the numerator need only be $8,100 rather than the current $8,400. So, there was an extra $300 that Seth had available for the purchase of the tech stock. The total price of the tech stocks was $100 \times \$121 = \$12,100$. The initial margin requirement for the tech stock was $.5 \times \$12,100 = \$6,050$, so Seth only had to contribute $5,750 of newly invested money to execute the tech stock purchase.

(2) Sitting in Seth's account were 200 shares of oil stock and 100 shares of tech stock, and he borrowed $7,800 for the oil stock and $6,050 for the tech stock. Now, consider the margin situation two months after the tech stock purchase, first looking at each of the types of stock separately. According to equation (7.2.3), if the tech stock had its own margin account, a margin call would go out if the price fell below $\frac{\$6,050}{1-.4} \approx \$10,083.3333$. Again by Equation (7.2.3), the oil stock (alone) would trigger a call if its price were below $\frac{\$7,800}{1-.3} \approx \$11,142.8571$. So, if we allocate Seth's assets to best serve him, he needed just $\$10,083.3333 + \$11,142.8571 \approx \$21,226.1904$ to avoid a margin call. If, as was the case after two months, the tech stocks had a price of $100 \times \$101 = \$10,100$, then there would have been a margin call unless Seth's holdings of oil stock had a price of at least $\$21,226.20 - \$10,100 = \$11,126.20$. Since $\frac{\$11,126.20}{200} = \$55.631$, per share, we needed a price of at least $55.64. ∎

In Examples (7.2.1) and (7.2.4), we've considered the prices stocks must command in order for margin calls to be avoided. But, sometimes stock prices experience a big drop, and the investor who purchased stocks on margin receives a maintenance call. Our next example concerns various responses an investor could make to such a call.

EXAMPLE 7.2.5   Responding to a margin call

Problem: Shannon Cox purchases $25,000 of a single stock using her margin account and borrowing to the full extent allowed. The account has a 30% maintenance margin requirement. Immediately after the sale, the value of the stocks plummets to $17,000.

(1) How much cash would she need to deposit in order for the account to meet the maintenance requirement without other changes?

(2) What must be the price of marginable securities added to satisfy the 30% margin requirement? Assume these securities again have a 30% maintenance margin and that no cash is added or stock sold.

(3) If the margin requirement is to be satisfied solely by liquidation of stocks, what is the total market value of the stocks that must be sold? The proceeds from the sale are put in the initial margin account.

Solution   The FRB initial margin requirement is 50%, so Shannon deposits $12,500 and borrows another $12,500. If the price of the purchased stock falls to $17,000, Shannon's debit balance remains $12,500 and her equity in the stock is just $17,000 − $12,500 = $4,500.

(1) If the security holdings held in the account are unchanged, the maintenance margin requirement requires her equity to be at least .3 × $17,000 = $5,100. Therefore, Shannon must deposit $5,100 − $4,500 = $600 to meet the maintenance requirement.

(2) We next consider what dollar value S of securities Shannon must add so that the new equity S + $4,500 = .3($17,000 + S). This equation expresses the requirement that the revised security holdings represent equity equal to 30% of the price of those holdings, and it is equivalent to .7S = $600. So, since $\frac{600}{.7} \approx$ $857.143, Shannon must deposit $857.15 of securities to meet the 30% margin requirement.

(3) Suppose that broker sells stocks from Shannon's account with market value Y. Then the market value of her remaining stocks is $17,000 − Y$ and her account also includes an additional amount Y of cash. In other words, the market value of her holdings is unchanged, and her debit balance remains $12,500. So, she still has $4,500 of equity. What changes is her ratio of equity to margined securities. In order that it equals 50%, we would want $4,500 = .5($17,000 − Y$)$, so Y = $8,000. Observe that under this scenario, if the share price returns to its original level, Shannon's equity would then be $\left[($8,000 + \frac{25}{17}($17,000 − $8,000)\right] − $12,500 \approx $8,735.29$. So, she would have lost $3,764.71. There is no such loss in (1) and in (2) where Shanon would gain equity if the newly purchased securities also experience a price rise.   ∎

Obviously, you would like to avoid the consequences of having a restricted account. In fact, if the value of the securities in your account rises before it falls, you may have the equity that exceeds the FRB's 50% transferred to another account, a so-called **special memorandum account (SMA)**, also called a **special miscellaneous account**. The purpose of an SMA account is to preserve the customer's buying power and ability to withdraw funds. Once money is credited to an SMA, it remains until used by the customer, even if the account becomes restricted. Here is a simple example of how an SMA may help a customer who has a margin account.

EXAMPLE 7.2.6   Special memorandum account

Problem: Uki Grissom opens a margin account and purchases 1,000 shares of pharmaceutical stock for $32.50 per share, borrowing 50% of the purchase price. An announcement that the FDA has granted approval of one of the company's drugs results in the stock soaring to $36.70 per share. However, the approval is withdrawn after several users die and lawsuits are feared. This results in the stock falling to $30 per share. In addition to the FRB's 50% margin requirement, the brokerage firm has a 35% maintenance margin for the stock. Trace the sequence of equity in Uki's account if there is no SMA and again if funds are transferred to an SMA to the full extent allowed, whenever equity exceeds the FRB's 50%.

Solution   The original price for the 100 shares was $32,500, and the initial debit amount and equity amount are each half of that amount, so $16,250. When the price jumps to $36.70 per share, for a total increase of $36,700 − $32,500 = $4,200, without an SMA the equity in the margin account increases to $16,250 + $4,200 = $20,450. The market value is now $36,700 so only $18,350 of equity is needed to satisfy the 50% initial margin requirement. If the excess $2,100 of equity is transferred to an SMA, then the margin account will now have equity of $18,350.

Without the SMA, following the price decline to $30,000 for 1,000 shares, Uki's margin account will have equity of $30,000 − $16,250 = $13,750, and Uki's account will have a margin of $\frac{\$13,750}{\$30,000} \approx .458333333$.

In the case where the $2,100 was moved to an SMA, Uki's equity is $13,750 − $2,100 = $11,650 and the margin is $\frac{\$11,650}{\$30,000} \approx .388333333$. In both cases the margin account is restricted, but if Uki has the SMA, he has $2,100 available: He may use this money as he wishes. The downside of using an SMA is that he now needs to be vigilant for continuing price declines that would make his account undermargined.   ■

7.3   GOING LONG: BUYING STOCK WITH BORROWED MONEY

How can you attempt to make money with common stocks? Since a stock is an equity, you could do so by determining a stock issuing company, or better yet several, whose future is better than the current price reflects. Of course this is a prediction about the future direction of a stock price (or prices) and is therefore at best an informed guess. If there is general agreement that a stock is underpriced, you would expect that market forces would quickly act to correct this undervaluing.

Beware that playing the stock market is a risky business!

If you believe a stock will rise in value, you are said to be **bullish** on the stock, while if you foresee a decline, you are **bearish**. [5] We defer discussion of how a bearish investor can try to make money on the stock market to Section (7.4).

---

[5]The terms "bear" and "bull" can be traced back to Thomas Mortimer's 1761 book "Every Man His Own Broker, or, A Guide to Exchange Alley."

Suppose you are confident that a stock will have a significant increase in value. You may wish to purchase shares of the stock now and sell them after their value increases. Purchasing an investment that will appreciate in value when the investment goes up in price is referred to as **taking a long position** or **going long**. When you take a long position on a stock, in addition to anticipating appreciation in the share price, you may foresee dividends being issued during the period in which you plan to hold the stock.

What should you consider when you make a decision as to whether to buy stock? If you are contemplating taking a long position on a stock using your own money, you should compare the anticipated gain from the stock purchase with the money you could earn by using your money for lower risk investments (e.g., U.S. Treasury securities). If borrowed money would be used to finance the stock purchase (or part of it), the interest rate on the loan must be considered. In both cases, it is wise to compare various stocks, evaluating the level of risk as well as the profit potential. You should also note dividend reinvestment options. If you make your purchase with a cash account, perhaps you have an option of having your dividend reinvested in the company without any transaction fee. When you purchase stock on margin, commonly dividends are credited to your account and reduce the debit balance.

In our next example, we illustrate that speculating using borrowed money can increase your yield, but it also magnifies the risk.

EXAMPLE 7.3.1   Going long with cash and on margin

Problem: Cong is confident that the price of shares of Promise Computer stock will soon rise dramatically. He pays $2,500 to purchase 1,000 shares. Cong's sister Shu Fang wishes to share in her brother's potential windfall, but only has $1,250 to invest. She opens a margin account with her broker and places her $1,250 in the account so that she may purchase $2,500 of the stock as her brother has done. So, she is taking out a $1,250 loan that is secured by the purchased Promise Computer shares. The stock pays dividends of $.20 per share at the end of the first six months. Calculate Cong's yield if he sells his 1,000 shares at the end of one year for $3,000. Also determine what his yield would be if he sells the stock at the end of one year for $2,000. If Shu Fang sells her shares at the same time as her brother, and for identical prices, what would her yield be in each case? Assume that Shu Fang pays $35 of interest for the year, and that there are no commissions.

Solution   The dividend paid halfway through the year on 1,000 shares is $.20 \times 1,000 = $200. So, if the price at the end of one year for 1,000 shares has jumped to $3,000, then Cong's yield rate i_C satisfies the equation of value

$$\$2,500 = \$200(1 + i_C)^{-\frac{1}{2}} + \$3,000(1 + i_C)^{-1}.$$

328 Chapter 7 Stocks and financial markets

Therefore, $i_C \approx 29.08940135\%$. (This may be found using the quadratic formula or the BA II Plus calculator **Cash Flow worksheet**.) On the other hand, if Cong only receives \$2,000 when he sells the 1,000 shares, his yield rate i_C satisfies

$$\$2,500 = \$200(1 + i_C)^{-\frac{1}{2}} + \$2,000(1 + i_C)^{-1},$$

and he has a negative yield rate $i_C \approx -12.51743063\%$. (Note that $1 + i_C$ is still positive.)

In Shu Fang's case, at the end of a year, the stock is sold and Shu Fang liquidates her margin account. The balance of her account at this time is the proceeds for the stock sale, less her debit balance. This debit balance is $\$1,250 - \$200 + \$35 = \$1,085$ since the dividends would be added to the account. So, if the sale price is \$3,000, she gets \$1,915, and if it is \$2,000 she gets \$915. Denoting her yield rate by i_{SF}, in the first case her time 1 equation of value is

$$\$1,250(1 + i_{SF}) = \$1,915,$$

while in the case that the stock goes down, it is

$$\$1,250(1 + i_{SF}) = \$915.$$

Therefore, in the case when her brother had a yield of about 29%, Shu Fang has a yield of $i_{SF} = 53.2\%$, and when Cong's yield is about -12.5%, hers is $i_{SF} = -26.8\%$. ∎

In Example (7.3.1), the yields have larger absolute value when borrowed money is used for part of the sale price. We shall see in Example (7.4.1) that positive and negative investment returns are also exaggerated if you anticipate a downturn in the price of a security and try and make a gain using borrowed securities.

> The term **leverage** is sometimes used to describe the use of borrowed assets to increase one's speculative capacity. Just as a lever can increase the force you apply, leveraging increases the force of your investment decisions.

With this language, Shu Fang's investment in Example (7.3.1) was leveraged; Cong's was not.

Whenever you purchase stock on margin, interest will be charged on the debit balance. In our example, the interest rate was roughly 3.4%. In practice, the **broker call rate** is the rate that the broker must pay to borrow money and the interest rate on a margin account is apt to exceed that by up to 2%. Interest rates tend to be lower on higher debit balances.

In addition to making money by charging interest, brokers collect more commissions and fees when they allow investors to buy stock on margin, because customers can then make transactions for the purchase of more shares.   This benefit is to some extent offset by client default, but the brokerages tend to have sufficiently high margin requirements to protect their interests.   Moreover, when an account is undermargined, the broker may choose not to give you time to respond to a margin call. If the broker makes a partial liquidation of your account, as allowed when it is undermargined, any choice of which stocks to liquidate may be made to lessen the brokerage's exposure to risks to which it is overexposed.

7.4   SELLING SHORT: SELLING BORROWED STOCKS

Recall that if you purchase stock on margin (take a long position), you borrow money so that you may buy more stock than you could without the added borrowed funds. This is done with the belief that the value of the additional purchased shares will increase at a rate greater than the interest rate on the debit balance. If your forecast is correct, you profit. But what do you do if you wish to make an investment based on a prediction that a stock is *overpriced* and the share price will *drop* substantially? In contrast with the situation where you were bullish on a stock and borrowed money so that you could buy more of it, you now **borrow stock** so you may sell more of it, planning on buying it back and returning shares when you can do so for less than you sold them for. Naturally, since the success of this strategy depends on your projection being correct, implementing it carries considerable risk. If you carry out the just outlined plan, you are a **short seller** and have made a **short sale** or established a **short position**.

When you **sell short** a stock, you borrow the stock shares from your broker. Perhaps they were in the brokerage's own inventory or the account of another one of the firm's customers. The shares are sold and the proceeds are credited to your margin account. Eventually you must buy back the same number of shares that you borrowed and return them to your broker.(This is called **closing** your position.) You should be aware that a broker may require a short seller to return the shares at any time, even when the price is high and it would be disadvantageous to the short seller. If dividends are paid to shareholders before you close your short position, these must be paid from your margin account (since dividends are owed both to the person who bought the stock and to the person from whom you borrowed it, and the company will only pay once). If the price of the shares falls sufficiently, you make money. If it rises or falls an amount insufficient to make up for dividends you had to pay and interest due on the borrowed stock, you lose money.

When you sell short, your broker will want you to place assets in your margin account to secure the loan of shares. The assets should have sufficient value so that, combining them with the cash from the stock sale, you can buy back the borrowed shares even if their price goes up significantly. We will suppose that your margin requirements are fullfilled by cash and that a modest rate of interest is credited on your margin balance. Treasury bills may sometimes be deposited in lieu of cash, and then the interest rate you receive is the Treasury-bill (T-bill) rate. If you deposit more volatile or less liquid assets, you may need them to have a higher deposited value since only a percentage of their total value is counted in figuring your margin balance. So, just as was the case for a margined stock purchase, a short sale has an initial margin requirement and a maintenance requirement [see Section (7.2)], the requirements made by the Federal Reserve Board and your broker.

EXAMPLE 7.4.1   Short sale

Problem: Dayita believes that stock in Confederated Foods is overpriced. The current price per share is $48.25. She decides to immediately sell 500 shares of Confederated Foods on margin and, as instructed by her broker, deposits the required FRB 50% margin in cash to her brokerage account. The annual effective interest rate on her cash margin deposit (but not on the funds from the stock sale) is 2.5%. If the stock declines by 20% over the next year and she closes out her short position after exactly one year, what will her yield be if no dividends were declared? How about if a dividend of $.15 was declared right before she covered her short sale? What yield would Dayita receive if the price rose by 20% and there were no dividends? How about if there were dividends of $.15 right before she bought back the stock?

Solution   The price of 500 shares at the time of the short sale is $48.25 × 500 = $24,125, and Dayita's margin deposit is $12,062.50, half of the value at the time the short position is initiated.   At the end of the year, the stocks are repurchased, any dividends due are paid out, and Dayita receives the balance from her margin account. This balance is the difference between $24,125 + (1.025)($12,062.50) ≈ $36,489.06, the sum of the proceeds from the short sale and Dayita's deposit to the margin account, and the money paid out to purchase the stock and pay any dividends. We were asked to calculate her yield in four different cases, and here are the results: In each case, the entry recorded in the "Money due to Dayita" column was found by calulating $36,489.06 − ("Repurchase Price" + "Dividends paid out").

Change in Stock Price	Repurchase Price	Dividends paid out	Money due to Dayita	Dayita's yield
−20%	.8($24,125)	$0	$17,189.06	$\dfrac{\$17,189.06}{\$12,062.50} - 1$
	= $19,300			$\approx 42.49997927\%$
−20%	.8($24,125)	500($.15)	$17,114.06	$\dfrac{\$17,114.06}{\$12,062.50} - 1$
	= $19,300	=$75		$\approx 41.87821762\%$
+20%	1.2($24,125)	$0	$7,539.06	$\dfrac{\$7,539.06}{\$12,062.50} - 1$
	= $28,950			$\approx -37.5000207\%$
+20%	1.2($24,125)	500($.15)	$7,464.06	$\dfrac{\$7,464.06}{\$12,062.50} - 1$
	= $28,950	=$75		$\approx -38.1217824\%$

We note that when the stocks go down as predicted, her yield is much higher than the percentage the stock prices fell, and when the movement was contrary to the bearish forecast, Dayita's percent loss was much worse than the percent gain by the stock.   ∎

The yield calculations illustrated in Example (7.4.1) might have been made by appealing to a formula. Specifically, let P_t denote the price of the short-sold stocks at time t. We suppose that the short sale is initiated at time 0 and terminated at time T. Suppose that the inital margin requirement is r [where $r = .5 = 50\%$ as in Example (7.4.1)], and J denotes the interest rate for the interval $[0, T]$ on the margin deposit. Also, let D denote the time T aggregate value of all dividends paid on $(0, T]$, where the values of the individual dividends are computed using the interest rate J and compound interest. Then the initial margin deposit is rP_0, and assuming that there are no intermediate margin deposits or withdrawals, the profit by the investor is $(P_0 - P_T) + J(rP_0) - D$, and the yield over the interval $[0, T]$ is

$$(7.4.2) \qquad i_{\text{shortsale}} = \frac{(P_0 - P_T) + J(rP_0) - D}{rP_0}.$$

EXAMPLE 7.4.3   Short sale with multiple dividend payments

Problem: On April 1, Zheng Wang initiates a short sale for 400 shares of Allied Machinery when the per share price is $43.13. She closes out her short position exactly two years later, buying back the stock for $38.95 per share. The inital margin requirement is 55%. Assume that there are no intermediate margin deposits except for quarterly withdrawals of $.36 per share to cover dividends. The annual effective interest rate earned by the margin deposit is 2.82%. Find Zheng's annual yield for this two-year transaction

Solution Zheng sells the stock for $P_0 = 400 \times \$43.13 = \$17,252$ and buys it back for $P_2 = 400 \times \$38.95 = \$15,580$. The margin requirement is $.55(\$17,252) = \$9,488.60$, and the interest rate for the two-year investment period is $J = (1.0282)^2 - 1$. We further note that at the end of each quarter (for eight quarters) dividends of $400 \times (\$.36) = \$144$ are paid out, so $D = \$144 s_{\overline{8}|((1.0282)^{.25}-1)}$. Therefore, Equation (7.4.2) gives

$i_{\text{short sale}}$

$$= \frac{(\$17,252 - \$15,580) + [((1.0282)^2 - 1)(\$9,488.60)] - \$144 s_{\overline{8}|((1.0282)^{.25}-1)}}{\$9,488.60}$$

$$\approx 10.8991522\%.$$

This is for the two-year period, so the annual effective yield is $1.0899152 2^{\frac{1}{2}} - 1 \approx 5.3086664\%$. ∎

Equation (7.4.2) is only uselful when the investor doesn't make intermediate deposits and withdrawals from the margin account. So, for instance, it is not applicable if a margin call is answered with a cash deposit. In general, you should use the basic principles introduced in Chapter 2 to find the yield.

EXAMPLE 7.4.4   Yield rate for a short sale with an intermediate cashflow

Problem: Dr. Amos Riesman short sells stock in Elwood Pharmaceutical. The purchase price is $38,300 and this requires Dr. Riesman to make a $19,150 margin deposit. Three months later, the price of the shorted shares is $56,200 and, since Dr. Riesman's account has a 35% maintenance margin, he receives a margin call that he answers with a cash deposit of $4,000 (although less was required). Six months after Dr. Riesman initiated the short sale, the price of the stock is just $28,800, and Dr. Riesman closes out his short position. If the interest rate paid on the margin account was a nominal rate of 2% convertible quarterly and no dividends were declared, what was Dr. Riesman's yield on the short sale?

Solution Note that, at the end of six months, Dr. Riesman's contributions to his margin account have accumulated to $\$19,150(1.005)^2 + \$4,000(1.005) \approx \$23,361.98$. Moreover, $38,300 was received from the sale of the stock. So, after the stock is repurchased for $28,800, Dr. Riesman liquidates $23,361.98 + $38,300 − $28,800 = $32,861.98. So, a time six months equation of value giving Dr. Riesman's short sale annual yield i for the short sale is

$$\$32,861.98 = \$19,150(1 + i)^{\frac{1}{2}} + \$4,000(1 + i)^{\frac{1}{4}}.$$

The quadratic formula may be used to find $(1 + i)^{\frac{1}{4}} \approx 1.20969138$. So, $i \approx 114.14\%$. This value of i may also be found using the BA II Plus calculator cashflow worksheet.

We note that there was considerable risk in Dr. Riesman's short sale. When the price of the stock soared to $56,200, it would not have been surprising if the investor, from whom the shares were borrowed, decided it was time to sell the shares. To accomplish the sale, the broker might have required Dr. Riesman to buy back the stocks, resulting in the doctor losing most of his $19,500 investment!   ∎

Why do brokers allow clients to execute short sales? A broker only offers a short sale if the stocks to be sold are readily available, either in the brokerage's own inventory (including the shares held in "street name") or in the inventory of another firm to which the broker has access. Should the owner of the borrowed stock shares require them back (most likely to sell them), the broker will try to borrow other shares. However, the broker may require the short seller to terminate the short position. While short sales are risky for investors, due to the margin requirements, the brokerage does not share substantially in that risk. Rather, it makes money from commissions and fees for the stock sale and from the later stock repurchase. In addition, the brokerage firm has access to the money from the sale of the shorted stocks, but does not pay interest on this money to either the investor or the party from whom the stocks were borrowed . Of course the brokerage firm can earn money on it, for instance by investing in Treasury bills. In some cases, there is an additional profit if the brokerage firm receives a higher rate of interest on the collateral than they pay the short seller.

You are not alone if you have concern that short selling might drive panic selling and that traders might attempt to create panic selling so as to drive down the price and thereby make a profit. Such concerns were the motivation behind the uptick rule. The **uptick rule** requires that on a short sale of a stock traded on the NYSE, the stock must be sold at a price no lower than the previous sale and higher than the last different price. Put another way, the uptick rule says you may not sell a stock short if it is already going down. There is a similar **bid rule** on NASDAQ trades.

Short selling can cause prices to rise as well as to fall. Specifically, when a number of short sellers simultaneously try to cover their positions, this drives up demand and therefore the price. This is called a **short squeeze**.

In general, a short position is taken when an investor (or trader) sells a borrowed security. That is to say, the investor borrows a security of which he was "short" and then sells it. This is done so that the investor will see a profit if the security's price declines sufficiently. An investor may wish to **hedge** — that is protect against — his risk by combining long and short positions. Hedging strategies are an active area of research in financial mathematics.

7.5   PROBLEMS, CHAPTER 7

(7.0) Chapter 7 writing problems

(1) Contact a brokerage house or go online to the website of a brokerage firm and learn what commissions or other fees that a broker would charge for buying and selling stock in a long position and also for a customer wishing to short sell a stock. Illustrate these with a specific stock (or stocks) and a stated number of shares.

(2) Pick a corporation that issues both preferred and common stock and a period of at least five years. Report on the price history of each type of stock and any dividends paid. Include mention of any special events such as a stock split.

(3) Market makers and specialists are important to the smooth running of the financial markets. Write a paragraph or two describing the role they play.

(4) Comment on the following statement: "As interest rates rise, stock prices tend to fall."

(7.1) Common and preferred stock

(1) At the beginning of a quarter, Bridget Dubois purchases 200 shares of preferred stock for a price to provide her with a yield rate equal to 6.2%. Assuming the stipulated dividends of $.28 per share are paid quarterly, what is Bridget's per share price?

(2) On July 1, Kevin Swisher purchases 6,000 shares of preferred stock for $25.13 per share. The stock is to pay a dividend of $.29(1.02)$^{k-1}$ at the end of each quarter of the k-th year Kevin holds the stock. That is to say, he first receives four successive end-of-quarter payments at a rate of $.29 per share, then four successive end-of-quarter payments at a rate of $.29(1.02) per share, etc. Find the yield rate for this purchase assuming all dividends are paid and Kevin holds the stock forever.

(7.2) Brokerage accounts

(1) Dr. Jennifer Rogowski opens a new margin account at Robertson and Hendricks brokerage firm, which requires a 65% initial margin and a

45% maintenance margin. She purchases $13,000 of stock in Bell Foods using her margin account to the full extent allowable. Prior to any interest being credited on the margin account, by what percentage may the stock fall without Dr. Rogowski's account violating the maintenance margin? Assume that she does not use the margin account for any other purpose and that no interest is credited.

(2) Pat Delmonico opens a new margin account with an initial margin requirement of m and a maintenance margin requirement of $(m - .2)$. He uses the account to purchase stock for $24,512, borrowing to the full extent allowable. At what price is the maintenance margin first violated, assuming that there is no interest or dividends paid to or from the margin account?

(3) Beverly Hibbert purchases $20,000 of Prosperity Industries stock using her margin account and borrowing to the full extent allowed. The account has a 35% maintenance requirement for all securities and a 50% initial margin requirement. Immediately after the sale, the value of the stocks falls to $15,000.

   (a) How much cash would she need to deposit in order for the account to meet the maintenance requirement without other changes?

   (b) What must be the price of marginable securities added to satisfy the 35% margin requirement? Assume no cash is added or stock sold.

   (c) If the margin requirement is to be satisfied solely by liquidation of stocks, what is the market value of the stocks that must be sold?

(7.3) Going long: buying stock with borrowed money

(1) Eleanor Michaels believes that the price of Continental Metals stock will go up. Its current price is $35.45 per share and she has $3,000 to invest. How many shares may she purchase with a cash brokerage account? How many may she purchase using a margin brokerage account with a 60% margin requirement? If the per share price is $38.45 three months later when she sells the shares, what is her quarterly yield in each case? Assume that there are no dividends and that in the case of the margin account, Eleanor deposits exactly $3,000 and interest charged to Eleanor for the three months reduces the funds received by $30.

(2) Fred Copeland decides to invest $10,000 in Supreme Foods stock. He chooses to purchase as much stock as he can with a margin account that has a 50% initial margin requirement. The price drops suddenly to only 30% of its initial value. At this point, Mr. Copeland receives a margin call and, fearing that the stock is about to lose its remaining value, sells it. How much of an additional payment must he make to his broker? Assume that there are no dividends or interest charged. (This problem

illustrates the fact that with a margined purchase, you may lose more than the amount of your initial investment.)

(7.4) Selling short: selling borrowed stocks

(1) Gregory sells a stock short at a price P and buys it back one year later for $.9P$. The required margin is 50% and interest on the margin deposit is paid at 5%. Gregory's yield for the year is 14%. Dividends are paid out at the end of the year. Find the amount of these dividends as a percent of Gregory's investment.

(2) Nadia sold short a stock. The price of the stock was $3,000 and she bought it back six months later for $3,400. The initial margin requirement was 60%, the maintenance margin requirement was 40%, and no interest was paid on her margin deposit. Dividends of $10 were declared just prior to her repurchasing the stock. Find Nadia's annual effective yield rate for her six-month investment.

(3) Ursula sells short $2,400 of a stock. Her broker requires a margin of q%. The stock does not pay any dividends for the year that she holds the stock, the repurchase price to cover the short sale is $2,012, and her yield is 34%. If the interest rate on the margin account is 6%, find the margin requirement.

(4) Byron Stewart believes that shares of Clever Computer are underpriced compared to shares of other high tech stocks. He therefore decides to purchase $10,000 of Clever Computer stock. However, Byron is a bit worried that high-tech stocks could generally be overpriced. To hedge the position he took on Clever Computer, he sells short $5,000 of Silly Chip stock, a stock he views as being overpriced compared to stocks in comparable companies. There is a 50% initial margin requirement on the short sale, and the margin account pays 5% annual effective interest. Suppose that Bryon's concern was justified and that on the average high-tech stocks lose 30% of their value over the year that Bryon maintains his long position with Clever Computer and his short position with Silly Chip. If Clever Computer loses 10% of its value during the year, Silly Chip loses 35% of its value during the year, and neither stock declares any dividends, what is Byron's yield for the year? You should assume that Byron sells the Clever Computer stock at the end of the year and also closes out his short position on Silly Chip at that time

(5) Warren sells short $3,000 of a common stock that pays dividends of $20 after six months and $15 after eighteen months. The margin requirement is 50%, and the annual effective interest rate on the margin account is 4%. Twenty months after he sold the stock, Warren repurchases the stock (to cover the short sale) for $2,650. Find his annual yield for this

twenty-month investment.

(6) Kim and Pauline each sell short a different stock for $2,000. They each face a 60% margin requirement, and the interest rate on their margin accounts is 5% annual effective. At the end of exactly one year, Kim purchases back his stock for X and Pauline repurchases hers for 8% more (that is to say, for $1.08X$). Kim's stock did not pay any dividends, but Pauline's declared a dividend of $10 just before she repurchased it. Kim's yield y is twice Pauline's. Find y.

Chapter 7 review problems

(1) Fernando Jones sells short common stock in Alpha Communications for $2,082. There is a 60% margin requirement, and the margin account pays interest at an annual effective rate of 3.2%. The stock price goes up by 2% during the next year, and dividends of $84 are declared at the end of six months. If Fernando closes out his short position at the end of the year, what is his yield for the one-year investment?

(2) What is the price of a preferred stock bought to yield 5% annually, if it pays level dividends of $40 semiannually, and the next dividend is due in exactly three months?

(3) Yuri Popescu purchases $16,850 of stock using his newly created margin account and borrowing to the full extent allowed. The account has a 40% maintenance margin requirement on all securities and a 50% initial margin requirement. Soon after the sale, the value of the stocks is $18,200 and Yuri withdraws the full amount of money allowable. The price then plummets to $13,600 and Yuri receives a margin call.

   (a) How much cash would he need to deposit in order for the account to meet the maintenance margin requirement without other changes?

   (b) What must be the price of marginable secuities added to satisfy the 40% margin requirement? Assume no cash is added or stock sold.

   (c) If the margin requirement is to be satisfied solely by liquidation of stocks, what is the market value of the stocks that must be sold?

(4) Ehud Katz anticipates that the price of Trinity Technology stock will go up. Its current price is $55.50 per share and he has $3,330 to invest. How many shares may he purchase with a cash brokerage account? How many may he purchase using a margin brokerage account with a 60% margin requirement? If the per share price is $60.20 six months later when Ehud sells the shares, what is his annual yield in each case? Assume that there are no dividends paid, and that in the case of the margin account, Ehud has no cashflows to or from the margin account except that interest for the six months reduced the funds he received by $32.

CHAPTER 8

Arbitrage, the term structure of interest rates, and derivatives

8.1 INTRODUCTION
8.2 ARBITRAGE
8.3 THE TERM STRUCTURE OF INTEREST RATES
8.4 FORWARD CONTRACTS
8.5 COMMODITY FUTURES HELD UNTIL DELIVERY
8.6 OFFSETTING POSITIONS AND LIQUIDITY OF FUTURES CONTRACTS
8.7 PRICE DISCOVERY AND MORE KINDS OF FUTURES
8.8 OPTIONS
8.9 USING REPLICATING PORTFOLIOS TO PRICE OPTIONS
8.10 USING WEIGHTED AVERAGES TO PRICE OPTIONS
8.11 SWAPS
8.12 PROBLEMS, CHAPTER 8

8.1 INTRODUCTION

Arbitrage refers to the possibility of making money with no outlay of capital or possibility of loss. Modern markets provide ready access to trading information and trades may be quickly transacted. Consequently, arbitrage opportunities should be short-lived. If you assume that there are no arbitrage opportunities, then two investments that have exactly the same cashflows must have the same price. This is the "law of one price" and is the fundamental principle of "no-arbitrage" pricing. Arbitrage is discussed in Section (8.2).

Market prices for noncallable bonds and other fixed-term investments determine interest rates for the term of the bond — so-called **spot rates**. "No-arbitrage pricing" will then allow you to infer interest rates for certain future periods. All this goes under the title **term structure of interest rates** and is the topic of Section (8.3).

A **derivative** is a security that provides its owner with cashflows that are a function of the value of other securities — that is, a derivative is an agreement that depends on the price of some other asset, called the **underlier**. The remainder of the chapter discusses various types of derivatives. Derivatives may be used to hedge against adverse price movements. In other words, a derivative may be a conservative investment used to provide insurance by capping expenses or guaranteeing a minimum revenue. Derivatives may also be used by speculators who are willing to take on risk in the hope of realizing significant profit. A desire to reduce transaction costs, to circumvent regulations, or to avoid tax liabilities may also motivate an individual or company to use derivatives.

A **forward contract** is a non-exchange-traded derivative that requires the parties to carry out a particular future sale at a specified price. The exchange-traded counterpart is a **future**. Forward contracts are discussed in Section (8.4), and futures are considered in Sections (8.5)–(8.7). The future contracts of Sections (8.5) and (8.6) all have commodities as underliers. Other types of futures are introduced in Section (8.7), which also includes a discussion of the importance of futures as a "price discovery" agent.

Unlike forward contracts and futures, options give the owner a choice: a **call option** gives the holder the right to buy a specified asset at a designated time and price, while a **put option** gives the holder the right to sell the underlier at a set time and price. Options are introduced in Section (8.8).

In general, it is difficult to theoretically determine the price of an option, although market forces do so. However, there is a simple model, the so-called binomial option-pricing model, that allows you to readily calculate the price; the model requires you to specify possible price movements, an interest rate, and that there is no arbitrage. In this simple model, you may find an easily priced portfolio that provides the same cashflows as the option. Then, by the "law of one price," the option must also have that price. This is explained in Section (8.9). In our binomial option pricing model, we do not specify probabilities of the stock going up or down. However, there is a probabilistic interpretation of the option pricing equation, and Section (8.10) is devoted to an elementary presentation of it. The section should be understandable even if you have not studied probability.

The last type of derivative we consider is a swap, presented in Section (8.11). **Swaps** are privately negotiated agreements, often arranged through a dealer, in which the parties trade sequences of cashflows arising from financial instruments.

8.2   ARBITRAGE

Arbitrage refers to a financial strategy that requires the instigator to expend no money, allows no possibility of a loss, and allows a possibility of a gain. It is important to note that if any money is to be spent, there must be a *simultaneous* inflow of at least as much money. Otherwise, you do not have a true arbitrage strategy. Also, the risk must truly be *zero*, not just slim. That is to say, in order that you be engaged in arbitrage, there must be *no* chance that you have a net outflow *at any time*.

You may think of an arbitrage opportunity as occurring when two portfolios are mispriced relative to each other. Imagine that portfolio *B* is better than (or at least as good as) portfolio *W* in the following sense: *B* guarantees all the same inflows as *W*, at least the possibility of some additional ones, and no outflows not also required with *W*. Further suppose that *B* has a lower price than *W*. An **arbitrageur** has the opportunity to simultaneously sell the overpriced portfolio *W* and buy the underpriced portfolio *B*. This may require the borrowing of *W*, which we assume may be done without cost, and the repaying of the inflows of *W* using the cashflows provided by *B*. Remember, *B* reproduces all the inflows of *W*! Now, if this opportunity is recognized by others, there should be an increased *demand* for *B* and an increased *supply* of *W*. The result is that the price of *B* should go up, and the price of *W* should go down. In other words, the fact that many people are happy to get something for nothing (or a possibility of something for nothing) along with the "law of supply and demand" should work to eliminate an arbitrage opportunity. In today's world, information is disseminated widely and quickly, so you would expect arbitrage opportunities to be fleeting.

If we *assume* that there are *no* arbitrage opportunities even though all assets may be bought and sold at will without borrowing or trading expenses, then we are said to have a **no-arbitrage model** (or **arbitrage-free model**). When the no-arbitrage model is used, any two portfolios that give exactly the same payments must have the same price. This is called the **"law of one price"**. Of course, the assumption that there are no transaction costs is clearly violated in the real world, so we cannot expect the no-arbitrage model to perfectly fit reality. On the other hand, large traders often have tiny costs relative to the prices of assets exchanged, and hence the model tends to be fairly good.

At several instances as we procede through this chapter, we will assume that we have the no-arbitrage model; for example, Sections (8.9) and (8.10) use this model when prices of "options" are derived.

We end this section by looking at some examples where there is an arbitrage opportunity.

EXAMPLE 8.2.1 An arbitrage opportunity: interest rates

Problem: Jennifer Atkins has an opportunity to borrow up to $123,000 for one year from Community Bank at an annual effective rate of 5.5%. She learns that she may purchase one-year $10,000 Treasury bills at an annual discount rate of 6%. How much money can Jennifer make if she purchases eight $10,000 Treasury bills using money borrowed at the 5.5% annual effective rate?

Solution   The $10,000 bonds each sell for $10,000$(1 - .06) = $9,400$, so Jennifer borrows $8 \times \$9,400 = \$75,200$. One year later, she will redeem her Treasury bills for a total of $80,000, and pay back the loan for $1.055 \times 75,200 = \$79,336$. This leaves her with $80, 000 - \$79,336 = \$664$ or $83 per bond purchased with no outlay or risk to Jennifer. It is likely that the decision makers at Community Bank did not know that Treasury bills would be offered for this price, since lending to the United States Treasury rather than Jennifer would have allowed the bank to make $664 more. That is to say, the Bank would not intentionally offer to lend the money at an interest rate lower than provided by Treasury bills. On the other hand, Treasury bills are sold on a competitive basis, so it is likely that others will spot the arbitrage opportunity and drive up prices. Rather quickly, we would expect this opportunity to disappear! ∎

EXAMPLE 8.2.2   An arbitrage opportunity: exchange rates

Problem: Hideki Kato often checks the exchange rates between dollars and yen. He observes that at First Exchange Bank, a dollar may be exchanged for 124 yen and that at Currency Bank, 100 yen may be exchanged for $.81. How might he use this information to make money, assuming he has $100,000 available for immediate use?

Solution   Mr. Kato simultaneously places an order for 12,400,000 yen at First Exchange Bank, and an order for $100,440 at Currency Bank, locking in the stated exchange rates. He then delivers the $100,000 to First Exchange Bank, receives 12,400,000 yen, and remits them to Currency Bank which gives him $100,440. A $440 profit has been made. This might conceivably all be done by wire in a matter of minutes. Of course the wire transfer could involve some fees that would cut into Mr. Kato's profit. Also, there is a concern as to the logistics so that the exchange rates would truly be locked in at the same instant, insuring that there is no risk from a change in exchange rates. In real life, it is hard to create a situation with absolutely no risk! Presumably, market forces would quickly cause the exchange rate difference between the banks to disappear. ∎

In Examples (8.2.1) and (8.2.2), the cashflows were predictable. However, this need not be the case. Our next example illustrates how a trader might develop an arbitrage strategy in a situation where the cashflows are not fixed but they take on known values if the state of the market is known.

EXAMPLE 8.2.3

Problem: The market includes two assets. Asset B has a current price of $46. One year from now, it will return $51 if the market is up and $44 if the market is down. Asset V sells for $23. One year from now it will return $25 if the market is up and \$22 if the market is down. Develop an arbitrage strategy involving these two assets.

Solution   We observe that asset B sells for twice the cost of asset V, and if the market is down, B returns twice as much as V does. However, if the market is up, B will return more than twice the amount an investor holding V receives. Therefore, asset V is overpriced relative to asset B, and an arbitrage strategy may be developed that involves selling V and using the proceeds to purchase B. More specifically, an investor can sell u units of V and simultaneously purchase $\frac{u}{2}$ units of B. One year from now, if the market is down, the investor's return is $-\$22u + \$44\left(\frac{u}{2}\right) = 0$ while an up market produces $-\$25u + \$51\left(\frac{u}{2}\right) = \$\left(\frac{u}{2}\right)$. No money is expended by the investor, the investor cannot lose money, and the investor makes money if the market is up. This is an arbitrage opportunity, and once again, you would expect it to vanish quickly. ∎

8.3   THE TERM STRUCTURE OF INTEREST RATES

Throughout this book, we have talked as if there is a single interest rate that governs compound interest investments, independent of the length of the investment. Occasionally, we have looked at problems involving a change in interest rates, but we have not recognized the fact, readily apparent to any investor, that the interest rates available depend on the length of time the investors are willing to tie up their money. The expression **term structure of interest rates** refers to the way interest rates vary with the investment term. This section presents various ways of describing this term structure, using such tools as yield rates, spot rates, forward rates, and discount factors.

There are a number of different theories that attempt to explain why interest rates vary with the investment period, even among investments with similar default risk. We wish to briefly consider some of these before describing mathematical ways to specify the term structure.

The **(pure) expectations theory**[1] has the underlying premise that investors view a sequence of short term bonds as a perfect substitute for a single longterm bond. For example, this theory states that investors are indifferent between a two-year bond and a one-year bond, followed by a one-year reinvestment *so long as they foresee the same overall yield.* Consequently, if y_n denotes the *annual* yield rate for an n-year investment, then the expectations theory says that $1 + y_n$ is the geometric mean of the growth factors for a sequence of one year bonds; that is, $1 + y_n = \left[\prod_{k=1}^{n}(1 + f_{[k-1,k]})\right]^{1/n}$ where $f_{[k-1,k]}$ is the

---

[1] This theory was first proposed by Irving Fischer (1867-1947), who was an American, pioneering mathematical economist. His 1930 book <u>Theory of Interest</u> is still studied.

yield rate for a bond with term $[k - 1, k]$. In order for the pure expectations theory to give an explanation of the fact that yield rates on longer term bonds commonly are greater than those on shorter term bonds of comparable default risk, you would have to believe that, on the average, investors tend to forecast interest rates to increase. Predictions of higher stated interest rates may be tied to beliefs that inflation will be higher [see Section (1.14)]. An argument in favor of the expectations theory is that it supports the documented phenomenon that yield rates on different term bonds tend to move up or down in unison.

Biased expectations theories are off-shoots of the pure expectations theory in which the investors are no-longer assumed to be indifferent about the length of their individual investments. The **liquidity preference theory**[2] asserts that investors prefer to invest their money for (a sequence of) short periods so that the money is frequently accessible. In addition, many investors are hesitant to commit to longer term investments, fearing that they will be missing out on higher rates available in the future. Consequently, market forces drive up the cost of long-term borrowing. To overcome this preference, issuers of long-term loan bonds must be prepared to offer a higher rate than is available on short-term bonds with similar default risk. In other word, they must offer a premium, called a **liquidity premium**, which is the amount the rate of return must be increased to compensate investors for holding a less liquid bond. More generally, the **preferred habitat theory**[3] asserts that investors each have their own preferred investment term, but may be induced to leave their "preferred habitat" if they are compensated by yields that are sufficiently higher than those for the investment period they view as being more desirable. This is called a **term premium**, and a liquidity premium is an example of a term premium. A term premium is different from a **risk premium**, which is the difference in yield rates offered on Treasury and non-Treasury bonds of comparable maturity.

The preferred habitat theory recognizes that investors have investment terms they prefer but allows them to shift to nonpreferred maturity segments if they are rewarded by sufficiently greater yield rates. In contrast, the so-called **market segmentation theory** states that the bond market is actually not one market but is rather divided into separate markets for each term, each subject to the laws of supply and demand within its own segment. According to the market segmentation theory, participants in one segment would not be influenced by interest rates in another segment.

Whatever may be the forces that influence the term structure, the **yield curve** is the classic way to visualize an investor's annual yield rate; for example, for a set of par-value bonds (or bonds trading at close to par) with various

---

[2]This theory was proposed by J.R. Hicks in 1939.

[3]The preferred habitat theory was proposed in the mid-1960's by Franco Modigliani and Richard Sutch.

times to maturity and each bond in the set having a similar default risk and coupon period. Bonds issued by the United States Treasury are viewed as having zero default risk, and the phrase "yield curve" often refers to the yield curve for these. Usually, the time until maturity of each bond is recorded as the x-coordinate, and the yield to maturity for the investment is the y-coordinate. As indicated above, often yields increase with the length of investment, and the yield curve is said to be a **normal yield curve** if it is increasing. You have an **inverted yield curve** if there are higher yield rates for shorter term investments, and a **flat yield curve** if the yield is independent of the investment term.

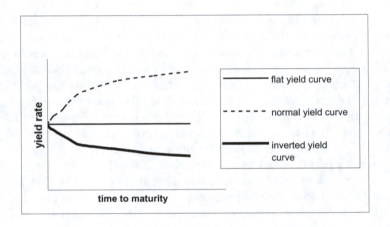

FIGURE (8.3.1)

Of course the yield rate may be increasing over some subintervals of time and decreasing over others. For instance, it may be increasing for most of the plotted period, but decrease for very long durations. Such a **humped yield curve** is fairly typical for United States Treasury securities.

A widely used set of yield rates is based on zero-coupon bonds. More precisely, for $t > 0$ we define the **spot rate** r_t to be the annual effective interest rate [4] earned by money invested now (time 0) for a period of t years. (Spot rates are sometimes called **zero-coupon rates**.) With notation as in Equation (1.3.5), the definition of the spot rate r_t may be expressed as

$$(8.3.2) \qquad (1 + r_t)^t = 1 + i_{[0,t]}.$$

Spot rates are most often reported for United States Treasury securities or other default-free zero-coupon securities. While **Treasury bills**, which have maturities of one year or less, are zero-coupon bonds, longer term obligations

---

[4]Elsewhere, you may occasionally see nominal rates, rather than effective rates, called spot rates.

pay semiannual coupons. [5] Therefore, if we want spot rates based on Treasury securities for terms longer than one year, we will have to determine them by removing the coupons. More specifically, as in Solution 2 in Example (8.3.3), you transform a coupon bond into a sequence of zero-coupon bonds with terms differing from one to the next by six months. These stripped securities are called **STRIPS** and have an active market. Their actual yields may be found in various publications.

When you do not have a market rate for zero-coupon bonds of a particular term, you may be able to determine a spot rate for that term by considering yield rates on available coupon bonds and assuming that there is no arbitrage. We first illustrate how this is done by considering a two-year bond having semiannual coupons. A more complicated example then follows.

EXAMPLE 8.3.3   Determining spot rates from yield rates

Problem: A two-year bond pays annual coupons of $30 and matures at $1,000. Its price is $984. The same corporation sells a one-year $1,000 zero-coupon bond for $975. Compute the two-year spot rate r_2.

Solution 1   The present value of the two-year coupon bond is $984. It is equal to the time 0 value of the bond's cashflows, namely $30 at the end of one year and $1,030 at the end of two years where we discount these using the spot rates r_1 and r_2, respectively. So, we have

$$\$984 = \frac{\$30}{1 + r_1} + \frac{\$1,030}{(1 + r_2)^2}.$$

But, the price of a one-year zero-coupon bond with redemption amount $1,000 is $975, so $\frac{1}{1+r_1} = .975$. Since $.975(\$30) = \$29.25$, $\$984 = \$29.25 + \frac{\$1,030}{(1+r_2)^2}$ and $r_2 = \left(\frac{1,030}{954.75}\right)^{\frac{1}{2}} - 1 \approx 3.866088985\%$.

Solution 2   Strip away the coupon at the end of one year, and view it as a separate asset. Thus, we have divided the original two-year bond into two pieces, namely a one-year zero-coupon bond with redemption $30 and a two-year zero-coupon bond with redemption $1,030. We know that the total price of these two (fictional) zero-coupon bonds is $984. Now what should the price of the one-year bond be? It is indistinguishable from 3% of the $1,000 one-year security, and since the $1,000 one-year security sells for

---

[5]**Treasury notes** is the term used to describe treasury securities issued for more than one year but no more than ten years, while **Treasury bonds** refers to securities issued for more than ten years. The Treasury suspended the issuance of Treasury bonds following the August 2001 auction, but their sale was resumed in February, 2006. Even when newly issued bonds were not available, many longterm securities were available on the secondary market, since bonds were previously issued with terms of twenty and thirty years.

$975, its price should be $.03 \times \$975 = \$29.25$. Then, the fictional zero-coupon two-year bond (the original security without its coupon at time 1) should have price $\$984 - \$29.25 = \$954.75$. Thus, $954.75(1 + r_2)^2 = \$1,030$, and $r_2 = (\frac{1,030}{954.75})^{\frac{1}{2}} - 1 \approx 3.866088985\%$, just as found in Solution 1.   ∎

In Example (8.3.3), we calculated the two-year spot rate to be about 3.866% by two different methods. The first one is a bit easier to apply, but the second is instructive as to what is going on. In fact, the perspective gained from the second solution is just what we need to see why the rate we found for the two-year spot rate is exactly what it must be in a no-arbitrage model.

To see this, suppose that there was a two-year zero-coupon bond B with a higher yield than the computed rate r_2, say 4%. Then, you could sell an original $1,000 two-year coupon bond and use the $984 proceeds to purchase a one-year zero-coupon bond with redemption amount $30 (cost $29.25) and a two-year zero-coupon bond B for the $954.75 balance. The first bond allows you to pay the $30 coupon due at the end of the first year on the sold bond. The second bond has a redemption value of $\$1,032.66 \approx \$954.75(1.04)^2$ which allows you to pay the $1,030 due on the sold bond and have an extra $2.66. As usual, we assume that there are no transaction charges. So, the higher two-year rate presents an arbitrage opportunity. In Problem (8.3.7), you are asked to explain how there is an arbitrage opportunity if there were a two-year zero-coupon bond with a lower yield than the rate we calculated for r_2.

The following table of bonds will be used in Examples (8.3.5) and (8.3.9).

Term	Yield
1 Year	$y_1 = 1.7568\%$
2 Year	$y_2 = 3.0153\%$
3 Year	$y_3 = 3.5463\%$
4 Year	$y_4 = 3.8616\%$
5 Year	$y_5 = 4.2984\%$

TABLE 8.3.4 (5% par-value bonds with annual coupons)

EXAMPLE 8.3.5   Determining spot rates from yield rates

Problem: Use Table (8.3.4) to determine the spot rate r_5. Note that the bonds are all 5% bonds with annual coupons.

Solution   The five-year bond has coupons at the end of each year, and we will have to discount each of them using a spot rate for the payment time. The one-year bond has its only coupon at the time of its maturity, so we may view it as a zero-coupon bond. Therefore, the given yield y_1 for the one-year bond is the one-year spot rate r_1. This rate may be used to determine r_2. Specifically, a two-year $100 bond has coupons of $5 at times 1 year and 2 years as well as its $100 redemption. So, its price to provide the yield 3.0153% of the table is $\frac{\$5}{1.030153} + \frac{\$105}{(1.030153)^2} \approx \$103.7968215.$ If you were to actually purchase the $100 bond, you would round to $103.80. However, for the purpose of computing spot rates, do not round. The reason for this is that we could have used a bond with any face value, say $(10)^s$ and been able to keep s of the places we now have behind the decimal point. Discounting each cashflow by the spot rate corresponding to its time, we have

$$\$103.7968215 \approx \frac{\$5}{1.030153} + \frac{\$105}{(1.030153)^2}$$
$$= \frac{\$5}{1 + r_1} + \frac{\$105}{(1 + r_2)^2} = \frac{\$5}{1.017568} + \frac{105}{(1 + r_2)^2}.$$

Therefore, $(1 + r_2)^2 \approx 1.061859431$ and $r_2 \approx 3.04656379\%$. Next, we find r_3 in a manner similar to that used to calculate r_2, only this time we use the two spot rates r_1 and r_2, as well as the yield rate $y_3 = .035463$. We work with a three-year $100 bond. From

$$\frac{\$5}{1 + r_1} + \frac{\$5}{(1 + r_2)^2} + \frac{\$105}{(1 + r_3)^3}$$
$$= \frac{\$5}{1.035463} + \frac{\$5}{(1.035463)^2} + \frac{\$105}{(1.035463)^3} \approx \$104.0691403,$$

we calculate $(1 + r_3)^3 \approx 1.111737654$ and $r_3 \approx 3.593881343\%$. Now for r_4. Thinking of a four-year $100 bond, and recalling that $y_4 = .038616$, we find

$$\frac{\$5}{1 + r_1} + \frac{\$5}{(1 + r_2)^2} + \frac{\$5}{(1 + r_3)^3} + \frac{\$105}{(1 + r_4)^4}$$
$$= \frac{\$5}{1.038616} + \frac{\$5}{(1.038616)^2} + \frac{\$5}{(1.038616)^3} + \frac{\$105}{(1.038616)^4}$$
$$\approx \$104.1457866.$$

So, $(1 + r_4)^4 \approx 1.166330699$ and $r_4 \approx 3.921504795\%$. Finally, consider a five-year $100 bond purchased to yield $y_5 = .042984$. From the equation

$$\left(\sum_{k=1}^{4} \frac{\$5}{(1.042984)^k} \right) + \frac{\$105}{(1.042984)^5} = \left(\sum_{k=1}^{4} \frac{\$5}{(1 + r_k)^k} \right) + \frac{\$105}{(1 + r_5)^5},$$

we obtain $(1 + r_5)^5 \approx 1.239807347$ and $r_5 \approx 4.392870872\%$.   ∎

Figure (8.3.6) is a yield curve depicting how the rates in Table (8.3.4) vary with their terms. The graph of the spot rates versus term is called the **spot rate curve** or **zero-coupon curve**; see Figure (8.3.7) and compare it to the yield curve [Figure (8.3.6)]. In each of these figures, we have interpolated linearly between the rates for integer terms, but the curves are sometimes drawn with other paths connecting known points.

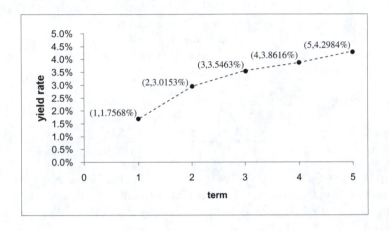

FIGURE (8.3.6) Yield Curve

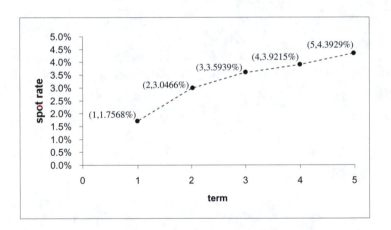

FIGURE (8.3.7) Spot Rate Curve

In Example (8.3.5), we determined the spot rate r_5 by finding a sequence of spot rates $\{r_k\}$ for times to the left of 5. This process is called **bootstrapping**. We note that the spot rate r_1 is equal to the yield rate y_1 but the spot rate for each later time exceeds the corresponding yield rate. In general, if the yield curve is normal, then the spot rate curve will lie above the yield curve. On the other hand, if there is an inverted yield curve, then the spot rate curve lies below the yield curve.

If $t > 0$, the spot rate r_t is the annual effective interest rate for money invested for the period $[0, t]$. That is to say, \$K invested at time 0 grows to $K(1 + r_t)^t$ at time t. Moreover, if $s > t$, then if there is to be a single accumulation function governing the growth of money, we would find that the investment continues to grow so as to be an amount $K(1 + r_s)^s$ at time s. So, money grows by a factor of $\frac{(1+r_s)^s}{(1+r_t)^t}$ over the interval $[t, s]$. This tells us that the annual effective interest rate for the interval $[t, s]$ is $f_{[t,s]}$ where

(8.3.8)
$$(1 + f_{[t,s]})^{s-t} = \frac{(1 + r_s)^s}{(1 + r_t)^t}.$$

If $t = 0$, we have not defined r_0, and this is why we asumed $t > 0$ in the above definition of the forward rate $f_{[t,s]}$. We extend our definition to include $t = 0$ by specifying

$$f_{[0,s]} = r_0 \quad \text{for all} \quad s > 0.$$

When $t = 0$, it is still true that money grows by an annual effective rate $f_{[t,s]}$ on the interval $[t, s]$.

We call $f_{[t,s]}$ the **theoretical forward rate** or the **implied forward rate**. The second name comes from the fact that it is the rate of interest for money invested for the period $[t, s]$ that is implied by the spot rates r_t and r_s.

Very often, an implied forward rate is referred to simply as a **forward rate**. However, this latter designation does not clearly distinguish the implied forward rate from the **market forward rate,** the rate that is actually offered to investors who invest money at time t for a period of length $s - t$. Market forward rates are quite often different from theoretical forward rates, because it is hard to predict what future interest rates are going to be. The implied forward rates are based on the spot rates, which in turn reflect *predictions* of interest rates in the future. If interest rates go up more rapidly than predicted, the implied forward rate will be lower than the market forward rate. On the other hand, the implied forward rates will be higher than the market forward rates if interest rates rise more slowly than anticipated. The term **future (interest) rate** may also be applied to an interest rate fixed today on a loan to be made at some future date.

Given a set of spot rates for times in a set $\mathcal{S}$, it is easy to determine the theoretical forward rate between any two dates in $\mathcal{S}$. A **forward curve** is a graph of the forward rates for intervals of a given length. In other words, one fixes the length w of the term and graphs pairs $(t, f_{t,t+w})$.

EXAMPLE 8.3.9   Implied forward rates curves

Problem: On a single coordinate axes, graph the implied forward rate curves for the bonds of Table (8.3.4) for investment periods of lengths 1, 2, 3, and 4.

Solution   In the solution to Example 8.3.5, we noted that $1 + r_1 = 1.017568$, $(1 + r_2)^2 \approx 1.061859431, (1 + r_3)^3 \approx 1.111737654, (1 + r_4)^4 \approx 1.166330699$, and $(1 + r_5)^5 \approx 1.239807347$. Equation (8.3.8) may now be used to find the forward rates for any pair of times chosen from $\mathcal{S} = \{0, 1, 2, 3, 4, 5\}$. For example,

$$1 + f_{[2,5]} = \left(\frac{(1 + r_5)^5}{(1 + r_2)^2}\right)^{\frac{1}{3}} \approx \left(\frac{1.239807347}{1.061859431}\right)^{\frac{1}{3}} \approx 1.053001668$$

and

$$f_{[2,5]} \approx 5.300166768\%.$$

In this manner, we derive the following table of forward rates for the bonds of Table (8.3.4).

Length 1	Length 2	Length 3	Length 4	Length 5
$f_{[0,1]} = 1.7568\%$	$f_{[0,2]} \approx 3.0466\%$	$f_{[0,3]} \approx 3.5939\%$	$f_{[0,4]} \approx 3.9215\%$	$f_{[0,5]} \approx 4.3929\%$
$f_{[1,2]} \approx 4.3527\%$	$f_{[1,3]} \approx 4.5248\%$	$f_{[1,4]} \approx 4.6533\%$	$f_{[1,5]} \approx 5.0625\%$	
$f_{[2,3]} \approx 4.6973\%$	$f_{[2,4]} \approx 4.8039\%$	$f_{[2,5]} \approx 5.3002\%$		
$f_{[3,4]} \approx 4.9106\%$	$f_{[3,5]} \approx 5.6029\%$			
$f_{[4,5]} \approx 6.2998\%$				

So, we have the following forward rate curves.

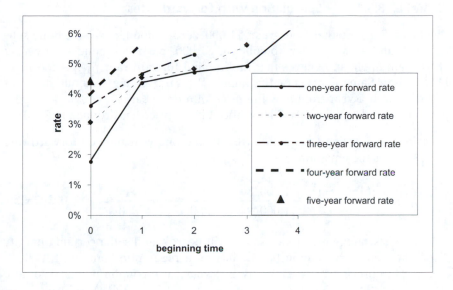

Forward rate curves for Table of bonds (8.3.4)

■

It is clear that the forward rate $f_{[t,s]}$ may be found using Equation (8.3.8), provided you know the spot rates r_t and r_s. You may also determine spot rates if you know a suitable sequence of forward rates. More precisely if $0 < t_1 < t_2 < \cdots < t_k = t$, then

$$(8.3.10) \quad (1 + r_t)^t = (1 + f_{[0,t_1]})^{t_1}(1 + f_{[t_1,t_2]})^{t_2 - t_1} \cdots (1 + f_{[t_{k-1},t_k]})^{t_k - t_{k-1}}.$$

As already mentioned, the forward rates $f_{[t,s]}$ are not usually equal to the actual market interest rate available to an investor at time t who wishes to lend his money from time t to time s. The market rate $i_{[t,s]}$ may be higher or lower than the implied forward rate $f_{[t,s]}$. If you believe that the market rate will be different from the theoretical forward rate, you may wish to act on that presumption. If your prediction is correct, you may profit. This does not contradict our arbitrage-free assumption, because *any speculation on forward interest rates only produces a profit if your prediction is correct. If you guessed incorrectly as to whether the market forward rate would be higher or lower than the implied, you would lose money.*

EXAMPLE 8.3.11 Speculating with forward rates

Problem: Suppose that a one-year, $1,000, zero-coupon, par-value bond sells for $940 and a two-year, zero-coupon, $1,000, par-value bond is priced at $860. Sam Spelman believes that the implied forward rate $f_{[1,2]}$ is too high. Discuss how Sam might act on this hunch. With the strategy you outline, what will happen if a year from now the price of a one-year, $1,000, zero-coupon, par-value bond remains $940? How about if the price has dropped to $900?

Solution According to equation (8.3.8), the one-year implied forward rate $f_{[1,2]}$ satisfies the equation

$$1 + f_{[1,2]} = \frac{(1 + r_2)^2}{1 + r_1} = \frac{1,000/860}{1,000/940} = \frac{940}{860} = 1 + \frac{80}{860} \approx 1 + .093023256.$$

Sam's strategy is as follows. He will borrow and sell a certain quantity X of one-year, zero-coupon bonds, having a redemption amount $1,000X$, and use the proceeds $940X$ to buy two-year, zero-coupon bonds. We note that the two-year bonds will have redemption amount $\left(\frac{\$1,000}{\$860}\right)(\$940X)$. Now, you may be concerned as to how Sam will cover the redemption payment of $1,000X$ that is due at the end of one year on the borrowed bonds, but Mr. Spelman has figured that out. He will borrow and sell one-year bonds at the time he needs to make the redemption payment, choosing the quantity so that their sale price will exactly equal the redemption that is due. Sam hopes that these bonds have a redemption amount, which he must again provide, that is less than the $\left(\frac{\$1,000}{\$860}\right)(\$940X)$ he will be getting from the two-year bonds. In this case, he will indeed realize a profit. In fact, Mr. Spelman will realize a profit from this scheme so long as the interest rate on one-year zero-coupon bonds at time 1 is less than the implied forward rate $f_{[1,2]} \approx .093023256$. For example, if the interest rate is the original $\frac{60}{940} \approx .063829787$, so that a one-year $1,000 bond still sells for $940, then at the end of one year he will borrow and sell $1,000X in bonds to get the $1,000X he needs in order to redeem the original X bonds. This $1,000X in new bonds is redeemable in one year for $(\$1,000X)\left(\frac{1,000}{940}\right)$. In this case his profit will be

$$\left(\frac{\$1,000}{\$860}\right)(\$940X) - (\$1,000X)\left(\frac{1,000}{940}\right) = (\$1,000X)\left(\frac{94}{86} - \frac{100}{94}\right) = \frac{\$236,000X}{8,084}.$$

Should the price instead drop to $900 for a one-year $1,000 par-value zero-coupon bond, with the higher yield rate $\frac{100}{900} \approx .11111111$, Sam's profit would be $\left(\frac{\$1,000}{\$860}\right)(\$940X) - (\$1,000X)\left(\frac{1,000}{900}\right) = -\frac{\$14,000X}{774}$. Poor Sam would have lost money! It is this last case, or one where the price of a one-year bond has fallen even more, that keeps Sam from letting X be too large! ∎

We have introduced three types of curves to show the term structure of interest rates, namely yield curves, spot rate curves, and forward rate curves. There is a fourth type of curve that you might see. It is called the **discount curve**, and you plot pairs (t, v_t) where $v_t = \dfrac{1}{1+r_t}$ is the discount factor associated with the spot rate r_t. The discount curve for the bonds of Table (8.3.4) is as follows.

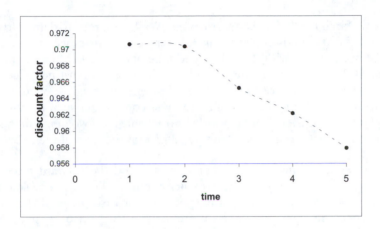

FIGURE (8.3.12) Discount curve for Table of bonds (8.3.4)

8.4   FORWARD CONTRACTS

There are several financial instruments that are **derivative** securities. A derivative is a financial investment whose worth is derived from that of some other asset or index. The remainder of this chapter discusses various derivatives, and in this section we consider the simplest of these, which are called **forward contracts**.

A **forward contract** is a contract that specifies that one party will sell a specified asset, the **underlier**, to the other at a designated date and price. Delivery details should be spelled out. The party contracting to buy the asset is said to take a **long position**, while the obligated seller assumes a **short position**. The holder of a long position, having locked in a price, hopes that the price of the asset will be higher at the agreed upon settlement date: then the buyer has contracted to get a bargain. Of course the contracted seller is hoping not to have to part with the asset at a price that is lower than the market price at the delivery date. In a forward contract, each party faces the risk that the price will change in an unfavorable direction. There is also a possibility of a party failing to live up to the contracted obligations. Sometimes, the agreement is collateralized with earnest money or assets securing the contract, thereby lessening **credit risk**, the risk of default. When two parties enter a forward contract, among the things they should spell out are whether the contract is transferable. If a party may trade its position in the contract, this will be a private (over-the-counter) transaction. Since forward contracts are individual agreements between two parties, there is much variation in the arrangements!

EXAMPLE 8.4.1   Forward contract

On March 18, 2005, fisherman Tommy Wallace agreed to deliver to Gavin McGraph 5,000 pounds of live Maine lobsters, each lobster weighing between one and five pounds, with the delivery to take place at 10AM on the morning of June 10th, 2005. The average weight of the lobsters must not exceed two pounds. The lobsters are to be delivered in one hundred wooden crates to Mr. McGraph's warehouse. Each crate must contain between forty-five and sixty pounds of lobster. The forward contract that the two gentlemen signed obligated Mr. McGraph to pay Tommy $25,500 at the time of the delivery. ■

The price set in a forward contract is called the **forward price**. It is frequently compared to the price at which the underlier could be sold at the time the contract is established, the **spot price** or **cash price**. If the forward price exceeds the spot price, the difference (forward price − spot price) is called a **forward premium**. If the spot price is higher than the forward price, you may hear the difference (spot price − forward price) called a **forward discount**. Storage costs and interest rates are among the factors that influence the relationship between forward prices and spot prices. The amount (possibly negative) by which the spot price exceeds the forward price is called the **basis**. So, the basis is equal either to the forward discount or to the negative of the forward premium.

We wish to consider how two parties might reasonably settle upon a forward price. For simplicity, we assume that there is a level term structure for risk-free interest rates, the annual effective risk-free rate being i. First suppose that the contract is for an asset with current price S_0 and that there are no costs associated with holding the asset and no income received by the holder of the asset. Then, the forward price K should equal $S_0(1 + i)^T$ where T is the length of time from contract initiation until maturity. The justification for this is a "no-arbitrage" argument that we discuss in the next paragraph.

There are two ways to now ensure that you hold the asset at time T. The first way is to purchase the asset now at a cost of S_0 and then hold on to it. The second way is to enter into a forward contract with forward price K and to invest $K(1 + i)^{-T}$ at the risk-free rate i, thereby ensuring you have the money required at time T to procure the asset under the provisions of the forward contract. Since neither of these involve any price uncertainty, they must have the same cost by the "law of one price" in the "no-arbitrage" model. So, $S_0 = K(1 + i)^{-T}$ and $K = S_0(1 + i)^T$.

This "no-arbitrage" argument may be modified if there are fixed cash-flows associated with holding the asset for the term of the forward contract, for instance, dividends paid or storage costs. If H denotes the value at the time the contract is being drawn up (determined using the risk-free accumulation function) of cashflows to the holder of the asset during the term of the forward contract, then the price of the forward contract should equal $K = (S_0 - H)(1 + i)^T$ if there is not to be an arbitrage opportunity [see Writing problem (8.0.3)]. Of course, in practice, forward contracts are entered into by a pair of individuals, and market forces do not necessarily dictate the price. Therefore, it is unrealistic to think that the price will necessarily be the "no-arbitrage" price.

Let's now consider just who might be a party to a forward contract. One possibility is a *speculator* who may enter into a forward contract, seizing a perceived opportunity to make a profit if the forward price is favorable. Another is someone wishing to *hedge risk*; if you currently feel you are too vulnerable to a downshift in the price of an asset, you may acquire a forward contract for that commodity in which you have the short position and hence would benefit from a price decrease. For example, a farmer who is growing wheat may wish to lock in a price for his crop, since if the price falls, he might have to sell off land to pay his creditors. Similarly, a jeweler, who has advertised that prices for his designs will be good through the end of the year, may wish to enter a forward contract to purchase a certain quantity of gold. There is a large range of assets for which forward contracts exist. These include agricultural products, currencies, and bonds.

Forward contracts are not a new invention. There is evidence, on Mesopotamian clay tablets, of such agreements having been made about 3,700 years ago. Seventeenth century Chinese feudal lords collected rice from their workers, which they sold on the Osaka rice market, and forward contracts for the delivery of rice were common. In Europe, there is evidence of forward trading in medieval times. Prior to the sixteenth century, forward contracts may just have involved the trading of that year's expected crops or catch. Trading became better organized and more sophisticated in Antwerp (where an exchange opened in 1531) and in Amsterdam (a principal market for grain, herring, and other bulk commodities). Still, the parties in forward contracts were not yet outside investors who viewed contracts as commodities in their own right. In the United States, forward contracts became common in the nineteenth century.

8.5   COMMODITY FUTURES HELD UNTIL DELIVERY

A **futures contract** is somewhat similar to a forward contract, but it differs in several ways. Firstly, a futures contract is *standardized* rather than designed to suit the participants. Secondly, futures contracts are traded on a **Futures**

Exchange, [6] [7] a derivatives marketplace (physical or online) with established rules and regulations. Among the rules that govern the operation of Futures Exchanges is the requirement that there be a **clearinghouse**. A clearinghouse is also called a **clearing organization** or **clearing association**. Thirdly, unlike forward contracts, which are each a contract between any two willing parties, the counterparty to each futures contract is the **clearinghouse** of a Futures Exchange. In this section we will begin to study these three distinctions and their consequences. One important consequence, namely that futures contracts tend to be highly liquid, is deferred until Section (8.6).

Note that like a forwards contract, a futures contract spells out a specific asset the (**underlier**), which is to be exchanged at the expiration date (usually a month or a portion of a month). Naturally the contract must clearly and fully describe the underlier, the price at delivery, and delivery instructions. In this section and in Section (8.6), the underliers we consider will be physical commodities. [8] Section (8.7) includes discussion of other types of underliers. In many contracts, speculators account for over 60% of the trades! The reason for this is that, as we shall illustrate in Examples (8.5.5) and (8.6.4), traders of futures may have a large amount of leverage.

EXAMPLE 8.5.1   Kansas City Board of Trade hard red wheat future

The **Kansas City Board of Trade** (**KCBT**) commenced trading a wheat future in 1876. The KW (ticker symbol) contract has as its underlier 5,000 bushels of #2 grade hard red winter wheat. Here, the grade is determined by government standards. Grade #1 may be substituted at a $1\frac{1}{2}$ cent premium per bushel, and grade #3 is allowable but results in a discount of five cents per bushel. Hard red wheat is grown primarily in the Great Plains, and it accounts for about 45% of the total United States wheat production.

The wheat can be delivered in Kansas City, Missouri. It can also be delivered in Hutchinson, Kansas, but then there is a nine cents per bushel discount. Allowable delivery dates are from the first business day of the liquidating month through the next to last business day of the month. There

---

[6]The **Chicago Board of Trade** (**CBOT**) opened in 1848 and was the first Futures Exchange. There were earlier markets that called themselves "Futures Exchanges," but they did not trade what we now call futures contracts. Rather, these grain exchanges, located in Buffalo and New York City among other places, traded individualized forward contracts. The advantages of modern futures trading quickly became apparent, and Futures Exchanges were established elsewhere in the United States. Chicago has, however, continued to house the most important exchanges.

[7]In the United States, the legal term for a Futures Exchange is a **designated contract market**.

[8]Early futures contracts had physical commodities as underliers, and it was fundamental that these were products with a cash market. So, the seller did not necessarily need to hold the product he promised to deliver, and it was not required that the buyer have a need for the product. This allowed speculators to enter the market as well as hedgers.

is no trading date after the fourteenth of the month. There are hard red wheat future contracts with delivery months March, May, July, September, and December.

Since it was established in 1913 by the KCBT, the KCBT Clearing Corportion has been the counterparty to each trade, and it has never defaulted. Trading hours are on business days from 9:30 AM until 1:15 PM, Central Time. During these hours, trading takes place by "open outcry" with traders shouting bids and offers in the "trading pit." The minimum price fluctuation from a previous bid is $12.50 per contract, and the maximum allowable fluctuation from the previous day's "settlement price" is $1,500.

In October 1984, hard red wheat futures with KE ticker symbol and identical contract specifications began trading electronically. Trading hours are 6:32 PM to 6 AM, Central Time Sunday through Friday.　■

EXAMPLE 8.5.2   Chicago Board of Trade gold future

The Chicago Board of Trade (CBOT) offers a futures contract (ticker symbol ZG) with the underlier being "100 troy ounces ($\pm 5\%$) of refined gold, assaying not less than .995 fineness, cast either in one bar or in three one-kilogram bars, and bearing a serial number and identifying stamp of a refiner approved and listed by the Exchange." The delivery requirements are physical delivery to specified New York area vaults. The trading is 100% electronic with trading from 6:16 PM to 4 PM Central Time, the next day.

CBOT also offers a "mini-sized" gold futures (ticker symbol YG) with the underlier being 33.2 ounces of refined gold. Delivery requirements and trading hours and platform are identical.　■

These examples included many of the specifications of the futures contract. In addition to the type of information given in Examples (8.5.1) and (8.5.2), specifications for a contract include the minimum price change per contract, called a **tick**. The fluctuation from the previous day's settlement price may be limited as may be the total number of contracts to be held by any one party.

It is natural to ask who might be interested in trading futures such as the above wheat and gold contracts. Just as for forward contracts, the traders fall into two categories, hedgers and speculators. Credit risk is a significant consideration if you enter a forwards contract. In contrast, credit risk is apt to be minimal if you take a position in a futures contract. This is because your counterparty is the futures' clearinghouse, and an exchange stands behind its clearinghouse. Furthermore, exchanges (and their clearinghouses) are subject to regulation that helps safeguard traders. [9]

There is also a monetary amount called the **futures price** associated with a futures contract. This is an amount agreed upon by the two parties, the holder of the **short position** (the seller) [10] and the holder of the **long position** (the buyer).

If you assume a **long position** in a futures contract, you are contracting the privilege (and duty) of paying the futures price in return for delivery to you of the underlier. As we shall see shortly [in Examples (8.5.4) and (8.5.5)], the timing of your payments depends upon the futures prices of later traded contracts having the same underlier and delivery requirements. In fact, your position may give rise to intermediate positive and negative cashflows! *However, your total outflows will exceed your total inflows by an amount equal to the contracted futures price.* So, the holder of a long futures position has, at least in some sense, locked in a purchase price. In hindsight, entering a long position was a profitable position for the long seller to have taken if the spot price at expiration of the underlier is above the contract's futures price.

If you establish a **short position** in a futures contract, you are contracting the right (and the obligation) to deliver the underlier at expiration and to receive the futures price. The sequence of cashflows that you will receive depends on subsequent prices on contracts with the same underlier and delivery instructions. However, if you hold the contract through expiration, you should receive more money in inflows than you pay in outflows, with the difference equaling the futures price you contracted. After the fact, a short position was valuable if the futures price at the time the position was established was greater than the spot price of the underlier at expiration.

If you enter into a long futures position, this may be described by saying you are **buying** the futures contract. Taking a short position is also referred to as **shorting** or **selling** the futures contract. In either case, you are said to have **opened a position**. But you are *not* actually buying or selling the *contract*; rather, you are contracting to buy or sell the *underlier*.

---

[9] The U.S. federal government first passed legislation pertaining to futures trading in 1923. The **Commodity Futures Trading Commission** (**CFTC**) was created by the Trade Commission Act of 1974 and oversees all futures trading. The **National Futures Association** (**NFA**), an industrywide self-regulatory body, also monitors futures brokers and their agents. Certain exchange functions are also regulated by the Securities and Exchange Commission, the Federal Reserve Bank, and the U.S. Treasury Board.

[10] Perhaps you will find it helpful to note that "sell" and "short" both begin with the letter "s".

A trader of a futures contract establishes either a **long position** or a **short position**. Since the holder of a long position has an obligation to accept delivery of the underlier at the expiration date and pay for it, the party with a long position is said to **buy** the contract. The holder of a short position has a commitment to deliver the underlier and receives money for doing this, so the party with an open **short position** may be referred to as the **seller of the contract.** Both buying and selling are described as **trading**. As was the case with stocks, if you have a long position you benefit from price appreciation. Once again, price decline is favorable to a short seller.

If you use the expressions "selling a futures contract" or "buying a futures contract," *it is important to remember that the sequence of cashflows received by the "seller" or "buyer" may be complicated.* As the following two examples show, *the cashflow sequence is not determined at the time the position is opened,* although the method for determining the cashflows as time passes has been established. However, if the futures price of futures contracts with the same underlier does not fluctuate between the time you open your position in the futures contract and expiration, then your cashflows will be just as they would have been had the contract been a forward contract with an earnest money deposit. We will consider the cashflows of the parties in a futures contract beginning with Example (8.5.4).

We now turn to the mechanics of trading futures contracts. A futures contract is traded through a brokerage firm that holds a **seat** on the Futures Exchange that trades the particular futures contract. Just as is the case with stocks, it will be necessary for you to open an account with a broker. Now, you will need to identify a security you would like to trade, and determine a position you wish to take, either long or short. No matter whether you buy or sell the contract, you will be asked to make a margin deposit. This is commonly between 2% and 15% of the current cash value of the asset due from the seller at expiration. Your margin deposit is quite different from a stock margin deposit, which is collateral on a loan. Here there is no loan. The purpose of the margin deposit is to cover any losses you may incur as a result of movement in the futures price, so the required amount will reflect the volatility of the contract. The initial margin is often set equal to the maximum one-day loss you might reasonably expect to incur on your contract. As is the case for stocks, there is an **initial margin requirement** and a **maintenance requirement**.

Each day, your margin balance is adjusted by any gain or loss in the trading value of your position for that day. This process is called **marking-to-market**, and the daily futures price used for this process is called the **settlement price**. Each day's settlement price is determined by a formula, and the settlement price may differ from the last futures price of the day. Commonly, the settlement price is some sort of average of futures prices near the end of the day's trading.

FACT 8.5.3:

• If the futures price of currently traded futures contracts with a specific underlier goes up by $U, then $U is added to the balance in every buyer's margin account and $U is subtracted from the balance in the account of every contract seller.

• If the futures price of currently traded futures contracts with a specific underlier goes down by $D, then $D is subtracted from the balance in every buyer's margin account and $D is added to the balance in the account of every contract seller.

For the clearinghouse, which handles the marking-to-market, there is no profit or loss from marking-to-market. The purpose of marking-to-market is to diminish default risk. Also, as we shall see in Example (8.6.2), marking-to-market makes the process of squaring accounts easy, even if a trader terminates his position before delivery: We will shortly explain how a trader can do this.

EXAMPLE 8.5.4 Marking-to-market

At the end of one day Klaus established a short position in a December 2005 CME milk class III future that had as underlier 200,000 pounds of "cheese milk." Such milk is used mainly in the manufacturing of cheddar cheese. The futures price when Klaus sold the contract was the same as the day's settlement price, namely $26,800. The next day, a similar contract had settlement futures price $26,940. This was a $140 price increase. Klaus naturally wished he had waited a day to open his position since the next day's settlement price was $140 better for a seller like Klaus than the one he contracted. This $140 disadvantage was reflected in a $140 deduction from Klaus's margin account. On the other hand, had Klaus established a long position rather than a short one, $140 would have been added to his margin balance. ∎

If a margin account balance falls below the maintenance margin, a margin call is made. (In fact, market conditions may be such that your account balance is negative, and you will be liable for any deficit in your account.) If you receive a margin call, you must immediately deposit money so that your account again satisfies the *initial* margin requirement. The required deposit is called the **variation margin**. Failure to respond to a margin call can result in your position being closed, [11] perhaps at a very disadvantageous time, and you may be left with considerable liability to the clearinghouse. For this reason, you may wish to make deposits to your margin account that are greater than required,

---

[11] A position is closed by the opening of an offsetting position. This is discussed in Section (8.6).

especially since margin accounts can pay competitive interest. In fact, clients (especially those with large accounts) may be allowed to keep their margin in Treasury bills.

EXAMPLE 8.5.5   Margins and marking-to-market

Problem: On May 1, 2005, a futures contract has futures price $40,000 and stipulates delivery of underlier A in exactly three months (August 1). The contract has an initial margin of $5,000 and a maintenance margin of $3,500. The futures price of contracts with underlier A and expiration date August 1 stays at $40,000 until May 25, when it falls to $37,000. Once again, there is a period with a level futures price (this time at $37,000) for underlier A and expiration date August 1. However, on June 16 it jumps to $42,000. There is a final change in the futures price on July 22 to $41,000. Consider two futures traders for this contract named Sam and Bob. Bob and Sam each open positions on May 1, Bob as a buyer and Sam as a seller. They each hold the futures through expiration and accept or make delivery as required by the contract. Suppose that Sam and Bob each only had outflows as required by the initial and maintenance margin requirements, except for Bob making a payment when accepting delivery. Furthermore, there was no interest earned on the margin accounts. (a) Describe Sam's cashflows. (b) Describe Bob's cashflows. (c) Find Bob's yield rate if he sells the underlier on August 1, immediately after taking delivery, for $41,000.

Solution   We begin with a helpful graphic representing the futures price over time.

$40,000 $42,000

Arrow up: Credit buyer's account, debit seller's account
Arrow down: Debit buyer's account, credit seller's account

$37,000 $41,000

(a) Sam sold the contract on May 1 at a futures price of $40,000. At the time of the sale, Sam deposited $5,000 into a margin account at the clearinghouse. On May 25, buyers were able to enter contracts, which granted identical rights and obligations to those Sam had contracted for, and the futures price was $3,000 lower than on Sam's contract. Consequently, Sam's margin account balance was increased by $3,000, resulting in a balance of $5,000 + $3,000 = $8,000. Sam's margin account then remained at $8,000 until June 16 when new contracts for underlier A and August 1 delivery had futures price $42,000. This $5,000 increase in the futures price led to a $5,000 decrease in Sam's margin account balance, so that it was now

$8,000 $-$ $5,000 = $3,000. Since his margin account had a balance below the $3,500 *maintenance* margin requirement, a margin call was made. Sam was required to bring his balance up to the $5,000 *initial* margin requirement, so he deposited $2,000. The July 22nd decrease in the futures price from $42,000 to $41,000 resulted in Sam's margin increasing by $1,000 to $6,000. On August 1, Sam withdrew the $6,000 margin and received $41,000 from the clearinghouse after showing that he had delivered the asset A. Note that Sam's cashflows total $(-\$5,000 - \$2,000 + \$6,000 + \$41,000) = \$40,000$. So, the sum of Sam's cashflows is equal to his contracted futures price.

(b) Bob bought the future on May 1 when the futures price was $40,000. To establish his long position, he deposited the required $5,000 to a new margin account. During the time Bob's position was open, his margin balance was adjusted to match changes in the futures price. So, when the price dropped from $40,000 to $37,000, Bob's balance in his margin account also dropped by $3,000 to $5,000 $-$ $3,000 = $2,000, below the $3,500 maintenance margin requirement. Bob deposited $3,000 on May 25 to once again attain the $5,000 initial margin. The $5,000 jump in the futures price on June 16 resulted in Bob's margin account's balance increasing to $5,000 + $5,000 = $10,000, and the July 22 drop in the futures price from $42,000 to $41,000 caused it to become $10,000 $-$ $1,000 = $9,000. On August 1, Bob used the $9,000 withdrawn from his margin account, along with another $41,000 $-$ $9,000 = $32,000 to pay for the underlier. Bob only had outflows, and the sum of these was $5,000 + $3,000 + $32,000 = $40,000, the futures price when Bob opened his long position.

(c) Read through the solution to part (b) and you will note that a complete list of Bob's *net* cashflows is as follows: On May 1, Bob paid $5,000; on May 25 Bob paid $3,000; on August 1, Bob received $41,000 $-$ $32,000 = $9,000. The period from May 1 to August 1 consists of $30 + 31 + 31 = 92$ days, and the period from May 25 to August 1 has $92 - 24 = 68$ days. Therefore, if j denotes Bob's *daily* yield rate, j satisfies the (August 1) equation of value

$$\$9,000 = \$5,000(1 + j)^{92} + \$3,000(1 + j)^{68}.$$

So, as you may discover by "guess and check," Newton's method, or using the BA II Plus cashflow worksheet, $j \approx .00141844852$. Since there were 365 days in the year 2005, Bob's annual yield was approximately $(1 + j)^{365} - 1 \approx$ .677604473 $\approx 67.76\%$. During the three-month period, Bob put in a total of $8,000 and received $9,000. Nice going Bob! ∎

8.6 OFFSETTING POSITIONS AND LIQUIDITY OF FUTURES CONTRACTS

Suppose you hold a short futures position *and* a long position in *identical* futures contracts and these positions share a common margin account. (When you have multiple positions, having a common margin account for all your positions is advantageous, because then you just need the total in the account to meet the sum of the margin requirements for the positions. This helps you avoid margin calls.) Recall that the clearinghouse is counterparty to each of your futures contracts. The long position obligates you to pay the expiration forward price \$P to the clearinghouse and to take delivery of the underlier. The short position mandates that you relinquish the underlier to the clearinghouse and that you receive payment of the amount \$P. But this leaves you with a net cashflow of \$0 and with no change in the amount of the underlying asset that you own. The same is true for the clearinghouse. So, it would be a waste of time (and perhaps of other resources) for you to carry out either delivery. Because of Fact (8.5.3) on marking-to-market and the maintenance margin requirement, you might be concerned that there would be other payments prior to delivery. However, by holding both the short position and the long position, with each daily movement of the futures price your margin account, which is used for both positions, has offsetting increases and decreases. Additionally, you would have money just sitting in a margin account, even though there is no risk in your combined short and long positions. The long position and the short position are said to be **offsetting**, and the clearinghouse cancels them both and allows you to withdraw the balance from your margin account if it is not needed for other positions you hold. A pair of contracts can only be offsetting if they have *exactly* the same underlier. The delivery date and requirements must be the same for the two contracts, and the commodity must also be identical.

Quite likely you are wondering why you might take both a short futures position and a long position on identical contracts. The answer involves the *times* when you acquire each of the two positions. These are almost surely different from each other. You take a position in a futures contract and then, when you wish to be rid of that position, you take the **offsetting position** on an identical contract. This is called **covering** or **closing** your position. If you have a short position in a futures contract, you cancel that contract by establishing a long position in the future. To cover a long position, you establish a short position.

An *individual* is unlikely to take a short position and simultaneously take a long position in an identical futures contract. However, that is what the *clearinghouse* does each time the exchange handles a futures transaction. The clearinghouse matches a buyer's and seller's offers, then splits them legally into two futures contracts, each with the clearinghouse as a counterparty. Therefore, if the clearinghouse holds exactly s short positions in a particular futures contract, it will also hold exactly s long positions in that future. Consequently, the portfolio of positions held by the clearinghouse consists entirely of offsetting positions! The only risk to the clearinghouse is a result of credit risk.

Now that we understand that if you establish a position in a futures contract, you need not hold it through delivery, we wish to consider some examples involving closing a position early.

EXAMPLE 8.6.1   Closing a position

Problem: On May 1, Sergio Reynolds bought a New York Board of Trade (NYBOT) domestic sugar contract for 112,000 pounds of sugar to be delivered in October. The futures price was $9,979.20 when Mr. Reynolds established his long position. The initial margin requirement was $700 and the maintenance margin requirement was $500. As the buyer of the contract, Mr. Reynolds is hoping that the price of October sugar goes up. On June 1, he notes that an October sugar contract, again for delivery of 112,000 pounds of sugar, has futures price $10,483.20. If he is content with a $504 gain, what should he do? If he still believes that the future is underpriced, what action is appropriate?

Solution   When the price goes up by $10,483.20 − $9,979.20 = $504, the clearinghouse accordingly adjusts the balance of his margin account upward by $504, and he is now effectively the buyer of a futures contract at $10,483.20. If Mr. Reynolds wishes to collect his $504 gain, he just needs to have his broker place a market order to sell the futures contract. Of course, he must realize that the price of the contract may change at any time: This is called **slippage**. Therefore, he is not guaranteed that the sale takes place at exactly $10,483.20. If Mr. Reynolds wishes, he may place a **limited order**, indicating that he only wishes to sell if the price is at least $10,483.20. If Mr. Reynolds believes the contract will experience significant further appreciation, he might want to hold on to it.   ∎

When we introduced marking-to-market, we mentioned that it allows easy reconciling of accounts even if a position is closed before delivery. The next example shows how this works.

EXAMPLE 8.6.2   Marking-to-market with offsetting positions

Problem: Return to the contract of Example (8.5.5). Suppose that instead of holding his long position until expiration, Bob covered it on July 10, at which time the futures price was $42,000. He did so by establishing a short position. This was possible because Brian simultaneously opened a long position. (a) Describe Bob's cashflows. (b) Describe Brian's cashflows. (c) How does Bob's behavior impact Sam?

Solution   (a) Bob's cashflows are unchanged prior to July 10. We recall that these were outflows of $5,000 on May 1 and $3,000 on May 25. Further recall that the balance in Bob's margin account on July 10 was $10,000. Bob covers his long position on July 10, so he is allowed to withdraw the $10,000. Note that the sum of Bob's cashflows is −$5,000 − $3,000 + $10,000 = $2,000 =

$42,000 − $40,000, the amount the futures price has risen during the time he was a buyer of the futures contract. Having held a long futures position, he benefited from the futures price going up.

(b) Brian's cashflows are very simple. On July 10, he deposited $5,000 to his margin account. On July 22, when the futures price has dropped to $41,000, his margin account balance has decreased to $4,000. On August 1, Brian paid $37,000 to Sam (or to the clearinghouse who passes the money on to Sam) and took delivery of the underlier.

(c) Sam makes delivery as instructed in the contract and gets the money he is owed from the clearinghouse. Sam is unaffected by Bob's behavior once Bob has established his initial position.   ■

We have discussed the possibility that you might terminate a futures position prior to delivery. To do so, you need to be able to obtain the offsetting position to the one you already hold, so you need the futures contract to be actively traded, that is, to be "liquid." We next consider how a cash market for an underlier works to make a futures contract more liquid.

If potential investors expect the price of the underlying commodity to be significantly above that of the futures contract at expiration, there would be a demand for long positions in the contract by individuals willing to take delivery and then sell the commodity at a profit. This demand would push up the price of the position. Likewise, investors who think the price of the underlier is apt to be significantly below the futures price would be interested in holding a short position. This would create a demand for the short futures position. As the time until contract expiration decreases, the uncertainty as to what price the delivered commodity would fetch usually lessens.

> **IMPORTANT FACT 8.6.3:** As the settlement date approaches, the futures price tends towards the cash price, an allowance **basis** being made for the costs of taking delivery and selling.

Physical delivery occurs in only a small percentage (less than 2%) of futures contracts. Usually, the delivery obligation is canceled out by the purchase of an offsetting (or covering) position. The veracity of Fact (8.6.3) is a major reason why this is true. You can usually obtain approximately the same amount of money by covering your position as you can obtain by taking delivery and then selling the underlier. Moreover, delivery may be rather inconvenient. If you opt for delivery, you may get a warehouse receipt for the product at a far-off location. So, even if you want the underlier, you may do better to cover your position and then purchase the commodity on the cash market.

Hedgers are more likely to take delivery than speculators, since they may have an actual need for the underlier. However, if you take a futures position as a hedge, you may not have found it possible or advantageous to take that position in a future for *exactly* the underlier you want. For example, a jeweler might purchase a futures contract for 99.99% pure silver but actually use sterling silver. A university system might use heating oil but buy a futures contract for crude oil. A wheat farmer with an anticipated July harvest might purchase an August wheat future. Trading a futures contract with one underlier to hedge against price fluctuations in another is called a **cross hedge**, and cross hedgers almost always cover their positions rather than taking physical delivery.

In Example (8.5.5), Bob had a high yield rate for his investment period. We end this section with another example, this time with no delivery, in which a futures trader has a very high yield rate.

EXAMPLE 8.6.4   Futures contracts and leverage

Problem: Michelle Schafer buys a July 2005 orange juice futures contract that trades on the New York Board of Trade. At the time her buy order is filled, the futures price for the 15,000 pound frozen concentrated orange juice contract (with origin Florida or Brazil) is $14,452.50. The initial margin is $1,050 and the maintenance margin is $750. She holds the contract for one month, during which time she does not receive any margin calls, and she sells the contract when the futures price is $15,067.50. By what percent does the price rise, and what is Michelle's monthly yield rate, assuming she deposits only the initial margin required and that the margin account pays interest at a nominal rate of 2.4% convertible monthly?

Solution   The price of the contract increased by a factor of $\frac{\$15,067.50}{14,452.50} \approx$ 1.042553191. So, the price rose by about 4.26%. Michelle's yield rate for the one month was much higher because of leveraging. (The initial margin was only about 7.27% of the purchase price.) Michelle sold the contract for $615 more than she bought it for and also earned $1,050\left(\frac{.024}{12}\right) = \$2.10$ interest on her margin deposit. Her initial deposit of $1,050 grew to $1,050 + $615 + $2.10 = $1,667.10. Therefore, her monthly yield rate is an impressive $\frac{\$1,667.10}{\$1,050} - 1 \approx 58.77\%$.   ∎

In Examples (8.5.5) and (8.6.4), the investors realized high yields thanks to leverage. Traders of futures have a tremendous amount of leverage available, since margin requirements reflect a small percentage of the market value of the underlier. However, you need to remember that *leverage exaggerates negative yield rates as well as positive yield rates*. Futures traders must be aware that their potential loss is greater than the amount of their margin deposit!

8.7   PRICE DISCOVERY AND MORE KINDS OF FUTURES

While futures originated with physical commodities as underliers, futures based on financial products now dominate, accounting for about three-quarters of the traded derivatives. The first **financial futures** were introduced in 1971 by the **Chicago Mercantile Exchange (CME)**, the largest United States Futures Exchange. [12] The introduction of these first financial futures, which were seven different foreign currency contracts, was spurred by the abolishment of the "gold standard." Their introduction was somewhat controversial. Viewing foreign currency as a traded asset whose value fluctuates requires a greater level of abstraction than that needed for understanding the trade in advance of corn or cattle!

Traded financial futures have values determined by stock indices, currency exchange rates, and debt instruments. The latter may be referred to as **interest rate futures**, and date back to 1976 when CME first offered a 90-day U.S. Treasury Bill contract. [13] The above list of types of financial futures is not exhaustive. The important thing to note is that *each financial future must have a certain rate that determines the payoff at settlement.* Recently introduced types of financial futures include products tied to the volatility of stock markets.

In contrast to commodity futures, some financial futures are **cash settled**, meaning that there is not an option for physical delivery: Instead, a settlement in cash is figured based on the spot prices at settlement.

Investing successfully in financial futures requires an economic sophistication well beyond that needed for commodity futures. However, financial futures take their roots from the contracts devised to allow farmers to manage price risk, and the mathematical considerations are the same. We include the following example for those of you who want an illustration of an important financial futures contract. It is not essential for what follows.

EXAMPLE 8.7.1   Eurodollars

In the 1960s Eurodollars originated when East European countries deposited U.S. dollars in West European banks that, in turn, made loans denominated in dollars. These were *not subject to U.S. banking regulation*, and hence the banks could operate on narrower spreads between lending and borrowing interest rates. This practice expanded beyond Europe, and the name Eurodollar is now used for CDs or other bank deposit liabilities denominated in U.S. dollars *but not subject to U.S. credit controls*. Since 1981, non-U.S. residents have been able to conduct business, without U.S. banking

---

[12]To give you some idea of how active this exchange is, in the first quarter of 2005 it averaged settlements of $1,500,000,000 (1.5 billion dollars) per day and its collateral deposits totaled over forty billion dollars.

[13]Recall, Treasury bills may be resold on the secondary market.

regulations, at International Banking Facilities (IBFs) located in the United States, and bank deposits at IBFs are also called Eurodollars. The United States does not stand behind Eurodollars. Instead, investors rely upon the issuing institutions not defaulting. When companies or individuals choose to deposit dollars in a bank not subject to U.S. regulations, they may be sacrificing a level of protection for a better interest rate.

In 1981, the Chicago Mercantile Exchange (CME) introduced the Eurodollars futures, the first futures contract that was settled in cash rather than by physical delivery. This contract has a very high volume of open contracts and has been described as "the most actively traded futures contract in the world." [14] The relevant interest rates, for the Eurodollars futures, are *three-month interbank rates* on Eurodollars for quarters ending in March, June, September, and December for up to ten years in the future and for the quarters ending in the nearest four additional months.

The value of a Eurodollar contract is quoted in a manner so that a quote of 100 corresponds to a cash settlement of 1,000,000 Eurodollars, and this corresponds to the relevant three-month Eurodollar interbank interest rate being 0%. For each .01% that the effective three-month interest rate exceeds 0, the payoff decreases by $25. So, if the rate were 1.22% at expiration, the contract would pay $1,000,000 - 25(122) = 996,950$ Eurodollars. A quote of Q corresponds to a payoff (in Eurodollars) of $750,000 + 2,500Q$. The final quote is $100 - I$ where I is the effective three-month interest rate.   ∎

Observe that a CME Eurodollars future has a cash payoff that is based on an interest rate, and a stock market index future has a payoff based on the value of a specified stock index at a designated time. The interest rate and the stock market index are numbers that seem distinctly financial, but a future could be marketed tied to the value of any variable. Recent successful entries into the market include CME weather futures using average temperature indices in various locations. Crucial to their success is the fact that there are hedgers who wish to diminish risk from extra energy costs necessitated by the weather as well as businesses whose volume is affected by the weather (e.g., hotels).

When futures are traded, all bids, offers, and sales are public. This results in exchanges serving as venues for "price discovery." The current price paid to buy or sell a contract may reasonably be viewed as resulting from the assimilation of the collective knowledge of all traders. The market price fluctuates so that supply and demand should balance, and both supply and demand reflect the available knowledge with regard to the many factors

---

[14]Statements found on www.tradecenterinc.com in August 2005 included the description of CME Eurodollars as "the most actively traded futures contract in the world with open interest recently surpassed the four million mark." This statement means that there were more than four million contracts open at once, and each contract has a principal value of $1,000,000.

that will determine the futures price. The myriad influences may include (but are not limited to) forecasts of weather, political developments, shifting popularity, and new technology developments. It is unreasonable to think any one person would be an expert on all these areas, and yet the futures price takes them all into account! This predictive power is what is meant by **price discovery**.

When we have talked about futures contracts, we have meant contracts that are traded on regulated exchanges. At the moment, this does not include "Event Markets" that sell contracts based on whether an event occurs (e.g., Howard Dean wins the 2008 Democratic presidential nomination). However, "event futures" are evolving and may someday trade on exchanges. They are noteworthy as a discovery instrument as well as for financial gain or protection [see Writing problem (8.0.7)].

8.8 OPTIONS

Forward contracts and future contracts *obligate* the holders to carry out a specified future sale. In contrast, an **option** gives the owner a *choice* as to whether to buy (or sell) a specified asset at a designated time and price. With **American** options, the option may be exercised at any time between the date of purchase and the expiration date. **European options** only allow for exercise at the expiration time. [15] The names do not indicate the geographic origin of the derivative although most exchange-traded options in the United States are American.

The price specified in an option contract is called the **strike price** or **exercise price**. If the option provides its owner with the right to *buy* an asset, it is called a **call option**. The owner of a call option may find it advantageous to exercise the option of the contract if the strike price is below the market price for the underlier. In contrast, a **put option** gives the holder the right to *sell* at a specified price. If the strike price is above the market price, then the buyer can make a profit by exercising the put option. For holders of options that trade on Options Exchanges, an alternative to exercising the option is taking an offsetting position. This is reminiscent of futures trading. Like forward and futures contracts, options may be used to hedge risk or to speculate.

The owner of an option is commonly referred to as an **option holder**. The option holder has paid money (the **premium** or **option price**) and in return has a *choice* as to whether the option is exercised. So, the option price is the amount the buyer of an option pays *initially* for the privilege of possibly

---

[15] In practice, there may be a short period just before maturity when option exercise is allowed. **Bermuda options** have a third type of styling with several discrete dates at which the option may be exercised.

exercising the option *later* at the strike price. [16] If you sell an option, you are an **option writer**. An option writer does not have any choice regarding whether the contract is exercised. An over-the-counter (non-exchange-traded) option has an option holder matched with an option writer.

The option writer is said to **hold the short position** (both on put and call options), and the option holder **goes long**. The option writer is the seller and his counterparty is the buyer, whether the option is a call option or a put option. So, the buyer in a put option can end up choosing to sell the underlier, while the seller in a put option can end up being forced to *buy* the underlier. In a call option, the seller sells and the buyer buys.

In exchange-traded options, which are commonly referred to as **listed options**, an option holder for a contract is only assigned an option writer for that contract if the holder declares a wish to exercise the option. This is because the **Option Clearing Corporation** (OCC) [17] serves as counterparty to every trade. So, while an options contract is only sold when there is also a buyer for the contract, the pairing of the buyer and seller is not ongoing.

Most exchange-traded options are never exercised. Instead they are sold. This is because, if an option can be profitably exercised, the market will set its premium accordingly. Over 60% of all listed option contracts each month are covered by an offsetting trade while only about 10% are exercised, most commonly in the week before expiration. The remaining 30% expire worthless.

EXAMPLE 8.8.1   Put option used for hedging

Problem: The owners of Sunnyvale Orchards are concerned that the price of peaches may be below average should weather conditions allow for a greater than usual national supply of peaches. Local weather conditions have caused the cost of producing the 5,000 pound crop of Sunnyvale peaches to be $2,950, which is $.59 per pound. The Sunnyvale owners are confident that the per pound price their crop will fetch will be between $.10 and $1.00. To hedge their possible losses if the national price turns out to be $x per pound, the owners pay $500 to purchase a put option to sell $5,000 pounds of peaches at $.60, at the time of picking. Explain how this helps control losses for x in the range $.1 \leq x \leq 1$.

Solution   First, consider Sunnyvale's situation *without* the put option. Whatever the price x per pound, Sunnyvale would sell its peaches for $5,000x$.

---

[16] **Capped style** options have a limit as to the profit that may be realized by the holder, and they are automatically exercised if the option holder can realize the maximum profit.

[17] The OCC was established in 1973, as was the **Chicago Board Option Exchange** (CBOE), the first U.S. options exchange. The CBOE is the second largest listed securities market in the U.S. following only the NYSE. Typically, over 1 million options, with a total contract value over $25 billion, change hands daily. The SEC oversees all trading on option exchanges.

Since it cost $2,950 to produce the peaches, Sunnyvale's profit would be $5,000x − $2,950, which for $.1 \leq x \leq 1$ could be as low as −$2,450 (a $2,450 loss) or as high as $2,050.

Second, consider Sunnyvale's situation *with* the put option. For $x \geq$ $.60, the strike price, Sunnyvale would *not* exercise the option, but would instead sell its peaches for $5,000x. In this case, since it cost $2,950 to produce the peaches and $500 to purchase the unused put option, Sunnyvale's profit would be $5,000x − $2,950 − $500 = $5,000x − $3,450. So, for $.60 \leq x \leq$ $1.00, Sunnyvale could at worst have a loss of $5,000($.60) − $3,450 = −$450 and at best have a gain of $5,000(1.00) − $3,450 = $1,550. On the other hand, if $x < $.60, Sunnyvale would exercise the option and sell its peaches for $5,000($.60) = $3,000. Once again, production costs were $2,950 and the option was purchased for $500, so Sunnyvale has a loss of $450.

By purchasing the put option, the owners hedged against a large loss, but they also reduced their maximum profit. *Without* the option, assuming that the per pound price they sell their peaches for is between $.10 and $1.00, Sunnyvale would have lost as much as $2,450 or gained as much as $2,050. *With* the put option, there is a narrower profit range. Sunnyvale might have a profit of as high as $1,550 or a loss of as much as $450.   ■

In Example (8.8.1), a put *commodity* option is used for hedging. Among the commonly listed options are *stock* options, in which the underlying commodity is 100 shares of a particular stock. In fact, options on more than 1,300 different stocks trade on the Chicago Board Option Exchange (CBOE).

Our next example concerns speculating with a call stock option.

EXAMPLE 8.8.2   Call option used for speculating

Problem: Darlene Leroux believes that she has identified a stock that is going to go up very soon. The stock is currently selling at $55.18 per share. Darlene purchases a call option, expiring in six weeks, for 100 shares of the stock. The option has a strike price of $55, and Darlene pays $165 to purchase it. Six weeks later the price of the stock per share is $60.97 and Darlene exercises the option, then sells the stock for $60.97. Pretending there are no commissions or other fees to be paid, what yield does Darlene achieve on her six-week investment? How does this compare to the yield she would have obtained had she purchased three shares of the stock using a cash account, assuming no dividends were declared during the six-week period? Again, assume there are no commissions or fees. How about the price for purchasing six shares of the stock on margin, assuming the initial margin requirement was 50%, there were no margin calls, dividends declared, or interest on the margin account during the six-week period?

Solution If Darlene purchases the option, then exercises it and trades the stock for cash, her initial \$165 investment has grown to $100(\$60.97 - \$55) = \$597$. This gives her a yield for the six-week period of about 262%. In contrast, had she bought three shares of stock, her yield on a \$165.54 investment would have been $\frac{\$60.97}{55.18} - 1 \approx 10.5\%$. The margined stock sale would have a yield that was 21% since the profit would be doubled (based on holding six shares instead of three) while for the margined as well as the unmargined sale, Darlene's investment would be \$165.54. ∎

Example (8.8.2) demonstrates that, thanks to *leverage*, an options trader has the potential to reap tremendous rewards. This is similar to what happens with futures. One difference is that if an investor buys a *call* option, the potential for loss is limited to the cost of the option. So, in Example (8.8.2), Darlene could have lost at most \$165. Had she instead purchased a stock future for 100 shares at \$55 per share and then had the company declared bankruptcy, her loss would have been \$5,500.

You may have tremendous risk if you are an option writer. For example, if you sell a *call* option on stocks that you do not already own, if the share price soars you will have a big loss! This may happen even if the price increase is short-lived because, as an option writer, you may be assigned to deliver the underlier. If you don't own the asset that you are required to deliver, you have to immediately buy it at the high market price, but you only get paid the strike price!

In Example (8.8.2), Darlene exercised her option at expiration and then immediately sold the stock she acquired. Since she evidently did not wish to hold on to the stock, she could alternatively have closed out her position as an option holder by purchasing an offsetting put option. The "no-arbitrage" principle indicates that the contract price should be equal to \$597, the amount she received by exercising the contract and then turning around and selling the stocks.

Suppose that an American call option has time left until maturity, but that a profit would result if the holder chose to immediately exercise the option. Then, the market price of this option (including commission) should exceed the profit obtainable by exercise and immediate sale of the underlier. This is because a buyer could either choose to exercise the option immediately or to hold on to it in hopes of even greater gain.

The gain, if any, that may be obtained by immediate exercise (and sale of the underlier) is called the **intrinsic value** of the option. (An option that may not immediately be exercised to realize a profit has zero intrinsic value.) The **time premium** is the amount by which the option's price (premium for the option) exceeds the intrinsic value.

EXAMPLE 8.8.3   Intrinsic price and time premium

Problem: Alan Jones purchases an American call option, expiring in four months, for 100 shares of Candy Cosmetics stock. He pays $225. The strike price is 5% below the current price of $18.50 per share. Find the intrinsic value and the time premium for the option at the time of Alan's purchase.

Solution   We are given that the current price of the stock is $18.50 per share, so the 100 underlying shares of Candy Cosmetics are worth $1,850. Moreover, we are told that the strike price is 5% below this price, so the intrinsic value of the option is .05($1,850) = $92.50. The time premium is the balance of the price Alan paid, namely $225 − $92.50 = $132.50. ∎

It is not easy to determine the price that the option market will assign to an American option, because the time premium is difficult to predict. There are, however, some general points worth noting. Firstly, the amount of time remaining until expiration, the volatility of the underlier, expected dividends, and interest rates should all influence the size of the time premium (and therefore the price of the option). A longer time until expiration or greater volatility should increase it, since each makes it more likely that the option will at some point have a large intrinsic value. Analogously, you pay a higher premium for a longer-term insurance policy, all other things being the same. Dividends tend to lower the share price and hence to increase the value of put options and decrease the value of call options. Naturally, if there are dividends, this will influence the optimal timing for exercising options. Premiums on call options usually move in the same direction as interest rates, while premiums on put options tend to move in the opposite direction.

You might wonder how the American or European styling of an option affects the premium. An American option allows the holder more freedom with regard to exercise time, hence may have a higher premium. However, before you conclude that you must pay more for an option to be American-styled, consider that the price for an option prior to expiration will always exceed its intrinsic value. Therefore,

> for a listed option with good liquidity and no dividends, the style (American or European) usually doesn't affect the price significantly.

We next introduce some notations for use as we study prices of *European* options. Each option contract has a contract origination date. At any one time, we will focus on one option contract, which may be a call option or a put option, and we denote the contract's origination date by 0 and its expiration date by T. For $t \in [0, T]$, S_t will denote the time t price of the underlying asset.

The contract's strike price is written as K. We use c_t to indicate the price of a call option and p_t to represent the price of a put option, each at time t.

We wish to determine how c_t and p_t are related in a *no-arbitrage model*. To do so, we shall use the "law of one price" that was introduced in Section (8.2). The law says that in order for arbitrage to be impossible, if two portfolios of investments give rise to exactly the same cashflows, they must have the same price or value at all times.

We introduce Portfolio P and Portfolio C, each giving rise to a single cashflow at time T. Portfolio P includes a single European put option, and Portfolio C contains a single European call option on the same underlier, for instance 75 shares of a specified stock or 1,000 lbs of sterling silver. The options have a common strike price K and expiration in T years. In addition to the put option, Portfolio P contains the underlier itself, so liquidation of Portfolio P at time T produces a single cashflow of value $\max(K, S_T)$. Portfolio C is completed by a zero-coupon T-year bond redeemable for K, so its value at time T is $\max(S_T - K, 0) + K = \max(K, S_T)$. So, the two portfolios produce identical total cashflows at time T and no other cashflows.

By the no-arbitrage model "law of one price," Portfolio P and Portfolio C must therefore have the same value at all times $t, 0 \leqslant t \leqslant T$. For Portfolio P, this is $p_t + S_t$, while for portfolio C it is $K(1 + i)^{t-T} + c_t$. Therefore,

(8.8.4)
$$p_t + S_t = K(1 + i)^{t-T} + c_t.$$

The no-arbitrage relationship expressed as Equation (8.8.4) is frequently referred to as **put-call parity**.

We remark that to derive Equation (8.8.4), we used an important technique that you will see again in Section (8.9).

> **TECHNIQUE (8.8.5)**: Begin with one portfolio of assets. Introduce a second portfolio that *exactly reproduces* the cashflows of the first. Under the assumption that there are no arbitrage opportunities, the original portfolio must have the same price as the newly introduced portfolio *at all times*.

8.9 USING REPLICATING PORTFOLIOS TO PRICE OPTIONS

Your intuition likely tells you that the price of an option, at a time prior to expiration, is probabilistic in nature — that is, it depends upon the likelihood of certain contingencies. For example, it seems evident that a call option at strike price K is more valuable if it is very likely that the underlier's price will be greater than K at maturity. The option is worth little if there is only a remote chance that the price will be above K at expiration. The mathematics dealing with the likelihood of random events is **probability theory.** However, and this may surprise you, there is an option pricing theory, the binomial option pricing model, that does not require a sophisticated analysis of random events. That is, it does not require probability theory. Although this model is overly simple, the binomial option pricing model is of considerable interest. In this section we introduce the model and explain how so-called "replicating portfolios" may be used to find option prices. We also give an example of how you may use a replicating portfolio in lieu of purchasing an option, provided you are prepared to practice "dynamic portfolio allocation." In Section (8.10), we present an alternative approach to the binomial option pricing method that involves weighted averages. In our discussion of the binomial option pricing model, we will focus on *European call* options.

Suppose that we wish to compute the no-arbitrage price c_0 of a portfolio P at the start of an interval of length T. (We will refer to the beginning of the interval as time 0.) We specify that P consists of units of some single underlier (for instance 100 units of IBM stock) and of options or other derivatives based on that underlier. Denote the unit price of the underlier at the start of the interval by S_0. (*Here you have a bit of choice. For example, you might think of the underlier as being one unit of the stock or as 100 units of the stock.*)

In the no-arbitrage **binomial option pricing model**, we assume that the unit price S_T of the underlier at the end of the interval can only be u (think of u for "up") or d where $d < u$; by assumption, there are no changes in the price of the underlier until the end of the interval nor any cashflows within the interval. Let V_u and V_d denote the no-arbitrage values or prices of the portfolio P at the end of the interval if the underlier's price S_T is u or d, respectively. Finally, suppose that there is a fixed effective interest rate i, the so-called **risk-free interest rate**, which applies to all risk-free investments (and loans) over the interval. We seek to find the no-arbitrage price or value c_0 of P at the start of the interval.

Note that the "no-arbitrage" assumption precludes the *possibility* that there is a gain with no risk or outlay of money. If $S_0(1 + i)^T \leq d$, then an investor could borrow S_0 at interest rate i at time 0 and use the funds to buy one share of stock. At time T, the investor would have to repay $S_0(1 + i)^T$, which is no larger, and possibly smaller, than what the investor could get by selling the stock (either for d or for u). This creates the outlawed arbitrage, so $S_0(1 + i)^T > d$. A similar argument shows that $S_0(1 + i)^T < u$. Therefore, $d < S_0(1 + i)^T < u$.

In our effort to find c_0, we **replicate** P with another portfolio P' that is easy to price. By "replicate," we mean that the new portfolio P' will consist of holdings that will again produce no midinterval cashflows and will have total value that is identical to that of P, no matter what happens to the price of the underlier. By the "law of one price" in a no-arbitrage model, the price of P must equal the price of P'. At the start of the interval, the portfolio P' consists of f in a risk-free investment and Δ units of the underlier; f or Δ may be negative, in which case the portfolio includes liabilities, namely dollars or units of the asset borrowed and to be repaid.

Graphically, we represent this situation over the interval by a diagram, commonly called a **tree**, with the option prices indicated below the underlier prices.

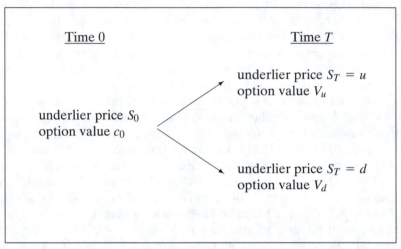

Time 0

Time T

underlier price $S_T = u$
option value V_u

underlier price S_0
option value c_0

underlier price $S_T = d$
option value V_d

TREE (8.9.1) Underlier and option prices at times 0 and T

The value of the portfolio P' (and hence also of P) at the end of the interval is

$$\begin{cases} f(1 + i)^T + u\Delta & \text{if } S_T = u, \\ f(1 + i)^T + d\Delta & \text{if } S_T = d. \end{cases}$$

So, in order that P' replicates the values V_u and V_d of P, we must have

$$(8.9.2) \quad \begin{cases} f(1 + i)^T + u\Delta = V_u, \\ f(1 + i)^T + d\Delta = V_d. \end{cases}$$

We can solve this system of equations for f and Δ, so we can determine the price of portfolio P'. Specifically, subtracting the second equation of System (8.9.2) from the first gives $u\Delta - d\Delta = V_u - V_d$ and

$$\Delta = \frac{V_u - V_d}{u - d}.$$

Substituting this value into the second equation of (8.9.2) and rearranging, we find

$$f(1 + i)^T = V_d - \left(\frac{V_u - V_d}{u - d}\right)d$$

$$= \frac{V_d(u - d) - (V_u - V_d)d}{u - d} = \frac{V_d u - V_u d}{u - d},$$

and

$$f = \frac{uV_d - dV_u}{(u - d)(1 + i)^T}.$$

So, the price c_0 of the replicating portfolio at the start of the interval, and therefore of the portfolio P if there is not to be an arbitrage opportunity, is

$$c_0 = f + \Delta S_0 = \frac{uV_d - dV_u}{(u - d)(1 + i)^T} + \left(\frac{V_u - V_d}{u - d}\right) S_0.$$

Note that the formula for Δ does not usually produce an integer, so it may not actually be possible to purchase Δ units of the underlier. However, we can still use the above fomula to determine the price c_0.

We summarize.

IMPORTANT FACT (8.9.3) (Binomial pricing model): Suppose that P consists of units of a single underlier and of options or other derivatives based on that underlier. Suppose that the price of the underlier at the start of the interval is S_0. Suppose that the price S_T of the underlier at the end of the interval can only be u or d where $d < S_0(1 + i)^T < u$; these are the only price changes for the underlier over the interval, and there are no cashflows within the interval. Let V_u and V_d denote the no-arbitrage values or prices of the portfolio P at the end of the interval if the underlier's price is u or d respectively. Finally, suppose that i is the risk-free rate of interest over the interval. Then the portfolio P is replicated by the portfolio P' consisting of f in a risk-free investment at the start of the interval with

$$f = \frac{uV_d - dV_u}{(u - d)(1 + i)^T},$$

and Δ copies of the underlier at the start of the interval with

$$\Delta = \frac{V_u - V_d}{u - d}.$$

Moreover, the no-arbitrage price of each of the portfolios is

$$c_0 = f + \Delta S_0 = \frac{uV_d - dV_u}{(u - d)(1 + i)^T} + \left(\frac{V_u - V_d}{u - d}\right) S_0.$$

As an illustration of the preceding, we look at how you find the time 0 price of a European call option with strike price K at time T, where the unit price of the underlier at time 0 is S_0 and at time T can only be either $S_T = u$ or

$S_T = d$. For this case, let the portfolio P consist of this call option alone. To express the values V_u and V_d of the option at the end of the interval, we use the standard mathematical notation $(x)_+$ for the **nonnegative part of** x:

$$(x)_+ = \begin{cases} x & \text{if } x \geqslant 0, \\ 0 & \text{if } x < 0. \end{cases}$$

If $S_T > K$, then the option is exercised and its immediate value is $S_T - K$; but if $S_T \leqslant K$, then the option is not exercised and its value is 0. That is, the value of the option is $(S_T - K)_+$. Thus, the values V_u and V_d at the end of the interval are $V_u = (u - K)_+$ and $V_d = (d - K)_+$. These values can be used to replicate the portfolio and price the call option at the start of the interval.

EXAMPLE 8.9.4   Binomial pricing model for a European call option

Problem: Consider the six-month call option from Example (8.8.2); the initial price of the stock was $S_0 = \$5,518$. (1) If Darlene believed that the price six months later would either be $6,097 or $5,200 and that interest was negligible, what price should she have been willing to pay? (2) In fact, in Example (8.8.2), the time 0 option price was $165 and the risk-free interest rate was 0. If there were just two possible prices $6,097 and an unknown d for the stock at the expiration of the contract, what must d equal to have $165 be the correct option price in the no-arbitrage binomial option pricing model?

Solution (1) We have $S_0 = \$5,518$, $K = \$5,500$, $u = \$6,097$, $d = \$5,200$, and $i \approx 0$. Then the values of the option at time 6 months are $V_u = (\$6,097 - \$5,500)_+ = \$597$ and $V_d = (\$5,200 - \$5,500)_+ = \$0$. The formula for c_0 in Important Fact (8.9.3) gives the fair price of the option to be

$$c_0 = \frac{(\$6,097)(\$0) - (\$5,200)(\$597)}{(\$6,097 - \$5,200)(1 + i)^{.5}} + \left(\frac{\$597 - \$0}{\$6,097 - \$5,200}\right)(\$5,518)$$

$$= \frac{\$597}{\$897}\left(\$5,518 - \frac{\$5,200}{(1 + i)^{.5}}\right).$$

Therefore, if $i = 0$, then $c_0 \approx \$211.645485$; if $i = .005$, $c_0 \approx \$220.2653478$. So, Darlene should have been willing to pay $211.65 if she believed the interest rate i truly was 0; she should have been willing to pay slightly more if she thought i was small but not zero (for instance, if i was .5%, then $220.27 is okay).

(2) Now we are given that $c_0 = \$165$, and d is unknown. Assuming, as is typical, that d is no larger than the strike price $5,500 so that $V_d = 0$, the

formula for c_0 in Important Fact (8.9.3) gives

$$\$165 = \frac{(\$6{,}097)(\$0) - d(\$597)}{(\$6{,}097 - d)(1 + 0)^{.5}} + \left(\frac{\$597 - \$0}{\$6{,}097 - d}\right)(\$5{,}518)$$

$$= \$597\left(\frac{\$5{,}518 - d}{\$6{,}097 - d}\right).$$

Therefore, $d \approx \$5{,}296.85$. ■

In the models of this section, so far there has been *one* time interval. We now want to consider two or more consecutive time intervals, where over each interval the price of the underlier can only either increase by a fixed percentage, say $100g\%$ ($g \geqslant 0$) or decrease by a fixed percentage, say $100b\%$ ($b \geqslant 0$), and where what happens in one interval is independent of what happens in the others. The case of just two intervals will be sufficient to demonstrate option pricing under the no-arbitrage binomial model.

If the initial price of the underlier is S_0, then at the end of the first interval its price is either $(1 + g)S_0$ or $(1 - b)S_0$. At the end of two intervals, the underlier's price must be one of $S_0(1 + g)^2$, $S_0(1 + g)(1 - b)$, or $S_0(1 - b)^2$; note that if the price decreases and then increases or increases and then decreases, the new price of the underlier at the end of two periods is the same, namely $S_0(1 + g)(1 - b)$. The following tree provides a graphical illustration of the evolution of the prices of the underlier.

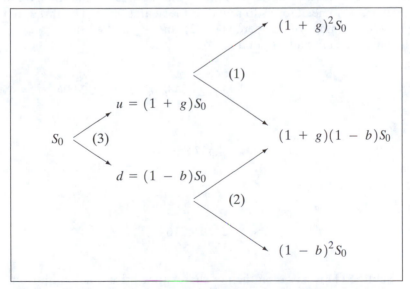

TREE (8.9.5) The evolution of underlier prices over two intervals

We want to find the no-arbitrage price of a two-interval European call option with strike price K. Our method is to use replicating portfolios and Important Fact (8.9.3) on the three two-branch subtrees [(1),(2), and (3)] of the above tree. Rather than present a general symbolic solution of this problem with lots of complicated formulas, we present two examples that illustrate the approach.

Our first example [Example (8.9.6)] concerns a two-interval call option, and the second [Example (8.9.7)] features a two-interval put option. In each case we will find the time 0 price of the option, assuming that there are no arbitrage opportunities. A broker might sell the option at time 0 for the determined price c_0, exclusive of any commissions or fees.

EXAMPLE 8.9.6   Two-interval binomial pricing model for a call option

Problem: The price of a stock is currently \$45.60 per share. A European call option for 100 shares of the stock has strike price \$44.50 per share and expires in two months. The risk-free rate of interest is an annual effective rate of 2.2%. The per share price will either go up by 4% or down by 6% each month. Use the two-period binomial option pricing model, with replicating portfolios, to determine the current price of the stock option.

Solution   We begin by making a tree showing stock and option prices for the two-month period, noting that the underlier is *one hundred* shares of the stock. We therefore have the following tree where, as usual, the price of the option is displayed below the price of the underlier. We note that since the prices listed are being used to derive the option price at time 0 rather than for actual sales, there is no need for us to round to the nearest cent. The three subtrees we plan to examine are numbered (1), (2), and (3), and we have used V_{3u}, for example, as the up-value in subtree (3).

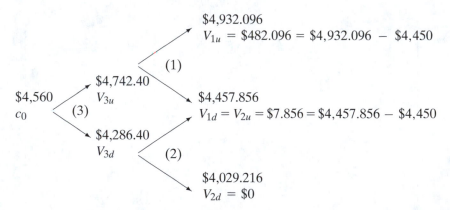

$4,932.096
$V_{1u} = \$482.096 = \$4,932.096 - \$4,450$

(1)

$4,742.40
V_{3u}

$4,560
c_0

(3)

$4,457.856
$V_{1d} = V_{2u} = \$7.856 = \$4,457.856 - \$4,450$

$4,286.40
V_{3d}

(2)

$4,029.216
$V_{2d} = \$0$

Subtree (1) is based on the per share stock price one month from now being \$47.424 = \$45.60(1.04). Imagine that at time one month, you wish to replicate the option payments of this subtree. Then, you should invest f_1 in the 2.2% annual effective risk-free account, and you should purchase $100\Delta_1$ shares of the stock (by paying $\$4,742.40\Delta_1$), where by Important Fact (8.9.3), with V_u and V_d replaced by V_{1u} and V_{1d}, we find

$$\Delta_1 = \frac{\$482.096 - 7.856}{\$4,932.096 - \$4,457.856} = 1,$$

and

$$\$f_1 = (1.022)^{-\frac{1}{12}} \left[\frac{(\$4,932.096)(\$7.856) - (\$4,457.856)(\$482.096)}{\$4,932.096 - \$4,457.856} \right]$$

$$\approx -\$4,441.937426242.$$

[In reporting the value of f_1, we note all the decimal places held in the calculator. We do this because the *stored* numerical value we subsequently calculate for Δ_3 will be used in Example (8.9.8).] Given our values for Δ_1 and f_1, by Important Fact (8.9.3), the value of the portfolio at time one month is

$$V \approx -\$4,441.937426242 + 1(\$4,742.40) = \$300.462573758.$$

This must also be the time one month price V_{3u} of the option, if the underlier price went up in the first month.

Next, we need to determine the price of the option at time one month if the per share price of the stock has dropped to $\$42.864 = \$45.60(.94)$. To do so, focus on subtree (2). The investments at time one month to replicate the option are f_2 in the risk-free account and $100\Delta_2$ shares of the stock (obtained for $\$4,286.40\Delta_2$), where by Important Fact (8.9.3), with V_u and V_d replaced by V_{2u} and V_{2d}; we find

$$\Delta_2 = \frac{\$7.856 - \$0}{\$4,457.856 - \$4,029.216} = \frac{7,856}{428,640},$$

and

$$f_2 = (1.022)^{-\frac{1}{12}} \left[\frac{(\$4,457.856)(\$0) - (\$4,029.216)(\$7.856)}{\$4,457.856 - \$4,029.216} \right]$$

$$\approx -\$73.71260403444.$$

Of course, you cannot actually purchase $100\Delta_2 = \frac{785,600}{428,640}$ shares of the stock (You can purchase one share or two shares but not an intermediate amount.), but for now we do not need to concern ourselves with that. The option value V_{3d}

is the value of this portfolio at time one month, namely $V_{3d} \approx -\$73.71260403 +$ $\$4,286.40(\frac{7,856}{428,640}) \approx \$4.84739596566$. We now redraw subtree (3), adding the possible time one month option values we just obtained,

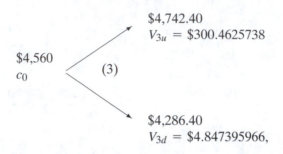

$4,742.40
$V_{3u} = \$300.4625738$

$4,560
c_0

(3)

$4,286.40
$V_{3d} = \$4.847395966,$

and we yet again solve a replication problem. This time, the investor, wishing to duplicate the payments that could be obtained by selling the option, should at time 0 invest f_3 at the risk-free interest rate and purchase $100\Delta_3$ shares of the stock; by Important Fact (8.9.3), with V_u and V_d replaced by V_{3u} and V_{3d}, we find that

$$\Delta_3 \approx \frac{\$300.4625738 - \$4.847395966}{\$4,742.40 - \$4,286.40} \approx .6482788986674,$$

and

$$\$f_3 = (1.022)^{-\frac{1}{12}} \left[\frac{(\$4,742.40)(\$4.847395966) - (\$4,286.40)(\$300.4625738)}{\$4,742.40 - \$4,286.40} \right]$$

$$\approx -\$2,768.90941607.$$

Therefore, the time 0 price of the option is $c_0 = f_3 + \$4,560\Delta_3 \approx \$187.2423583$. So, the no-arbitrage price of the option is $\$187.24$. ■

Important Fact (8.9.3) can also be used repeatedly for a multiple-interval put option. Once again, we illustrate with two periods.

EXAMPLE 8.9.7   Two-interval binomial pricing model for a put option

Problem: The price of a stock is currently $25 per share. A European put option for 100 shares of the stock has strike price $24 per share and expires in six months. The risk-free rate of interest is 4% convertible quarterly. The per share price will either go up by 10% or down by 20% each quarter. Use the two-period binomial pricing model, with replicating portfolios, to determine the current price of the stock option.

Solution This problem concerns a put option rather than a call option, but our method is similar. We again begin with a tree showing the underlier prices. As usual, we also include the values of the option, and we number the subtrees.

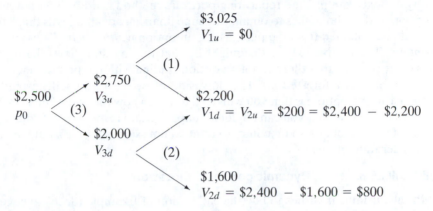

$3,025
$V_{1u} = \$0$

(1)

$2,750
V_{3u}

$2,500
p_0

(3)

$2,000
V_{3d}

(2)

$2,200
$V_{1d} = V_{2u} = \$200 = \$2,400 - \$2,200$

$1,600
$V_{2d} = \$2,400 - \$1,600 = \$800$

We note that if the underlier price rises twice, its price at time 2 is above the strike price of the option, so the put option will not be exercised. This accounts for its price V_{1u} being 0. The interest rate for a three-month period is 1%, so to find replicating portfolios for subtrees (1) and (2), use Important Fact (8.9.3), finding $\Delta_1 = -\frac{200}{825}$, $f_1 = -\frac{\$3,025\Delta_1}{1.01} \approx \$726.0726073$, $\Delta_2 = -1$, and $f_2 = \frac{\$2,400}{1.01} \approx \$2,376.2376238$. Therefore, the price of the option at time three months is $V_{3u} = f_1 + \$2,750\Delta_1 \approx \$59.40594059$ if the underlier price went up in the first quarter and $V_{3d} = f_2 + \$2,000\Delta_2 \approx \$376.2376238$ if it went down. So, if we redraw subtree (3) including these contingent option prices, it looks like

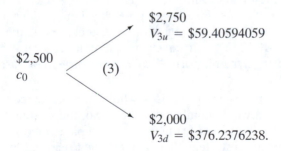

$2,750
$V_{3u} = \$59.40594059$

$2,500
c_0

(3)

$2,000
$V_{3d} = \$376.2376238.$

Using Important Fact (8.9.3) , we obtain $\Delta_3 \approx -.422442244$ and $f_3 \approx \$1,209.03$. Therefore, the price of the *put* option is $p_0 = f_3 + \$2,500\Delta_3 \approx \$152.93$.   ∎

As we have seen, replicating portfolios can be useful for pricing options so long as there is a single underlier whose price shift each period is either up by a specified percentage or down by another named percentage. As illustrated by our next example, the replicating portfolio method is useful to a potential option trader who wishes to obtain the same financial advantages as those that would be obtained from the purchase of an option *without actually buying an option*. The motivation for this might be that the investor views the option as overpriced or that there is not an option trading with the precise underlier the investor is interested in. The downside of doing it yourself, instead of purchasing an option, is that you will need to watch how the underlier's market price changes, so that you can implement appropriate responses. Of course, in real life you should also take into account commissions and fees, but we omit consideration of these.

EXAMPLE 8.9.8   Dynamic portfolio allocation

Problem: Jamaar wishes to buy the call option of Example (8.9.6) to cover a call option he has previously sold. The call option is for 100 shares of a stock that currently sells for $4,560, and the strike price is $4,450. The option expires in two months. Jamaar agrees with the assertion of Example (8.9.6) concerning the price movements of the underlier; the price will either go up by 4% or down by 6% each month. As in Example (8.9.6), the risk-free rate of interest is 2.2%. So, Jamaar believes that a fair price for the call option is $187.2423583, the value we calculated in Example (8.9.6). However, the option is selling for $250.

(a) Suppose that, at any time, Jamaar may sell any number of shares, including fractional numbers, and that at any time he may borrow any amount he wishes at the risk-free interest rate. Further suppose that the prices of a share of stock and the amounts borrowed are not limited to whole number of cents. Finally, suppose that Jamaar does not pay any commissions or fees when he buys or sells stock. Explain how Jamaar may use only $187.2423583 to cover the call option, thereby saving more than $62 by using dynamic portfolio allocation instead of purchasing the option for $250.

(b) In reality, the per share price is a whole number of cents and you may only trade a whole number of shares. So, the strategy outlined in item (1) of the solution to (a) may not be implemented. Given these limitations, suggest a strategy that Jamaar might reasonably implement, and compare it with the option selling for $250.

Solution   (a) Let Δ_1, Δ_2, and Δ_3 be as in the solution to Example (8.9.6).

(1) Jamaar should immediately purchase $100\Delta_3 \approx 64.827889867$ shares of the stock. The cost of this is $\$4,560\Delta_3 \approx \$2,956.151778$, so he needs to borrow $\$2,956.151778 - \$187.242423583 \approx \$2,768.90942$ at the 2.2% annual effective interest rate.

(2) At the end of one month, Jamaar should rearrange his portfolio. The way he should do this depends on whether the stock has gone up or down.

(3) If the stock has gone up to $\$47.424$ per share, Jamaar's portfolio's value is $(100\Delta_3)(\$47.424) - \$2,768.90942(1.022)^{\frac{1}{12}} \approx \$300.4625738$. He should then rearrange his assets so as to produce a portfolio with $100\Delta_1 = 100$ shares of the stock. So, Jamaar needs to purchase an additional $100 - 100\Delta_3 \approx 35.17211013$ shares at a cost of $(35.17211013)(\$47.424) \approx \$1,668.002151$. Towards this end, he borrows $\$1,668.002151$ at an annual effective rate of 2.2%. At the end of the second month, since the price of the stock is above the $\$4,450$ strike price of the call option, the option will be exercised, and Jamaar (the seller of the call option he wishes to cover) will be required to deliver the stocks for payment of $\$4,450$. He uses the $\$4,450$ to repay the bank the money he owes, namely $\$2,768.90942(1.022)^{\frac{2}{12}} + \$1,668.002151(1.022)^{\frac{1}{12}} \approx \$4,500$. Jamaar is left with no money, but he has met his obligation under the call option he sold.

(4) If the stock goes down to $\$42.864$ per share, Jamaar's portfolio is worth $(100\Delta_3)(\$42.864) - \$2,768.90942(1.022)^{\frac{1}{12}} \approx \$4.8474$. To determine how he should reallocate the holdings of his portfolio, this time, the relevant delta is $\Delta_2 = \frac{7,856}{428,640}$. So, he now should only continue to hold $100\Delta_2 = \frac{785,600}{428,640}$ shares. So, Jamaar should sell $100\Delta_3 - \frac{785,600}{428,640}$ shares, thereby getting $\left(100\Delta_3 - \frac{785,600}{428,640}\right)(\$42.864) \approx \$2,700.222671$. He should use this money to pay off most of his debt. Assuming he does so immediately, at the end of two months his debt will only be $\$2,768.90942(1.022)^{\frac{2}{12}} - \$2,700.222671(1.022)^{\frac{1}{12}} \approx \$73.8464$. How does this compare to the value of his assets at time 2 months? If the stock's price after two months is $\$40.29216$, the call option will *not* be exercised and the portfolio is hence worth $\left(\frac{785,600}{428,640}\right)\$40.29216 \approx \$73.8464$. If the per share price is $\$44.57856$, then Jamaar will first spend $\left(100 - \frac{785,600}{428,640}\right)\$44.57856 \approx \$4,376.1536$ to buy the balance of the

100 shares he needs to sell since the call option will be exercised. He would then receive $4,450 when the option is exercised. So, deducting the cost of the just purchased shares, he will again find himself with $73.8464. This is exactly the amount he needs to pay off the loan.

(b) The per share price of the stock is now constrained to be a whole number of cents. Therefore, the possible prices of the stock at the end of one month are taken to be $47.42 or $42.86, while the possible share price at the end of two months should be $49.32, $44.58, or $40.29. If we were to repeat the calculation of the deltas of the binomial pricing method using these new rounded prices, we would obtain minor changes in the values of the deltas. In fact, we actually used $100\Delta_1 = 100$, $100\Delta_2 = \frac{785,600}{428,640} \approx 1.832773423$, and $100\Delta_3 \approx 64.8278899$. These were numbers of shares to be held (or potentially to be held depending on the direction the stock price moved), so it is reasonable to think of using 65 for the number of shares initially purchased, then planning on holding either 100 or 2 shares depending upon whether the stock price has risen to $47.42 or fallen to $42.86.

So, the plan is to buy 65 shares initially for $65 \times \$45.60 = \$2,964$. Then, if the price goes up at the end of the month to $47.42, you purchase another 35 shares, so as to be holding 100. This is clearly the right strategy since, with our assumptions about the price movement in the second month, if the stock price went up in the first month, the call option Jamaar sold is now certain to be exercised. The cost of purchasing the additional 35 shares is $35 \times \$47.42 = \$1,659.70$. On the other hand, if the price of the stock went down to $42.86, Jamaar should sell 63 shares so that he is only left with 2. This seems appropriate since the call option Jamaar sold will either not be exercised or will be exercised at $4,458 − \$4,450 = \$8$ below the cost of the shares. There is therefore little incentive to stockpile the shares. When Jamaar sells the 63 shares, he receives $63 \times \$42.86 = \$2,700.18$ which he can use to pay down his debt.

Let's go ahead and look at the cost to Jamaar of this strategy. Unlike part (a) when the cost in all cases was exactly $187.24, this time the costs will be somewhat different depending on the price history of the underlier.

If the price first goes up to $47.42, then no matter what happens in the next month, the call option Jamaar sold would be exercised. Jamaar would forfeit his 100 shares and receive $4,450. So, the cost to him in time 0 dollars would be

$$\$2,964 + \$1,659.70(1.022)^{-\frac{1}{12}} - \$4,450(1.022)^{-\frac{2}{12}} \approx \$186.8034714.$$

If the price first goes down to $42.86 and then goes up to $44.58, then the call option that Jamaar sold would be exercised. Therefore, since the strategy was to only continue to hold two shares if the share price fell to $42.86, Jamaar would have to purchase 98 shares for 98($44.58) = $4,368.84. He would then hand over the newly acquired shares, along with the 2 shares he was previously holding; he would receive $4,450. So, the cost to Jamaar in time 0 dollars would be

$$\$2,964 \ - \ \$2,700.18(1.022)^{-\frac{1}{12}} \ + \ (\$4,368.84 \ - \ \$4,450)(1.022)^{-\frac{2}{12}}$$
$$\approx \$187.8460521.$$

If the price first goes down to $42.86 and then goes down again to $40.29 then the call option that Jamaar sold would *not* be exercised. In this case, Jamaar would sell the 2 shares that he still held for $40.29 apiece, so he would take in $80.58. In this case, the total cost of Jamaar's position, in time 0 dollars, would be

$$\$2,964 \ - \ \$2,700.18(1.022)^{-\frac{1}{12}} \ - \ \$80.58(1.022)^{-\frac{2}{12}} \approx \$188.4239523.$$

The cost to Jamaar in all cases (in time 0 dollars) was between $186.80 and $188.43. This is at least $61.57 less than the $250 for the option.

By the strategy outlined, Jamaar has considerably reduced the variation in the cost. Had he done nothing, the cost of the call option (in time 0 dollars) would have ranged from a high of ($4,932 $-$ $4,450)(1.022)^{-\frac{2}{12}} \approx $480.2549932 to a low of $0. Just as in (a), you can consider Jamaar's strategy to be a hedge against a loss of approximately $480. ∎

When you actively manage your portfolio, moving assets as desirable, you are implementing **dynamic portfolio allocation.** In contrast, if you do not make changes, you have a **static position**. The dynamic portfolio allocation of Example (8.9.8) is called a **delta portfolio allocation**, since it is accomplished by holding a number of copies of the underlier (100 shares of the stock) equal to the delta of the relevant subtree. The delta portfolio allocation of part (a) of Example (8.9.8) is **self-financing**, meaning no additional money needs to be added or withdrawn after the initial portfolio is purchased. Instead, already committed funds are reallocated.

Dynamic portfolio allocation that is used to accomplish a hedge is called **dynamic hedging**. This term is more common than our "dynamic portfolio allocation."

8.10   USING WEIGHTED AVERAGES TO PRICE OPTIONS

If you use a little elementary probability theory, there is an interesting and computationally useful way to interpret the binomial model option pricing equation of Important Fact (8.9.3). Define

(8.10.1)
$$p^* \stackrel{\text{def}}{=} \frac{S_0(1 \ + \ i)^T \ - \ d}{u \ - \ d}.$$

Then, the formula for c_0 in Important Fact (8.9.3) may be rewritten as

$$(8.10.2) \qquad \boxed{c_0 = \left[V_u(1 + i)^{-T}\right]p^* + \left[V_d(1 + i)^{-T}\right](1 - p^*).}$$

As we discussed in Section (8.9), under our assumption that there is no arbitrage, we have the string of inequlities $d < S_0(1 + i)^T < u$. Combining this with Equation (8.10.1) gives

$$(8.10.3) \qquad\qquad 0 < p^* < 1,$$

and p^* *could* be thought of as the **probability** of some event — that is, the proportion of time that the event would happen on average. Moreover, if p^* is thought of as if it were the probability that the value of the underlier rises, then $1 - p^*$ represents the probability that the value of the underlier falls. Furthermore,

$$1 - p^* = \frac{u - d}{u - d} - \frac{S_0(1 + i)^T - d}{u - d} = \frac{u - S_0(1 + i)^T}{u - d},$$

and

$$
\begin{aligned}
(8.10.4) \qquad & \left[u(1 + i)^{-T}\right]p^* + \left[d(1 + i)^{-T}\right](1 - p^*) \\
&= \frac{u}{(1+i)^T}\left(\frac{S_0(1+i)^T-d}{u-d}\right) + \frac{d}{(1+i)^T}\left(\frac{u-S_0(1+i)^T}{u-d}\right) \\
&= S_0.
\end{aligned}
$$

Thus, the price S_0 is a weighted average of the time 0 values of u and d, the weights being the numbers we are treating as though they are the probabilities that the underlier has time T price u and d. Note that the first and third lines of Equation (8.10.4) completely determine p^*. In the language of probability theory, Equation (8.10.4) tells us that S_0 is the **expectation** (or **expected value**) of the time 0 value of the payoff of the underlier under the artificial probability distribution we've introduced. This fabricated probability p^* is called the **risk-neutral probability** and we say that the time 0 underlier price S_0 is equal to the **risk-neutral** expectation of the time 0 value of the time T value of the underlier. (This is a bit of a mouthful, but is conceptually simple. You start with the possible time T values, discount them to time 0, and then average them using risk-neutral probabilities as weights.)

FACT (8.10.5): In the no-arbitrage model, there is an artificial probability distribution, the risk-neutral probability distribution, with respect to which the initial price c_0 of the European call option is the expectation as in Equation (8.10.2) of the present value of the time T option value. The probability function that gives the risk neutral probabilities of the two possible outcomes u and d is given by

$$p(x) \overset{\text{def}}{=} \text{the probability the new price is } x$$
$$= \begin{cases} p^* & \text{if } x = u, \\ 1 - p^* & \text{if } x = d, \\ 0 & \text{otherwise} \end{cases}$$

where p^* is defined by (8.10.1). The value p^* is determined by the fact that the initial price S_0 of the underlier is the expectation of the present value of the time T underlier price:

$$S_0 = p^*\left[u(1 + i)^{-T}\right] + (1 - p^*)\left[d(1 + i)^{-T}\right].$$

Our intuition tells us that the value of the call option must depend upon how likely it really is that the underlier's price moves up or down — that is, on the *true* probability distribution for these outcomes. Such a dependence is not apparent in the option pricing formula (8.10.2). However, the probability distribution for the underlier's time T price is presumably reflected in its time 0 price S_0, and according to Definition (8.10.1) and Fact (8.10.5), p^* depends on S_0. So, our intuition was indeed correct, even though *you do not need the real probability distribution for the underlier's time T price in order to price an option.* All you need is the risk-neutral probability p^*.

METHOD (8.10.6) Risk-neutral method (One period):
(1) In the no-arbitrage model, the risk-neutral probability distribution p^* is completely determined as follows. Set the risk-neutral expectation of the time 0 value of the time T price of the underlier equal to the time 0 price S_0 of the underlier. Solve for p^*.
(2) Compute the risk-neutral expectation of the time 0 values of the option's possible payoffs V_u and V_d. This is equal to the time 0 price c_0 of the call option.

EXAMPLE 8.10.7   Risk-neutral method

Problem: Suppose that a European call option with expiration in one month has a strike price $37.75. The current price for the underlier is $36.50, one month from now it will be $34 or $39, and the risk-free nominal interest rate convertible monthly is 3%. Determine the risk-neutral probability p^* of an upward movement of the underlier's price and the no-arbitrage price of the option at time 0.

Solution We use Method (8.10.6). The risk-neutral probability p^* must make the risk-neutral expectation of the time 0 value of the outcomes of the underlier equal the time 0 value of the underlier; that is,

$$p^*\left[\$39\left(1 + \frac{.03}{12}\right)^{-1}\right] + (1 - p^*)\left[\$34\left(1 + \frac{.03}{12}\right)^{-1}\right] = \$36.50,$$

which gives $p^* = .51825$. Then the risk-neutral price of the option at time 0 is the risk-neutral expectation of time 0 values $V_u = \$39 - \$37.75 = \$1.25$ and $V_d = \$0$. That is,

$$c_0 = p^*\left(\frac{\$1.25}{1.0025}\right) + (1 - p^*)\left(\frac{\$0}{1.0025}\right) = .51825\left(\frac{\$1.25}{1.0025}\right) \approx \$.646197007.$$

∎

The same approach works to price options over more than one time interval.

EXAMPLE 8.10.8   Risk-neutral method

Problem: A European call option, with expiration in three years, has strike price $4,100. The current price of the underlier is $4,000, and this price is guaranteed to either be up by exactly 10% or down by exactly 5% at the end of the first two years. Moreover, the final year will independently have either a 10% increase or a 5% decrease in the underlier's price. Use the risk-neutral method to find the no-arbitrage price of the option, assuming interest rates remain at an annual effective rate of 3% throughout the three-year period.

Solution We have strike price $K = \$4,100$, initial price $S_0 = \$4,000$, and $i = .03$. The price S_2 of the underlier at time 2 is either $\$4,000(1 + .1) = \$4,400$ or $\$4,000(1 - .05) = \$3,800$. Therefore, with respect to the yet to be calculated risk-neutral probability p_1^* of an increase in the underlier price during the first period $[0, 2]$, the expectation of the time 0 value of the time 2 price of the underlier is

$$p_1^*\left[\$4,400(1.03)^{-2}\right] + (1 - p_1^*)\left[\$3,800(1.03)^{-2}\right].$$

According to Fact (8.10.5), p_1^* may be found by setting this expectation equal to the time 0 underlier price $S_0 = \$4,000$. In fact, $p_1^* = [4,000(1.03)^2 - 3,800]/(4,400 - 3,800) = 443.6/600 \approx .739333333$.

The risk-neutral probability p_2^* for the second interval $[2, 3]$ is again given by looking at expectations. If the time 2 price is \$4,400, then the time 3 price will be either $\$4,400(1 + .1) = \$4,840$ or $\$4,400(1 - .05) = \$4,180$. According to Fact (8.10.5), p_2^* must satisfy

$$\$4,400 = p_2^*[\$4,840(1.03)^{-1}] + (1 - p_2^*)[\$4,180(1.03)^{-1}],$$

which gives $p_2^* = \frac{8}{15} \approx .533333333$. Alternatively, if the time 2 price is \$3,800, then the time 3 price would be either $\$3,800(1 + .1) = \$4,180$ or $\$3,800(1 - .05) = \$3,610$. Then p_2^* would have to satisfy

$$\$3,800 = p_2^*[\$4,180(1.03)^{-1}] + (1 - p_2^*)[\$3,610(1.03)^{-1}],$$

which again gives $p_2^* = \frac{352}{660} \approx .533333333$, since this second equation for p_2^* is just $(1.1)^{-1}(1 - .05)$ times the first equation for p_2^*. (With constant percentage increases or decreases over each time interval, the same value of p_2^* will be found regardless of the price at the start of the interval.)

At time 3 there are three possibilities for the underlier's price, namely \$4,840, \$4,180, and \$3,610. The corresponding time 3 values of the call option are $V_{uu} = \$4,840 - \$4,000 = \$740$, $V_{ud} = \$4,180 - \$4,000 = \$80$, and $V_{dd} = 0$. But the time 0 price of the call option is equal to the risk-free expectation of the time 0 value of these time 3 values of the option. The risk-neutral probability of two increases is $p_1^* p_2^*$, of one increase and one decrease (in either order) is $(1 - p_1^*)p_2^* + p_1^*(1 - p_2^*)$, and of two decreases is $(1 - p_1^*)(1 - p_2^*)$. So,

$$c_0 = p_1^* p_2^* \left[\$740(1.03)^{-3}\right] + \left[(1 - p_1^*)p_2^* + p_1^*(1 - p_2^*)\right]\left[\$80(1.03)^{-3}\right]$$

$$+ (1 - p_1^*)(1 - p_2^*)\left[\$0(1.03)^{-3}\right]$$

$$= \left(\frac{443.6}{600}\right)\left(\frac{352}{660}\right)\left[\$740(1.03)^{-3}\right]$$

$$+ \left[\left(1 - \frac{443.6}{600}\right)\left(\frac{352}{660}\right) + \left(\frac{443.6}{600}\right)\left(1 - \frac{352}{660}\right)\right]\left[\$80(1.03)^{-3}\right]$$

$$\approx \$302.47.$$

∎

The approach illustrated above in one-interval and two-interval binomial models is a powerful and general one for pricing options in a no-arbitrage setting. First, find the risk-neutral probabilities using the prices of the underlier;

second, compute the option price as a risk-neutral expected value. In fact, under appropriate hypotheses one can prove the Arbitrage Theorem: The existence of such risk-neutral probabilities and this method of pricing options is *equivalent* to the impossibility of arbitrage.[18]

We next consider two more examples of how to price options using the risk-neutral method. The first concerns a European call option and the second a European put option, where the same approach used for call options can be applied. In each case, we will find the time 0 price of the option under the assumption that there are no arbitrage opportunities.

EXAMPLE 8.10.9   Example (8.9.6) revisited: risk-neutral method

Problem: The current price of a stock is \$45.60 per share. A European call option for 100 shares of the stock has strike price \$44.50 per share and expires in two months. The risk-free rate of interest is an annual effective rate of 2.2%. The per share price will either go up by 4% or down by 6% each month. Use the risk-neutral method to find the no-arbitrage price of the stock option.

Solution   We have, for 100 shares, $S_0 = \$4,560$ and $K = \$4,450$. The stock price $S_{\frac{1}{12}}$ at the end of one month is either $(1 + .04)\$4,560 = \$4,742.40$ or $(1 - .06)\$4,560 = \$4,286.40$.

The risk-neutral probability p_1^* of an increase in the first month can be found using Fact (8.10.5) and Method (8.10.6):

$$\$4,560 = p_1^* \left[\$4,742.40(1.022)^{-\frac{1}{12}} \right] + (1 - p_1^*) \left[\$4,286.40(1.022)^{-\frac{1}{12}} \right].$$

Therefore, $p_1^* = \left[4,560(1.022)^{\frac{1}{12}} - 4,286.4 \right]/456 \approx .61815103$.

The risk-neutral probability p_2^* of an increase in the second month may be found similarly. If the stock price after the first month is $S_{\frac{1}{12}}$, then after the second it is either $(1 + .04)S_{\frac{1}{12}}$ or $(1 - .06)S_{\frac{1}{12}}$. So, p_2^* must satisfy

$$S_{\frac{1}{12}} = p_2^* \left[(1.04)S_{\frac{1}{12}}(1.022)^{-\frac{1}{12}} \right] + (1 - p_2^*) \left[.94 S_{\frac{1}{12}}(1.022)^{-\frac{1}{12}} \right],$$

which yields $p_2^* = \left[(1.022)^{\frac{1}{12}} - .94 \right]/.1 \approx .61815103$. Note that $p_1^* = p_2^*$; this happens whenever the time intervals have equal length and the percentage changes are the same each period.

---

[18]The Arbitrage Theorem is discussed in Sheldon Ross's *Introduction to Probability Models*, (seventh edition) Academic Press, 2000. Ross's theorem (10.1) is more general than what we use here, but it is still in a discrete setting. Also see Section (11.2.9) of Panjer's *Financial Economics*.

Now, if the price after two months is $\$4,932.096 = (1 + .04)^2\$4,560$, the option is exercised with value $V_{uu} = \$4,932.096 - \$4,450 = \$482.096$. Likewise, if the price is $\$4,457.856 = (1 + .04)(1 - .06)\$4,560$, the option is exercised with value $V_{ud} = \$4,457.856 - \$4,450 = \$7.856$. On the other hand, if that price is $\$4,029.216 = (1 - .06)^2\$4,560$, the option is not exercised since the strike price is $\$4,450 > \$4,029.216$. Method (8.10.6) gives the no-arbitrage price as

$$c_0 = p_1^* p_2^* \left[V_{uu}(1.022)^{-\frac{1}{6}} \right] + \left[(1 - p_1^*)p_2^* + p_1^*(1 - p_2^*) \right] \left[V_{ud}(1.022)^{-\frac{1}{6}} \right]$$

$$\approx (.61815103)^2 \left[\$482.096(1.022)^{-\frac{1}{6}} \right]$$

$$+ \; 2(.38184897)(.61815103) \left[\$7.856(1.022)^{-\frac{1}{6}} \right]$$

$$+ \; (1 - .61815103)^2 \left[\$0(1.022)^{-\frac{1}{6}} \right]$$

$$\approx \$187.24.$$

Our answer agrees with the one found in Example (8.9.6) using replicating portfolios. ∎

We made a useful observation in the course of our solution to the last example. It should be displayed, for easy reference.

> **FACT (8.10.10):** The risk-neutral probability p_1^* for an increase in the underlier price over the interval $[0, T]$ and the risk-neutral probability p_2^* for an increase in the underlier price over the interval $[T, T']$ are equal whenever $T' = 2T$ – that is, whenever the time intervals are equal.

EXAMPLE 8.10.11   Example (8.9.7) revisited: risk-neutral method

Problem: The current price of a stock is $\$25$ per share. A European put option for 100 shares of the stock has strike price $\$24$ per share and expires in six months. The risk-free rate of interest is 4% convertible quarterly. The per share price will either go up by 10% or down by 20% each quarter. Use the risk-neutral method to find the no-arbitrage price of the stock option.

Solution   We measure time in quarters, so the risk-free interest rate per time period is $i = \frac{.04}{4} = .01$. We also have $S_0 = \$2,500$. The stock price at time 1 is either $(1 + .1)\$2,500 = \$2,750$ or $(1 - .2)\$2,500 = \$2,000$. As usual, p_1^* may be found using Fact (8.10.5):

$$\$2,500 = p_1^* \$2,750(1.01)^{-1} + (1 - p_1^*)\$2,000(1.01)^{-1}.$$

Therefore, $p_1^* = .7$. Since both intervals are one quarter and the percentage changes in the stock prices are the same each quarter, we may apply Fact (8.10.10) to get $p_2^* = p_1^* = .7$.

The price at time two quarters (six months) has three possible values, namely $\$3,025 = (1 + .1)^2\$2,500$, $\$2,200 = (1 + .1)(1 - .2)\$2,500$, and $\$1,600 = (1 - .2)^2\$2,500$. The only difference from our previous analysis occurs now, when we have to compute the values of the *put* option at time 2 corresponding to each of the possible stock prices. For a stock price of $\$3,025$, we do not exercise the option since it exceeds the strike price $K = \$2,400$. That is, $V_{uu} = \$0$. If the price is $\$2,200$, we do exercise the option; the option value is $V_{ud} = \$2,400 - \$2,200 = \$200$. Likewise, for $\$1,600$, the option value is $V_{dd} = \$2,400 - \$1,600 = \$800$.

We now determine the option price p_0 as the risk-neutral expected value of the time 0 values of these option values:

$$
\begin{aligned}
p_0 &= p_1^* p_2^* \left[V_{uu}(1.01)^{-2} \right] + \left[p_1^*(1 - p_2^*) + (1 - p_1^*)p_2^* \right]\left[V_{ud}(1.01)^{-2} \right] \\
&\quad + (1 - p_1^*)(1 - p_2^*)\left[V_{dd}(1.01)^{-2} \right] \\
&= (.7)^2[0] + [2(.7)(.3)]\left[\$200(1.01)^{-2} \right] + (.3)^2\left[\$800(1.01)^{-2} \right] \\
&\approx \$152.93.
\end{aligned}
$$

This agrees with the price p_0 that we computed in Example (8.9.7).   ∎

8.11   SWAPS

A **swap** is a contract in which the parties agree to trade sequences of cashflows arising from financial instruments. Usually, there are two participants in the contract, and the cashflows exchanged may be in one or more currencies. Most often, the payment times for each party in the swap are the same and, if there is a common currency for the cashflows, the payments are **netted**. This means that, at any time, only the difference in the cashflows changes hands. This is important if there is default risk.

One reason to use a swap is to accomplish a change in the nature of a liability or asset without liquidating it. This is useful since liquidation may be impossible or costly. For example, a life insurance company may be obligated to make inflation-indexed payments to a beneficiary, and prefers to exchange these for a fixed stream of payments. An investor, who faces substantial tax liability if stock is sold, may wish to replace volatile stock dividends with a more reliable income.

Borrowers find swaps useful as a way to obtain lower payments. The reason for this is that not all borrowers qualify for the same loan rates. This is due to differences in creditworthiness and to regulations. So, you may have an advantage compared to another potential borrower on a particular type of loan, and you may be able to benefit by entering into a contract that transfers your borrowing advantage. Likewise, investors may find swaps helpful as they attempt to maximize their return.

A swap is called an **interest-rate swap** if both cashflow sequences are defined as the interest due on some specified amount, called the **notional principal**. Usually, one of the cashflows consists of the periodic interest due at some **fixed** rate on the stated principal, while the other payment sequence is made up of the interest, calculated using a **floating** rate, on the same notional principal. An interest-rate swap of this sort is often referred to as a **pure vanilla** swap. You may also encounter interest swaps with two different floating-interest rates. Such a swap may be used by hedgers concerned with the movement of particular rates. A **basis swap** is an exchange based on forward rates at two different points on the yield curve. Swaps may be directly negotiated by two corporations or individuals, but it is more common for an investment or commercial bank to be the counterparty. The bank earns money by means of a bid-ask spread for similar swaps. Swap dealers facilitate many swaps, warehousing swaps until they can find counterparties and engineering multiparty arrangements.

EXAMPLE 8.11.1   Interest-rate swap

Problem: On February 1, 1999, Value Enterprises (VE) has a low credit rating and therefore has trouble qualifying for loans longer than six months. In fact, VE cannot find a lender who will offer it a four-year loan. Therefore, in order to finance a four-year $3,600,000 building project, VE anticipates needing to take out a sequence of eight six-month $3,600,000 loans. (VE plans to make the interest payments at the end of each six months with its own money, so the loan amount will remain at $3,600,000.) So, unless VE enters an interest rate swap, it is subject to the risk that the interest rate rises. With uncertain interest rates, VE would not know ahead of time what their building project would cost, even though the contractor has promised to produce the building in return for a February 1, 1999 payment of $3,600,000.

(a) Suppose that VE took out a sequence of eight six-month loans, each of which was at an interest rate that is numerically equal to the then current six-month LIBOR[19] rate plus 2%. These LIBOR rates, reported chronologically on an "actual/360" basis were 5.168%, 5.913%, 6.328%, 6.831%, 4.955%, 3.479%, 2.068%, and 1.815% so VE's rates were 7.168%, 7.913%, 8.328%, 8.831%, 6.955%, 5.479%, 4.068%, and 3.815%. What interest payments did VE have to make?

(b) Suppose that on February 1, 1999, VE also entered into a fixed-to-floating rate interest-rate swap with Swapland Bank, with cashflows computed using a notional principal of $3,600,000. VE contracted to pay semiannual interest at a nominal rate of 5% convertible semiannually and to receive interest calculated using the six-month LIBOR rate. Assuming there was no default by either VE or Swapland, what netted cashflows did they exchange?

(c) Show that VE's total semiannual cashflows are nearly level during the four-year period.

Solution   (a) Interest payments took place at the end of each six-month period, hence on each August 1 and each February 1. In the non-leap-years 1999, 2001, and 2002, there are 181 days in the six-month period beginning on February 1, and this period has 182 days in the leap year 2000. In all four years, the six-month period beginning on August 1 has 184 days. Therefore, VE's interest payments are

$3,600,000(.07168)(181/360) = $129,740.80 on 8/1/1999,
$3,600,000(.07913)(184/360) = $145,599.20 on 2/1/2000,
$3,600,000(.08328)(182/360) = $151,569.60 on 8/1/2000,
$3,600,000(.08831)(184/360) = $162,490.40 on 2/1/2001,
$3,600,000(.06955)(181/360) = $125,885.50 on 8/1/2001,
$3,600,000(.05479)(184/360) = $100,813.60 on 2/1/2002,
$3,600,000(.04068)(181/360) = $73,630.80 on 8/1/2002,
$3,600,000(.03815)(184/360) = $70,196.00 on 2/1/2003.

(b) Under the terms of the swap, at the end of each six-month period, VE owes Swapland a payment of $3,600,000\left(\frac{.05}{2}\right) = $90,000. However, only netted cashflows exchange hands, and Swapland owes VE interest at the LIBOR rate. That is, Swapland owes VE

$3,600,000(.05168)(181/360) = $93,540.80 on 8/1/1999,
$3,600,000(.05913)(184/360) = $108,799.20 on 2/1/2000,
$3,600,000(.06328)(182/360) = $115,169.60 on 8/1/2000,
$3,600,000(.06831)(184/360) = $125,690.40 on 2/1/2001,

---

[19]Recall that LIBOR stands for London interbank offer rate, and is an average of the interest rates that the largest international banks charge each other for loans. It is quoted on an "actual/360" basis, and the sequence of rates used in this example was found on www.erate.com/six_month_libor_index.6-months-libor.html.

$3,600,000(.04955)(181/360) = \$89,685.50$ on 8/1/2001,
$3,600,000(.03479)(184/360) = \$64,013.60$ on 2/1/2002,
$3,600,000(.02068)(181/360) = \$37,430.80$ on 8/1/2002,
$3,600,000(.01815)(184/360) = \$33,396.00$ on 2/1/2003.

The netted cashflows are therefore

$\$93,540.80 - \$90,000 = \$3,540.80$ from Swapland to VE on 8/1/1999,
$\$108,799.20 - \$90,000 = \$18,799.20$ from Swapland to VE on 2/1/2000,
$\$115,169.60 - \$90,000 = \$25,169.60$ from Swapland to VE on 8/1/2000,
$\$125,690.40 - \$90,000 = \$35,690.40$ from Swapland to VE on 2/1/2001,
$\$90,000 - \$89,685.50 = \$314.50$ from VE to Swapland on 8/1/2001,
$\$90,000 - \$64,013.60 = \$25,986.40$ from VE to Swapland on 2/1/2002,
$\$90,000 - \$37,430.80 = \$52,569.20$ from VE to Swapland on 8/1/2002,
$\$90,000 - \$33,396.00 = \$56,604.00$ from VE to Swapland on 2/1/2003.

Note that the first four cashflows are inflows to VE, while the last four were outflows. Of course, there is no reason that the number of inflows need equal the number of outflows. We note that, under the swap, there is considerable fluctutation in the amount of money changing hands from one period to the next.

(c) In parts (a) and (b) of this solution, VE's cashflows [interest in (a), netted in (b)] were far from being level. However, if we combine the interest outflows [from (a)] with the swap cashflows [from (b)], so as to obtain VE's total outflows on each of the eight dates, we see that these total outflows show much less variation. In fact, VE's total outflows are

$\$129,740.80 - \$3,540.80 = \$126,200.00$ on 8/1/1999,
$\$145,599.20 - \$18,799.20 = \$126,800.00$ on 2/1/2000,
$\$151,569.60 - \$25,169.60 = \$126,400.00$ on 8/1/2000,
$\$162,490.40 - \$35,690.40 = \$126,800.00$ on 2/1/2001,
$\$125,885.50 + \$314.50 = \$126,200.00$ on 8/1/2001,
$\$100,813.60 + \$25,986.40 = \$126,800.00$ on 2/1/2002,
$\$73,630.80 + \$52,569.20 = \$126,200.00$ on 8/1/2002,
$\$70,196.00 + \$56,604.00 = \$126,800.00$ on 2/1/2003.

So, the payments are all close to $126,500. We note that $\$126,000 = \$3,600,000(7\%/2)$. This is as we would expect since the fixed rate was a nominal rate of 5% convertible semiannually, the floating rate was 2% above LIBOR, and 7% = 5% + 2%. Of course the LIBOR rates were six-month rates, rather than nominal rates convertible semiannually, and this is why things only work out approximately.

Without the Swap, VE's interest payments in (a) were subject to variation, from about $70,000 to more than $162,000. This variation could make it difficult for VE to plan, and could even lead to default. With the swap, VE's net outflows in (c) were nearly level, making it easier for VE to plan. The price for this convenience was over $51,000 more in outflows with the swap than without it. The LIBOR rates beyond the first six-month period were not known when the swap agreement was made, so at the outset VE might have imagined the cost would have been much less, but it could have been even more.   ∎

As illustrated in Example (8.11.1), the notional principal is *not* exchanged. As a result, the credit risk inherent in an interest-rate swap is much less than the default risk for a loan of the named principal. This is why swaps may be used by parties whose credit ratings are compromised. Differences in the borrowing opportunities available to individuals (or companies) fuel the interest-rate swap market. Desires to skirt regulations or taxes also play a significant role.

Currency swaps are another common type of swap. [20] To understand currency swaps, it is helpful to first give a little thought to foreign exchange rates. A **foreign exchange rate** (**exchange rate**, **FX rate**) is the price, expressed in the currency of one country, that you would have to pay for the basic unit of the currency of another country. For example an exchange rate of 5 Egyptian pounds to the U.S. dollar means that 5 Egyptian pounds are worth the same as one dollar. With this exchange rate, you are making a fair exchange if you trade X U.S. dollars for $5X$ Egyptian pounds. On the other hand, with the stated exchange rate, each Egyptian pound is worth .20 U.S. dollars. So, just as you may translate English to Spanish and Spanish to English, you may *translate* between two currencies using the exchange rate as your dictionary. [21]

An accurate Spanish/English–English/Spanish dictionary written one

---

[20] An early, and much described, currency swap took place between the World Bank and IBM in August 1981. IBM had issued callable bonds in Swiss Francs and Deutsche Marks, and the dollar had appreciated against these currencies, making it tempting for IBM to call the bonds and finance the repayment with new U.S. dollar issues. However, this would involve a number of costs including the cost of exchanging currencies, call premiums, and costs of reissue. The World Bank (also known as the International Bank for Reconstruction) borrows money internationally and lends it to developing countries to finance construction projects. At the time, the interest rate the World Bank faced for borrowing Swiss Francs was about 8% and the rate for loans of Deutshche Marks was 11%–12%. These rates may seem high to you, but they were atttractive compared to a 16%–17% rate for U.S. dollars. But the World Bank was near the borrowing limits imposed by the Swiss and German governments for the amounts of these currencies they could borrow in Switzerland and Germany. The agreement reached between IBM and World Bank, with Salomon Brothers as intermediary, called for World Bank to borrow U.S. dollars in the U.S. market with IBM then assuming the repayment while World Bank took over their Swiss Francs and Deutsche Marks bond obligations.

[21] If you actually attempt to exchange currencies at a bank, you are apt to encounter service fees so that you experience a slight loss in value.

day is apt to be very similar to a Spanish/English–English/Spanish dictionary recording the next day's language. However, the exchange rate, our "foreign currency dictionary" between two specified currencies, may show much greater day-to-day changes! Therefore, when we describe a (two-party) currency swap, which is basically an interest-rate swap in two different currencies, we must be very careful about the translating.

As was the case for an interest rate swap, a currency swap has a notional principal. Using the exchange rate that is in effect at initiation, the notional principal may be given in two different currencies, say as an amount N_1 of currency 1 or as an amount N_2 of currency 2. (It is optional whether these notional amounts are exchanged at the start of the swap, since they then have the same value.) The currency swap will also have a term to maturity and dates given for interest cashflows to be exchanged. Additionally, there will be two different interest rates specified; a rate j_1 to be used for the principal denominated in currency 1, and a rate j_2 to be used for the principal denominated in currency 2. The rates j_1 and j_2 may each be floating or fixed.

One party in the currency swap makes interest payments, whose amount is determined using the interest rate j_1 on a principal amount N_1 : This is all in currency 1. The other party makes interest payments in currency 2, these being the interest due on a loan of amount N_2 of currency 2 if the interest rate on the loan is j_2. Each swap party has one additional payment it must make to its counterparty. The party that made payments in currency $k, k \in \{1, 2\}$, (as if it had borrowed an amount N_k of currency k,) must make an additional payment of N_k in currency k. Since our "exchange rate dictionary" has likely changed, we do not expect these additional payments at maturity to have the same value.

EXAMPLE 8.11.2   Currency swap

Uncle Sam Industries (USI) and Iwasawa Technology (IT) are counterparties in a currency swap that involves U.S. dollars and Japanese yen. The exchange rate at initiation is 109.44 yen per dollar, and the notional principal is \$10,000,000 = 1,094,400,000 yen. Uncle Sam Industries will make payments in yen as if it has borrowed the 1,094,400,000 yen for one year, and Iwasawa Technology will have outflows in dollars as if it has borrowed \$10,000,000, again for one year. (In fact, USI may have sent IT \$10,000,000 in return for 1,094,400,000 yen.) The interest rate to be paid on the dollar loan is the three-month Treasury bill rate, and interest on the borrowed yen is a fixed rate of 3.5% convertible quarterly. Describe all cashflows after the swap is initiated if the four three-month nominal Treasury bill rates convertible quarterly are 2.75%, 2.87%, 3.05%, and 3.11%.

Solution   Iwasawa Technology is required to make quarterly payments for one year on the \$10,000,000 loan using four different quarterly interest rates, namely $\frac{.0275}{4}$ = .006875, $\frac{.0287}{4}$ = .007175, $\frac{.0305}{4}$ = .007625, and $\frac{.0311}{4}$ = .007775. So, Iwasawa Technology makes payments of .006875($10,000,000) = $6,875,000 in three months, .007175($10,000,000) = $7,175,000 in six months, .007625($10,000,000) = $7,625,000) in nine months, and a final payment of .007775($10,000,000) + $10,000,000 = $7,775,000 + $10,000,000 = $17,775,000 in one year. Uncle Sam Industry's payments are quarterly payments for one year on a loan of 1,094,400,00 at a quarterly interest rate of $\frac{.035}{4}$ = .00875. Since (.00875)(1,094,400,000) = 9,576,000, USI makes three payments of 9,576,000 yen, and then a final payment of 1,103,976,000 yen + 1,094,400,000 yen = 1,103,976,000 yen.   ∎

Interest-rate swaps and currency swaps each involve an exchange of loan payments. However, a currency swap does not use netting and this may increase credit risk.

There are other types of swaps where the payments involved do not behave like loan payments. Foremost among these are equity swaps. An **equity swap** is a swap in which at least one of the cashflows is determined by the performance of a single stock, portfolio of securities, or stock index. In a **total return equity swap**, the periodic security-based cashflows due from one party are the sum of the capital gains (possibly negative) and the value of any dividends that have been paid. If for some period the value of the losses exceeds the value of the dividends paid on the stock, then the party that makes payments based on the security makes a negative payment — that is, it receives money from the counterparty. In a **commodity swap**, at least one stream of cashflows is based on a commodity index.

EXAMPLE 8.11.3   Total return equity swap

Problem: Dr. Rebecca Rubinstein has owned telephone stock for many years and wishes to continue to do so, even though she believes that the stock is going to go down soon. Dr. Rubinstein enters a two-year total return equity swap with Swapmasters, exchanging her position in the stock for fixed semiannual payments at a nominal rate of 3.25%. At the time she does so, the value of her stock is $22,240. At the end of the first six months, the stock's price has fallen to $20,600, and no dividends are paid. The price falls farther the following year, namely to $20,000, and a $100 dividend is paid. Six months later the stock price is $21,800 and a $300 dividend is paid, and one year later it is $23,100, but there is no dividend. Describe all payments that Dr. Rubinstein receives or must make.

Solution   Each six-month period, the retun on the stock is the sum of the dividends and the appreciation of the stock. Therefore, in the successive periods, we have returns of $0 + ($20,600 − $22,400) = −$1,800, $100 + ($20,000 − $20,600) = −$500, $300 + ($21,800 − $20,000) =

$2,100$, and $0 + (\$23,100 - \$21,800) = \$1,300$. These are the amounts that Dr. Rubinstein is responsible for paying to Swapmasters, but we use netting to determine the cash exchanges. The payments she is entitled to receive from Swapmasters are each for $\$22,240\left(\dfrac{.0325}{2}\right) = \$361.40$. So, after six months Dr. Rubinstein receives a payment for $\$361.40 - (-\$1,800) = \$2,161.40$, and after one year she gets $\$364 - (-\$500) = \$861.40$. In the second year, in each six-month period the stock has returns that exceed $\$361.40$, and Dr. Rubinstein must pay the excess. So, eighteen months after the swap is initiated, Dr. Rubinstein pays $\$2,100 - \$361.40 = \$1,738.60$, and at the end of the two years she pays $\$1,300 - \$361.40 = \$938.60$. ∎

Swaps are over-the-counter, tailor-made financial products, and are not limited to those we have described. Circumventing regulations or taxation, utilizing competitive advantage, hedging, and speculating all contribute to the popularity of swaps.

8.12   PROBLEMS, CHAPTER 8

(8.0) Chapter 8 writing problems

(1) [following Section (8.3)] Write a paragraph or two concerning the evolution of coupon stripping. Include the Tax Equity and Fiscal Responsibility Act of 1982 (TEFRA), Treasury Investment Growth Receipts (TIGRs), and Separate Trading of Registered Interest and Principal of Securities (STRIPS).

(2) [following Section (8.3)] Discuss how the yield curve for Treasury Securities has changed over a twenty-year period of your choosing. Mention major political developments that occurred during your period in (or affecting) the United States. Note any relations you hypothesize between changes in the term structure of interest rates and the political events.

(3) [following Section (8.4)] Assume that there is a level term structure for risk-free interest rates, the annual effective risk-free rate being i. A forward contract is initiated at time 0 for an asset whose time 0 price is S_0, the contract calling for the sale of the asset at time T for forward price K. Suppose there are fixed cashflows associated with holding the asset for the term of the forward contract, and these have time 0 value H if you use the risk-free accumulation function $a(t) = (1 + i)^t$. Explain why with "no-arbitrage" pricing, the time 0 price of the forward contract is $K = (S_0 - H)(1 + i)^T$.

(4) [following Section (8.5)] As we did in Examples (8.5.1) and (8.5.2), write a short description of a currently traded futures contract.

(5) [following Section (8.11)] Write a paragraph or two describing some reasons someone might wish to become a party in an equity swap.

(6) [following Section (8.11)] Write a paragraph describing how various types of derivatives may be used to benefit from a predicted stock appreciation. Include discussion of what happens if your prediction is wrong.

(7) [following Section (8.7)] Write a paragraph describing an event futures. Include your thoughts on how (if at all) it is useful or beneficial.

(8.2) Arbitrage

(1) Mr. Ralbracht qualifies for his company's low interest loans. This allows him to borrow up to $20,000 for two years at an effective interest rate of 2.5%. Mr. Ralbracht does not have any need to borrow money. However, he observes that two-year Treasury notes are currently selling so as to provide an investor an annual effective yield of 3.72%. Treasury notes pay interest semiannually. Describe in detail the arbitrage opportunity available to Mr. Ralbracht.

(2) Lorenzo notes that the market includes securities A and B whose prices in one-year depend only upon whether the market is "up," "steady," or "neutral." A, which has a current share price of $25, will sell for $32, $25, or $22 respectively, while B, with a current share price of $14, will sell for $16, $14, or $12, respectively. Explain how Lorenzo has an arbitrage opportunity, assuming there are no costs associated with trading A and B.

(3) Leonard Talbot observes that there are three options, all with identical prices. The price of the first will increase by 10% if the market goes up, and decrease by 10% if the market goes down. The second will increase in value by 20% if the market is up, and decrease by 15% if the market is down. The third will increase by 16% if the market is up and decrease by 20% if it is down. Describe an arbitrage opportunity.

(8.3) The term structure of interest rates

(1) The current prices on one-year, two-year, and three-year $10,000 zero-coupon bonds are $9,765, $9,428, and $8,986.82, respectively. Find all forward rates implied by these prices.

(2) A three-year, $1,000, 4%, par-value bond with annual coupons sells for $990, a two-year, $1,000, 3% bond with annual coupons sells for $988, and a one-year, zero-coupon, $1,000 bond sells for $974. Determine the spot rates $r_1, r_2,$ and r_3.

(3) Assuming the market is arbitrage-free, if a six-month pure discount bond yields 1.9%, a one-year pure discount bond yields 2.3%, an eighteen-month pure discount bond yields 2.65%, and a two-year discount bond

yields 3.05%, what should be the price of a two-year $1,000 6% par-value bond with semiannual coupons?

(4) Alan Jones observes that two-year zero-coupon bonds yield 3.2%, two-year 10% bonds with annual coupons yield 3%, and the one-year spot rate is 1.8%. Describe how he might use this to make money without tying up any of his own money.

(5) Consider the following table of 4% par-value bonds having annual coupons.

Term	Yield
1 Year	$y_1 = 1.435\%$
2 Year	$y_2 = 2.842\%$
3 Year	$y_3 = 3.624\%$
4 Year	$y_4 = 3.943\%$
5 Year	$y_5 = 4.683\%$

Determine the forward rate $f_{[3,5]}$.

(6) Spot rates associated with a four-year, par-value, $3,000, 6% bond with annual coupons are $r_1 = 4.5\%$, $r_2 = r_3 = 5.5\%$, and $r_4 = 6\%$. Calculate the value of the bond and its yield if it is sold at a price equal to its value.

(7) In Example (8.3.3) we calculated the two-year spot rate and found $r_2 \approx 3.866088985\%$. Explain how there would be an arbitrage opportunity if there was a two-year zero-coupon bond available with a yield rate of 3.5%.

(8.4) Forward contracts

(1) Nadia Sobolev has entered a forward contract for the purchase of 10,000 ounces of high grade copper in six months at $16,800. The risk-free rate of interest is 3% convertible quarterly, and the present value of the cost of holding the copper for six months is $40. Find the cost per ounce of copper at the time the contract was entered, assuming that the forward price was determined on a no-arbitrage basis.

(2) Consider a forward contract for an asset that does not have any cashflows associated with holding it over the contract period $[0, T]$. Denote the asset's price at time t by S_t. The contract's forward price was determined using no-arbitrage pricing, assuming a level risk-free effective interest rate i, so that the forward price is $K = S_0(1 + i)^T$. Show that the no-arbitrage value of the contract to the buyer at time t, where $0 \leq t \leq T$, is $S_t - K(1 + i)^{t-T}$.

(8.5) Commodity futures held until delivery

(1) Boyd Harkey takes a long position on a futures contract and maintains the position until he takes delivery, at which time he sells the underlier at the spot price. We denote his profit (possibly negative) by B. At exactly the same time, his wife Linda takes a short position on the same contract, and at expiration she purchases the underlier for cash so that she may make the required delivery. We denote her profit (possibly negative) by L. Assume that there are no fees, delivery expenses, or costs associated with taking delivery and reselling. Find $B + L$.

(2) Travis Scott took a short position on a July 2005 NYBOT cotton futures contract. This contract calls for the delivery of 50,000 pounds net weight of cotton. He established his position when the futures price was $.4785 per pound, and at expiration, the futures price was $.5225 per pound. How much has Travis earned or lost?

(3) Katrina takes a long position on a futures contract when the futures price is $6,150. She deposits $800 in a margin account. The next day's settlement price for a contract with identical underlier, delivery instructions, and delivery date is $6,225. How is her margin balance adjusted?

(4) On April 15, 2003, a futures contract had futures price $32,500 and stipulated delivery of underlier A on July 31. The contract had an initial margin of $3,000 and a maintenance margin of $2,400. The futures price of contracts with underlier A and expiration date July 31 stayed at $32,500 until May 1, when it jumped to $35,000. It then remained at $35,000 until May 27, when it fell to $31,000. The next change in the futures price came on May 30, when the futures price was $34,000. There is one final change in the futures price prior to delivery: On June 20, it was $36,000. Consider two futures traders for this contract named Serena and Barbara. Barbara and Serena each open positions on April 15, Barbara as a buyer and Serena as a seller. Suppose that Serena and Barbara each only had outflows as required by the initial and maintenance margin requirements, and there was no interest on the margin account. They each held the futures through expiration and accepted or made delivery as required by the contract. Aside from Barbara's payment at delivery, the only outflows made were those mandated by the initial and maintenance margin requirements. Furthermore, there was no interest earned on the margin accounts. Describe Serena's cashflows and Barbara's cashflows.

(8.6) Offsetting positions and liquidity of futures contracts

(1) Trevor Osterman purchases a September 2005 palladium futures contract. At the time his buy order is filled, the price for the 100 troy ounce palladium contract is $19,120. The initial margin is $2,700 and the

maintenance margin is $2,000. Trevor holds the contract for two months, during which time he does not receive any margin calls, and sells the contract for $18,310. By what percent does the price fall, and what is Trevor's annual effective yield rate, assuming he deposits only the initial margin required and that the margin account pays interest at a nominal rate of 2.7% convertible monthly?

(2) Sam Goodman bought a futures contract on August 14, 2005 for $72,850 and sold it on August 23rd for $76,550. The contract had an inital margin requirement of $5,500 and a $3,000 maintenance requirement. Sam made margin deposits to his account only as required and he did not make any withdrawals prior to settlement. If the prices for the futures contract used for marking-to-market were $72,850 on 8/14, $72,200 on 8/15, $71,700 on 8/16, $73,400 on 8/17, $75,200 on 8/18, $74,500 on 8/19, $75,900 on 8/22, and $76,400 on 8/23, make a chart showing the marking-to-market values, Sam's deposits, and Sam's margin account balances. Assume that no interest is paid on the margin account. Repeat, assuming that Sam took a short position instead of a long position; the trading times and futures prices are assumed to be unchanged.

(3) Rhonda Stallings established a long position on a 5,000 bushel CBOT November 2005 soybean future, and twelve days later she closed the position. The initial margin requirement was $1,485, and the maintenance margin was $1,100. At the time she established her position, the futures price was $30,500 (610 cents per bushel), and she closed her position when the futures price was $29,300. In addition to her initial deposit of $1,485, she made a deposit of $600 to her margin account six days after she established her position. Find Rhonda's annual effective yield for the twelve-day period during which her long position was open.

(4) Jose took a short position on a CME random-length lumber futures contract whose price was $46,805. At the same time, his cousin Rodrigo took a long position on the contract. The initial margin requirement was $1,898, and the maintenance margin was $1,265. Ten days later, there have been no margin calls, and the futures price drops to $45,760. One day later, the futures price rebounds, soaring to $47,685. It stays level at this price for ten more days, then drops suddenly to $44,770. Just after this second drop, and before there are any further price changes, Jose and Rodrigo each close their positions. Suppose that Jose and Rodrigo each only had outflows as required by the initial and maintenance margin requirements, and there was no interest on the margin account. Describe Jose's cashflows and Rodrigo's cashflows, and then calculate the *daily* yield rate for each of the cousins.

(8.7) Price discovery and more kinds of futures

(1) Tuscany Flores takes a long position on a Eurodollar futures contract when the price is quoted as 98.75. She holds the contract until delivery, at which time the relevant three-month Eurodollar interest rate is 1.55%. What is the value of Tuscany's position at delivery?

(2) A CBOT five-year U.S. Treasury note futures has as underlier U.S. Treasury notes with face value at maturity of $100,000. These notes must have a maturity of not more than five years and three months and a remaining maturity of not less than four years and two months as of the first day of the delivery month. The invoice price equals the futures price times a conversion factor plus accrued interest. The conversion factor is the price of the delivered note ($1 par value) to yield 6 percent. Drake Chamberlain buys a five-year U.S. Treasury note futures with December 31 delivery when the futures price is $99,687.50, and he sells the contract one month later when the contract is $98,031.25. His initial margin deposit was $675 and the maintenance margin was $500. Assuming that there were no margin calls for the one month that Drake held the contract, find the ratio of Drake's gain (possibly negative) to his investment.

(8.8) Options

(1) Joao Teixeira buys a call option to purchase 10,000 gallons of 82 octane gasoline at $2.35 per gallon in three months. The premium for the option is $1,000. Let $f(x)$ give the amount that Joao's option is worth in three months if x is the dollar price per gallon of 82 octane gasoline. Graph $f(x)$.

(2) Rita Gipson owns 500 shares of Explore stock. The current price per share is $22.45. Rita is counting on having at least $10,000 from these shares six months from now to pay her college tuition, so she decided to purchase five European put options, each with 100 shares of Explore stock as underlier, expiration in six months, and a strike price of $2,000. The options each cost $30, and she withdraws the $150 to pay for them from her savings account that has a 4% annual effective interest rate. Six months from now, the price of a share of Explore stock is X, and Rita observes that buying the put option ended up increasing the total value of her holdings at time 6 months by $127.03. (a) Find X. (b) What is the largest amount by which buying the put options might have decreased the total value of her holdings at time 6 months?

(3) Sasha purchased a call option to buy 100 shares of Search Computer stock six months from the option purchase date at $54 per share. The cost of the option was $325.50. The per share price of the stock was

$54.25 when Sasha bought the option, and the price six months later was $55.80. Find Sasha's annual yield if he sold the option for its value right before its expiration. How does this compare with the annual yield he would have received if he had used his $325.50 to purchase six shares of the stock, which he sold at the end of the six months?

(4) Cherie Coleman wishes to purchase a European call option with expiration in one year and strike price $6,840. The underlier is 100 shares of Eureka Energy stock. The stock's current price is $67.90 per share, and the annual effective interest rate is 4.2%. The price of a European put option, with expiration in one year and strike price $6,840, is $32.17. How much should Cherie expect to pay for the desired option, assuming that pricing is by a no-arbitrage model?

(8.9) Using replicating portfolios to price options

(1) Easylife stock currently sells for $23.80 per share. Three months from now, the price will either have gone up to $25.20 or have fallen to $22.20. A call option to purchase 100 shares of the stock in three months for $24 per share sells for $72. With what risk-free annual effective rate of interest is this consistent, if there is to be no arbitrage opportunity?

(2) The current price of a share of Excel Computers is $47.20. One month from now, the price of a share will be either $48 or $46.50. The nominal risk-free interest rate is 3% convertible monthly. In order that there not be an arbitrage opportunity, what should be the price of a European put option for 100 shares of this stock if it allows the holder to purchase the stock for $47 per share in one month?

(3) A European call option has strike price $20,000 and exercise in three years. The current price of the underlier is $18,400 and this will either go up by 10% or down by 8% each year, independent of what happens in the other years. Find the no-arbitrage price of the option, by using replicating portfolios, if the annual effective risk-free effective interest rate is 4%.

(4) Tan Yoon has agreed to sell 50 shares of a stock to his brother in two months for $2,500, if his brother so desires. The current price of the shares is $2,430, and each month that price will either go up by 8% or down by 5%, independent of what happens in the other month. If the price ends up going up each month, then Tan's brother will accept the offer and Tan will receive only $2,500 for an asset worth $2,834.35. Otherwise, the stocks will have a market price of less then $2,500, so Tan's brother will not accept the offer to buy the stock from Tan. (a) Suppose that Tan Yoon notes that he could purchase a call option having as underlier 100 shares of the same stock, expiration in two months, and strike price

$5,000, and that it has a no-arbitrage price X, based on a risk-free annual effective interest rate of 5%. Find X. Do not round your answer to the nearest cent. (b) The option in part (a) is for twice as many shares as Tan is obligated to offer to his brother, so Tan does not feel that it fits his needs. However, he would like to protect himself from a possible $334.35 loss in two months. Explain how he may do so, at a cost of $X/2$, using dynamic portfolio allocation. As in Example (8.9.8), suppose that any number of shares of the stock, including fractional numbers, may be traded at any time, and that there are no fees or commissions. Further suppose that Tan may borrow money at any time at an annual effective interest rate of 5% and that the loan amounts are not limited to whole numbers of cents.

(8.10) Using weighted averages to price options: risk-neutral probabilities

(1) Axle Automotive Stock currently sells for $64 a share. Each quarter, its price will either increase by 7% or fall by 13%. The risk-free interest rate is 6% convertible quarterly. Use risk-neutral probabilities to find the no-arbitrage price of a European call option to purchase 100 shares of the stock at its current $64 per share price if its expiration is in two quarters.

(2) A put option is to be sold for 100 shares of a security whose current price is $76.40 per share; it is for exercise in one year. The security's price will change (up or down) by 10% during the year, and the risk-free annual effective interest rate is 5%. The no-arbitrage price of the option is $100. Use risk-neutral probabilities to find the exercise price for the option.

(3) Use risk-neutral probabilities to solve Problem (8.9.1).

(4) As in Example (8.10.7), consider a European call option with expiration in one month and strike price $37.75. The current price of the underlier is $36.50, one month from now it will be $34 or $39, and the risk-free nominal interest rate convertible monthly is 3%. Let p denote the real-world probability that the value of the underlier goes up to $39 and X the value of the underlier at the end of the month.

(a) Let J be the nominal rate-of-return convertible monthly for a purchaser of the call option assuming the option's price is $.51825\frac{\$1.25}{1.0025} \approx .646197$, the price found in Example (8.10.7). Then J is a function of X, say $J = g(X)$, and (by a standard result in probability theory), the expected value of J with respect to the real-world probability p is $E(J) = p[g(39)] + (1 - p)[g(34)]$. Find p if the expected value $E(J)$ of the nominal rate of return convertible monthly is equal to 8.4%.

(b) Let J' be the nominal rate-of-return convertible monthly for a purchaser of the underlier who sells it at the end of one month. Like J, J' is a function of X, say $J' = h(x)$. Moreover, the expected value of J with respect to the real-world probability p is $E(J') = p[h(39)] + (1 - p)[h(34)]$. Find $E(J')$ assuming p has the numerical value found in (a).

(c) Compare $E(J)$ and $E(J')$. Can you justify their relative sizes?

(5) A European call option has strike price $20,000 and exercise in three years. The current price of the underlier is $18,400 and this will either go up by 10% or down by 8% each year, independent of what happens in the other years. Find the no-arbitrage price of the option, by using risk-neutral probabilities, if the annual effective risk-freeinterest rate is 4%. (If you did Problem (8.9.3), compare your answer to the one found using replicating portfolios.)

(8.11) Swaps

(1) On April 1, 1994, Great Savings Bank had issued many fixed-rate CDs, and had loaned out much of the money it obtained from these at variable rates. The board of directors is concerned that the Bank faces too much interest-rate risk and decides to take the fixed-leg in a fixed-for-floating interest rate swap — that is, lock in a fixed interest rate. The notional balance is $35,000,000, the fixed rate is $i^{(4)} = 5.25\%$, and the floating rate is the three-month LIBOR. The swap is for one year and the LIBOR rates, given on an "Actual/360 basis," turn out to be 4.250% for the period beginning 4/01/94, 4.875% for the period beginning 7/01/94, 5.688% for the period beginning 10/01/94, and 6.328% for the period beginning 1/01/95. Detemine the amount and time of all payments.

(2) Kennedy Contemporary Design had outstanding par-value bonds with face values totaling 12,200,000 Danish krone (DKK). They had two years until maturity. The company took in dollars and its management was fearful that the dollar would become weaker against the krone. However, the par-value bonds were issued at a lower rate than the company would have had to offer on newly issued bonds so, rather than issuing new bonds in dollars and calling in the krone bonds, Kennedy decided to enter a currency swap in which it would pay out dollars and take in krone. At the time, the exchange rate for krone is 6.1 DKK per dollar. Suppose that the company entered a two-year currency swap with a notional principal of 12,200,000 DKK or $2,000,000, the nominal dollar interest rate convertible semiannually was 8%, the nominal krone interest rate convertible semiannually was 6.5%, and that payments were to be made each six months. Further suppose that the exchange rate

in six months was 6.2DKK/$1, in one year was 6.3DKK/$1, in eighteen months was 6.45DKK/$1, and in two years was 3.68DKK/$1. (a) Based on a nominal interest rate of 5% convertible semiannually, what is the time 0 value K, in krone, of the payments received by Kennedy Contemporary Design from the currency swap? (b) Based on a nominal interest rate of 5% convertible semiannually, what is the time 0 value D, in dollars, of the payments paid by Kennedy Contemporary Design from the currency swap? (c) Find $K - 6.1D$. In hindsight, and based on discounting by compound interest at a nominal interest rate of 5% convertible semiannually, this was the time 0 value, in krone, of the cashflows from the swap.

(3) Mrs. Eleanor Markov has owned stock in her brother's business for thirty years and would have to pay taxes on a large capital gain if she sold it. Its current market price is $56,400. Eleanor's brother has warned her that the company's financial outlook for the next couple of years is discouraging, but that he sees a bright financial picture thereafter. Based on this information, Mrs. Markov decides to enter a two-year total return equity swap, exchanging her position in the stock for fixed semiannual payments at a nominal rate of 2.5%. At the end of the first six months, the stock pays dividends of $150 and the stock's price has risen to $57,600. At the end of each of the next three six-month periods, there are no dividends paid out and the selling prices are $57,900, $55,500, and $54,000. Describe all payments that Mrs. Markov receives or must make.

Chapter 8 review problems

(1) Assuming the market is arbitrage-free, if a three-month zero-coupon bond yields 2.25%, a six-month zero-coupon bond yields 2.45%, a nine-month zero-coupon bond yields 2.95%, and a one-year zero-coupon bond yields 3.35%, what should be the price of a one-year $1,000 5% par-value bond with quarterly coupons?

(2) The market includes two assets. Asset A has a current price of $85. One year from now, it will return $95 if the market is up and $70 if the market is down. Asset B sells for $21. One year from now it will return $24 if the market is up and $19 if the market is down. Develop an arbitrage strategy involving these two assets.

(3) No-arbitrage pricing, with an annual effective risk-free interest rate of 5.8%, was used to price a European call option with expiration in six months and $4,200 strike price. The option price is $173.51. The underlier has current price $4,238 and at expiration the price will either be $4,424 or X. Find X. Explain why you could not determine X if the option price were $154.75.

(4) On June 1, 1998, Confidence Life and Casualty (CLC) entered into an interest rate swap with Great Commercial Bank (GCB). The swap lasted five years, and had interest cashflows at the end of each year. The notional principal was $50,000,000. CLC was assured interest calculated using the one-year LIBOR and was obligated to pay interest at an annual effective rate of 6%. If the five consecutive June 1 LIBOR rates used were 5.940%, 5.803%, 7.214%, 4.055%, and 2.251%, what was the June 1, 2003 total value, calculated using the compound interest accumulation function $a(t) = (1.04)^t$, of the netted payments that CLC received from the swap?

(5) A European option on 100 shares of Omega Industries stock has an exercise price of $2,300 and expiration in six months. The risk-free annual effective interest rate for investing is 3.2%. The current price of the stock is $24 per share. Each three months, the price of the stock will either rise by 7% or fall by 5%. Find the price of the option if it is a call option and again if it is a put option. In each case, show how dynamic portfolio allocation may be used to replicate the option.

(6) Xu Chang notes that the yields on 6% par-value bonds with annual coupons are 2.825% for a one-year bond, 3.745% for a two-year bond, 4.212% for a three-year bond, and 4.730% for a four-year bond. Determine the forward rate $f_{[2,4]}$.

(7) Marjorie Majeski bought a futures contract on Monday, January 5, 2003 when the futures price was $32,460. The contract had an initial margin requirement of $2,500 and a $2,000 maintenance requirement. Marjorie held the account for one week. She made margin deposits to her account only as required and did not make any withdrawals prior to settlement. The prices for the futures contract used for marking-to-market were $32,075 on 1/5, $31,200 on 1/6, $29,800 on 1/7, $30,600 on 1/8, $32,975 on 1/9, and $31,400 on 1/12. Marjorie sold the contract for $31,800. Make a chart showing the marking-to-market values and any deposits required of Marjorie. Assume that no interest is paid on the margin account.

(8) Farmer Escobar purchases a March 2006 pork bellies futures contract. At the time his buy order is filled, the price of the 40,000 pound contract is 83.2 cents per pound. The initial margin is $1,863 and the maintenance margin is $1,380. Farmer holds the contract for three months, during which time he does not receive any margin calls, and he sells the contract for 82 cents per pound. By what percent does the price fall, and what is Farmer's three month effective yield rate, assuming he deposits only the initial margin required and that the margin account is credited with $10.40 of interest?

CHAPTER 9

Interest rate sensitivity

9.1 OVERVIEW

For our readers who know calculus and wish a more comprehensive treatment, this section serves as an introduction to the issues that we discuss in greater depth, using calculus, in Sections (9.2) through (9.5). For others, we hope to provide a brief survey of problems and possible solutions that face an investor who lives in a world where interest rates may change in an unpredictable fashion.

To get started, imagine that you are obligated to pay $50,000 five years from now. To cover this liability, you decide to buy bonds now. You believe that the prices for five-year zero-coupon bonds are not as attractive as those for three-year zero-coupon bonds or seven-year zero-coupon bonds. However, you note that should you buy a five-year zero-coupon bond with maturity $50,000, you are *certain* to meet your obligation. On the other hand, the attractively priced three-year bond that sells for the same price as the five-year $50,000 par-value bond will mature for less than $50,000, so you only accumulate $50,000 at time five if you can reinvest at a sufficiently good rate. In other words, there is some **reinvestment risk**, because you are hoping to have high interest rates three years from now but could possibly have very low rates.[1] Realizing this, you wonder whether you can avoid risk caused by uncertain rates in the future by purchasing the seven-year bond, then selling it after five years so as to realize the needed $50,000. But of course you once again have a problem, since the price your seven-year bond commands at time

---

[1] Some of you may be wondering about whether you should speculate on the forward rate $f_{[3,5]}$. That is an interesting possibility, but in this motivating example, it is not what we wish to consider, nor are other possible ways of purchasing a guaranteed rate at time 3.

5 is based on the rates of return that are available then. If interest rates have risen, you may again have a problem meeting your obligation. So, under this scenario, you are hoping to see low rates at time five.

We note that if you invest for less than five years, you hope for *high* rates at reinvestment time, while if you start with an investment of more than five years, you hope to see *low* interest rates at time five. So, either low or high interest rates can cause problems. It is natural to wonder whether there might be some way that you could split your investment, buying three-year zero-coupon bonds and seven-year zero-coupon bonds, and put yourself in a good position no matter whether rates go up or down. This is the basic idea behind **immunization** strategies. Of course, you could forget this puzzle and just invest in a five-year $50,000 zero-coupon par bond. This would be an example of **asset matching** or producing a **dedicated portfolio**. However, sometimes it pays to **immunize** your position, even though it may require you to be vigilant with respect to interest rates over time.

We now have established the idea that you may wish to follow how the price of a set of *predetermined* future cashflows changes as the interest rate fluctuates. Actually, as you know from Section (8.3), the term "the interest rate" may not be appropriate since you probably have a nonflat yield curve, but in Sections (9.2)–(9.4) we *assume that the yield curve is flat and remains so*. In Section (9.5), we will consider the problem when the cashflows are interest-sensitive (so not predetermined), and in that section we also briefly address how you might take into consideration the existence of a nonflat yield curve.

EXAMPLE 9.1.1   Immunization

Problem: As in the preceding paragraphs, suppose that you need to pay $50,000 in five years and that you can finance this with zero-coupon bonds with terms of three years and seven years. Imagine that you buy a $22,675.74 three-year bond and a $27,562.51 seven-year bond, each priced to yield 5%. Suppose also that, at the end of three years, no matter what the yield rate i (from a flat yield curve) may be, you sell the remaining bond at a purchase price to yield i, combine the proceeds with the $22,675.74 from the redeemed bond, and use the total to buy a two-year zero-coupon bond. Illustrate that this immunizes against interest rate risk by showing that it produces the needed $50,000 five years after your initial bond purchases if $i = 20\%$ (a high rate), $i = 5\%$ (the present moderate rate), or $i = 1\%$ (a low rate).

Solution   If $i = 20\%$, then the original seven-year $27,562.51 bond that has four years until maturity can be sold for $\frac{\$27,562.51}{(1.2)^4} \approx \$13,292.11$. Combining this with the $22,675.74 from the redeemed three-year bond, you have $13,292.11 + $22,675.74 = $35,967.85 to invest for two years at 20%, producing $\$35,967.85(1.2)^2 \approx \$51,793.70$. You have $\$1,793.70$ more than you needed.

On the other hand, if $i = 5\%$, then the original seven-year bond can be sold for $\frac{\$27,562.51}{(1.05)^4} \approx \$22,675.75$. Combining this with the $\$22,675.74$ from redemption, you have $\$22,675.75 + \$22,675.74 = \$45,351.49$ to invest for two years at $i = 5\%$. So, five years after your initial bond purchases, you have $\$45,351.49(1.05)^2 \approx \$50,000.02$. You have only two cents more than you need for your $\$50,000$ obligation.

Finally, if $i = 1\%$, then the original seven-year bond can be sold for $\frac{\$27,562.51}{(1.01)^4} \approx \$26,487.03$. Putting this with the $\$22,675.74$ from redemption, you have $\$26,487.03 + \$22,675.74 = \$49,162.77$ to invest for two years at $i = 1\%$. in this case, five years after your initial bond purchases, you have $\$49,162.77(1.01)^2 \approx \$50,150.94$. This is $\$150.94$ more than you needed.

Note: Section (9.4) uses calculus to develop the tools needed to determine this immunization strategy.   ∎

Another important concept we can introduce here is the sensitivity of the market value of a set of cashflows — that is, how changes in the interest rate change the value, or price, of the cashflows. Fix a set of cashflows, and let $P(i)$ denote their total price (present value) using compound interest at an effective interest rate i. If all the cashflows are positive, $P(i)$ is a decreasing function of i. The graph of $P(i)$ is called the **price curve**. The price curve is discussed further in Section (9.2), and Figure (9.2.2) depicts a price curve corresponding to a set of positive cashflows.

The **duration** of a set of cashflows is a measure of how sensitive the price of the portfolio is to a shift in the interest rate. If the portfolio has duration 5, then for small h it will lose about the fraction $5h$ of its value if the annual effective interest rate goes up from i_0 to $(i_0 + h)$ and go up in value by approximately the fraction $5h$ of its value if the interest rate decreases to $(i_0 - h)$. This is only an approximation since the price curve is not a straightline, but the approximation should be good if h is small. Actually, we will define several types of duration and the above statement refers to the **modified duration** $D(i_0, 1)$ defined by (9.2.14). The above description may be interesting, but it does not tell us how to precisely calculate duration; that requires calculus.

Another form of duration is the **Macaulay duration** $D(i_0, \infty)$, which is given by $D(i_0, \infty) = (1 + i_0)D(i_0, 1)$. Historically, the concept now known as Macaulay duration predates the modified duration, and it was introduced in a 1938 paper by Frederick Macaulay concerning the volatility of Unites States bond yield rates and stock prices to changes in interest rates.[2] The Macaulay duration is a weighted average of the times of the cashflows, the weights being the associated proportion of the total present value that is attributable to the cashflow at the corresponding time. You should view $D(i_0, \infty)$ as some sort of average length of the investment. The Macaulay duration of an n-year zero-coupon bond turns out to be n, but an n-year bond with coupons has a smaller Macaulay duration.

Modified duration captures the rate of change of the price relative to the price. It does not, however, tell you anything beyond that about the way the price curve is shaped. More precisely, as we shall explain in Section (9.2), a duration D at i_0 can be used to approximate the price curve $P(i)$ near $i = i_0$ by a linear function as in $P(i_0 + h) \approx P(i_0)(1 - Dh)$, but this does not indicate whether the price curve is decreasing more or less rapidly near $i = i_0$ as i increases. The concept of **convexity** is introduced in Section (9.3) to address this. If the convexity is positive, the curve bends upward and decreases less rapidly as i increases, and if it is negative it bends downward and decreases more rapidly as i increases.

Asset-liability management refers to techniques for maintaining a firm's ability to meet its cashflow obligations in a world of changing interest rates. Tools used to counter interest rate risk include immunization procedures such as **Redington immunization** and **full immunization**. These are methods that involve the ideas of duration and convexity, and immunization is discussed in Section (9.4). Redington immunization and full immunization are based on a hypothetical flat yield curve, but they give a strategy for protecting a portfolio of cashflows from interest rate changes: The strategy might be useful even if the yield curve is not flat. Redington immunization focuses on small interest rate shifts while full immunization addresses arbitrary parallel shifts of a flat yield curve. For Redington immunization, you structure your holdings so that the **assets** (promised incoming cashflows) and **liabilities** (promised outgoing cashflows) have equal present values and durations while the convexity of the assets is greater than the convexity of the liabilities. The immunization in Example (9.1.1) is an example of full immunization.

We end this section with a simple example of **asset-liability matching**[3] which requires that the outflows from liabilites are exactly offset by inflows from assets. These inflows may include bond coupons and redemption payments.

EXAMPLE 9.1.2   Asset-liabilty matching

Problem: Sandy is obligated to pay $10,000 in six months, $15,000 in twelve months, and $25,000 in eighteen months. He wishes to purchase bonds to

[2]Macaulay, F. R. *Some Theoretical Problems Suggested by Movements of Interest Rates, Bond Yields, and Stock Prices in the United States Since 1856* New York: National Bureau of Economic Research, 1938.
[3]Asset-liability matching is also called **dedication**.

exactly match these liabilities — that is, to provide inflows so that the net cashflow at the three payment times, as well as at all other times times, is zero. The bonds available for purchase by Sandy are of the following three types:

(a) Six-month zero-coupon bonds, sold to yield the investor 6% nominal interest convertible semiannually;

(b) Twelve-month 6% par-value bonds with semiannual coupons;

(c) Eighteen-month 5% par-value bonds with semiannual coupons.

How much of each of these should Sandy purchase? Assume that each may be purchased for any par value that Sandy would like.

Solution Sandy should first figure out how much of the eighteen-month bonds to purchase to match his $25,000 obligation, since the other types of bonds will not help with this liability. If he purchases face amount F_c of the eighteen-month bond, then Sandy has the rights to coupon payments of $(.025)F_c$ every six months and also to the redemption payment of F_c. Therefore, at time eighteen months, he stands to receive $(1.025)F_c$, the total from the time-eighteen-months coupon payment and the redemption payment. Setting this equal to $25,000, we find $F_c \approx \$24,390.2439$. We round to the nearest penny, so $F_c = \$24,390.24$, and the coupon payments every six months are each $(.025)(\$24,390.24) \approx \$609.76$. Sandy wishes to receive a total of $15,000 at time twelve months and since he is already scheduled to be paid a $609.76 coupon, he needs to purchase a one-year bond that will pay him $15,000 - \$609.76 = \$14,390.24$ at the end of the year. So, he should purchase face amount F_b of the twelve-month 6% bond where $(1.03)F_b = \$14,390.24$. Rounding the face value to the nearest cent, we obtain $F_c = \$13,971.11$. Since $(.03)F_b \approx \$419.13$ and the six-month bond is bought to yield 3% per coupon period, Sandy should purchase a face amount F_a satisfying $(1.03)F_a = \$10,000 - \$609.76 - \$419.13$. Therefore, $F_a = \$8,709.82$. ∎

9.2 DURATION

(calculus needed here)

We have many times looked at sequences of cashflows. The important concept of the net present value of a set of cashflows was introduced in Section (1.7). We recall that the net present value depends on the choice of a particular accumulation function and that usually we specify a compound interest accumulation function. We once again assume that accumulation is by compound interest. Moreover, we add two assumptions. Firstly, we *assume a flat term structure of interest rates*. That is, we assume that spot rates are all equal to some common value. Secondly, we *assume that if one spot rate changes, all other spot rates change by an equal amount* so that we again have a flat yield curve. This second assumption is often expressed by saying that "the term structure only experiences parallel shifts." We wish to understand how the value of a series of cashflows shifts as the interest rate (the common value of the spot rates) changes. [In Section (9.5) we will address what might be done if we drop our two assumptions about the spot rates. We recognize that our spot rate assumptions rarely jibe with reality, but our analysis will still be helpful in real-life situations.]

Fix a set of cashflows $\{C_t : t \geqslant 0\}$ with only finitely many of the C_t being nonzero, and a reference time zero at which we will value the set. [4] Given an interest rate i, the price of the set that will provide the buyer with a yield rate equal to i is just the present value

$$(9.2.1) \qquad P(i) = \sum_{t \geqslant 0} C_t(1 + i)^{-t}.$$

We let i_0 denote the initial interest rate. If $P(i_0) > 0$, then the continuous function $P(i)$ remains positive for i close to i_0, and a typical **price curve** might look as in Figure (9.2.2).

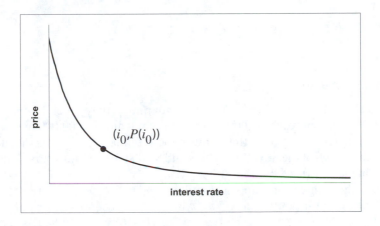

FIGURE (9.2.2)

The function $P(i)$ is an infinitely differentiable function, so it has a Taylor series

$$(9.2.3) \quad \sum_{n=0}^{\infty} \frac{P^{(n)}(i_0)}{n!}(i - i_0)^n = P(i_0) + P'(i_0)(i - i_0) + \frac{P''(i_0)}{2}(i - i_0)^2 + \cdots .$$

The first two terms of this series $P(i_0) + P'(i_0)(i - i_0)$ give us the tangent-line

---

[4]We have fixed the cashflows. In particular, we do not allow them to be sensitive to the interest rate i. This would not, for instance, be true of the cashflows of a callable or putable bond. We will consider such interest rate-dependent cashflows in Section (9.5).

approximation to $P(i)$ at interest rate i_0, namely

$$(9.2.4) \qquad P(i) \approx P(i_0) + P'(i_0)(i - i_0),$$

which is of course only a good approximation for i very close to i_0. The first three terms give us the second-Taylor-polynomial approximation

$$(9.2.5) \qquad P(i) \approx P(i_0) + P'(i_0)(i - i_0) + \frac{P''(i_0)}{2}(i - i_0)^2,$$

and we can expect this to be a better approximation for i close to i_0. We note that the derivatives involved in Approximations (9.2.4) and (9.2.5) are straightforward to compute; taking successive derivatives in Equation (9.2.1), we have

$$(9.2.6) \qquad P'(i) = -\sum_{t \geq 0} C_t t(1 + i)^{-t-1}$$

and

$$(9.2.7) \qquad P''(i) = \sum_{t \geq 0} C_t t(t + 1)(1 + i)^{-t-2}.$$

EXAMPLE 9.2.8

Problem: Excel Finance holds loans that guarantee they receive repayments of $1,000,000 in one year, $2,000,000 in two years, and $2,000,000 in six years. They also have issued bonds that require them to make payments of $300,000 at the end of each of the next three years and $3,000,000 in four years. Calculate Excel Finance's price function at 2%, 4%, 4.9%, 5%, 5.1%, 6%, and 10%. Then use the tangent line approximation at $i_0 = 5\%$ to estimate the prices at each of these rates, excluding 5%. Also use the second Taylor polynomial around $i_0 = 5\%$ to estimate the prices at 2%, 4%, 6%, and 10%. Discuss your findings.

Solution Denote the amount of any positive inflow at time t by A_t, and let L_t designate the amount of any liability you must pay at t. Then $A_1 = \$1,000,000$, $A_2 = A_6 = \$2,000,000$, $L_1 = L_2 = L_3 = \$300,000$, and $L_4 = \$3,000,000$. Therefore, the price, when figured at interest rate i, is

$$P(i) = (\$1,000,000 - \$300,000)(1 + i)^{-1} + (\$2,000,000 - \$300,000)(1 + i)^{-2}$$
$$- \$300,000(1 + i)^{-3} - \$3,000,000(1 + i)^{-4} + \$2,000,000(1 + i)^{-6}.$$

Computing this net present value seven times, with i equaling .02, .04, .049, .05, .051, .06, and .10, we obtain approximately $1,041,971.22, $994,340.06,

$975,755.73, \$973,788.87, \$971,840.87, \$955,125.61, and \$895,835.37 respectively. [Equipped with a BA II Plus calculator, you can most easily check these values using the **Cash Flow worksheet** and $\boxed{\text{NPV}}$. The contribution registers should have nonzero entries CF0 $= 0$, C01 $= 700{,}000$, C02 $= 1{,}700{,}000$, C03$=$ $-300{,}000$, C04$=-3{,}000{,}000$, C05$=0$, C06 $= 2{,}000{,}000$, and the corresponding frequency registers should all contain the numeral 1. Remember to enter the interest rates as percents after keying $\boxed{\text{NPV}}$ (Don't forget to press $\boxed{\text{ENTER}}$), and then to press $\boxed{\downarrow}$ $\boxed{\text{CPT}}$.] Note that

$$P'(i) = -\$700{,}000(1+i)^{-2} - \$3{,}400{,}000(1+i)^{-3} + \$900{,}000(1+i)^{-4}$$
$$+ \$12{,}000{,}000(1+i)^{-5} - \$12{,}000{,}000(1+i)^{-7},$$

and therefore $P'(.05) \approx -\$1{,}957{,}398.21$. (Again, the BA II Plus calculator **Cash Flow worksheet** and $\boxed{\text{NPV}}$ allow you to check this calculation efficiently.) Therefore, the tangent line at $i = .05$ gives

$$P(i) \approx \$973{,}788.87 - \$1{,}957{,}398.21(i - .05).$$

In particular, using this repeatedly, we obtain $P(.02) \approx \$1{,}032{,}510.82$, $P(.04) \approx \$993{,}362.85$, $P(.049) \approx \$975{,}746.27$, $P(.051) \approx \$971{,}831.47$, $P(.06) \approx$ $\$954{,}214.89$, and $P(.10) \approx \$875{,}918.96$. We note that these approximations *underestimate* the exact values by \$9,460.40 ($\approx$.91%), \$977.21 ($\approx$.098%), \$9.46 ($\approx$.00097%), \$9.40 ($\approx$.00097%), \$910.72 ($\approx$.095%), and \$19,916.41 ($\approx 2.2\%$), respectively. So, they are all very good approximations! Of course, those for interest rates very close to 5% are spectacular.

Now for the quadratic approximations. Note that

$$P''(i) = \$1{,}400{,}000(1+i)^{-3} + \$10{,}200{,}000(1+i)^{-4}$$
$$- \$3{,}600{,}000(1+i)^{-5} - \$60{,}000{,}000(1+i)^{-6}$$
$$+ \$84{,}000{,}000(1+i)^{-8},$$

and $P''(.05) \approx \$18{,}861{,}826.29$. Therefore,

$$P(i) \approx \$973{,}788.87 - \$1{,}957{,}398.21(i - .05) + \$9{,}430{,}913.15(i - .05)^2.$$

Using this approximation, we find $P(.02) \approx \$1{,}040{,}998.64$, $P(.04) \approx \$994{,}305.94$, $P(.06) \approx \$955{,}125.98$, and $P(.10) \approx \$899{,}496.24$. The first two understate the values by \$972.58($\approx$.09334%) and \$34.12 ($\approx$.00343%), and the last two overstate it by \$32.37 ($\approx$.00339%) and \$3,660.87 ($\approx$.40865%). These are improved estimates from our previous ones. ∎

In Example (9.2.8) we worked first with the tangent approximation (9.2.4), and this approximation may be rewritten as

(9.2.9)
$$\frac{P(i) - P(i_0)}{P(i_0)} \approx \frac{P'(i_0)}{P(i_0)}(i - i_0).$$

The approximation (9.2.9) allows us to estimate the **fractional price change** or **relative price change** — the price change divided by the original price — as the interest rate increases.

It is best to avoid talking about a "one-percent increase of the interest rate" i, since the phrase is ambiguous. Instead, we introduce **basis points**. If the yield increases from $i = q\% = .01q$ to $i = (q + 1)\% = .01(q + 1)$, we say that the yield has increased by **one hundred basis points**, and more generally an increase of $100b$ basis points means that the yield has increased from one value to another value that is numerically $(.01)b$ more. That is, 100 basis points equals $.01 = 1\%$. Substituting $i - i_0 = .01 = 100$ basis points into the approximation (9.2.9) gives $\frac{P(i)-P(i_0)}{P(i_0)} \approx \frac{P'(i_0)}{P(i_0)}(.01)$. With this new language, the approximation (9.2.9) gives us the following statement.

> If the yield is initially equal to i_0 and then it increases by one hundred basis points, the approximate relative price change is $\frac{P'(i_0)}{P(i_0)}$ percent.

Another way to write Approximation (9.2.9) is as

(9.2.10)
$$\left[\frac{P(i) - P(i_0)}{P(i_0)}\right] \Big/ (i - i_0) \approx \frac{P'(i_0)}{P(i_0)}.$$

We may view the right-hand side of Approximation (9.2.10) as a normalized or relative rate of change of the price. This is reminiscent of the way we viewed $\left[\frac{a(t+h)-a(t)}{a(t)}\right]/h$ when we discussed force of interest [see Section (1.12)]. By Approximation (9.2.9), multiplying the ratio $\frac{P'(i_0)}{P(i_0)}$ by the numerical change in the interest rate gives an estimate of the relative change in the price $P(i)$. You expect this to be a good estimate if i is close to i_0.

EXAMPLE 9.2.11

Problem: Use Approximation (9.2.9) to estimate the relative price change in Example (9.2.8) when i changes from .05 to .051.

Solution In Example (9.2.8), we calculated $P(.05) \approx \$973,788.87$, and $P(.051) \approx \$971,840.87$. Therefore, the true relative change is $\frac{P(.051)-P(.05)}{P(.05)} \approx$

$\frac{\$971,840.87-\$973,788.87}{\$973,788.87} \approx -.002000434$. Approximation (9.2.9) estimates this to be $\frac{P'(.05)}{P(.05)}(.051 - .05)$, which from Example (9.2.8) equals

$$\frac{-\$1,957,398.21}{\$973,788.87}(.001) \approx -.002010085$$

rather than the true $-.002000434$.   ∎

The factor $\frac{P'(i_0)}{P(i_0)}$ plays a key role in both Approximations (9.2.9) and (9.2.10) and leads to our first formal definition of a duration. For any number $i > -1$, we define the **modified duration** $D(i,1)$ by

(9.2.12)
$$D(i,1) = -\frac{P'(i)}{P(i)}.$$

Approximation (9.2.9) indicates that the prices of investments with larger modified durations are more sensitive to interest rate changes.

Recalling Equation (9.2.6), we see that

(9.2.13)
$$D(i,1) = \sum_{t\geqslant 0} C_t t(1+i)^{-t-1} \Big/ \sum_{t\geqslant 0} C_t(1+i)^{-t}.$$

The "1" in the notation $D(i,1)$ reminds us that we were considering the price as a function of the effective interest rate, which is a nominal interest rates convertible *one* time per year. More generally, we may define a **modified duration** for a nominal interest rate $i^{(m)}$ convertible m times per year. To do this, we begin by observing that the price of a cashflow is a function of the nominal interest rate $i^{(m)}$, since i is a function of $i^{(m)}$, namely $i^{(m)} = m[(1+i)^{\frac{1}{m}} - 1]$. We define the **modified duration** $D(i,m)$ by

(9.2.14)
$$D(i,m) = -\frac{\frac{dP}{di^{(m)}}}{P(i)} \quad \text{where} \quad i^{(m)} = m[(1+i)^{\frac{1}{m}} - 1].$$

Since $i = \left(1 + \frac{i^{(m)}}{m}\right)^m - 1$, the chain rule gives

$$\frac{dP}{di^{(m)}} = \frac{dP}{di}\frac{di}{di^{(m)}} = P'(i)\left(1 + \frac{i^{(m)}}{m}\right)^{m-1}$$

$$= P'(i)\frac{\left(1 + \frac{i^{(m)}}{m}\right)^m}{1 + \frac{i^{(m)}}{m}} = P'(i)\frac{1+i}{1 + \frac{i^{(m)}}{m}}.$$

Therefore,

(9.2.15)

$$D(i, m) = \frac{1 + i}{1 + \frac{i^{(m)}}{m}} D(i, 1) \quad \text{or equivalently}$$

$$D(i, m)\left(1 + \frac{i^{(m)}}{m}\right) = D(i, 1)(1 + i).$$

It follows that

$$D(i, m)\left(1 + \frac{i^{(m)}}{m}\right) \quad \text{is independent of } m.$$

Define

(9.2.16)

$$D(i, \infty) = -\frac{\frac{dP}{d\delta}}{P(i)} \quad \text{where} \quad \delta = \ln(1 + i).$$

Then, since $i = e^{\delta} - 1$,

(9.2.17)

$$\frac{dP}{d\delta} = \frac{dP}{di}\frac{di}{d\delta} = \frac{dP}{di}(1 + i),$$

and

(9.2.18)

$$D(i, \infty) = D(i, 1)(1 + i).$$

Combining (9.2.15) and (9.2.18), we have

IMPORTANT FACT (9.2.19): If m is a positive integer, then

$$D(i, \infty) = D(i, m)\left(1 + \frac{i^{(m)}}{m}\right).$$

Also note that

$$D(i, \infty) = D(i, 1)(1 + i) = D(i, 1)\left(\lim_{m \to \infty} \frac{1 + i}{1 + \frac{i^{(m)}}{m}}\right)$$

$$= \lim_{m \to \infty} D(i, m).$$

We call $D(i, \infty)$ the **Macaulay duration.** Important Fact (9.2.19) shows that what we call modified duration $D(i, 1)$ is obtained from Macaulay duration by a simple modification. This is the origin of the term modified duration. Another name for modified duration is **volatility**, but the word "volatility" has other meanings too.

It is instructive to write a formula for $D(i, \infty)$ that does not involve derivatives or other durations. Specifically, the price of the cashflows is given by $\sum_{t \geqslant 0} C_t e^{-\delta t}$ and the derivative of this expression *with respect to* δ is $-\sum_{t \geqslant 0} C_t t e^{-\delta t}$. Therefore,

(9.2.20)

$$
\begin{aligned}
D(i, \infty) &= -\frac{\frac{dP}{d\delta}}{P(i)} = \sum_{t \geqslant 0} C_t t e^{-\delta t} \Big/ \sum_{t \geqslant 0} C_t e^{-\delta t} \\
&= \sum_{t \geqslant 0} \left(\frac{C_t e^{-\delta t}}{\sum_{t \geqslant 0} C_t e^{-\delta t}} \right) t \\
&= \sum_{t \geqslant 0} \left(\frac{C_t (1 + i)^{-t}}{P(i)} \right) t.
\end{aligned}
$$

We therefore have

IMPORTANT FACT (9.2.21): The Macaulay duration $D(i, \infty)$ of a set of cashflows is a weighted average of the payment times at which the cashflows occur. If the cashflow at time t is C_t, the weight given to the time t is $C_t(1 + i)^{-t}/P(i)$, the proportion of the total price attributable to the time t cashflow.

We note that if the interest rate i were zero, then the Macaulay Duration $D(i, \infty)$ would be the weighted average used in the method of equated time [see Section (2.3)]. Macaulay and modified duration each provide a measure of the average length of an investment. Consequently, they are important in considering risk arising from the uncertainty of future interest rates when money needs to be reinvested. This risk is often referred to as **reinvestment risk.**

The simplest set of cashflows is a single payment — for instance, the redemption payment of a zero-coupon bond.

EXAMPLE 9.2.22   **Duration of a zero-coupon bond**

Problem: An N-year zero-coupon bond is purchased to provide an annual effective yield i. Find the Macaulay duration $D(i, \infty)$ and the modified duration $D(i, 1)$.

Solution   The bond has a single cashflow at its redemption time, so the average of its payment times is simply N. Therefore, Important Fact (9.2.21) gives $D(i, \infty) = N$. Equation (9.2.18) tells us that $D(i, 1) = (1 + i)^{-1} D(i, \infty) = \dfrac{N}{1+i}$.   ∎

Calculating duration is more complicated for a bond with coupons, as we see in our next examples.

EXAMPLE 9.2.23   **Duration of a coupon bond**

Problem: Find the Macaulay duration $D((1.03)^2 - 1, \infty)$ of a ten-year 8% $15,000 bond with semiannual coupons and redemption amount $16,500.

Solution   This bond has twenty semiannual $\left(\frac{.08}{2}\right)$ $15,000 = $600 coupon payments, occurring at times $\frac{1}{2}, \frac{2}{2}, \dots, \frac{20}{2}$, and a redemption amount of $16,500 at time $\frac{20}{2}$. The given annual effective interest rate is $i = (1.03)^2 - 1$, so the semiannual interest rate is $(1 + i)^{\frac{1}{2}} - 1 = .03$. Therefore, the basic price formula tell us that the price of the bond is $P(.03) = $600 a_{\overline{20}|3\%} + $16,500(1.03)^{-20} \approx $18,062.13486$. The Macaulay duration of the bond is, according to Important Fact (9.2.21), the weighted average of the payment times at which the cashflows occur: The weight given to time $t \in \left\{\frac{1}{2}, \frac{2}{2}, \dots, \frac{19}{2}\right\}$ is

$$\frac{$600(1 + i)^{-t}}{P(i)} = \frac{$600}{$18,062.13486}(1.03)^{-2t},$$

and the weight given to time $\frac{20}{2}$ (the maturity time) is

$$\frac{($600 + $16,500)(1 + i)^{-t}}{P(i)} = \frac{$600 + $16,500}{$18,062.13486}(1.03)^{-20}.$$

Thus, the Macaulay duration is

$$D((1.03)^2 - 1, \infty)$$

$$= \left[\left(\sum_{k=1}^{20} \$600(1.03)^{-k} \left(\frac{k}{2} \right) \right) + \$16,500(1.03)^{-20} \left(\frac{20}{2} \right) \right] \Big/ \$18,062.13486$$

$$= \left[\$300 \left(\sum_{k=1}^{20} (1.03)^{-k} k \right) + \$91,356.49944 \right] \Big/ \$18,062.13486$$

$$= \left[\$300(Ia)_{\overline{20}|3\%} + \$91,356.49944 \right] \Big/ \$18,062.13486.$$

We can now use Equation (3.9.6) to compute $\$300(Ia)_{\overline{20}|3\%}$. We find

$$\$300(Ia)_{\overline{20}|3\%} = \$300 \left(\frac{\ddot{a}_{\overline{20}|3\%} - 20(1.03)^{-20}}{.03} \right) \approx \$42,502.84023.$$

Therefore,

$$D((1.03)^2 - 1, \infty) \approx \frac{\$42,502.84023 + \$91,356.49944}{\$18,062.13486} \approx 7.411047515$$

for this ten-year coupon bond.   ∎

We next consider the Macaulay duration of a par-value bond with coupons, bought for its redemption amount. In this situation, the Macaualy duration, calculated at the bond's yield rate, turns out to be given by a simple formula.

EXAMPLE 9.2.24   Duration of a coupon bond

Problem: A noncallable par-value bond has m coupons per year and matures at the end of n years. As usual, the coupons are level. Find the Macaulay duration $D(i, \infty)$ and the modified duration $D(i, m)$ where i is the effective annual yield realized by a buyer who purchases the bond for its redemption amount and then holds it to maturity.

Solution   The coupon amount is $P \left((1 + i)^{\frac{1}{m}} - 1 \right) = P \left(\frac{i^{(m)}}{m} \right)$, because the bond is purchased for its redemption amount and the resulting effective yield on the investment of $P = C$ is i. Let $N = mn$. The cashflows provided by the bond are the level coupon payments at times $\frac{1}{m}, \frac{2}{m}, \ldots, \frac{N}{m}$, as well as a redemption payment $C = P$ at time $n = \frac{N}{m}$, the end of the last coupon period. Recalling that the Macaulay duration is the weighted average of the payment times, the weight at time t being the amount of the total present value contributed at time t [see Equation (9.2.20)], since $P = C$ we have

$$D(i, \infty) = \left[\left[\sum_{k=1}^{N} P\left(\frac{i^{(m)}}{m}\right) (1 + i)^{-\frac{k}{m}} \left(\frac{k}{m}\right) \right] + P(1 + i)^{-n} n \right] \Big/ P$$

$$= \left[\sum_{k=1}^{N} \left(\frac{i^{(m)}}{m}\right) (1 + i)^{-\frac{k}{m}} \left(\frac{k}{m}\right) \right] + (1 + i)^{-n} n$$

$$= i^{(m)} \left[\sum_{k=1}^{N} (1 + i)^{-\frac{k}{m}} \left(\frac{k}{m^2}\right) \right] + (1 + i)^{-n} n$$

$$\overset{(*)}{=} i^{(m)} \left[(I^{(m)}a)_{\overline{n}|i}^{(m)} \right] + (1 + i)^{-n} n$$

$$= i^{(m)} \left[\frac{\ddot{a}_{\overline{n}|i}^{(m)} - n(1 + i)^{-n}}{i^{(m)}} \right] + (1 + i)^{-n} n$$

$$= \ddot{a}_{\overline{n}|i}^{(m)}.$$

In obtaining the above equality marked "$\overset{(*)}{=}$", we have used the definition of $(I^{(m)}a)^{(m)}$ [see Figure (4.5.1) and Equation (4.5.2)]. To obtain the modified duration $D(i, m)$, we use Equation (9.2.19). We have,

$$D(i, m) = \frac{\ddot{a}_{\overline{n}|i}^{(m)}}{1 + \frac{i^{(m)}}{m}} = a_{\overline{n}|i}^{(m)}.$$

■

The result of Example (9.2.24) is significant, and we record it here so that it is easy to reference, along with a helpful observation.

> **IMPORTANT FACT (9.2.25):** The Macaulay duration of a par-value bond bought for its redemption amount, with m coupons per year and a fixed n-year term, is $\ddot{a}_{\overline{n}|i}^{(m)}$ where i is the yield rate to the purchaser. So, the duration increases with n and decreases if i or m increase.

If a bond is bought at a discount, there is a larger redemption payment than the price would indicate. The redemption payment is made at the end of the term, so this larger redemption payment results in an increased duration compared to that in Important Fact (9.2.25). On the other hand, if a bond is bought at a premium, the price exceeds the redemption amount, and the duration is decreased.

We next consider the duration for a mortgage that does not allow early repayment. How one should measure the length of time until reinvestment for a mortgage allowing repayment or for a callable or putable bond is discussed in Section 9.5.

EXAMPLE 9.2.26   Duration of an amortized loan with no early repayment option

Problem: On the first day of the year, Friendly Mortgage Company counts among its assets the rights to all future payments on a mortgage having N level payments of X remaining, the payments due at the end of each m-th of a year. Friendly wishes to determine the Macaulay duration $D(i, \infty)$ of this asset. Find the numerical value if $N = 180$, $m = 12$, and $i = 6\%$.

Solution   If we let time zero be the first day of the year referred to in the statement of the problem, then payments are to occur at times $\frac{1}{m}, \frac{2}{m}, \ldots, \frac{N}{m}$, and the present value of the payment at time $\frac{k}{m}$ is $X(1 + i)^{-\frac{k}{m}}$. So, the total present value is $\sum_{k=1}^{N} X(1 + i)^{-\frac{k}{m}}$, and this is the outstanding loan balance if the loan was made at the annual effective rate i. Let $n = \frac{N}{m}$. The Macaulay duration is

$$D(i, \infty) = \sum_{k=1}^{N} \left(\frac{k}{m}\right) X(1 + i)^{-\frac{k}{m}} \Big/ \sum_{k=1}^{N} X(1 + i)^{-\frac{k}{m}}$$

$$= \sum_{k=1}^{N} \left(\frac{k}{m^2}\right)(1 + i)^{-\frac{k}{m}} \Big/ \sum_{k=1}^{N} \left(\frac{1}{m}\right)(1 + i)^{-\frac{k}{m}}$$

$$= (I^{(m)}a)_{\overline{n}|i}^{(m)} \Big/ a_{\overline{n}|i}^{(m)} = \frac{\ddot{a}_{\overline{n}|i}^{(m)} - nv^n}{i^{(m)}} \Big/ \frac{1 - v^n}{i^{(m)}}$$

$$= \left(\frac{\ddot{a}_{\overline{n}|i}^{(m)} - nv^n}{1 - v^n}\right) \left(\frac{(1 + i)^n}{(1 + i)^n}\right) = \frac{\ddot{s}_{\overline{n}|i}^{(m)} - n}{(1 + i)^n - 1}$$

$$= \frac{\ddot{s}_{\overline{n}|i}^{(m)}}{(1 + i)^n - 1} - \frac{n}{(1 + i)^n - 1} = \frac{1}{d^{(m)}} - \frac{n}{(1 + i)^n - 1}.$$

Here we have used Equations (4.3.11) and (4.5.2). Now suppose $N = 180$, $m = 12$, and $i = 6\%$. Then $n = \frac{N}{m} = 15$, and $d^{(m)} = d^{(12)} = 12\left[1 - (1.06)^{-\frac{1}{12}}\right]$.

So,

$$D(i, \infty) = \frac{1}{d^{(m)}} - \frac{n}{(1 + i)^n - 1} = \frac{1}{12\left[1 - (1.06)^{-\frac{1}{12}}\right]} - \frac{15}{(1.06)^{15} - 1}$$

$$\approx 6.462820597.$$

The mortgage lasts for 15 years, but the Macaulay duration is less than half the length of the mortgage. ■

We next consider the duration of a portfolio of assets, each with known Macaulay duration at interest rate i. We suppose that the portfolio consists of s assets and that the k-th asset has present value $P^{[k]}(i)$ and Macaulay duration $D^{[k]}(i, \infty)$ when figured using compound interest at effective interest rate i or the equivalent continuously compounded rate δ. Note that

$$P^{[k]}(i) = \sum_t C_t^{[k]}(1 + i)^{-t},$$

where $C_t^{[k]}$ is the cashflow of the k-th asset at time t. In addition,

$$D^{[k]}(i, \infty) = \sum_t \left(\frac{C_t^{[k]}(1 + i)^{-t}}{P^{[k]}(i)}\right) t.$$

Then, with $P(i)$ as in Equation (9.2.1), the Macaulay duration of the whole portfolio $D^{\text{portfolio}}(i, \infty)$ is given by

(9.2.27)

$$\boxed{\begin{aligned} D^{\text{portfolio}}(i, \infty) &= \sum_t \left(\sum_{k=1}^s C_t^{[k]}(1 + i)^{-t} t\right) / P(i) \\ &= \sum_{k=1}^s \left(\sum_t C_t^{[k]}(1 + i)^{-t} t\right) / P(i) \\ &= \sum_{k=1}^s D^{[k]}(i, \infty) \frac{P^{[k]}(i)}{P(i)}. \end{aligned}}$$

Therefore,

IMPORTANT FACT 9.2.28: The Macaulay duration of a portfolio is a weighted average of the Macaulay durations of the individual assets, the weight given to the duration of an individual asset being the proportion of the total price attributable to that asset. Moreover, Important Fact (9.2.19) guarantees that you may replace "Macaulay" with "modified" in this statement.

EXAMPLE 9.2.29   Duration of a portfolio of bonds

Problem: Julian Bradley's bond portfolio consists of two bonds. Specifically, there is a \$10,000 five-year zero-coupon bond and a \$2,000 6% par-value two-year bond with semiannual coupons. Compute the Macaulay duration $D^{\text{portfolio}}(.045, \infty)$ for the portfolio.

Solution   Unfortunately, we can not just use Important Fact (9.2.25), because the nominal coupon rate, namely 6%, is not equivalent to the interest rate $i = 4.5\%$ at which we want to calculate the Macaulay duration. However, we can determine the individual Macaulay durations $D^{\text{two}}(.045, \infty)$ and $D^{\text{five}}(.045, \infty)$ of the two-year and five-year bonds respectively, and then use Important Fact (9.2.28) to find the Macaulay duration $D^{\text{portfolio}}(.045, \infty)$ of the portfolio of bonds.

The two-year bond has semiannual coupons that are each for an amount $\$2{,}000\left(\frac{.06}{2}\right) = \$60$ and an additional redemption payment of \$2,000 at the end of the second year. So, calculated at 4.5%, its price is

$$\$60(1.045)^{-.5} + \$60(1.045)^{-1} + \$60(1.045)^{-1.5} + \$2{,}060(1.045)^{-2}$$
$$\approx \$2{,}058.680315,$$

and by Important Fact (9.2.21),

$$D^{\text{two}}(.045, \infty)$$

$$= \frac{1}{\$2{,}058.680315}\left(\$60\frac{.5}{(1.045)^{.5}} + \$60\frac{1}{(1.045)^{1}} + \$60\frac{1.5}{(1.045)^{1.5}} + \$2{,}060\frac{2}{(1.045)^{2}}\right)$$
$$\approx 1.915703104.$$

This is a bit less than 2, the Macaulay duration of a two-year zero-coupon bond.

The price of the five-year zero-coupon bond to yield 4.5% is the present value of the redemption amount $\$10{,}000(1.045)^{-5} \approx \$8{,}024.510465$. The bond has a single payment at the end of five years, and therefore $D^{\text{five}}(.045, \infty) = 5$.

The price of the portfolio, to yield 4.5%, is the sum of the prices of the individual bonds, namely $\$2{,}058.680315 + \$8{,}024.510465 \approx \$10{,}083.19078$. So, using Important Fact (9.2.28) or Equation (9.2.27), we calculate

$$D^{\text{portfolio}}(.045, \infty) \approx \left(\frac{\$2{,}058.680315}{\$10{,}083.19078}\right)(1.915703104)$$

$$+ \left(\frac{\$8{,}024.510465}{\$10{,}083.19078}\right)(5) \approx 4.370280555.$$

■

We end this section by noting that (Macaualy or modified) duration is a measure of the length of an investment. However, the duration does not always decrease proportionally to the elapsed time. Rather, the way it decreases depends on the type of investment. To see this, consider Examples (9.2.24) and (9.2.26). This observation will be important in connection with immunization in Section (9.4), since it implies that immunized portfolios do not generally stay immunized without active management.

9.3   CONVEXITY

(calculus needed here)

Modified duration $D(i,1)$ was introduced when we discussed how the tangent line could be used to estimate the value of the price function $P(i)$ as interest rates change. As we witnessed in Example (9.2.8), such estimation tends to work extremely well for very *small* changes in the interest rate, but using the quadratic approximation (9.2.5) in place of the tangent-line approximation (9.2.4) may give you a significantly better estimate for larger interest rate changes. The quadratic approximation (9.2.5) may be rewritten as

(9.3.1)
$$\frac{P(i) - P(i_0)}{P(i_0)} \approx \frac{P'(i_0)}{P(i_0)}(i - i_0) + \frac{P''(i_0)}{2P(i_0)}(i - i_0)^2.$$

Define the **modified convexity** by

(9.3.2)
$$\boxed{C(i_0, 1) = \frac{P''(i_0)}{P(i_0)}.}$$

(We recall that the sign of the second derivative of a function governs whether it is concave upward or concave downward. But a function being concave is equivalent to a certain associated region of the xy-plane being convex,[5] and

---

[5] A subset S of the xy-plane is **convex** if whenever points $P = (x_1, y_1)$ and $Q = (x_2, y_2)$ lie in S, the line segment joining them consists entirely of points from S.

Let $a < b$ and suppose $f(x)$ is a function that is differentiable on (a, b) and continuous on $[a, b]$. Further suppose that $S_{[a,b]}$ is the subset of the xy-plane bounded by the graph $\{(x, f(x)) | a \leq x \leq b\}$ along with the line segment joining the points $(a, f(a))$ and $(b, f(b))$. The function $f(x)$ is **concave upward** on the interval $[a, b]$ if its graph $\{(x, f(x)) | a \leq x \leq b\}$ curves in a counterclockwise direction and it is **concave downward** on the interval $[a, b]$ if its graph $\{(x, f(x)) | a \leq x \leq b\}$ curves in a clockwise direction. The function $f(x)$ is concave, either upward or downward, on $[a, b]$ if and only if the subset $S_{[a,b]}$ is convex.

It is a standard fact of calculus that $f(x)$ is concave upward on $[a, b]$ exactly when the derivative $f'(x)$ is an increasing function on $[a, b]$, and the condition for $f(x)$ to be concave downward on $[a, b]$ is that the derivative $f'(x)$ be a decreasing function on $[a, b]$. Therefore, for a twice differentiable function on (a, b), the second derivative $f''(x)$ being positive on (a, b) is equivalent to the function being concave upward on $[a, b]$, while $f''(x)$ being negative on (a, b) is exactly what is needed for the function to be concave downward.

this is why the term "convexity" is used.) Using Definitions (9.2.12) and (9.3.2), we may rewrite the approximation (9.3.1) as

$$(9.3.3) \qquad \boxed{\frac{P(i) - P(i_0)}{P(i_0)} \approx -D(i_0, 1)(i - i_0) + C(i_0, 1)\frac{(i - i_0)^2}{2}.}$$

Approximation (9.3.3) indicates that if two investments have the same modified duration, then the one whose convexity $C(i_0, 1)$ has the larger absolute value is apt to have the price that is more sensitive to interest rate changes.

The sign of the modified convexity $C(i_0, 1)$ is positive if $P(i_0)$ and $P''(i_0)$ are both positive. Recalling Equation (9.2.7), we see that this is the case for a set of nonnegative cashflows, at least one of which is positive — for example, for the set of cashflows obtained by the holder of a bond. Comparing Approximations (9.2.9) and (9.3.3), if the convexity is positive, we would guess that the estimate of the relative price change given by Approximation (9.2.9) is too low.

EXAMPLE 9.3.4

Problem: A five-year zero-coupon bond redeemable at C is purchased to provide an annual effective yield of 6%. Find the modified convexity $C(i, 1)$. Use this, along with the modified duration $D(i, 1)$, to estimate the relative price change if the interest rate goes up by 100 basis points.

Solution   The bond provides a single cashflow C at time 5, so has price function $P(i) = C(1 + i)^{-5}$. Therefore, $P'(i) = -5C(1 + i)^{-6}$, $P''(i) = 30C(1 + i)^{-7}$,

$$D(i, 1) = -\left[\frac{-5C(1 + i)^{-6}}{C(1 + i)^{-5}}\right] = \frac{5}{1 + i} \text{ and } C(i, 1) = \frac{30C(1 + i)^{-7}}{C(1 + i)^{-5}} = \frac{30}{(1 + i)^2}.$$

Approximation (9.3.3) thus gives us

$$\frac{P(i) - P(.06)}{P(.06)} \approx -D(.06, 1)(i - .06) + C(.06, 1)\frac{(i - .06)^2}{2}$$

$$= -\frac{5}{1.06}(i - .06) + \frac{30}{(1.06)^2}\left[\frac{(i - .06)^2}{2}\right].$$

Thus, if the interest rate goes up by 100 basis points, so that it is 7%, the relative price change is estimated by (9.3.3) to be

$$-\frac{5}{1.06}(.01) + \frac{30}{(1.06)^2}\left[\frac{(.01)^2}{2}\right] \approx -.047169811 + .001334995 \approx -.045834817.$$

Approximation (9.2.9) would yield the estimate $-.047169811$. In fact, the issue price of the bond to yield 6% is $C/(1.06)^5$, and the issue price to yield 7% is $C/(1.07)^5$. So, the exact relative price change is

$$\frac{C/(1.07)^5 - C/(1.06)^5}{C/(1.06)^5} = \left(\frac{1.06}{1.07}\right)^5 - 1 \approx -.045863658.$$

Approximation (9.3.3) does indeed give a better approximation than (9.2.9). ∎

So far, we have introduced the modified convexity $C(i_0, 1)$, which arose naturally as we considered the price as a function $P(i)$ of the annual effective interest rate, and the second Taylor polynomial of $P(i)$. Of course, for any positive number m, the price of a fixed set of cashflows may also be thought of as a function of the nominal interest rate $i^{(m)}$ or of the force of interest δ. For any positive number m, we define the **modified convexity** $C(i, m)$ and **Macaulay convexity** $C(i, \infty)$ by

(9.3.5)
$$C(i, m) = \frac{\frac{d^2 P}{di^{(m)^2}}}{P(i)},$$

and

(9.3.6)
$$C(i, \infty) = \frac{\frac{d^2 P}{d\delta^2}}{P(i)}.$$

Since $P(i) = \sum_{t \geqslant 0} C_t(1 + i)^{-t} = \sum_{t \geqslant 0} C_t e^{-\delta t}$, the Macaulay convexity is given by

(9.3.7)
$$C(i, \infty) = \frac{\frac{d^2 P}{d\delta^2}}{P(i)} = \frac{\sum_{t \geqslant 0} C_t t^2 e^{-\delta t}}{\sum_{t \geqslant 0} C_t e^{-\delta t}}$$
$$= \sum_{t \geqslant 0} \left(\frac{C_t e^{-\delta t}}{\sum_{t \geqslant 0} C_t e^{-\delta t}}\right) t^2$$
$$= \sum_{t \geqslant 0} \left(\frac{C_t(1 + i)^{-t}}{P(i)}\right) t^2.$$

So, just as we can view the Macaulay duration as a weighted sum of the times of the cashflows [see Important Fact (9.2.21)], we see the Macaulay convexity as a weighted sum of the *squares* of the times of the cashflows. Specifically, we have

> **IMPORTANT FACT (9.3.8):** The Macaulay convexity $C(i,\infty)$ of a set of cashflows is a weighted average of the squares of the times at which the cashflows occur. If the cashflow at time t is C_t, the weight given to t^2 is $C_t(1 + i)^{-t}/P(i)$, the proportion of the total price attributable to the time t cashflow — the same weight as is used to compute the Macaulay duration.

We now know that the Macaulay convexity is easy to compute, as is the Macaulay duration. Moreover, you can quickly find the modified convexity $C(i, m)$ if you know both the Macaulay convexity and the Macaulay duration. In fact,

$$(9.3.9) \qquad C(i, m) = \frac{C(i,\infty) + \frac{1}{m}D(i,\infty)}{\left(1 + \frac{i^{(m)}}{m}\right)^2}.$$

The derivation of Equation (9.3.9) is a calculus exercise, and Problem (9.3.5) guides you through it.

EXAMPLE 9.3.10

Problem: As in Example (9.3.4), a five-year zero-coupon bond is purchased to provide an annual effective yield of 6%. Find the Macaulay convexity $C(i,\infty)$, and verify that in this case, Equation (9.3.9) with $m = 1$ is correct.

Solution   There is a single cashflow made at $t = 5$. So, according to Important Fact (9.3.8), we have Macaulay convexity $C(.06, \infty) = 5^2 = 25$. The Macaulay duration $D(.06, \infty) = 5$, the time of the lone payment. Thus, equation (9.3.9) gives $C(i, 1) = \frac{25+5}{(1.06)^2}$, the same value for the modified convexity as we found in Example (9.3.4). ∎

It is worth noting conditions that guarantee that the Macaulay duration $D(i,\infty)$ is a decreasing function of i. The reason for our interest is that, if $D(i,\infty)$ is a decreasing function of i, when the interest rate increases, the average length of the investment decreases. This may be important if you are trying to match the duration of your assets to the duration of your liabilities, in an effort to protect yourself from interest rate risk. Remember, not all durations change with an interest rate shift: In particular, the duration of a zero-coupon bond is equal to the time until the single redemption payment, no matter what the interest rate may be.

We note the following assertion that gives conditions for $D(i, \infty)$ being a decreasing function of i.

> **CLAIM (9.3.12)**: If all the cashflows are *positive* and there is *more* than one time at which there is a nonzero cashflow, then the Macaulay duration $D(i, \infty)$ [and hence each of the modified durations $D(i, m)$] is a decreasing function of the interest rate i. When one increases the interest rate, the present values of all of the cashflows are affected. However, there is a greater change in the later cashflows and consequently the duration decreases.

Our proof of Claim (9.3.12) will use a measure of how spread out the payment times are around the Macaulay duration. Define the **dispersion**

$$(9.3.11) \qquad V(i) = \frac{1}{P(i)} \sum_{t \geqslant 0} (t - D(i, \infty))^2 C_t e^{-\delta t}.$$

Note that $(t - D(i, \infty))^2$ is always nonnegative, so there is no chance that values less than $D(i, \infty)$ compensate for values greater than $D(i, \infty)$, as there would be if we dropped the exponent 2.

Proof of Claim (9.3.12): Let's take a closer look at the dispersion. Note that expansion of the squared term in Equation (9.3.11), the definition of $V(i)$, gives

$$V(i) = \frac{1}{P(i)} \left(\sum_{t \geqslant 0} t^2 C_t e^{-\delta t} - 2D(i, \infty) t C_t e^{-\delta t} + (D(i, \infty))^2 C_t e^{-\delta t} \right)$$

$$= \left(\frac{1}{P(i)} \sum_{t \geqslant 0} t^2 C_t e^{-\delta t} \right) - 2D(i, \infty) \left(\frac{1}{P(i)} \sum_{t \geqslant 0} t C_t e^{-\delta t} \right)$$

$$+ (D(i, \infty))^2 \left(\frac{1}{P(i)} \sum_{t \geqslant 0} C_t e^{-\delta t} \right).$$

But Equations (9.3.7) and (9.2.19), respectively, tell us that $\frac{1}{P(i)} \sum_{t \geqslant 0} t^2 C_t e^{-\delta t} = C(i, \infty)$ and $\frac{1}{P(i)} \sum_{t \geqslant 0} t C_t e^{-\delta t} = D(i, \infty)$, and of course by Equation (9.2.1) $\frac{1}{P(i)} \sum_{t \geqslant 0} C_t e^{-\delta t} = \frac{1}{P(i)} \sum_{t \geqslant 0} C_t (1 + i)^{-t} = 1$, so we have

$$(9.3.13) \qquad V(i) = C(i, \infty) - (D(i, \infty))^2.$$

Now compute the derivative of $D(i, \infty)$ with respect to the force of interest δ. Using the product rule and chain rule, from Equation (9.2.16) we obtain

(9.3.14)
$$\frac{d}{d\delta} D(i, \infty) = \frac{d}{d\delta} \left(-(P(i))^{-1} \frac{dP}{d\delta} \right)$$
$$= \left((P(i))^{-2} \frac{dP}{d\delta} \right) \frac{dP}{d\delta} - (P(i))^{-1} \frac{d^2 P}{d\delta^2}$$
$$= (D(i, \infty))^2 - C(i, \infty) = -V(i).$$

Equation (9.3.14) tells us that the derivative $\frac{d}{d\delta} D(i, \infty)$ is negative when the dispersion $V(i)$ is positive. By Definition (9.3.11) this happens when there are cashflows at more than one time and all the cashflows are positive.

We end the section with an example of convexities of a bond with coupons, followed by a discussion of the Macaulay convexity of a portfolio of bonds.

EXAMPLE 9.3.15   Convexity of a bond

Problem: An N-year zero-coupon bond has the same duration as a two-year 8% bond with annual coupons, when figured at the current yield rate of 5%. Calculate the Macaulay convexity of the two-year bond and of the N-year zero-coupon bond. Which is more sensitive if the interest rate increases from 5% to 6%?

Solution   We begin by calculating the Macaulay duration and the Macaulay convexity of the two-year bond. To do so, we note that if the two-year bond has face value F, then it provides the holder of the bond with a payment of $.08F$ at the end of one year and a payment of $1.08F$ (coupon and redemption amount) at the end of two years. Therefore, at the current interest rate of 5%, the two-year bond sells for $.08F(1.05)^{-1} + 1.08F(1.05)^{-2} \approx 1.055782313F$. Therefore, applying Important Fact (9.2.21), we obtain the following Macaulay duration for the two-year bond at the current interest rate:

$$D(.05, \infty) \approx 1 \left(\frac{.08F(1.05)^{-1}}{1.055782313F} \right) + 2 \left(\frac{1.08F(1.05)^{-2}}{1.055782313F} \right) \approx 1.927835052.$$

In addition, Important Fact (9.3.8) tells us that the Macaulay convexity of the two-year bond is

$$C(.05, \infty) \approx 1^2 \left(\frac{.08F(1.05)^{-1}}{1.055782313F} \right) + 2^2 \left(\frac{1.08F(1.05)^{-2}}{1.055782313F} \right) \approx 3.783505155.$$

Since the duration of a zero-coupon bond is its term until maturity, and we require that the N-year zero-coupon bond has the same duration as the two-year bond, we must have $N \approx 1.927835052$. But then, the zero-coupon bond has Macaulay convexity $C(.05, \infty) = N^2 \approx (1.927835052)^2 \approx 3.716547986$ which is *slightly* less than the Macaulay convexity of the two-year bond. The convexities are very close, so we would expect that the sensitivity to small interest rate shifts would also be close.

We next wish to determine which bond's price is more sensitive to an interest rate shift from 5% to 6%. If the interest rate goes up to 6%, then the price of the two-year bond is $.08F(1.06)^{-1} + 1.08F(1.06)^{-2} \approx 1.036667853F$, and the relative price change is about $\frac{1.036667853F - 1.055782313F}{1.055782313F} \approx -.018104546$. On the other hand, the relative price change of the zero-coupon bond is

$$\frac{(1.06)^{-1.927835052} - (1.05)^{-1.927835052}}{(1.05)^{-1.927835052}} \approx -.018107508.$$

The values of the relative price changes are very close, as we might expect from the fact that the convexities are close. It is worth noting that convexity of the zero-coupon bond is slightly smaller, but the relative price change of the zero-coupon bond has absolute value that was slightly larger; this alerts us to the fact that the quadratic approximation does not tell the whole story, and that this would require the whole Taylor series! ∎

We have previously noted that by stripping off the coupons of a bond with coupons, you may view a bond with coupons as a portfolio of zero-coupon bonds. Important Fact (9.2.21) explained how to find the duration of a portfolio of cashflows from the duration of the component cashflows, and there is a similar statement for how one might compute the convexity of a portfolio of cashflows from the individual convexities.

> **IMPORTANT FACT 9.3.16:** The Macaulay convexity of a portfolio is a weighted average of the Macaulay convexities of the individual assets, the weight given to the convexity of an individual asset being the proportion of the total price attributable to that asset.

The verification of Fact (9.3.16) is left as an exercise [Problem(9.3.6)].

EXAMPLE 9.3.17   Convexity of a portfolio of bonds

Problem: As in Example (9.2.29), Julian Bradley's bond portfolio consists of a $10,000 five-year zero-coupon bond and a $2,000 par-value two-year 6% bond with semiannual coupons. Compute the Macaulay convexity $C^{\text{portfolio}}(.045, \infty)$ for the portfolio.

Solution   We first find the individual Macaulay convexities $C^{\text{two}}(.045, \infty)$ and $C^{\text{five}}(.045, \infty)$ of the two-year and five-year bonds respectively, and then use Important Fact (9.3.16) to find the Macaulay convexity $C^{\text{portfolio}}(.045, \infty)$ of the portfolio of bonds. For our solution, it is useful to recall the prices, to yield 4.5%, that were found in the solution to Example (9.2.29): The two-year bond has price $2,058.680315, the five-year bond has price $8,024.510465, and the portfolio has as its price the sum of these individual prices, namely $10,083.19078.   The two-year bond has semiannual $60 coupons and an additional $2,000 redemption payment at the end of two years, so Important Fact (9.3.8) tells us that

$$C^{\text{two}}(.045, \infty)$$

$$= \frac{1}{\$2,058.680315}\left(\$60\frac{(.5)^2}{(1.045)^{.5}} + \$60\frac{1^2}{(1.045)^1} + \$60\frac{(1.5)^2}{(1.045)^{1.5}} + \$2,060\frac{2^2}{(1.045)^2}\right)$$

$$\approx 3.761671472,$$

which is a little less than the convexity $2^2 = 4$ of a two-year zero-coupon bond. The Macaulay convexity of the five-year zero-coupon bond is 25, the square of the term of the bond. Therefore, Important Fact (9.3.16) tells us that the Macaulay convexity of the portfolio is

$$C^{\text{portfolio}}(.045, \infty) \approx \left(\frac{\$2,058.680315}{\$10,083.19078}\right)(3.761671472)$$

$$+ \left(\frac{\$8,024.510465}{\$10,083.19078}\right)(25) \approx 20.66378046.$$

∎

9.4   IMMUNIZATION

(calculus needed here)

A set of cashflows may be broken up into inflows and outflows. The promised inflows are *assets* and the required outflows are *liabilities*. The total inflow at time t will be denoted by A_t and the total outflow at time t will be called L_t; so $A_t \geq 0$ and $L_t \geq 0$.[6] Throughout this section we assume that the cashflows *do not depend on the interest rate i*. There may be offsetting inflows and outflows, and the net cashflow inward at time t is the difference $A_t - L_t$. We denote the **surplus** valued at interest rate i by $S(i)$. That is,

$$(9.4.1) \quad S(i) = \sum_{t \geq 0}(A_t - L_t)(1+i)^{-t} = \sum_{t \geq 0} A_t(1+i)^{-t} - \sum_{t \geq 0} L_t(1+i)^{-t}.$$

---

[6] You have already seen this notation used in Example (9.2.8).

Note that $S(i)$ is just the price function $P(i)$ of the portfolio that consists of the assets *and* the liabilities.

> A nonnegative surplus means that the time 0 value of the assets is greater than the time 0 value of the liabilities; you are *ahead* by an amount $S(i)$. If you have a negative surplus, you are *behind* and this may have significant consequences.

We will next show that if a portfolio of cashflows is such that $S(i_0) = 0$, $S'(i_0) = 0$, and $S''(i_0) \geqslant 0$ where i_0 is the current interest rate, then *for sufficiently small* changes in the interest rate, the surplus is positive; that is, if i_0 is your current yield rate, making $S(i_0) = 0$, and you also have $S'(i_0) = 0$, and $S''(i_0) \geqslant 0$, then *nearby interest rates would make you no worse off.*[7]

Note that the surplus is zero when the value of the set of assets is equal to the value of the set of liabilities. Moreover, by Definition (9.2.12), the condition that the derivative $S'(i)$ must be zero is equivalent to the duration of the assets equaling the duration of the liabilities. (Here it doesn't matter whether we use modified duration or Macaulay duration, but we do need to use the same kind of duration for the assets as we do for the liabilities.) Further note that by Equation (9.3.2), the second derivative $S''(i)$ is nonnegative precisely when the convexity of the assets is at least as large as the convexity of the liabilities.

Assume that we have the following conditions:

(9.4.2) $\boxed{S(i_0) = 0, \ S'(i_0) = 0, \quad \text{and} \quad S''(i_0) \geqslant 0.}$

Then, if h is very small, the first three terms of the Taylor series expansion for $S(i)$ around i_0 give us the approximation

$$S(i_0 + h) \approx S(i_0) + S'(i_0)h + \frac{S''(i_0)}{2}h^2 \approx \frac{S''(i_0)}{2}h^2 \geqslant 0,$$

which is just Equation (9.2.5) with $i = i_0 + h$. Thus,

---

[7]We remind you that we continue to assume that there is a flat yield curve and that any shift in the yield curve results in a new parallel yield curve. You may encounter the term "C3 risk" or "C-3 risk." This refers to risk that is due to change in the shape of the yield curve as well as to the risk due to parallel shifts that we allow. In addition, "C3 risk" must deal with the fact that many cashflows are interest rate sensitive, a serious complication that we are not permitting.

> **IMPORTANT FACT (9.4.3): (Redington immunization)**
> • *If* Condition (9.4.2) holds, if the yield curve is flat, and if all shifts in the yield curve are parallel, *then* $S(i_0 + h) \geqslant 0$ for h sufficiently small.
> • Condition (9.4.2) is equivalent to the assets and liabilities having equal present values and equal Macaulay durations, and the Macaulay convexity of the assets being at least as large as the Macaulay convexity of the liabilities.

While Redingon immunization (9.4.3) does not tell you how small h needs to be for you to be certain to have a surplus, it is still quite useful.

EXAMPLE 9.4.4   Redington immunization

Problem: Alan and Peabody Insurance is obligated to make a payment of $120,000 in exactly four years. In order to provide for this obligation, their financial officer decides to purchase a combination of two-year zero-coupon bonds and five-year zero-coupon bonds. Each of these is sold to yield an annual effective yield of 4.5%. How much of each type of bond should be purchased so that, together with the $120,000 outflow in four years, the zero-coupon bonds constitute a portfolio satisfying Redington immunization Condition (9.4.2) at an annual effective interest rate of 4.5%? How does this immunization help?

Solution   Let a denote the amount spent on two-year bonds and b the amount spent on five-year bonds. Then, the present value of the assets, when figured using the annual effective interest rate 4.5% is $a + b$, and Important Fact (9.4.3) says that for Redington immunization this must equal the present value of the $120,000 liability to be paid in four years, again figured at 4.5%. Therefore,

$$a + b = \frac{\$120,000}{(1.045)^4} \approx \$100,627.3612.$$

We will settle for $a + b = \$100,627.36$ because, as usual, there is no way to pay fractional numbers of cents.

Next, according to Important Fact (9.4.3) again, we need the Macaulay durations of the set of two assets to equal 4, the Macaulay duration of the single liability at time 4. Recalling Important Fact (9.2.21), the portfolio of assets has Macaulay duration $2\left(\frac{a}{a+b}\right) + 5\left(\frac{b}{a+b}\right)$, and therefore the duration condition for Redington immunization is

$$\frac{2a}{a + b} + \frac{5b}{a + b} = 4.$$

Multiplying this condition by $a + b$ we have $2a + 5b = 4a + 4b$. Thus, we must have $b = 2a$ as well as $a + b = \$100,627.36$. Therefore, we want

$3a = \$100,627.36$. So $a \approx \$33,542.4533$ and $b = \$67,084.90667$. Since these purchase amounts must have whole numbers of cents, we take $a = \$33,542.45$ and $b = \$67,084.91$.

Now, let us compute the Macaulay convexity of the portfolio of assets and see whether it exceeds $4^2 = 16$, which is the Macaulay convexity of the single liability to be paid at time 4. Using Important Fact (9.3.8), we see that the set of assets has Macaulay convexity

$$\left(\frac{\$33,542.45}{\$100,627.36.} \right) 2^2 + \left(\frac{\$67,084.91}{\$100,627.36.} \right) 5^2 \approx 18.$$

This is greater than 16, so the portfolio satisfies the Redington immunization Condition (9.4.2) at interest rate 4.5%.

Let us now consider how paying $\$33,542.45$ for two-year bonds and $\$67,084.91$ for five-year bonds might help you if these bonds were purchased to yield 4.5%. The two-year bonds will provide $\$33,542.45(1.045)^2 \approx \$36,629.19$ in two years, and the five-year bonds produce a redemption payment of $\$67,084.91(1.045)^5 \approx \$83,600.00$ in five years. If interest rates rise 50 basis points to 5%, at time two years, then the five-year bond, with three years before it may be redeemed for $\$83,600$, may be sold for $\$83,600(1.05)^{-3} \approx \$72,216.82$. Combine this with the $\$36,629.19$ redemption from the two-year bond to obtain $\$36,629.19 + \$72,216.82 = \$108,846.01$. You may invest this total for two years at 5%, giving you $\$108,846.01(1.05)^2 \approx \$120,002.73$, $\$2.73$ above the amount you need to cover your liability. On the other hand, if interest rates go down by 50 basis points to 4%, you should sell the five-year bonds at the end of two years for $\$83,600.00(1.04)^{-3} \approx \$74,320.10$. Once again, combine the money obtained from the sale of the five-year bond with the redemption amount $\$36,629.19$, this time realizing a total of $\$36,629.19 + \$74,320.10 = \$110,949.29$, and invest this for two years. This will result in $\$110,949.29(1.04)^2 \approx \$120,002.75$. Once again, you have a little money left after you take care of the $\$120,000$ obligation: This time you are left with $\$2.75$. You immunized yourself against small interest rate changes, either up or down. ∎

We now modify the above example so that the two-year bond has coupons.

EXAMPLE 9.4.5 Redington immunization

Problem: Alan and Peabody Insurance is still obligated to make a payment of $\$120,000$ in exactly four years. In order to provide for this obligation, their financial officer now decides it is better to purchase a combination of two-year par-value 6% bonds *with semiannual coupons* and five-year zero-coupon bonds. Each of these is sold to yield an annual effective yield of 4.5%. How much of each type of bond should be purchased so that, together with the $120,000 outflow in four years, the zero-coupon bonds constitute a portfolio satisfying Redington immunization condition (9.4.2) at interest rate 4.5%?

Solution   We begin, just as we did in Example (9.4.4), by letting a denote the amount spent on two-year bonds and b the amount spent on five-year bonds. Once again, the fact that the present value of the assets must equal the present value of the liabilities implies that $a + b = \$100,627.36$.

Observe that we have seen such two-year and five-year bonds in Examples (9.2.29) and (9.3.17), and that we calculated that the two-year bond has Macaulay duration equal to 1.915703104 and its Macaulay convexity is 3.761671472. The zero-coupon bond has Macaualy duration equal to 5 and Macaulay convexity 25.

Using Important Fact (9.2.21), we find that the Macaulay duration of the assets is

$$(1.915703104)\left(\frac{a}{a+b}\right) + 5\left(\frac{b}{a+b}\right).$$

But this must equal the Macaulay duration of the liability, namely 4. Therefore,

$$(1.915703104)\left(\frac{a}{a+b}\right) + 5\left(\frac{b}{a+b}\right) = 4,$$

and so $1.915703104a + 5b = 4a + 4b$, and $b = 2.084296896a$.

The two-year coupon bond has slightly lower duration than the zero-coupon bond of Example (9.4.4). To accommodate this in our Redington immunization duration condition, we need to devote a little less of our total spending to the two-year bond. Then $a + b = \$100,627.36$ gives us $3.084296896a = \$100,627.36$. So, $a \approx \$32,625.70479$ and $b \approx \$68,001.65521$. The purchase amounts a and b must each be a whole number of cents, so $a = \$32,625.70$ and $b = \$68,001.66$. Finally, we need to check the convexity condition. The liability once again has Macaulay convexity 16. To determine the Macaulay convexity $C(4.5\%, \infty)$ of the portfolio of assets, we use Important Fact (9.3.16), along with the convexities taken from Example (9.3.17). Then, the portfolio of assets has Macaulay convexity that is approximately

$$\left(\frac{\$32,625.70}{\$100,627.36.}\right)(3.761671472) + \left(\frac{\$68,001.66}{\$100,627.36.}\right)5^2 \approx 18.114.$$

Again this is greater than 16, so we have an immunized portfolio. We note that the calculation with coupon bonds is significantly longer. Sometimes, it may be a good idea to first use zero-coupon bonds to get a basic feel of your situation. ∎

Let us now consider how you might try to structure a portfolio of cashflows so that you are protected from *large* interest rate changes as well as small ones. Not surprisingly, this is not generally easy to implement.

To begin with, suppose that you have a set of three cashflows consisting of a single outflow of amount L to be paid at time T and inflows U due at times $T - u$ and W due at time $T + w$. This was the situation in our Redington immunized portfolio of Example (9.4.4).

CASHFLOW:		U	$-L$	W
TIME:	0	$T - u$	T	$T + w$

Let

$$S(\delta) = Ue^{-\delta(T-u)} - Le^{-\delta T} + We^{-\delta(T+w)} = e^{-\delta T}\left(Ue^{\delta u} - L + We^{-\delta w}\right),$$

the net present value (or surplus) of the set of cashflows. By δ_0 we indicate the current force of interest, and we stipulate that $S(\delta_0) = 0$ and $S'(\delta_0) = 0$. It may be established by an argument given in an optional technical note at the end of this section that these two assumptions are enough to force $S(\delta)$ to be positive for any $\delta \neq \delta_0$. So, the pair of conditions $S(\delta_0) = 0$ and $S'(\delta_0) = 0$ are enough to guarantee that a change in the interest rate from the current rate $S(\delta_0)$ will *increase* the value of the surplus.

IMPORTANT FACT (9.4.6): (Full immunization) Let δ_0 denote the current force of interest. Suppose a portfolio consists of a single liability L, to be paid at time T, along with a pair of assets U and and W, to be paid at times $T - u$ and $T + w$, respectively, where $0 < u < T$ and $w > 0$. Further suppose that the net present value $S(\delta_0)$ of the portfolio is 0, as is the derivative $S'(\delta_0)$. Then, if $\delta \neq \delta_0$, the net present value $S(\delta)$ is positive.

We now know that the portfolio we found in Example (9.4.4) is *fully* immunized since Important Fact (9.4.6) applies!

EXAMPLE 9.4.7   Full immunization

Problem: Consider again the situation of Example (9.4.4). As previously considered, Alan and Peabody Insurance is obligated to make a payment of $120,000 in exactly four years and purchase $33,542.45 of two-year zero-coupon bonds and $67,084.91 of five-year zero-coupon bonds, priced to yield an annual effective yield of 4.5%. Suppose that we have a rather large change in the interest rate, say up to 10% or down to 1%. Illustrate that Alan and Peabody still can cover the $120,000 liability, as is guaranteed by full immunization.

Solution   If interest rates are now 10% at the end of two years, the amount Alan and Peabody gets by selling the five-year bond at time two years is only $83,600.00(1.10)^{-3} \approx $62,809.92. So, at time two years, they only have $36,629.19 + $62,809.92 = $99,439.11 to invest. But, thanks to the relatively high 10% interest rate, the money will still accumulate to more than the $120,000 needed at time four — specifically to $99,439.11(1.10)^2 \approx $120,321.32.

Should the interest rate at time two years be a modest 1%, then at time two years the five-year bond, with redemption in three years, will fetch $83,600.00(1.01)^{-3} \approx $81,141.34 so Alan and Peabody would have a combined $36,629.19 + $81,141.34 = $117,770.53 to invest for two years at 1%. This would give them $117,770.53(1.01)^2 \approx $120,137.72 so, again, Alan and Peabody can take care of the $120,000 liability. ∎

Having a fully immunized portfolio seems to be rather fantastic, so we wish to further analyze the pair of constraint equations $S(\delta_0) = 0$ and $S'(\delta_0) = 0$. Since $e^{-\delta_0 T} \neq 0$, the condition $S(\delta_0) = 0$ is equivalent to

$$(9.4.8) \qquad Ue^{\delta_0 u} - L + We^{-\delta_0 w} = 0.$$

Moreover, using Equation (9.4.8), we find

$$0 = S'(\delta_0) = -(T - u)Ue^{-\delta_0(T-u)} + TLe^{-\delta_0 T} - (T + w)We^{-\delta_0(T+w)}$$
$$= \left(uUe^{-\delta_0(T-u)} - wWe^{-\delta_0(T+w)}\right) - T\left(Ue^{-\delta_0(T-u)} - L + We^{-\delta_0(T+w)}\right)$$
$$= \left(uUe^{-\delta_0(T-u)} - wWe^{-\delta_0(T+w)}\right) - Te^{-\delta_0 T}\left(Ue^{\delta_0 u} - L + We^{-\delta_0 w}\right)$$
$$= uUe^{-\delta_0(T-u)} - wWe^{-\delta_0(T+w)} = e^{-\delta_0 T}\left(uUe^{\delta_0 u} - wWe^{-\delta_0 w}\right).$$

Therefore, since $e^{-\delta_0 T} \neq 0$, we have

$$(9.4.9) \qquad uUe^{\delta_0 u} - wWe^{-\delta_0 w} = 0.$$

FACT (9.4.10): A portfolio consisting of a single outflow of amount L at time T and inflows U at time $T - u$ and W at time $T + w$ is fully immunized. If the system of equations

$$\begin{cases} Ue^{\delta_0 u} + We^{-\delta_0 w} = L \\ uUe^{\delta_0 u} - wWe^{-\delta_0 w} = 0 \end{cases}$$

is satisfied, the portfolio will *never* force default.

Of course you may wish to fully immunize a portfolio with multiple liabilities rather than a single one. You may adopt the strategy of selecting a pair of assets to immunize each of the liabilities. Just as was true for Redington immunization, you will need to frequently rebalance your portfolio to maintain an immunized status, because the derivative condition $S'(i) = 0$ will not continue to hold with the passing of time.

The assumption that you have a flat yield curve that remains flat is not a realistic one. In fact, a consequence of Redington immunization is that if assets and liabilities can be structured so as to make condition (9.4.2) hold, then there is an arbitrage opportunity. You could short-sell the liabilities and use the profits to purchase the assets. When the interest rate had a slight shift, you could sell your assets, cover the short sale, and make a profit. Similarly, full immunization is contrary to the market being essentially arbitrage-free. These arbitrage opportunities result from our unrealistic hypothesis that the yield rate curve is flat. We explained in Section (8.3) that the term structure of interest rates should be considered on the premise that there is no arbitrage. This points toward more advanced immunization strategies in which many different spot rates are used. Asset management remains an active area of research.

Optional technical note:

We end this section with the promised somewhat technical argument establishing Important Fact (9.4.6) that the net present value $S(\delta)$ will be positve if $\delta \neq \delta_0$. ($S(\delta) \geqslant 0$ is the important condition that you can cover your liability.) We begin by noting that Equation (9.4.9) may be rewritten as

$$(9.4.11) \qquad\qquad We^{-\delta_0 w} = \frac{u}{w} U e^{\delta_0 u}.$$

Equations (9.4.8) and (9.4.11), derived from the conditions $S(\delta_0) = 0$ and $S'(\delta_0) = 0$ respectively, together give us

$$(9.4.12) \qquad\qquad U e^{\delta_0 u} - L + \frac{u}{w} U e^{\delta_0 u} = 0.$$

Now, if δ is any force of interest, we wish to show that $S(\delta)$ is positive. With this goal in mind, we note that

$$S(\delta) = e^{-\delta T}\left(Ue^{\delta u} - L + We^{-\delta w}\right)$$

$$= e^{-\delta T}\left[\left(Ue^{\delta u} - L + We^{-\delta w}\right) - \left(Ue^{\delta_0 u} - L + \frac{u}{w}Ue^{\delta_0 u}\right)\right] \quad \text{by (9.4.12)}$$

$$= e^{-\delta T}\left(Ue^{\delta u} + We^{-\delta w} - Ue^{\delta_0 u} - \frac{u}{w}Ue^{\delta_0 u}\right)$$

$$= e^{-\delta T}\left(Ue^{\delta u} + \frac{u}{w}Ue^{\delta_0(u+w)}e^{-\delta w} - Ue^{\delta_0 u} - \frac{u}{w}Ue^{\delta_0 u}\right) \quad \text{by (9.4.11)}$$

$$= e^{\delta_0 u - \delta T}U\left(e^{(\delta-\delta_0)u} + \frac{u}{w}e^{(\delta_0-\delta)w} - 1 - \frac{u}{w}\right).$$

So, our objective is to prove that $e^{(\delta-\delta_0)u} + \frac{u}{w}e^{(\delta_0-\delta)w} - 1 - \frac{u}{w}$ is positive. To accomplish this, we introduce the function

$$f(x) = e^{ux} + \frac{u}{w}e^{-wx} - 1 - \frac{u}{w}.$$

What we must show is that $f(\delta - \delta_0) > 0$ for $\delta \neq \delta_0$, and because $f(0) = 0$, this is equivalent to the function $f(x)$ having a minimum at 0. It is therefore enough to show that the derivative $f'(x)$ has the same sign as x, that is $f'(x) > 0$ if $x > 0$ and $f'(x) < 0$ if $x < 0$, a fact which is immediate from

$$f'(x) = ue^{ux} - ue^{-wx} = u(e^{ux} - e^{-wx}),$$

since the constants u and w are positive.

9.5   OTHER TYPES OF DURATION

(calculus needed here)

In Section (9.2) we introduced Macaulay duration and modified duration for a set of cashflows. To do so, we postulated that the cashflows were *not* interest rate sensitive, and we also supposed that we had a flat yield curve at all times. Of course, we know that these assumptions are not necessarily appropriate in real life applications. We conclude our discussion with a brief look at what might be done if these assumptions are dropped. This is still an active area of work with competing suggestions: You may even find "duration derbies" where competing types of durations face off, using actual real life examples, to see how well the models predict the price response of a bond or other financial instrument to changes in interest rates.[8]

[8]If you want an example, see Allen, D. E., Thomas, L. C., and Zheng, H. "The Duration Derby: A Comparison of Duration Stategies in Asset Liability Management," *J. Bond Trading and Management* 1, 371-380, 2003.

Consider investments with interest-sensitive cashflows. Examples of such cashflows include the payments to be received by the holder of a callable bond or a mortgage-backed security. If interest rates fall sufficiently, then a callable bond is likely to be called early and, thanks to refinancing options, a mortgage is more apt to be repaid early.

A set of interest-sensitive cashflows may not have an everywhere differentiable price function, and since the definitions of durations use the derivative of P, the durations we have considered so far may not be defined. To illustrate why we might have interest rates i_0 at which the price function $P(i_0)$ is not differentiable, let us consider a par-value bond, with an American call option and no lockout period, that is purchased at a discount. Recall that such a bond gives the buyer a yield rate that exceeds the coupon rate, assuming that the bond is held to maturity. If, immediately after the bond is purchased, the interest rate drops *below* the coupon rate, the bond's current price would be above the redemption amount. So, from the issuer's viewpoint, it is advantageous to call in the bond and reissue it: The new purchase price, yielding the new buyer a lower interest rate, allows the issuer to recoup the price paid out to the holder of the old bond and also to keep the premium. But, the bond being called results in a new cashflow (the redemption payment) and cancels the old ones. This *sudden* change makes the price function have an interest rate at which it fails to be differentiable, namely the interest rate numerically equal to the coupon rate. Here we have considered an interest rate shift *immediately* after purchase; however, subsequent changes can also lead to the bond being called, and the price function failing to be differentiable at particular interest rates.

Now, recall that the modified duration was used to estimate the price change that would result from a small shift in the interest rate, and we had the basic definition

$$D(i_0, 1) = -\frac{P'(i_0)}{P(i_0)}.$$

How might we approximate this if $P'(i_0)$ were not necessarily defined? Well, it is standard to use the slope of a secant line in place of the slope of a tangent line. Specifically, we define

(9.5.1) $$m_h(i_0) = \frac{P(i_0 + h) - P(i_0 - h)}{2h},$$

which is the slope of the line containing the points $(i_0 - h, P(i_0 - h))$ and $(i_0 + h, P(i_0 + h))$,[9] and we set

---

[9] Quite likely you have considered secant lines through $(i_0, P(i_0))$ and $(i_0 + h, P(i_0 + h))$, having slope $\frac{P(i_0+h)-P(i_0)}{h}$, but then you are focusing on what happens to the right of i_0. We are concerned with shifts in the interest rate in either direction, hence adopt this two-sided approach.

(9.5.2)
$$E_h(i_0, 1) = -\frac{m_h(i_0)}{P(i_0)}.$$

Note that if $D(i_0, 1)$ is defined, if h is small, and if the cashflows are fixed, then, since the slope of the secant line $m_h(i_0)$ approximates $P'(i_0)$, the **effective duration** $E_h(i_0, 1)$ approximates $D(i_0, 1)$.

> The **effective duration** $E_h(i_0, 1) = -\frac{m_h(i_0)}{P(i_0)}$ may be used in place of the modified duration. Do this when you have interest-sensitive cashflows, using a natural choice of h suggested by the problem. *Unlike modified duration, effective duration takes into account the fact that the interest rate may influence cashflows.* If the yield increases by 100 basis points and cashflows are sensitive to interest rate changes, the approximate relative price change is $-E_h(i_0, 1)$ percent.

EXAMPLE 9.5.3   Using effective duration to estimate prices

Problem: The price of a callable bond, purchased when interest rates are 7%, is $970.00. The bond has effective duration $E_{.01}(.07, 1)$ that turns out to equal 4.4. Estimate the price if interest rates drop to 6%. Also, estimate the price if there is a more modest drop to 6.5%.

Solution   If the interest rate decreases from 7% to 6%, we have a 100 basis point decrease. The price then increases by about 4.4%. This is an increase of about .044 × $970 = $42.68, so we estimate a price of $970 + $42.68 = $1,012.68. Should we have a rate increase of 50 basis points (to 6.5%), the price increase is estimated to be half as large, hence $21.34. So, the estimated price if the interest rate falls to 6.5% is $970 + $21.34 = $991.34.   ∎

Let's now look at an example where we calculate the effective duration.

EXAMPLE 9.5.4   Modified duration and effective duration, bond purchased at a discount

Problem: Consider a three-year $1,000 5% par-value bond with annual coupons whose current price is $986.51. You are given that the bond has an American call option with no lockout period, so it may be called at any time. Calculate the investor's yield j if the bond is held to maturity, and then find $E_{.01}(j, 1)$. Also, for comparison, calculate the modified duration $D(j, 1)$ and effective duration $E_{.01}(j, 1)$ of a three-year $1,000 5% par-value noncallable bond with annual coupons whose price at issue is $986.51.

Solution The bonds have annual $50 coupons and a $1,000 redemption amount. You may check that the basic price formula $986.51 = $50a_{\overline{3}|j}$ + $1,000(1 + j)^{-3}$ forces $j \approx 5.500012463\%$. If interest rates go up 100 basis points to 6.5%, since the coupon rate is only 5%, it is unlikely the bond will be called. We therefore figure the price as if the bond were noncallable. The basic price formula gives

$$P(.065) = \$50a_{\overline{3}|j+.01} + \$1,000(1 + (j + .01))^{-3}$$
$$\approx \$50a_{\overline{3}|6.500012463\%} + \$1,000(1.06500012463)^{-3}$$
$$\approx \$960.2725464.$$

On the other hand, should the interest rate fall by 1%, the bond would most likely be called for $1,000. The reason for this is that the interest rate is then less than the coupon rate, so the bond should be priced at a premium. So, calling in the bond at its coupon rate gives the issuer a bargain. Therefore, with the bond able to be called in at $1,000, we figure $P(j - .01) = \$1,000$. (Knowing that the bond is apt to be called in, a buyer would not want to pay more than the redemption amount!) Recalling Equations (9.5.1) and (9.5.2), we calculate that $m_{.01}(j) = (\$960.2725464 - \$1,000)/2(.01) \approx -1,986.372682$ and $E_{.01}(j, 1) \approx \frac{1,986.372682}{986.51} \approx 2.0135353.$

If we have a noncallable three-year $1,000 5% par-value bond with annual coupons, then Example (9.2.24) tells us that the modified duration $D(j, 1)$ is $D(j, 1) = a_{\overline{3}|j} \approx 2.697932752$. The noncallable bond, having modest coupons, has a modified duration that is reasonably close to its term. As for the calculation of the effective duration $E_{.01}(j, 1)$, once again $P(j + .01) \approx \$960.2725464$. However, if the interest rate drops to $j - .01$, with no call option the price rises to $1,013.744476. The effective duration is then $(\$1,013.744476 - \$960.2725464)/[2(.01)(\$986.51)] \approx 2.71015649$. We note that this is quite close to the value 2.697932752 just calculated for $D(j, 1)$.

The values of the durations are higher for the noncallable bond than the effective duration was for the callable bond. This will always be true! ∎

The idea behind the definition of effective duration was to replace the first derivative, which gives the slope of the tangent line to the price curve at i_0, by the slope of a secant line. We replaced $\frac{dP}{di}\big|_{i=i_0}$ by $\frac{P(i_0+h)-P(i_0-h)}{h^2}$. Similarly, you may define **effective convexity** by replacing the second derivative $\frac{d^2P}{di^2}\big|_{i=i_0}$ by $\frac{P(i_0+h)+P(i_0-h)-2P(i_0)}{2h}$ [see Problem (9.5.4)]. Like effective duration, effective convexity may be useful when the cashflows are interest-sensitive.

As we remarked in Section (9.2), there are problems with the durations we have defined so far even if we have fixed cashflows because it is unrealistic to think that the yield curve will be flat. **Keyrate duration** or **partial duration** is defined using several different spot rates. In fact, it is common to use eleven different spot rates for Treasury securities, although any finite set of spot rates does not determine the yield curve. Partial derivatives with respect to the spot rates take the place of the derivatives we have discussed.

Recall that the motivation for considering duration is to manage your portfoilo so as to be reasonably protected from interest rate volatility. As you might now guess, no one type of duration has been found that clearly does this best for all classes of portfolios.

9.6   PROBLEMS, CHAPTER 9

(9.0) Chapter 9 writing problems

(1) [following Section (9.1)] C. L. Trowbridge classified risks as C1, C2, C3, and C4 risks when he was president of the Society of Actuaries. C3 risks are addressed by asset-liability management. Describe Trowbridge's classification. Include several examples of each type of risk.

(2) Write a short essay on the history of cash-flow management and immunization. Names you might look for in your research include Tjalling C. Koopmans and Frank M. Redington.

(9.1) Overview

(1) Leland Price must pay $5,000 six months from now and $10,000 one year from now. He wishes to purchase bonds so that together they form a portfolio of assets that exactly match his liabilities. Available bonds are a six-month zero-coupon $1,000 bond that has a 3.0225% annual yield, and a one-year $1,000 par-value 6% bond with semiannual coupons and a 4% nominal yield (convertible semiannually). How much must Leland pay to purchase the bonds? Assume that he may buy any quantity he wishes of each bond.

(2) Suppose that you need to pay $35,000 in three years and that you can finance this with zero-coupon bonds yielding 5.5% with terms of two years and six years. Imagine that you spend $22,354.86 purchasing a two-year bond and $7,451.62 for a six-year bond, and these are each priced to yield 5.5%. Suppose also that, at the end of two years, no matter what the yield rate i may then be, you sell the remaining bond at a purchase price to yield i, combine the proceeds with the $24,881.52 from the redeemed bond, and use the total to buy a one-year zero-coupon bond. Illustrate that this immunizes against interest rate risk by showing that it produces the needed $35,000 three years after your initial bond purchases if $i = 18\%$ (a high rate) or $i = 1\%$ (a low rate).

(9.2) Macaulay and modified duration

(1) Compute the Macaulay duration of a ten-year 6% $1,000 bond having annual coupons and a redemption of $1,200 if the yield to maturity is 8%.

(2) The current price of a bond having annual coupons is $1,312. The derivative of the price function of the bond with respect to the yield to maturity is $-$7,443.81$ when evaluated at the current annual yield, which is 7%. Calculate the Macaulay duration $D(.07, \infty)$ and the modified duration $D(.07, 1)$ of the bond.

(3) A zero-coupon bond matures in eight years. It is sold to yield 5% annually. Find the modified duration $D(.05, 1)$.

(4) The current price of a noncallable bond with annual coupons is $1,120.58, and the current annual yield is 4.25%. The modified duration $D(.0425, 2)$ is 3.58. Estimate the price of the bond if the yield increases to 4.4%. About what would the price be if the discount rate decreases to an annual effective rate of 3.7%?

(5) Calculate the Macaulay duration $D(.05, \infty)$ and the modified duration $D(.05, 2)$ of a preferred stock that pays dividends forever of $50 each six months, with the next dividend in exactly six months.

(6) Calculate the Macaulay duration $D(.06, \infty)$ and the modified duration $D(.06, 1)$ of a stock that pays annual dividends forever, assuming that the first dividend, payable in exactly one year, is $100 and then, each subsequent dividend is 2% more than the previous one.

(7) On January 1, 1978, Jonathan Linden is obligated to pay annual level payments for sixteen years, beginning with a payment on January 1, 1979. His financial advisor tells him that this liability has Macaulay duration 7.39. Determine the annual effective interest rate that the advisor used to figure the duration.

(8) Santosh inherits a perpetuity with annual payments. The first payment is $1,400 and payments increase by 3% each year. Obtain a formula for the Macaulay duration $D(.05, \infty)$ of the remaining payments an instant after the k-th payment.

(9) Mustafa's portfolio consists of an annuity with monthly payments of $1,000 each month for five years and a $20,000 8% eight-year par-value bond bearing semiannual coupons. Calculate the Macaulay duration of the portfolio at 9%.

(10) A thirty-year mortgage with no early repayment option is repaid with level monthly payments. The interest rate on the mortgage is 6.8% nominal convertible monthly. Calculate the Macaulay duration on the mortgage at the equivalent annual effective interest rate.

(9.3) Convexity

(1) Calculate the Macaulay convexity of a ten-year 6% $1,000 bond having annual coupons and a redemption of $1,200 if the yield to maturity is 8%. [This is the same bond considered in Problem (9.2.1).]

(2) A $2,000 7% three-year bond has semiannual coupons and redemption amount $2,225. It is purchased to provide the investor with a 5% annual effective yield. Find the Macaulay convexities $C(4\%, \infty)$, $C(5\%, \infty)$, and $C(6\%, \infty)$. Also compute the modified convexity $C(5\%, 2)$.

(3) A bond has Macaulay duration $D(i, \infty) = 5.8$ and Macaulay convexity $C(i, \infty) = 1.2$. Determine $C(i, 4)$ as a function of i.

(4) A perpetuity has a level payment at the end of each year. The annual effective interest rate is 5.2%. Calculate the Macaulay duration and the Macaulay convexity. Use these to estimate the percentage decline in the market value of the perpetuity if the annual effective interest rate increases by 50 basis points.

(5) This problem guides you through a derivation of Equation (9.3.9).

(a) Show that $\dfrac{d\delta}{di^{(m)}} = \left(1 + \dfrac{i^{(m)}}{m}\right)^{-1}$.

(b) Show that $\dfrac{d^2\delta}{d(i^{(m)})^2} = -\dfrac{1}{m}\left(1 + \dfrac{i^{(m)}}{m}\right)^{-2}$.

(c) Use the chain rule and the product rule to establish

$$\frac{d^2 P}{d(i^{(m)})^2} = \frac{d}{di^{(m)}}\left(\frac{dP}{d\delta}\right) \cdot \frac{d\delta}{di^{(m)}} + \frac{dP}{d\delta} \cdot \frac{d^2\delta}{d(i^{(m)})^2}.$$

(d) Show

$$\frac{d^2 P}{d(i^{(m)})^2} = \frac{d^2 P}{d\delta^2}\left(1 + \frac{i^{(m)}}{m}\right)^{-2} - \frac{dP}{d\delta}\cdot\frac{1}{m}\left(1 + \frac{i^{(m)}}{m}\right)^{-2}.$$

(e) Use the result of (d) to demonstrate the following;

$$C(i, m) = \left(C(i, \infty) + \frac{1}{m}D(i, \infty)\right)\Big/\left(1 + \frac{i^{(m)}}{m}\right)^2.$$

(6) Verify Fact (9.3.16).

(9.4) Immunization

(1) Providence Health Care is obligated to make a payment of $300,000 in exactly three years. In order to provide for this obligation, their financial officer decides to purchase a combination of one-year zero-coupon bonds and four-year zero-coupon bonds. Each of these is sold to yield an annual effective yield of 4%. How much of each type of bond should be purchased so that the present value and duration conditions of Redington immunization are satisfied? Is the convexity condition also satisfied at $i = 4\%$.?

(2) A court has ordered Security Enterprises to pay $200,000 in two years and $500,000 in five years. In order to meet this important liability, they wish to invest in a combination of two-year 10% par-value bonds with annual coupons and five-year zero-coupon bonds. Each of these is sold to yield an annual effective yield of 4%. How much of each type of bond should be purchased so that the present value and duration conditions of Redington immunization are satisfied at the current 4% rate? Is the convexity condition also satisfied?

(3) Tomorrow Financial Associates is required to make a $500,000 payment in exactly four years. To cover this liability, they have purchased a two-year zero-coupon bond with redemption amount $343,398.73 and a ten-year zero-coupon bond with redemption amount $162,782.52. We assume that there is a flat yield curve and that the current interest rate is 4.5%. Is this portfolio fully immunized? In one year, will it be fully immunized? What will the surplus be if interest rates rise immediately to 8%? What will the surplus be in one year if interest rates are 8%?

(9.5) Other types of duration

(1) A four-year callable bond's current market value is $95.40. If the interest rate increases by 100 basis points, the market value is anticipated to be $92.50 and if the interest rate decreases by 100 basis points, the price is expected to be $96.60. Compute the effective duration.

(2) A 4% $2,000 six-year par-value bond with annual coupons is purchased for $1,895.26. It may be called at any time. If interest rates fall by 1%, the bond should be called, while if they rise, it will not be called. Calculate the original yield i that an investor would receive if the bond were held to maturity, and then calculate the effective durations $E_{.01}(i, 1)$ and $E_{.02}(i, 1)$.

(3) Simeon Burnitz holds a callable bond with cashflows of $2,500 in one year and again in two years and $54,200 in three years. If the annual yield is 5%, then the price is expected to be $49,200 (because of the call feature). The current yield is 6%, and the bond's price is $48,392.75. Should the interest rate rise to 7%, the price will be $47,181.90. Calculate the effective duration $E_{.01}(.06, 1)$ of the bond and use it to approximate the price if the interest rate falls to 5.4%. What if it falls to 4%? Compare your answers to the actual prices, and also to the estimated price obtained using the modified duration $D(.06, 1)$, of a noncallable bond with cashflows of $2,500 in one year and $54,200 in three years.

(4) Let $P(i)$ denote the market value of a set of cashflows if it is purchased to yield i. Define the effective convexity $F_h(i_0, 1)$ by $F_h(i_0, 1) = \frac{P(i_0+h)+P(i_0-h)-2P(i_0)}{h^2 P(i_0)}$. Discuss how $F_h(i_0, 1)$ and $C(i_0, 1)$ are related. In particular, show that $F_h(i_0, 1) \approx C(i_0, 1)$ if the cashflows are not interest rate sensitive and h is sufficiently small.

(5) An eight-year 7% bond pays annual coupons. Its current price is $102.43, and its current yield is 6.6%. If interest rates rise to an effective rate of 6.8%, then its price will fall to $101.20. If interest rates fall to an effective rate of 6.4%, then its price will rise to $103.21. Find the bond's effective duration and effective convexity, and use the calculated values to estimate the price if interest rates rise to an effective rate of 7.2%.

Chapter 9 review problems

(1) A fifteen-year 7.5% par-value bond has annual coupons. Compute the modified durations $D(9\%, 1)$ and $D(7.5\%, 1)$.

(2) The current price of a noncallable bond with semiannual coupons is $1,020.60, and the current annual yield is 5.35%. The Macaulay duration $D(5.35\%, \infty) = 6.26$. Estimate the price of the bond if the annual effective interest rate increases to 5.6%. Approximate the bond's price if the annual effective interest rate decreases to 4%. How should these estimates be refined if you know that the Macaulay convexity $C(5.35\%, \infty)$ is equal to 1.64?

(3) A portfolio consists of three $1,000 par-value bonds. The first is a two-year zero-coupon bond, the second is a three-year zero-coupon bond, and the last is a five-year 6% bond with annual coupons. At interest rate $i = 5\%$, calculate the Macaulay duration of each bond and of the portfolio. Also calculate the Macaulay convexities.

(4) Calculate the Macaulay duration $D(.06, \infty)$ on January 1 of a preferred stock that pays dividends forever of $40 each June 30th and $60 each December 31st . What is the modified duration $D(.06, 2)$? — Use the 30/360 method of counting time.

(5) National Reliance Insurance (NRI) is obligated to make a payment of $250,000 in exactly five years. Its managers wish to fund this liability with a combination of two-year zero-coupon bonds and seven-year zero-coupon bonds, purchased today. The current yield rate for bonds is 5%. How much should the insurance company invest in each type of bond in order to immunize its portfolio? Assuming NRI allocates its resources so as to immunize its position, what is the time zero value of the portfolio of asset and liabilities if yield rates fall from 5% to 4%?

(6) A 5% $10,000 five-year par-value bond with annual coupons is purchased for $9,736.11. It may be called at any time. If interest rates fall by 1%, the bond should be called, while if the interest rate rises, it will not be called. Calculate the original yield i that an investor would receive if the bond were held to maturity, and then calculate the effective duration $E_{.01}(i, 1)$.

APPENDIX A

Some useful formulas

SPECIAL ACCUMULATION FUNCTIONS

Simple interest at rate r: $a(t) = 1 + rt$

Simple discount at rate q: $a(t) = \dfrac{1}{1-qt}$

Compound interest at effective interest rate i: $a(t) = (1 + i)^t$

RATES

Effective interest rate: $i_{[t_1,t_2]} = \dfrac{a(t_2)-a(t_1)}{a(t_1)}$;

Effective discount rate: $i_{[t_1,t_2]} = \dfrac{a(t_2)-a(t_1)}{a(t_1)}$;

Spot rate: $r_t = (1 + i_{[0,t]})^{\frac{1}{t}} - 1$

Forward rate: $f_{[t,s]} = \dfrac{(1+r_s)^s}{(1+r_t)^t} - 1$

Inflation rate: $r_{[t_1,t_2]} = \dfrac{p(t_2)-p(t_1)}{p(t_1)}$; real interest rate $j_{[t_1,t_2]} = \dfrac{i_{[t_1,t_2]}-r_{[t_1,t_2]}}{1+r_{[t_1,t_2]}}$

Under compound interest, $i, d, i^{(m)}$, and $d^{(m)}$ are *equivalent* if

$$\left(1 + \frac{i^{(m)}}{m}\right)^m = 1 + i = (1 - d)^{-1} = \left(1 - \frac{d^{(m)}}{m}\right)^{-m}.$$

Rate i is *equivalent* to constant force of interest δ if $\delta = \ln(1 + i)$.

Under *a general accumulation function* $a(t)$,

$$\delta_t = \frac{a'(t)}{a(t)} = \frac{d}{dt}\ln a(t) \quad ; \quad a(t) = e^{\int_0^t \delta_r \, dr}$$

FORMULAS NOT SPECIFIC TO INTEREST THEORY

SUM OF GEOMETRIC SERIES

$$c + cr + cr^2 + \cdots + cr^{n-1} = c\left(\frac{1-r^n}{1-r}\right) \quad \text{for} \quad r \neq 1.$$

$$c + cr + cr^2 + cr^3 + \cdots = \frac{c}{1-r} \quad \text{for} \quad |r| < 1.$$

NEWTON'S METHOD

A root is sought using the recursion formula $x_{k+1} = x_k - \dfrac{f(x_k)}{f'(x_k)}$.

QUADRATIC FORMULA

$ax^2 + bx + c = 0$ has solutions $x = \dfrac{-b \pm \sqrt{b^2 - 4ac}}{2a}$.

ANNUITIES AND PERPETUITIES WITH LEVEL PAYMENTS

Present values: payment period = conversion period

$$a_{\overline{n}|} = v(1) + v(2) + \cdots + v(n); \quad a_{\overline{\infty}|} = \sum_{k=1}^{\infty} v(k)$$

$$a_{\overline{n}|i} = v + v^2 + \cdots v^n = \frac{1-v^n}{i}; \quad a_{\overline{\infty}|} = \lim_{n \to \infty} \frac{1-v^n}{i} = \frac{1}{i}$$

$$\ddot{a}_{\overline{n}|} = 1 + v(1) + v(2) + \cdots + v(n-1); \quad \ddot{a}_{\overline{\infty}|} = \sum_{k=0}^{\infty} v(k)$$

$$\ddot{a}_{\overline{n}|i} = 1 + v + v^2 + \cdots v^{n-1} = \frac{1-v^n}{d}; \quad \ddot{a}_{\overline{\infty}|} = \lim_{n \to \infty} \frac{1-v^n}{d} = \frac{1}{d}$$

Present values: payment period ≠ conversion period

$$a_{\overline{n}|}^{(m)} = v\left(\frac{1}{m}\right) + v\left(\frac{2}{m}\right) + \cdots + v\left(\frac{nm}{m}\right); \quad a_{\overline{\infty}|}^{(m)} = \sum_{k=1}^{\infty} v\left(\frac{k}{m}\right)$$

$$a_{\overline{n}|i}^{(m)} = v^{\frac{1}{m}} + v^{\frac{2}{m}} + \cdots + v^{\frac{nm}{m}} = \frac{1-v^n}{i^{(m)}}; \quad a_{\overline{\infty}|i}^{(m)} = \sum_{k=1}^{\infty} v^{\frac{k}{m}} = \frac{1}{i^{(m)}}$$

$$\ddot{a}_{\overline{n}|}^{(m)} = 1 + v\left(\frac{1}{m}\right) + v\left(\frac{2}{m}\right) + \cdots + v\left(\frac{nm-1}{m}\right); \quad \ddot{a}_{\overline{\infty}|}^{(m)} = \sum_{k=0}^{\infty} v\left(\frac{k}{m}\right)$$

$$\ddot{a}_{\overline{n}|i}^{(m)} = 1 + v^{\frac{1}{m}} + v^{\frac{2}{m}} + \cdots + v^{\frac{nm-1}{m}} = \frac{1-v^n}{d^{(m)}}; \quad \ddot{a}_{\overline{\infty}|i}^{(m)} = \sum_{k=0}^{\infty} v^{\frac{k}{m}} = \frac{1}{d^{(m)}}$$

$$\overline{a}_{\overline{n}|} = \int_0^n v(t)\, dt; \quad \overline{a}_{\overline{n}|i} = \frac{1-v^n}{\delta} = \frac{1-e^{\delta n}}{\delta}$$

Basic relations between $a_{\overline{n}|}$ and $\ddot{a}_{\overline{n}|}$, $a_{\overline{\infty}|}$ and $\ddot{a}_{\overline{\infty}|}$

$$\ddot{a}_{\overline{n}|i} = (1+i)a_{\overline{n}|i} \quad \ddot{a}_{\overline{\infty}|i} = (1+i)a_{\overline{\infty}|i}$$

$$\ddot{a}_{\overline{n}|} = 1 + a_{\overline{n-1}|}; \quad \ddot{a}_{\overline{\infty}|} = 1 + a_{\overline{\infty}|};$$

Basic relations between $a_{\overline{n}|}$ and $s_{\overline{n}|}$

$s_{\overline{n}|} = a(n)a_{\overline{n}|}$. In particular, $s_{\overline{n}|i} = (1 + i)^n a_{\overline{n}|i}$.

$\dfrac{1}{a_{\overline{n}|i}} = \dfrac{1}{s_{\overline{n}|i}} + i$.

Accumulated values (*Obtain these, and other formulas, by multiplying expressions for the present values by $a(n)$.*)

$s_{\overline{n}|} = \dfrac{a(n)}{a(1)} + \dfrac{a(n)}{a(2)} + \cdots + \dfrac{a(n)}{a(n)}$

$s_{\overline{n}|i} = (1 + i)^{n-1} + (1 + i)^{n-2} + \cdots (1 + i)^0 = \dfrac{(1+i)^n - 1}{i}$

$\ddot{s}_{\overline{n}|} = (1 + i)^n + (1 + i)^{n-1} + \cdots (1 + i)^1 = \dfrac{(1+i)^n - 1}{d}$

Basic relations between $s_{\overline{n}|}$ and $\ddot{s}_{\overline{n}|}$

$\ddot{s}_{\overline{n}|i} = (1 + i)s_{\overline{n}|i}$

$s_{\overline{n+1}|} = \ddot{s}_{\overline{n}|} + 1$

ANNUITIES WITH SPECIAL PAYMENT PATTERNS

Present values: payment period = conversion period

$(I_{P,Q}a)_{\overline{n}|i} = Pv + (P + Q)v^2 + (P + 2Q)v^3 + \cdots + [P + (n - 1)Q]v^{n-1}$

$\qquad = Pa_{\overline{n}|i} + \dfrac{Q}{i}\left(a_{\overline{n}|i} - nv^n\right)$

$(Ia)_{\overline{n}|i} = (I_{1,1}a)_{\overline{n}|i} = 1v + 2v^2 + \cdots + nv^n = \dfrac{\ddot{a}_{\overline{n}|i} - nv^n}{i}$;

$(Da)_{\overline{n}|i} = (I_{n,-1}a)_{\overline{n}|i} = nv + (n - 1)v^2 + \cdots + 1v^n = \dfrac{n - a_{\overline{n}|i}}{i}$

Present values: payment period ≠ conversion period

$(Ia)^{(m)}_{\overline{n}|i} = \dfrac{\ddot{a}_{\overline{n}|i} - nv^n}{i^{(m)}}$ (*payments increase $\frac{1}{m}$ each interest period*)

$(Ia^{(m)})^{(m)}_{\overline{n}|i} = \dfrac{\ddot{a}^{(m)}_{\overline{n}|i} - nv^n}{i^{(m)}}$ (*payments increase $\frac{1}{m^2}$ each payment period*)

$(\overline{Ia})_{\overline{n}|} = \int_0^n tv(t)\,dt$;   (*pays continuously at rate t at time t*)

$(\overline{Ia})_{\overline{n}|i} = \dfrac{\overline{a}_{\overline{n}|i} - nv^n}{\delta}$

Note: *If you have compound interest , formulas for the symbol giving the* **present value of an annuity-due** *may be obtained by multiplying the expressions for the corresponding annuity-immediate symbol by* $(1 + i)^T$, *where T is the time between annuity payments. Alternatively, you may view the annuity as the sum of the first cashflow and an annuity-immediate with one fewer payment. For example,*

$$(I_{P,Q}\ddot{a})_{\overline{n}|i} = P\ddot{a}_{\overline{n}|i} + \frac{Q}{d}\left(a_{\overline{n}|i} - nv^n\right); (I_{P,Q}\ddot{a})_{\overline{n}|i} = P + (I_{P+Q,Q}a)_{\overline{n-1}|i}$$

Note: *For an annuity, formulas for the symbol giving the* **accumulated values at time** *n may be obtained from those giving the present value by multiplying by* $a(n)$. *When you have compound interest, you multiply by* $a(n) = (1 + i)^n$. *For example,*

$$(\overline{I}\overline{s})_{\overline{n}|} = a(n)\int_0^n tv(t)\,dt;$$

$$(I_{P,Q}s)_{\overline{n}|i} = Ps_{\overline{n}|i} + \frac{Q}{i}\left(s_{\overline{n}|i} - n\right);$$

$$(Is)_{\overline{n}|i} = (I_{1,1}s)_{\overline{n}|i} = \frac{\ddot{s}_{\overline{n}|i}-n}{i} = \frac{s_{\overline{n+1}|i}-(n+1)}{i};$$

Note: *If you want the* **present value of a perpetuity**, *look at a limit as n* $\to \infty$ *of the present value of the first n payments. The following limit is useful to note:* $\lim_{n \to \infty} nv^n = 0$. Formulas obtained in this manner include

$$(I_{P,Q}a)_{\overline{\infty}|i} = \frac{P}{i} + \frac{Q}{i^2}$$

$$(Ia)_{\overline{\infty}|i} = \frac{i}{id}$$

LOANS WITH LEVEL PAYMENTS

Amortized loan: n payments, payment amount Q, payment period = conversion period

$$B_k = Qa_{\overline{n-k}|i}, \ I_k = B_{k-1}i, \ Q = I_k + P_k, \ P_k = Qv^{n-k+1},$$

Loan by the sinking-fund method: n payments, loan amount L, i = rate on loan, j = rate on sinking-fund account

$$\text{total payment} = iL + \frac{L}{s_{\overline{n}|j}}$$

BONDS

Basic notation: Rates: $r = \frac{\alpha}{m}; j = \frac{I}{m}; i = (1 + j)^m - 1$

Coupon amount: $Fr = Cg = Gj$

Basic Price formula: $P = (Fr)a_{\overline{n}|j} + Cv_j^n$

Premium-discount formula: $P = C(g - j)a_{\overline{n}|j} + C$

Bond amortization: $B_t = C(g - j)a_{\overline{n-t}|j} + C; B_t - B_{t-1} = P_t$

$$Cg = I_t + P_t; I_t = jB_{t-1}; P_t = C(g - j)v_j^{n-t+1}$$

Valuing at yield rate j: [see Table (6.6.9)]

$$\mathcal{D}_T = (1 + j)^f B_{\lfloor T \rfloor}$$

$$\mathcal{C}_T = \mathcal{D}_T - Cg\left(\frac{(1+j)^f - 1}{j}\right) = \mathcal{D}_T - Cgs_{\overline{f}|j}$$

$$\mathcal{D}_T^{\text{prac}} = (1 + fj)B_{\lfloor T \rfloor}$$

$$\mathcal{C}_T^{\text{prac}} = \mathcal{D}_T^{\text{prac}} - Cgf$$

Pricing at rate $\tilde{j}$: [see Table (6.7.8)]

$$\mathcal{D}_T^{\tilde{j}} = \mathcal{D}_{\lfloor T \rfloor}^{\tilde{j}}(1 + \tilde{j})^f$$

$$\mathcal{C}_T^{\tilde{j}} = \mathcal{D}_T^{\tilde{j}} - Cgs_{\overline{f}|\tilde{j}}$$

$$\mathcal{C}_T^{\tilde{j},\,\text{semiprac}} = \mathcal{D}_T^{\tilde{j}} - Cgf$$

$$\mathcal{A}_T^{\tilde{j}} = Cg\left(\frac{(1+\tilde{j})^f - 1}{\tilde{j}}\right) = Cgs_{\overline{f}|\tilde{j}}$$

$$\mathcal{A}_T^{\text{prac}} = Cgf$$

SHORTSALE OF STOCKS

$$i_{shortsale} = \frac{(P_0 - P_T) + i_{[0,T]}rP_0 - D}{rP_0}$$

where D is the time-T accumulated value of any dividends paid.

NO-ARBITRAGE PRICING OF OPTIONS

Put-Call parity: $P_t + S_t = K(1 + i)^{t-T} + c_T$

One-period binomial pricing model: [See (8.9.3)]

Replicating portfolio:

$c_0 = f + \Delta S_0$ where $f = \frac{uv_d - dV_u}{(u-d)(1+i)^T}$, the amount in the portfolio placed in a risk-free investment

$\Delta = \frac{V_u - V_d}{u-d}$, the number of copies of the underlier in the portfolio

Risk-neutral method: [See (8.10.5)]

$p^* = \frac{S_0(1+i)^T - d}{u-d}$; $c_0 = p^*V_u(1 + i)^{-T} + (1 - p^*)V_d(1 + i)^{-T}$

DURATION AND CONVEXITY

Price function: $P(I) = \sum_{t \geq 0} C_t(1 + i)^{-t}$

Macaulay duration: $D(i, \infty) = -\frac{dP}{d\delta} / P(i) = \sum_{t \geq 0} \left(\frac{C_t(1+i)^{-t}}{P(i)} \right) t$

Modified duration: $D(i, m) = -\frac{dP}{di^{(m)}} / P(i)$

Relationships between $D(i, m)$ and $D(i, \infty)$:

$D(i, \infty) = D(i, m) \left(1 + \frac{i^{(m)}}{m} \right)$

$D(i, \infty) = \lim_{m \to \infty} D(i, m)$

Macaulay convexity: $C(i, \infty) = \frac{d^2 P}{d\delta^2} / P(i) = \sum_{t \geq 0} \left(\frac{C_t(1+i)^{-t}}{P(i)} \right) t^2$

Modified convexity: $C(i, m) = \frac{d^2 P}{di^{(m)^2}} / P(i)$

Relationship between $C(i, m)$ and $C(i, \infty)$: $C(i, m) = \dfrac{C(i, \infty) + \frac{1}{m} D(i, \infty)}{\left(1 + \frac{i^{(m)}}{m} \right)^2}$

Effective duration: $E_h(i_0, 1) = -\left(\dfrac{P(i_0 + h) - P(i_0 - h)}{2h} \right) \Big/ P(i_0)$

APPENDIX B

Answers to end of chapter problems

Solutions are only given for those problems for which there is a short answer. Monetary amounts are given to the nearest cent. Rates are expressed as percents. We give five significant places after the decimal point, or after leading zeros, unless the problem specifies otherwise or the number may be expressed exactly with fewer digits displayed.

PROBLEMS, CHAPTER 0

(1) 9-formatting and rounding to nearest hundredth gives 8.51; 2-formatting gives 8.51; 2-formatting with rounding each product gives $3.53 + 4.97 = 8.50$.

(3) (a) 1,508.7335
(b) 212.9612
(c) 1.8577

(4) (a) The calculator computes $542(1.06)^5 + (768)^{\frac{1}{4}}$ and stores the result, 730.5825591, in register 7.

(b) The sequence clears the memories, enters 1 in memory 0, 2 in memory 1, and 3 in memory 2. Note that 4 is keyed, but it is not entered in a register.

(c) The calculator computes $(625 \times 52) + 4{,}089 = 36{,}589$.

(d) The calculator computes $2 \times 1.04 = 2.08$, $3 \times 1.04 = 3.12$, and $58 \times 1.04 = 60.32$. It first stores 60.32 in register 4, then multiples it by 1.04 and adds 54 to it , storing the displayed result, 116.7328, in register 3. The final $\boxed{2}\boxed{=}$ does not accomplish anything.

(e) The calculator first adds 10% of 5 to 5, displaying 5.5, then adds 10% of 100 to 100, displaying 110, and finally starts with 200, increases it by 10%, multiplies it by .01, and again by .01; the final display is 0.022.

(f) Here you are making a sequence of errors. All you end up accomplish-
ing is having the display show "RST". (If you now push ENTER ,
the calculator will be reset to factory default settings.)

PROBLEMS, CHAPTER 1

(1.3) Accumulation and amount functions

(1) $K = 10; a(20) = 1.25$

(2) $2,400

(3) $\alpha = .01; \beta = 1; i_1 \approx 7.14285\%$

(4) $2,057.14

(5) $(n + 1)\frac{n}{2}$

(6) $2^{n+1} - 2$

(7) $i_n = \frac{6n-1}{3n^2-4n+801}$; If $f(x) = \frac{6x-1}{3x^2-4x+801}$, then $f'(x) = \frac{-18x^2+6x+4,802}{(3x^2-4x+801)^2} < 0$
for $x > 17$.

(1.4) Simple interest

(1) 4th year interest is $50; balance at end of 4th year is $1,200.

(2) 10

(3) 6%

(4) 8.75%

(5) $600

(6) (a) the fourth period [3,4]
    (b) $i_{[4,6]} \approx 8.33333\%$

(7) 27,793

(1.5) Compound interest

(1) $T \approx 32.91588$

(2) $14,716.53 (This actually accumulates to $32,168.01.)

(3) $i \approx 8.00597$

(4) $57.88

(5) $a = 3, b = 1, c = -1$

(6) $1,245.96

(7) 6.03988%

(8) (a) no, $4,250 > $4,207.66
    (b) yes, $5,000 < $5,154.56

(9) 24.56361 years

(10) $5,257.17

(1.6) Effective discount rates/ Interest in advance

(1) $2,760
(2) $2,553.19
(3) $d \approx 8.96555\%, i \approx 9.84848$
(4) $3,020.80
(5) $d_{[1,3]} \approx 13.57189\%$

(1.7) Discount functions/ The time value of money

(1) $2,628.57
(2) $3,318.42
(3) $121,262.40
(4) $X = 2,462.75$ (This repays a debt of $6,000.01.)
(5) $3,084.81
(7) $i = 21\%$
(8) $31.94

(1.8) Simple discount

(1) $2,378.75
(2) 2.22222%
(3) 4.16667 years
(4) $22.45
(5) (a) 8.69231 years
     (b) 9.30272%

(1.9) Compound discount

(1) $1,391.94
(2) (a) $4,953.32
     (b) 3.62694%
(3) 1.63124%
(4) $66.27
(5) $33.16
(6) 17.49966%
(7) $I \approx 5.31137\%, X \approx 2,934.48$

(1.10) Nominal rates of interest and discount

(1) $d \approx 7.76318\%, d^{(3)} \approx 7.97321, i \approx 8.41658\%, i^{(6)} \approx 8.13575\%$

(2) $d^{(12)} \approx 6.37434\%$ and the monthly rate is approximately $.53120\%$

(3) $i^{(12)} = 6\%, i \approx 6.16788\%, d \approx 5.80947\%$

(4) $3,932.32

(5) 10.00002

(6) 42

(7) (a) $i^{\left(\frac{1}{m}\right)} = \frac{1}{m}[(1 + i)^m - 1]$

   (b) 5.91342%

   (c) $d^{\left(\frac{1}{m}\right)} = i^{\left(\frac{1}{m}\right)} / \left[1 + mi^{\left(\frac{1}{m}\right)}\right], d^{\left(\frac{1}{m}\right)} = \frac{1}{m}\left[1 - (1 - d)^m\right]$

(1.11) A friendly competition (Constant force of interest)

(1) 3.21287%

(2) $366.42

(3) B is the best, with an annual effective rate of 5.40967%. A is the worst with an annual effective rate of 5.2%. C's annual effective rate is 5.29545%.

(1.12) Force of interest

(1) (a) $358.09

   (b) $384.82

(2) 29.72973%

(3) $74.61

(4) 3.8%

(5) 10.61491%

(6) .57658 years

(7) In both case, shift the money after 4.78008 years.

(8) (a) $\frac{r}{1+rt}$

   (b) $\frac{s}{1-st}$

   (c) $\frac{r-s}{2rs}$

(9) 5.13178%

(10) $12,140.26

(1.13) Note for those who skipped Section (1.11) and (1.12)

(1) 3.82120%

(2) 4.02013%

(1.14) Inflation

(1) (a) 1.16505%
    (b) −.38241%
(3) 1.79485%
(4) $Y = X(1 + r)^n$
(5) $p = .40024\%$
(6) The inflation-adjusted force of interest is 1.78399%. The stated rate of interest is $i = 4.1414\%$, so the force of interest is 4.05794. The force of interest exceeds the inflation-adjusted force of interest by $\ln(1 + r)$, where r is the rate of inflation.

Chapter 1 review problems

(1) $8,353.30
(2) .15277%
(3) $1,060.89
(4) If you assume $K must be an integer number of cents, then $K = $1,082 and the answer is $1,060.89. Without the assumption on $K, you get $1,060.88.

   (a) $d - 1$
   (b) $d - 1$

(5) $N \approx 49.57218$.
(6) $P_{Javier} \approx \$3,126.73, i_{Chan} \approx 7.84342\%, i_{Javier} \approx 13.08566\%,$
(7) (a) $a = .002, b = .04, c = 1$
    (b) $714.47
(8) $i \approx 4.51953\%$
(9) $f(n) = \left(\frac{n}{n-1}\right)^4$

PROBLEMS, CHAPTER 2

(2.2) Equations of value for investments involving a single deposit made under compound interest

(1) 868.81
(2) 5.22998
(3) 2.75003 years
(4) By the rule of 72, you find 14.4 years (at $i = 5\%$) and 7.2 years (at

$i = 10\%$). More exactly, the answers are 14.20670 years and 7.27254 years.

(5) $n = 114$; A rule of 114 gives answers of 28.5 years (at $i = 4\%$) and 9.5 years at (at $i = 12\%$). More exactly, the answers are 28.01102 years and 9.69404 years.

(2.3) Equations of value for investments with multiple contributions

(1) $6,505.49

(2) 2.20768%

(3) 1.06369

(4) 5.37687%

(5) $\overline{T} \approx 2.86487$, $T \approx 2.82481$

(6) 4.51430 years after the loan is made

(7) (a) $y - \ln A = \frac{1}{A}(x - A)$

(9) $T \approx .019706$ years, and $\overline{T} = 0$; so $T > \overline{T}$.

(10) .45338 years after the loan is made

(11) (a) $950.06
     (b) $964
     (d) $1,016.28 for both revised (a) and revised (b)

(12) .342

(2.4) Investment return

(1) 11.46372%

(3) 8.28581%

(4) $i_A \approx 8.14837\%$, $i_B \approx 9.75123\%$, $i_C \approx 7.23805\%$, $i_D \approx 7.49501\%$

(5) yield rates 6.38298%, 5.26316%, and 4.16667%

(6) 10%

(8) 7.19257%

(9) 18.92%

(2.5) Reinvestment considerations

(1) $i_{\text{Angela}} \approx 9.65463\%$, $i_{\text{Kathy}} \approx 11.37510\%$,

(2) $i_{\text{Randy}} \approx 6.54709\%$, $i_{\text{Kurt}} \approx 5.46623\%$,

(3) $i_A \approx 7.06134\%$, $i_B \approx 8.01234\%$, $i_C \approx 7.25891\%$

(2.6) Approximate dollar-weighted yield rates

(1) (a) $\$10{,}832(1 + i)^2 + \$2{,}000(1 + i)^{\frac{21}{12}} - \$7{,}000(1 + i)^{\frac{1}{12}} = \$12{,}566$
     (b) By (2.6.5), $i \approx 39.87454\%$, and by (2.6.8), $i \approx 34.46967\%$.
(2) -2.69544%
(3) September 26, 1996
(4) The annual dollar-weighted yield is 6.56422% or 6.90450%, depending
    upon whether the payment is after six months or one year; the difference
    is .34028%.

(2.7) Fund performance

(1) $\left(\dfrac{15{,}264}{163{,}748{,}000} P \right)^{\frac{1}{4}} - 1$
(2) (a) 1.2829%
     (b) -2.09097%
(3) \$2,306,938.21

Chapter 2 review problems

(1) (a) A time 0 equation of value is $-\$20{,}000 + \dfrac{\$4{,}000}{1+r} + \dfrac{\$18{,}000}{1+3r} = 0,$
         and $r \approx 3.82448\%$.
     (b) A time 3 equation of value is $-\$20{,}000(1 + i)^3 + \$4{,}000(1 + i)^2 +$
         $\$18{,}000 = 0$, and $i \approx 3.70\%$.
     (d) 3.697017%
(2) (a) $g(n) = K[(1 + i)^n - 1]^2/(1 + i)^n$
     (b) yes
(3) 6.36499%
(4) 5.60349%
(5) (a) 8.42664%
     (b) 10.72970%
     (c) 10.68650%
(6) (a) \$528.90
     (b) unique positive yield rate

PROBLEMS, CHAPTER 3

(3.2) Annuities-immediate

(1) \$166,957.07 (using $i^{(12)} = 3\%$), \$166,751.66 (using $d = 3\%$) When
    $d = 3\%$, the equivalent $i^{(12)}$ is greater than 3%. This results in a smaller

present value than the first case where $i^{(12)} = 3\%$.

(2) $v^n \approx .85801$

(3) Based on the 5% rate, the first 17 payments should each be for $5,331.94, and the last payment should be for $5,331.75. If the rate changes to 4.5% immediately after the tenth payment, the next 7 payments should be for $5,823.82, an increase of $491.88, and the new final payment should be increased by $491.81.

(4) $111,514.93

(5) $i \approx 6.32000\%$, $X = \$1,585.00$

(6) 5.57790%

(7) $23,380.84

(8) 1.89171 years

(3.3) Annuities-due

(1) $38,068.76

(2) $11,141.62

(3) (a) $a_{\overline{n+1}|} - a_{\overline{n}|}$ measures the value of a payment of 1 made $n + 1$ periods later

  (b) 38

(4) $5,922.83

(5) 9.09432

(6) 32.12891

(7) $i \approx 3.26453\%$, $n = 237$

(8) $i = 8.5\%$, $n = 43$

(10) (c) 412

(3.4) Perpetuities

(1) $1,040,000

(2) $245,695.34

(3) (a) 1990

  (b) $2,109.79

(4) $31,155.52

(5) 41.89597

(6) $62,035.20

(3.5) Deferred annuities and annuity values on any date

(1) At $i = 5\%$, the perpetuity has present value of $172,767.52 > \$160,000$.
    At $i = 6\%$, the perpetuity has present value of $139,936.55 < \$160,000$.

(2) $n = 12$, At $i = 6\%$, the charity aiding children of veterans gets 49.69694% of the total.

(3) $6,328.00

(4) $n \approx 17.99132$ The closest integer is $n = 18$.

(5) (a) $50,673.92
    (b) $5,659.45

(3.6) Outstanding loan balances

(1) (a) $5,253.10
    (b) $6,607.82

(2) $7,165.81

(3) $146,105.22

(4) $12,560.43

(5) (a) $234.07
    (b) $10,147.36

(6) The first 359 payments are each for $1,008.20. He gets a check for $33,463.18.

(7) $678.95

(3.7) Non-level annuities

(1) $409.86

(2) $4,758.51

(3) 84

(4) 7.38007%

(5) .62659

(6) $1,070.44

(7) (a) $5,817.13
    (b) 7.32%

(8) (a) $132,439.18
    (b) $4.16

(3.8) Annuities with payments in geometric progressions

(1) The common present value is $821.39 and $n = 14$.

(2) $33,954.85

(3) $196,614.90
(4) $99,501.73
(5) $397,388.55

(3.9) Annuities with payments in arithmetic progressions

(1) (a) $704.37
     (b) $1,178.78
     (c) $1,963.80
(2) 13.27815%
(3) $19,591.87
(4) 35.29412%
(5) $132,817.13
(6) $69,042.81
(7) $\frac{1}{i}[1 + (I_{3,2}a)_{\overline{n-1}|i} - n^2 v^n]$, 3,967.63808

(3.10) Yield rate examples involving annuities

(1) 8.43260%
(2) $59,619.29
(3) 7.94921%
(4) 4.60462% and 14.80726%
(5) $i_A \approx 5.41616\%$, $i_B \approx 5.10476\%$, $i_C \approx 5.58379\%$
(6) 7.32773%
(7) -4.89686%
(8) (a) 5.86144%
    (b) 5.77861%
    (c) 24.32561
    (d) $1,096.12
(9) 13.39%

(3.11) Annuity-symbols for non-integral terms

(1) There are 9 payments. The drop payment is $9,523.87.
(2) $34,429.58
(3) $T \approx 11.04724$ and the amount of the last payment is $27.69.

(3.12) Annuities governed by general accumulation functions

(1) (a) $654.99
    (b) 3.49845%

(2) $a_{\overline{5}|} \approx 4.41612,\ s_{\overline{5}|} \approx 5.63367,\ s_{\overline{5}|}/a_{\overline{5}|} = a(5)$

(3) 675

(3.13) The investment Year Method

(1) $3,713.16

(2) 5.25%

Chapter 3 review problems

(1) $37,920.31

(2) $410,714.79 10, HW2, 10, Spring 2004, h4S4

(3) $2,688.48

(4) 19

(5) 6.48384%

(6) $222,186.62

(7) $519,729.09

PROBLEMS, CHAPTER 4

(4.2) Level annuities with payments less frequent than each interest period

(1) (a) 1.21741%, $5,114.05

(2) $11,806.30

(3) $3,765.30

(4) $3,099.00

(5) The term is calculated to be 6.00062. Assuming it is an integer, it is 6.

(7) $X = [(1 + i)^3 - 1]\left(\$1,000 + \dfrac{\$1,000}{(1+i)^2 - 1}\right)$

(4.3) Level annuities with payments more frequent than each interest period

(1) $\ddot{s}_{\overline{23}|}^{(4)}$ denotes the accumulated value at the end of 23 interest periods of an annuity with a payment of $\frac{1}{4}$ at the beginning of each quarter of an interest period, and $\ddot{s}_{\overline{23}|2.25\%}^{(4)} \approx 30.11567$. $a_{\overline{\infty}|}^{(12)}$ denotes the present value of a perpetuity with a payment of $\frac{1}{12}$ at the end of each twelfth of an interet period and $a_{\overline{\infty}|}^{(12)4\%} \approx 25.45509$.

(2) $\dfrac{1-v}{i^{(12)}} \approx .94258$

(3) $640\left[\left(1 + \frac{i^{(4)}}{4}\right)^{84} - 1\right] \Big/ \left[1 - \left(1 + \frac{i^{(4)}}{4}\right)^{-\frac{4}{3}}\right]$

(4) $\$400a^{(4)}_{\overline{4}|i}(1 + j)^6 + \$420a^{(12)}_{\overline{6}|j}$; $\$5,082.45$

(5) There are various correct expressions including. Among these are $T = \frac{1}{4} + \ln\left(\dfrac{750}{13,520[(1+i)^{\frac{1}{4}} - 1]}\right) \Big/ \ln(1 + i)$ and $T = \ln\left(\dfrac{750/13,520}{1-(1+i)^{-\frac{1}{4}}}\right) \Big/ \ln(1 + i)$.

(4.4) Annuities with payments less frequent than each interest period and payments in arithmetic progression

(1)   (a) The effective quarterly interest rate is about 1.50751%, and the accumulated value is $251,477.70.

   (b) Again, the accumulated value is $251,477.70.

(2) $\dfrac{\$160,806.60}{s_{\overline{4}|i}} + \dfrac{1}{is_{\overline{4}|i}}\left(\dfrac{\$20,100.825}{a_{\overline{4}|i}} - \dfrac{\ln(1+9.332i)}{\ln(1+i)}\right)$

(3) $125,325.91

(4.5) Annuities with payments more frequent than each interest period and payments in arithmetic progression

(1) $(I^{(4)}\ddot{a})^{(4)}_{\overline{\infty}|}$ gives the value at the time of the first payment of a perpetuity having quarterly payments, the k-th of which is for an amount $\frac{k}{16}$.
   $(I^{(4)}\ddot{a})^{(4)}_{\overline{\infty}|3.2\%} \approx \$1,015.86$

(2) $43,290.39

(3) $665,782.44

(4) $82,984.31

(5) $198,058.59

(6) $142.879.58

(7) $15(Is)^{(3)}_{\overline{40}|j}(1 + j)^8 + \$600s^{(3)}_{\overline{8}|j} + \$40,000$

(4.6) Continuously paying annuities

(1) $\overline{a}_{\overline{18}|}$ denotes the present value of an annuity which pays levelly and continuously for eighteen interest periods, the total per interest period being 1, and $\overline{a}_{\overline{18}|3.2\%} \approx 13.739\%$.

(2) $119.89

(3) $82.40

(4) $837.82

(5) $70.67

(6) 17.10632 years, including the 12 years prior to any withdrawals

(4.7) A yield rate example

(2) 5.85509%

(3) $7,985.21, 5.24160%

Chapter 4 review problems

(1) (a) $\bar{s}_{\overline{30}|}$ denoes the accumulated value at the end of 30 interest periods of an annuity which pays levelly and continuously for thirty interest periods, the total per interest period being 1, and $\bar{s}_{\overline{30}|3\%} \approx 48.28553$

(b) $(I^{(12)}a)_{\overline{\infty}|}^{(12)}$ denotes the present value of a perpetuity with payments at the end of each month; the k-th payment is for $\frac{k}{144}$, and $(I^{(12)}a)_{\overline{\infty}|3\%}^{(12)} \approx 1,144.52720$

(c) $(Ia)_{\overline{\infty}|}^{(4)}$ denotes the present value of a perpetuity with payments at the end of each quarter of an interest period; the payments in the k-th interest period are each $\frac{k}{4}$, and $(Ia)_{\overline{\infty}|3\%}^{(4)} \approx 1,157.24016$.

(2) $52,253.29

(3) $1,400\ddot{a}_{\overline{15}|i}^{(12)} + (\$43,200)(I^{(12)}a)_{\overline{1}|}^{(12)}(1 + i)^{12}\ddot{a}_{\overline{15}|i}$. This is approximately $486,386.33 when $i = 3\%$.

(4) $500 \left[\left(1 + \frac{i^{(4)}}{4}\right)^{80} - 1 \right] \left(1 + \frac{i^{(4)}}{4}\right)^{16} \Big/ \left[1 - \left(1 + \frac{i^{(4)}}{4}\right)\right]^{-\frac{4}{3}}$

(6) $1,936.33

(8) $P\ddot{a}_{\overline{12}|j}\ddot{a}_{\overline{N}|I}$ where $I = \frac{(1+.01j)^{12}}{1-.01q} - 1$; This is $129,222.89 when $N = 15$, $P = \$1,200$, $q = 3$, and $j = .4$.

PROBLEMS, CHAPTER 5

(5.2) Amortized loans and amortization schedules

(1)

TIME	PAYMENT	INTEREST	PRINCIPAL	BALANCE
0	—	—	—	$29,119.00
1	$8,000.00	$1,223.00	$6,777.00	$22,342.00
3	$14,350.14	$1,916.14	$12,434.00	$9,908.00
4	$10,324.14	$416.14	$9,908.00	$0.00

(2) $604.59
(3) $530.18
(4) (a) $77,884.78
  (b) $489.90
  (c) $12,168.43, so the total paid at the end of fifteen years is $13,156.88
(5) $73,797.79

(5.3) Loans by the sinking fund method

(1)

TIME IN YEARS	INTEREST ON LOAN	SINKING FUND DEPOSIT	INTEREST ON S.F.	S.F. BAL. AFTER DEPOSIT	NET BALANCE ON LOAN
0	$0.00	$0.00	$0.00	$0.00	$14,000.00
1	$889.00	$5,200.00	$0.00	$5,200.00	$8,800.00
2	$889.00	$3,000.00	$218.40	$8,418.40	$5,581.60
4	$1,834.45	$4,859.60	$722.00	$14,000.00	$0.00

(2) $7,980.98
(3) $2,835.71
(4) $1,123.07
(5) 5.0285%

(5.4) Loans with other repayment patterns

(1) The balance just after the 20th payment is $20,147.74, the interest in the 20th payment is $217.23, and principal in the 20th payment is $1357.73.
(2) $n = 24$, $1,758.02
(3) $460.31
(4) fifth = $3,528, tenth = $2,500, fifteenth = $16,346.58
(5) $79,068.68
(6) $26,106.43

(5.5) Yield rate examples and replacement of capital

(1) (a) 6.39218%
  (b) 7.00773%
(2) $9,508.90
(3) (a) 7.05723%; (b) 6.56436%; (c) 8.93246%; (combined) 8.26688%
(4) 4.78740%

(5) .97384%
(6) 5.26394%

Chapter 5 review problems

(1) $11,801.14
(2) 33
(3) 4.34380%
(4) 15.75169%
(5) $3,778.99
(6) $677.43
(7) $4,098.13
(8) 5.60430%
(9) $113.66
(10) $7,456.04

PROBLEMS, CHAPTER 6

(6.2) Bond alphabet soup and the basic price formula

(1) $50; 6.10925%
(2) $2,370.69
(3) 21.5 years
(4) 8.5%; $n = 28$
(5) $4,081.54
(6) 5.21616%
(7) bond B, since it has a higher inflation adjusted yield, namely 4.13625%;
    A: 4%, B: 4.69667%.

(6.3) The premium-discount formula

(1) $2,997.95
(2) $4,438.18
(3) $20.71

(6.4) Other pricing formulas for bonds

(1) $780.68 by Makeham; $780.69 by other formulas using $n = 34$ and
    $K = 355.40$, i.e. $P = \$20 a_{\overline{34}|(1.06)^{1/2}-1} + \$355.40$.
(2) $15,814.29
(3) 31 since the number of coupons must be an integer

(6.5) Bond amortization schedules

(1) The price is $1,771.45, the amount for accumulation of discount in the tenth is $8.34, and the amount of interest in the tenth is $73.34.

(2) $119.00

(3) $48,739.16

(4) $1,863.73

(5) The following charts were made using the actual yields resulting from fact you must pay an integer number of cents.

5% $P = \$1,062.03$ so $j \approx 2.500043113\%$

TIME	COUPON	I_t	P_t	B_t
0	—	—	—	$1,062.03
1	$30	$26.55	$3.45	$1,058.58
2	$30	$26.46	$3.54	$1,055.05
3	$30	$26.38	$3.62	$1,051.42
4	$30	$26.29	$3.71	$1,047.71
5	$30	$26.19	$3.81	$1,043.90
6	$30	$26.10	$3.90	$1,040.00

6% $P = \$1,033.50$ so $j \approx 2.999988779\%$

TIME	COUPON	I_t	P_t	B_t
0	—	—	—	$1,033.50
1	$30	$31.00	−$1.00	$1,034.50
2	$30	$31.04	−$1.04	$1,035.54
3	$30	$31.07	−$1.07	$1,036.60
4	$30	$31.10	−$1.10	$1,037.70
5	$30	$31.13	−$1.13	$1,038.83
5	$30	$31.16	−$1.17	$1,040.00

7% $P = \$1,005.90$ so $j \approx 3.499949549\%$

TIME	COUPON	I_t	P_t	B_t
0	—	—	—	$1,005.90
1	$30	$35.21	−$5.21	$1,011.11
2	$30	$35.39	−$5.39	$1,016.49
3	$30	$35.58	−$5.58	$1,022.07
4	$30	$35.77	−$5.77	$1,027.84
5	$30	$35.97	−$5.97	$1,033.82
6	$30	$36.18	−$6.18	$1,040.00

(6.6) Valuing a bond after its date of issue

(1) In the table below, the values for times $T = 3.25$, $T = 3.5$, and $T = 3.75$ are based on $B_3 \approx \$2,973.612445$ with $\$2,427.36$ as redemption value.

T	$\lfloor T \rfloor$	f	$\mathcal{D}_T$	$\mathcal{C}_T$	$\mathcal{D}_T^{\text{prac}}$	$\mathcal{C}_T^{\text{prac}}$
3.25	3	.25	$3,017.25	$2,931.65	$3,018.22	$2,930.72
3.50	3	.5	$3,061.52	$2,889.07	$3,062.82	$2,887.82
3.75	3	.75	$3,106.45	$2,845.87	$3,107.43	$2,844.93
4.00	4	0	$2,802.03	$2,802.03	$2,802.03	$2,802.03

(2) $\mathcal{D}_T = \$1,093.95$, $\mathcal{C}_T = \$1,072.72$, $\mathcal{D}_T^{\text{prac}} = \$1,094.14$, and $\mathcal{C}_T^{\text{prac}} = \$1,072.68$.

(3) In the following chart, the time T is measured in half-years.

T	$\mathcal{D}_T$	$\mathcal{C}_T$	$\mathcal{D}_T^{\text{prac}}$	$\mathcal{C}_T^{\text{prac}}$
.5	$1,023.35	$1,008.48	$1,023.50	$1,008.50
1.0	$1,011.10	$1,011.10	$1,011.10	$1,011.10
1.5	$1,028.65	$1,013.77	$1,028.80	$1,013.80
2.0	$1,016.49	$1,016.49	$1,016.49	$1,016.49
2.5	$1,034.13	$1,019.26	$1,034.28	$1,019.28
3.0	$1,022.07	$1,022.07	$1,022.07	$1,022.07
3.5	$1,039.80	$1,024.93	$1,039.96	$1,024.96
4.0	$1,027.84	$1,027.84	$1,027.84	$1,027.84
4.5	$1,045.67	$1,030.80	$1,045.83	$1030.83
5.0	$1,033.82	$1,033.82	$1,033.82	$1,033.82
5.5	$1,051.75	$1,036.88	$1,051.91	$1,036.91
6.0	$1,040.00	$1,040.00	$1,040.00	$1,040.00

(6.7) Selling a bond after its date of issue

(1) $B_5 -$ Invoice price $\approx \$1,614.16$, $i \approx 5.88258\%$

(2) $\mathcal{A}_T \approx \$14.84$, $\mathcal{A}_T^{\text{prac}} = \$15$

(3) $\tilde{\jmath} \approx 10.05172\%$; $\mathcal{C}_T^{\tilde{\jmath}} \approx \$18,395.84$

(4) $\mathcal{A}_T \approx \$3,261.51$, $\mathcal{A}_T^{\text{prac}} = \$3,306.67$. The theoretically calculated $\$3,261.51$ includes $-\$147.09$ of principal and $\$3,408.60$ of interest.

(5) The maximum occurs when $f = \ln\left(\dfrac{\tilde{\jmath}}{\ln(1+\tilde{\jmath})}\right) \Big/ \ln(1 + \tilde{\jmath})$ If $\tilde{\jmath} = .01$, then $f \approx .50041$, if $\tilde{\jmath} = .07$, then $f \approx .50289$, and if $\tilde{\jmath} = .21$, then $f \approx .50794$.

(6.8) Yield rate examples

(1) $2,341.67
(2) 5.99141% (5.99139% if you do not assume reinvestment is liquidated and hence do not round it to the nearest cent)
(3) $11.19, 8.98066%
(4) $14.66
(5) 9.37777%
(6) $2,202.06
(7) The price at issue was $10,529.85, and the price paid by Pierre was $10,407.12.

(6.9) Callable bonds

(1)   (a) The lowest is 7.63414%, and the highest is 7.79235%.
      (b) The lowest is 6.89884%, and the highest is 7.20108%.
(2) $p(T) = \left(\$20,939.51 - \$900a_{\overline{[T]}4\%} \right)(1.04)^T - \$18,500$
(3)   (a) 6.6%
      (b) 6.17807%

(6.10) Floating rate bonds

(1) 6.45198%
(2) $9,140.43

(6.11) The BA II Plus calculator Bond worksheet

(1) The $1,665.08 February 23rd settlement price includes $42.08 of accrued interest.
(2) The $5,664.56 November 13th settlement price includes $268.33 of accrued interest
(3) $28.03 accrued interest (practical method with exact day count); $3,042.86

Chapter 6 review problems

(1) $57.86
(2) $\mathcal{D}_T \approx \$958.89$, $\mathcal{C}_T \approx \$938.80$
(3) $4,408.04, 34 coupons
(4) $985.77
(5) $15,531.17
(6) Using the "30/360" method: the price is $27,246.90, the theoretically determined accrued interest is $680.23, and accrued interest by the

practical method is $683.33.

Using the "actual/actual" method: the price is $27,243.80, the theoretically determined accrued interest is $679.35, and accrued interest by the practical method is $676.23.

(7) $I > 6.86391\%$

PROBLEMS, CHAPTER 7

(7.1) Common and preferred stock

(1) $18.48

(2) 6.73093%

(7.2) Brokerage accounts

(1) 36.36370%

(2) It is first violated as soon as it is below $24,512(1 - m)/(1.2 - m)$.

(3) (a) $250

    (b) $384.62

    (c) $5,000

(7.3) Going long: buying stock with borrowed money

(1) With the cash account, she purchases 84 shares and has a three-month yield of 8.46262%. With the margin account, she purchases 141 shares and has a three-month yield of 13.1%.

(2) $4,000

(7.4) Selling short: selling borrowed stocks

(1) 11%

(2) −40.36728%

(3) 57.73810%

(4) 7%

(5) 15.79407%

(6) 25.115% (with rounding of Kim's repurchase price to $1,758.62)

Chapter 7 review problems

(1) −6.96438%

(2) $1,639.63

(3) (a) $940
     (b) $1,566.67
     (c) $2,350
(4) With the cash account, she purchases 60 shares and has an annual yield of 17.65409%. With the margin account, she purchases 100 shares and has an annual yield of 28.03636%.

PROBLEMS, CHAPTER 8

(8.3) The term structure of interest rates

(1) $f_{0[,1]} \approx 2.40655\%$, $f_{[0,2]} \approx 2.98885\%$, $f_{[0,3]} \approx 3.62503\%$, $f_{[1,2]} \approx 3.57446\%$, $f_{[1,3]} \approx 4.23969\%$, $f_{[2,3]} \approx 4.90919\%$,
(2) $r_1 \approx 2.66940\%$, $r_2 \approx 3.64757\%$, $r_3 \approx 4.4062\%$,
(3) $1,057.82
(5) 6.49879%
(6) $3,006.12, 5.94121%

(8.4) Forward contracts

(1) $1.65

(8.5) Commodity futures held until delivery

(1) 0
(2) lost $2,200
(3) $75 credited to her account, resulting in $875 balance
(4) Serena: April 15: $-$3,000, May 1: $-$2,500, June 20: $-$1,000, July 31: +$39,000
     Barbara: April 15: $-$3,000, May 27: $-$1,500, July 31: $-$28,000

(8.6) Offsetting positions and liquidity of futures contracts

(1) price falls 4.23640%, annual yield $\approx -87.77357\%$
(2) The following table applies to Sam's long position:

Date	Price	Change	Deposit	Post-deposit balance
8/14	$72,850	—	—	$5,500
8/15	$72,200	−$650	—	$4,850
8/16	$71,700	−$500	—	$4,350
8/17	$73,400	+$1,800	—	$6,050
8/18	$75,200	+$1,700	—	$7,850
8/19	$74,500	−$700	—	$7,150
8/22	$75,900	+$1,400	—	$8,550

following table applies if Sam took a short position instead:

Date	Price	Change	Deposit	Post-deposit balance
8/14	$72,850	—	—	$5,500
8/15	$72,200	−$650	—	$6,150
8/16	$71,700	−$500	—	$6,650
8/17	$73,400	+$1,800	—	$4,950
8/18	$75,200	+$1,700	—	$3,150
8/19	$74,500	−$700	—	$3,850
8/22	$75,900	+$1,400	$3,050	$2,450

(3) −100%

(4) Jose has an initial outflow of $1,898. Eleven days later, he has a $880 outflow. On the twenty-firt day, he receives a $4,813 liquidation payment. His daily yield rate is 3.11796%. Rodrigo has the same initial outflow. ten days later, he has a $1,045 outflow, and his liquidation onflow is $908. Rodrigo's daily yield is -6.83999%.

(8.7) Price discovery and more kinds of futures

(1) −$750

(2) 2.45370

(8.8) Options

(1) $f(x) = \begin{cases} 0 & \text{if } x < 2.35 \\ 10,000x - 23,500 & \text{if } x \geqslant 2.35 \end{cases}$

(2) (a) $19.44

   (b) $152.97

(3) −64.41961%, compared with 5.79592%

(4) $257.87

(8.9) Using replicating portfolios to price options

(1) 3.68625%
(2) $22.68
(3) $1,373.59
(4) (a) $114.76

(8.10) Using weighted averages to price options: risk-neutral probabilities

(1) $473.14
(2) $7,296
(3) 3.68625%
(4) (a) .52058
    (b) 3.38241%
(5) $1,373.59 as in Problem (8.9.3)

(8.11) Swaps

(1) GSB received $83,368.06 on 7/01/94 and $23,333.33 on 10/01/94. GSB paid $49,385 on 1/01/05 and $94,325 on 4/01/95.
(2) (a) 12,544,220.64 DKK
    (b) $2,112,859.23
    (c) $-344,220.64$
(3) Mrs Yolente pays $645 at the end of 6 months. She receives $405, $3,105, and $2,205 at the end of 12 months, 18 months, and 24 months respectively.

Chapter 8 review problems

(1) $1,016.69
(3) $4,105.04
(4) $-2,375,263.66$
(5) $c_0 \approx \$171.26$, $p_0 \approx 35.32$
(6) 5.86420%

(7)

Time	Balance	Deposit	New Balance
1/5 at purchase	$0	$2,500	$2,500
1/5	$2,115	$0	$2,115
1/6	$1,240	$1,260	$2,500
1/7	$1,100	$1,400	$2,500
1/8	$3,300	$0	$3,300
1/9	$5,675	$0	$5,675
1/12 at sale	$4,500	−$4,500	$0

(8) Price falls by 1.46341%; Three-month effective yield is −25.20665%

PROBLEMS, CHAPTER 9

(9.1) Overview

(1) $14,536.39
(2) If $i = 18\%$, there is a surplus of $613.64 at time 3. If $i = 1\%$, there is a surplus of $102.77 at time 3.

(9.2) Macaulay and modified duration

(1) 7.84562
(2) $D(.07, \infty) = 6.07079$, $D(.07, 1) = 5.67364$
(3) $D(.05, 1) \approx 7.61905$
(4) At $i = 4.4\%$, the price is $1,114.69. At $d = 3.7\%$, it is $1,136.60.
(5) $D(.05, \infty) \approx 20.74695$, $D(.05, 2) \approx 20.24695$
(6) $D(.06, \infty) \approx 26.5$, $D(.06, 1) = 25$
(7) 5.42791
(8) Independent of k, the Macaulay duration is 52.5.
(9) 3.39072
(10) 10.27553

(9.3) Convexity

(1) 71.32585
(2) $C(4\%, \infty) \approx 8.12137$, $C(5\%, \infty) \approx 8.10828$, $C(6\%, \infty) \approx 8.09513$, $C(5\%, 2) \approx 9.04837$
(3) $\dfrac{2.65}{\sqrt{1+i}}$
(4) $D(i, \infty) \approx 20.23077$, $C(i, \infty) \approx 798.337278$, down by about 8.69083%

(9.4) Immunization

(1) $88,899.64 of one-year bond, $177,799.27 of four-year bond, yes

(2) Spend $179,736.12 to purchase the two-year coupon bond (face value $161,463.94) and $416,138.68 to purchase the five-year zero-coupon bond. The convexity of the assets exceeds the convexity of the assets as is required for Redington immunization.

(3) We say that "the portfolio is fully immunized" since the price of the assets "equals" the price of the liabilities and the Macaulay duration of the assets "equals" the Macaulay duration of the liabilities; both prices round to $419,280.67, but they differ by about $.00126, and the durations are each about 4.0000000. The surplus now is $2,293.94, and the surplus in about one year is $2,477.45.

(9.5) Other types of duration

(1) 2.14885

(2) $i \approx 5.03287\%$, $E_{.01}(, i, 1) \approx 5.26890$, $E_{.02}(i, 1) \approx 3.80993$

(3) $E_{.01}(.06, 1) \approx 2.08513$, so if $i = 5.4\%$ the price of the callable bond is approximately $48,998.18, and if $i = 4\%$ this price is approximately $50,410.85. For a non-callable bond, the prices are $50,911.27 at 5.4% and $52,898.84 at 4%. Using $D(.06, 1)$, you should get estimates $50,902.15 and $52,795.20 respectively.

(5) $E_h(i_0, 1) \approx 4.90579$; $F_h(i_0, 1) \approx -1098.31104$; Estimated price at 7.2% is $97.39.

Chapter 9 review problems

(1) $D(7.5\%, 1) \approx 8.82712$, $D(9\%, 1) \approx 8.83574$

(2) $P(5.6\%) \approx \$1,005.44$, and $P(4\%) \approx \$1,102.47$. The refined estimates are $P(5.6\%) \approx \$1,005.46$, and $P(4\%) \approx \$1,103.13$.

(3) The two year bond has $D(.05, \infty) = 2$ and $C(.05, \infty) = 4$. The three year bond has $D(.05, \infty) = 3$ and $C(.05, \infty) = 9$. The five year bond has $D(.05, \infty) \approx 4.47775$ and $C(.05, \infty) \approx 21.36936$. The portfolio has $D(.05, \infty) \approx 3.22554$ and $C(.05, \infty) \approx 11.97415$.

(4) $D(.06, \infty) = 17.46316$, $D(.06, 2) \approx 16.96172$

(5) Purchase $78,352.62 of the two-year bonds and $117,528.92 of the seven-year bonds. At 4%, the time zero value is $56.27.

(6) $i \approx 5.6200\%$, $E_{.01}(i, 1) \approx 3.44617$

Bibliography

1. Cvitanić, Jakša and Zapatero, Fernando, *Introduction to the Economics and Mathematics of Financial Markets*, The MIT Press, Cambridge, Massachusetts, 2004.

2. Hull, John C., *Options, Futures, and Other Derivatives* (Sixth Edition), Pearson Prentice Hall, Upper Saddle River, New Jersey, 2006.

3. Kellison, Stephen G., *The Theory of Interest* (Second Edition), Irwin, Homewood, Illinois, 1991.

4. McDonald, Robert L., *Derivative Markets* (Sixth Edition), Pearson Addison Wesley, Boston, Massachusetts, 2006.

5. Panjer, Harry H. (editor), *Financial Economics: with Applications to Investments, Insurance, and Pensions*, The Actuarial Foundation, Schaumberg, Illinois, 1998.

6. Ruckman, Chris, Francis, Joe, *Financial Mathematics: A Practical Guide for Actuaries and other Business Professionals*, BPP Professional Education, Weatogue, Connecticut, 2004.

7. Sharpe, William F., Alexander, Gordon J., Bailey, Jeffrey V., *Investments* (Sixth Edition), Prentice Hall, Upper Saddle River, New Jersey, 1999.

Index

cash settled future contract, 367–368
CD, *see* certificate of deposit
certificate of deposit, 25
Chicago Board of Trade (CBOT), 356
Chicago Board Option Exchange (CBOE), 370, 371
Chicago Mercantile Exchange (CME), 367
Chn, 2
clearing association, *see* clearinghouse
clearing organization, *see* clearinghouse
clearinghouse, 356–357, 363
closed-end fund, 319
closing a position, 363
commissions, 320, 328, 333, 372, 380
Commodity Futures Trading Commission (CFTC), 358
common stock, 316
competitive advantage (interest rates), 395
compound discount, 41–46
compound interest, 25
    comparison with simple interest, 22
    introduction to, 20–25
    varying rates, 24–25
compound interest accumulation function, 21
compounded, 46
concavity and convex sets, 430
Consumer Price Index, 61
continuous compounding, 56, 60
convertible, 46
convetible continuously, 55
convexity, 415, 430–437
    effective, 448, 453
    Macaulay, 432–437
        portfolio, 436–437
        weighted average, 433
    modified, 430–433
counting days, 17
    Date worksheet, 19–20
covering a position, *see* closing a position

D
Date worksheet, 19–20, 303

dealer, 320
debit balance, 321
dedicated portfolio, 413
dedication, 415–416
default, 11, *see* risk, credit
delta portfolio allocation, 387
derivative, 338–339, 353
designated contract market, 356
discount, 25–29
    amount of, 26
    forward, 354
discount curve, 353
discount factor, 31
discount function, 29
    linear, *see* simple discount
discount rate, 26
    compound interest, 29
    effective, 26–29, 42
    nominal, 48–56
dispersion, 434–435
dividend, 316–317
    reinvestment options, 327
    short sale, 329–332
dividend discount model, 318
dollar-weighted yield rate, 87
Dow Jones Industrial Average
    Dow divisor, 318
Dow Jones Industrial Average (DJIA), 318–319
drop payment, 162
duration, 414, 416–430, 433–435, 447–448
    effective, 447–448
    keyrate, 449
    Macaulay, 414–415, 422–430, 433–435
        portfolio, 428–430
        weighted average, 423
    modified, 414, 421–422, 424, 426
    partial, 449
dynamic portfolio allocation, 384–387

E
effective discount rate, 26–29, 42
effective interest rate, 14–15
equation of value
    multiple deposits, 79–87
    single deposit, 77–79
equity, 316, 322